T0221137

RF and Microwave Circuit Design

Microwave and Wireless Technologies Series

Series Editor: Professor Steven (Shichang) Gao, Chair of RF and Microwave Engineering, and the Director of Postgraduate Research at School of Engineering and Digital Arts, University of Kent, UK.

Microwave and wireless industries have experienced significant development during recent decades. New developments such as 5G mobile communications, broadband satellite communications, high-resolution earth observation, the Internet of Things, the Internet of Space, THz technologies, wearable electronics, 3D printing, autonomous driving, artificial intelligence etc. will enable more innovations in microwave and wireless technologies. The Microwave and Wireless Technologies Book Series aims to publish a number of high-quality books covering topics of areas of antenna theory and technologies, radio propagation, radio frequency, microwave, millimetre-wave and THz devices, circuits and systems, electromagnetic field theory and engineering, electromagnetic compatibility, photonics devices, circuits and systems, microwave photonics, new materials for applications into microwave and photonics, new manufacturing technologies for microwave and photonics applications, wireless systems and networks.

RF and Microwave Circuit Design: Theory and Applications
Charles E. Free, Colin S. Aitchison
September 2021

Low-cost Smart Antennas
Qi Luo, Steven (Shichang) Gao, Wei Liu
March 2019

RF and Microwave Circuit Design

Theory and Applications

Charles E. Free
UK

Colin S. Aitchison
UK

This edition first published 2022
© 2022 John Wiley & Sons Ltd

All rights reserved. No part of this publication may be reproduced, stored in a retrieval system, or transmitted, in any form or by any means, electronic, mechanical, photocopying, recording or otherwise, except as permitted by law. Advice on how to obtain permission to reuse material from this title is available at http://www.wiley.com/go/permissions.

The right of Charles E. Free and Colin S. Aitchison to be identified as the authors of this work has been asserted in accordance with law.

Registered Offices
John Wiley & Sons, Inc., 111 River Street, Hoboken, NJ 07030, USA
John Wiley & Sons Ltd, The Atrium, Southern Gate, Chichester, West Sussex, PO19 8SQ, UK

Editorial Office
The Atrium, Southern Gate, Chichester, West Sussex, PO19 8SQ, UK

For details of our global editorial offices, customer services, and more information about Wiley products visit us at www.wiley.com.

Wiley also publishes its books in a variety of electronic formats and by print-on-demand. Some content that appears in standard print versions of this book may not be available in other formats.

Limit of Liability/Disclaimer of Warranty
While the publisher and authors have used their best efforts in preparing this work, they make no representations or warranties with respect to the accuracy or completeness of the contents of this work and specifically disclaim all warranties, including without limitation any implied warranties of merchantability or fitness for a particular purpose. No warranty may be created or extended by sales representatives, written sales materials or promotional statements for this work. The fact that an organization, website, or product is referred to in this work as a citation and/or potential source of further information does not mean that the publisher and authors endorse the information or services the organization, website, or product may provide or recommendations it may make. This work is sold with the understanding that the publisher is not engaged in rendering professional services. The advice and strategies contained herein may not be suitable for your situation. You should consult with a specialist where appropriate. Further, readers should be aware that websites listed in this work may have changed or disappeared between when this work was written and when it is read. Neither the publisher nor authors shall be liable for any loss of profit or any other commercial damages, including but not limited to special, incidental, consequential, or other damages.

Library of Congress Cataloging-in-Publication Data

Names: Free, Charles E., author. | Aitchison, Colin S., author.
Title: RF and microwave circuit design : theory and applications / Charles
   E. Free, UK, Colin S. Aitchison, UK.
Description: Hoboken, NJ, USA : Wiley, 2022. | Series: Microwave and
   wireless technologies series | Includes bibliographical references and
   index.
Identifiers: LCCN 2020028781 (print) | LCCN 2020028782 (ebook) | ISBN
   9781119114635 (hardback) | ISBN 9781119114673 (adobe pdf) | ISBN
   9781119114666 (epub)
Subjects: LCSH: Radio circuits–Design and construction. | Microwave
   circuits–Design and construction.
Classification: LCC TK6560 .F68 2022 (print) | LCC TK6560 (ebook) | DDC
   621.3841/2–dc23
LC record available at https://lccn.loc.gov/2020028781
LC ebook record available at https://lccn.loc.gov/2020028782

Cover Design: Wiley
Cover Images: Circuit board background © Berkah/Getty Images, Inset sketch by Charles
Free

Set in 9.5/12.5pt STIXTwoText by Straive, Chennai, India
Printed and bound by CPI Group (UK) Ltd, Croydon, CR0 4YY

C9781119114635_260821

Contents

Preface

In recent years, the rapid expansion of communication applications at RF and microwave frequencies has created significant interest in this area of high-frequency electronics, both in industry and academia. This textbook provides a rigorous introduction to the theory of modern-day circuits and devices at RF and microwave frequencies, with an emphasis on current practical design.

One of the themes of the book is that of the design of high-frequency hybrid integrated circuits in which individual passive and active components are interconnected within a planar circuit structure. The traditional method of making such circuits is to assemble the components on a low-loss copper-clad printed circuit board, which has been etched with the required interconnection pattern. In recent years, the development of new materials and new fabrication techniques has created greater scope for the circuit designer, with the opportunity to use hybrid circuit structures at much higher frequencies, well into the millimetre-wave region. In particular, the use of a photoimageable thick-film and low-temperature co-fired ceramic (LTCC) materials has enabled low-cost, high-performance multilayer structures to be fabricated. The properties of these materials, and their application to RF and microwave circuits, are discussed in the book.

The book has been organized to provide a cohesive introduction to RF and microwave technology at undergraduate and Master's degree level. It assumes only a basic knowledge of electronics on the part of the reader, with the bulk of the high-frequency material being developed from first principles. Many worked examples have been included in the text to emphasize the key points, and supplementary problems with answers are provided at the end of most chapters.

The material presented in the book is based largely on courses in RF and microwave communications taught by both authors at several UK Universities.

Chapter 1 introduces the theory of RF and microwave transmission lines, which are fundamental to high-frequency circuits. The basic theory is expanded to include the principles and applications of the Smith chart, which is a graphical tool that can be used to represent and solve transmission line problems. This chapter includes summaries of the properties of some common high-frequency transmission lines, namely coaxial cable, microstrip, coplanar waveguide, and hollow metallic waveguide.

Chapter 2 expands on the theory and applications of microstrip, which was introduced in the first chapter. Microstrip is by far the most common type of interconnection used for RF and microwave applications, and the chapter discusses the properties of this transmission medium, and introduces some typical passive microstrip components.

Chapter 3 presents information on modern circuit materials and the associated fabrication techniques. This is an important chapter in that good electrical design of planar circuits at high frequencies requires a good understanding of the properties and capabilities of the circuit materials. In addition to the traditional etched circuit techniques, the chapter discusses newer fabrication approaches using photoimageable thick-film, LTCC, and ink jet printing. The chapter concludes with a discussion of the various methods available for characterizing circuit materials at high frequencies.

Chapter 4 continues the theme of planar circuit design through a discussion of the discontinuities associated with microstrip components. Also in this chapter are the equivalent circuits of packaged lumped-element passive components, as well as miniature planar components that can be used in single layer and multilayer formats.

Chapter 5 introduces the concept of *S*-parameters. These are network parameters used to characterize RF and microwave devices, and an understanding of their meaning and usage is essential for microwave design. The chapter includes information on power gain definitions, and the use of flow graphs in high-frequency network analysis.

Chapter 6 presents information on microwave ferrites. Ferrite materials have a well-established place in microwave technology, primarily for providing non-reciprocal components. The properties of ferrite materials, as well as their use in traditional metallic waveguide components, are explained. More recent uses of ferrite material in multilayer planar circuits are also discussed.

Chapter 7 is concerned with measurements at RF and microwave frequencies. The chapter focuses in particular on the use of the vector network analyzer (VNA), which is the main item of test instrumentation in all high-frequency laboratories. As well as describing the functions and use of the VNA, the chapter addresses the important issues of calibration and measurement errors.

Chapter 8 provides an introduction to RF filters, which are essential circuits in most high-frequency sub-systems. This chapter commences with a review of the principal filter responses, and then extends the theoretical discussion into practical design detail. A number of worked examples of the design of microstrip filters have been included, and the chapter ends with a brief review of the advantages of multilayer formats for producing high-performance filters at RF and microwave frequencies.

Chapter 9 presents information on the design of hybrid microwave amplifiers in which a packaged transistor, usually a metal semiconductor field effect transistor (MESFET), is positioned between input and output matching networks. The chapter focuses on the design of the matching networks, and considers the effects of these networks on transducer gain, noise, and stability. Worked examples are included to show the design strategies employed for microstrip implementation.

Chapter 10 introduces the function and design of switches and phase shifters in planar circuits. The functions of four devices commonly used as switches in RF and microwave circuits are described, namely PIN diodes, field effect transistors (FETs), microelectromechanical switches (MEMS), and inline phase change switches (IPCS). The use of these switches in various types of phase shifters are described, with supporting worked examples to show how microstrip phase shifters are designed.

Chapter 11 gives information on the various types of oscillators used in high-frequency circuits. Starting with the criteria for oscillation in a feedback circuit, the three main types of transistor oscillators are described, namely the Colpitts, Hartley, and Clapp–Gouriet oscillators. Leading on from the basic oscillators the concept of the voltage-controlled oscillator (VCO) is introduced. Many RF and microwave receivers, and most test instrumentation, derive the required frequency from a very stable low-frequency crystal oscillator, and a frequency synthesizer. The chapter includes a discussion of crystal oscillators, together with a review of the main types of frequency synthesizers using phase-locked loops. Also given are descriptions of several types of oscillators used specifically at microwave frequencies; these include the dielectric resonator oscillator (DRO), and oscillators using the Gunn and Impatt principles. Oscillator noise is a significant issue in many situations, particularly for low-noise receivers, and the chapter concludes with a discussion of oscillator noise, together with a method for measuring this noise.

Chapter 12 presents information on RF and microwave antennas. The chapter commences with a discussion of the theoretical aspects of electromagnetic radiation from simple wire structures, including the half-wave dipole. The analysis is then extended to consider the behaviour of wire arrays. The most notable array used for RF and low microwave frequency communication is the Yagi–Uda array, and this is considered in some detail. The short wavelengths associated with microwave frequencies offer more scope for the antenna designer, and the chapter gives design details of microwave antennas using planar patch structures and also those using radiation from apertures.

Chapter13 provides an introduction to power amplifiers, and distributed amplifiers. Power amplifiers are commonly positioned in as the last device in a transmitter before the signals are transmitted, and consequently any distortion introduced by the power amplifier cannot be corrected. Distortion is therefore one of the main themes of this chapter. Also described in this chapter are distributed amplifiers, which provide an attractive combination of high available gain and very wide bandwidth at microwave frequencies.

Chapter 14 brings together a number of devices introduced in earlier chapters with a discussion of RF and microwave receivers. Noise is an important issue in any receiver, particularly when the received signal levels are very low, and this chapter includes a discussion of typical sources of noise, and how they affect the performance of a receiver.

About the Companion Website

This book is accompanied by a companion website:

www.wiley.com/go/free/rfandmicrowave

The website includes:
- Teaching PDF Slides (by chapter)
- Microstrip Design Graphs

1

RF Transmission Lines

1.1 Introduction

Transmission lines, in the form of cable and circuit interconnects, are essential components in RF and microwave systems. Furthermore, many distributed planar components rely on transmission line principles for their operation. This chapter will introduce the concepts of RF transmission along guided structures, and provide the foundations for the development of distributed components in subsequent chapters.

Four of the most common forms of RF and microwave transmission line are shown in Figure 1.1.

(i) *Coaxial cable* is an example of a shielded transmission line, in which the signal conductor is at the centre of a cylindrical conducting tube, with the intervening space filled with lossless dielectric. The dielectric is normally solid, although for higher-frequency applications it is often in the form of dielectric vanes so as to create a semi-air-spaced medium with lower transmission losses. A typical coaxial cable is flexible with an outer diameter around 5 mm, although much smaller diameters are available with 1 mm diameter cable being used for interconnections within millimetre-wave equipment. Also, for very high-frequency applications, the cable may have a rigid or semi-rigid construction. Further data on coaxial cables are provided in Appendix 1.A.

(ii) *Coplanar waveguide* (CPW), in which all the conductors are on the same side of the substrate, is also shown in Figure 1.1. This type of structure is very convenient for the mounting of active components, and also for providing isolation between signal tracks. Coplanar lines are widely used in compact integrated circuits for high-frequency applications. Further data on coplanar lines are given in Appendix 1.B.

(iii) *Waveguide*, formed from hollow metal tubes of rectangular or circular cross-section, is a traditional form of transmission line used for microwave frequencies above 1 GHz. For many circuit and interconnection applications, waveguide has been superseded by planar structures, and its use in modern RF and microwave systems is restricted to rather specialized applications. It is the only transmission line that can support the very high powers required in some transmitter applications. Another advantage of an air-filled metal waveguide is that it is a very low loss medium and therefore can be used to make very high-Q cavities, and this application is discussed in more detail in Chapter 3 in relation to dielectric measurements. A more recent application of traditional waveguides is in substrate integrated waveguide (SIW) structures for millimetre-wave applications, and this is explained in more detail in Chapter 4 in the context of emerging technologies. Further data on the theory of waveguides are given in Appendix 1.C.

(iv) *Microstrip* is the most common form of interconnection used in planar circuits for RF and microwave applications. As shown in Figure 1.1, it consists of a low-loss insulating substrate, with one side completely covered with a conductor to form a ground plane, and a signal track on the other side. Further data on microstrip are given in Appendix 1.D. This is a particularly important medium for high-frequency circuit design and so Chapter 2 is devoted to an in-depth discussion of microstrip and the associated design techniques.

1.2 Voltage, Current, and Impedance Relationships on a Transmission Line

In its simplest form, a transmission line can be viewed as a two-conductor structure with a go and return path for the current. For the purpose of analysis we may regard any transmission line as made up of a large number of very short lengths (δz), each of which can be represented by a lumped equivalent circuit, as shown in Figure 1.2. In the equivalent circuits, R and L

RF and Microwave Circuit Design: Theory and Applications, First Edition. Charles E. Free and Colin S. Aitchison.
© 2022 John Wiley & Sons Ltd. Published 2022 by John Wiley & Sons Ltd.
Companion website: www.wiley.com/go/free/rfandmicrowave

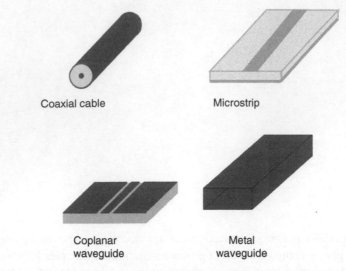

Coaxial cable

Microstrip

Coplanar
waveguide

Metal
waveguide

Figure 1.1 Common types of high-frequency transmission line.

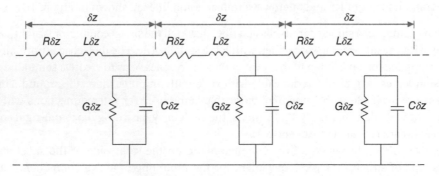

Figure 1.2 Representation of a transmission line in terms of lumped components.

represent the series resistance and inductance per unit length of the conductors, respectively, C represents the capacitance between the lines per unit length, and G is the parallel conductance per unit length, and represents the very high resistance of the insulating medium between the conductors.

It should be noted that it is legitimate to represent a continuous transmission line by the lumped equivalent circuit shown in Figure 1.2 providing that δz is small compared to a wavelength. R, L, G, and C are normally referred to as the primary line constants, and have the units of Ω/m, H/m, S/m, and F/m, respectively.

In order to establish relationships between the voltage and current on a transmission line we need first to specify a line excited by a sinusoidal voltage at the sending end whose angular frequency is ω. If we then let the voltage and current at some arbitrary point on the line be V and I, respectively, we can consider the effect on an elemental length at this point. The voltage drop across the elemental length will be δV and the parallel current will be δI, as shown in Figure 1.3.

Using standard AC circuit theory, we can relate the change in voltage, δV, to the components of the equivalent circuit as

$$-\delta V = (R\delta z)I + (L\delta z)\frac{\partial I}{\partial t}$$
$$= (R\delta z)I + (L\delta z)j\omega I,$$

i.e.

$$\frac{\delta V}{\delta z} = -(R + j\omega L)I.$$

Considering the limit, as $\delta z \to 0$, $\frac{\delta V}{\delta z} \to \frac{dV}{dz}$ giving

$$\frac{dV}{dz} = -(R + j\omega L)I. \tag{1.1}$$

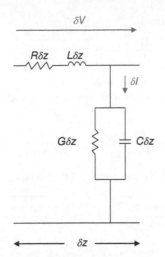

Figure 1.3 Equivalent circuit of an elemental length, δz, of a transmission line.

Considering the parallel current, δI, we have

$$-\delta I = (G\delta z)V + (C\delta z)\frac{\partial V}{\partial t}$$
$$= (G\delta z)V + (C\delta z)j\omega V,$$

i.e.

$$\frac{\delta I}{\delta z} = -(G + j\omega C)V.$$

As $\delta z \to 0$, $\dfrac{\delta I}{\delta z} \to \dfrac{dI}{dz}$ giving

$$\frac{dI}{dz} = -(G + j\omega C)V. \tag{1.2}$$

Differentiating Eq. (1.1) with respect to time gives

$$\frac{d^2V}{dz^2} = -(R + j\omega L)\frac{dI}{dt}.$$

Substituting for $\dfrac{dI}{dz}$ from Eq. (1.2) gives

$$\frac{d^2V}{dz^2} = (R + j\omega L)(G + j\omega C)V,$$

which can be written as

$$\frac{d^2V}{dz^2} = \gamma^2 V, \tag{1.3}$$

where

$$\gamma = \sqrt{(R + j\omega L)(G + j\omega C)}. \tag{1.4}$$

Similarly

$$\frac{d^2I}{dz^2} = \gamma^2 I. \tag{1.5}$$

To determine the variation of V along the line, we have to solve the differential Eq. (1.3) for V. This is a second-order differential equation with a standard solution in the form

$$V = V_1 e^{-\gamma z} + V_2 e^{+\gamma z}. \tag{1.6}$$

The two terms on the right-hand side of Eq. (1.6) show how the peak amplitudes and phases of waves travelling in the forward and reverse directions vary with distance. The values of the amplitudes and phases of these waves are determined by the value of γ, which is defined as the propagation constant (this is considered in more detail in Section 1.3).

Differentiating the expression in Eq. (1.6) gives

$$\frac{dV}{dz} = -\gamma V_1 e^{-\gamma z} + \gamma V_2 e^{\gamma z}.$$

(1.7)

Combining Eqs. (1.7) and (1.1) gives

$$-(R + j\omega L)I = -\gamma V_1 e^{-\gamma z} + \gamma V_2 e^{\gamma z},$$

i.e.

$$I = \frac{\gamma}{(R + j\omega L)} V_1 e^{-\gamma z} - \frac{\gamma}{(R + j\omega L)} V_2 e^{\gamma z}.$$

(1.8)

Remembering that $\gamma = \sqrt{(R + j\omega L)(G + j\omega C)}$ we can rewrite Eq. (1.8) as

$$I = \sqrt{\frac{(G + j\omega C)}{(R + j\omega L)}} V_1 e^{-\gamma z} - \sqrt{\frac{(G + j\omega C)}{(R + j\omega L)}} V_2 e^{\gamma z}$$

or

$$I = \frac{V_1}{Z_O} e^{-\gamma z} - \frac{V_2}{Z_O} e^{\gamma z},$$

(1.9)

where

$$Z_O = \sqrt{\frac{(R + j\omega L)}{(G + j\omega C)}}.$$

(1.10)

The impedance, Z_O, is termed the characteristic impedance of the transmission line. Characteristic impedance is an important property of any transmission line and it is useful to have an appreciation of its physical significance. Theoretically, it is the ratio of the voltage to current at an arbitrary position on an infinitely long transmission line that supports a wave travelling in one direction. If the line is lossless, i.e. $R = 0$ and $G = 0$, then we see from Eq. (1.10) that $Z_O = \sqrt{L/C}$ and has a constant value that is independent of frequency. It follows that if such a line is terminated by an impedance equal to the characteristic impedance, there will be no reflections from the termination. Moreover, if a transmission line is terminated with its characteristic impedance, then the impedance at the input of the line will be equal to the characteristic impedance; under these conditions the line is said to be matched.

Considering the sending end of the line, i.e. $z = 0$, then from Eqs. (1.6) and (1.9) we obtain

$$V = V_S = V_1 + V_2,$$
$$I = I_S = \frac{V_1 - V_2}{Z_O},$$

(1.11)

where V_S and I_S are the voltage and current at the sending end of the line, respectively.

Rearranging Eq. (1.11) to obtain V_1 and V_2 gives:

$$V_1 = \frac{V_S + Z_O I_S}{2},$$
$$V_2 = \frac{V_S - Z_O I_S}{2}.$$

(1.12)

The voltage, V, and current, I, at any distance, z, along the transmission line can now be found in terms of the voltage and current at the sending end by substituting V_1 and V_2 from Eq. (1.12) into Eqs. (1.6) and (1.9) giving

$$V = \frac{V_S + Z_O I_S}{2} e^{-\gamma z} + \frac{V_S - Z_O I_S}{2} e^{\gamma z}$$
$$= V_S \left(\frac{e^{\gamma z} + e^{-\gamma z}}{2} \right) - I_S Z_O \left(\frac{e^{\gamma z} - e^{-\gamma z}}{2} \right).$$

(1.13)

Equation (1.13) may be written in terms of hyperbolic functions as

$$V = V_S \cosh(\gamma z) - I_S Z_O \sinh(\gamma z).$$

(1.14)

Similarly,

$$I = I_S \cosh(\gamma z) - \frac{V_S}{Z_O} \sinh(\gamma z).$$

(1.15)

The impedance, Z_z, at any distance z from the sending end of the line can now be found by dividing Eq. (1.14) by Eq. (1.15) giving

$$Z_z = \frac{V}{I} = \left(\frac{Z_S \cosh(\gamma z) - Z_O \sinh(\gamma z)}{Z_O \cosh(\gamma z) - Z_S \sinh(\gamma z)} \right) Z_O, \tag{1.16}$$

where $Z_S = \frac{V_S}{I_S}$ is the impedance at the sending end of the line.

If we now consider a transmission line of finite length, l, terminated by an arbitrary impedance, Z_L, then $Z_z = Z_L$ when $z = l$, and Eq. (1.16) can be rewritten as

$$Z_L = \left(\frac{Z_S \cosh(\gamma l) - Z_O \sinh(\gamma l)}{Z_O \cosh(\gamma l) - Z_S \sinh(\gamma l)} \right) Z_O. \tag{1.17}$$

The input impedance, Z_{in}, of a transmission line terminated by an impedance, Z_L, can be found by rearranging Eq. (1.17) to give

$$Z_{in} = Z_S = \frac{Z_O(Z_L \cosh(\gamma l) + Z_O \sinh(\gamma l))}{(Z_O \cosh(\gamma l) + Z_L \sinh(\gamma l))}. \tag{1.18}$$

This is an important, but complicated, expression giving the input impedance, Z_{in}, of a transmission line terminated in Z_L in terms of the propagation constant, γ, the characteristic impedance, Z_O, and the line length, l. If the line is low loss, Eq. (1.18) can be significantly simplified, as is shown later in the chapter.

1.3 Propagation Constant

The propagation constant was introduced in Eq. (1.4). This constant determines the amplitude and phase of a wave propagating along a transmission line at a particular frequency, and may conveniently be expressed as

$$\gamma = \alpha + j\beta, \tag{1.19}$$

where α is the attenuation constant and β is the phase propagation constant. Considering the first term on the right-hand side of Eq. (1.6) we have

$$V_F = V_1 e^{-\gamma z} = V_1 e^{-(\alpha + j\beta)z} = V_1 e^{-\alpha z} e^{-j\beta z},$$

where V_F represents the voltage of the forward wave at a distance l along the line. The magnitude of this wave is given by

$$|V_F| = |V_1| e^{-\alpha z}.$$

Rearranging this equation gives

$$\alpha = -\log_e \left| \frac{V_F}{V_1} \right| = -\ln \left| \frac{V_F}{V_1} \right|. \tag{1.20}$$

Taking the natural logarithm of the ratio of two voltages gives the ratio in the units of Nepers; so α will have the units of Np/m (Nepers/metre). Although Nepers are not in common use as a unit in RF work, it is important to be able to convert a voltage ratio from Neper to the more usual power unit of dB.

Considering a voltage ratio, α, we have

$$\alpha_{Np} = \log_e \alpha \quad \Rightarrow \quad \alpha = e^{\alpha_{Np}}, \tag{1.21}$$

$$\alpha_{dB} = 20 \log_{10} \alpha \quad \Rightarrow \quad \alpha = 10^{\alpha_{dB}/20}. \tag{1.22}$$

Therefore,

$$e^{\alpha_{Np}} = 10^{\alpha_{dB}/20},$$

i.e.

$$\alpha_{Np} \log_e e = \frac{\alpha_{dB}}{20} \log_e 10,$$

$$\alpha_{Np} = 0.115 \times \alpha_{dB}, \tag{1.23}$$

i.e.

$$1 \ Np \equiv 8.686 \ dB. \tag{1.24}$$

The imaginary part of the propagation constant gives the transmission phase change experienced by the wave in travelling a distance, z. Since there are 2π radians in one wavelength, the phase propagation constant is always given by

$$\beta = \frac{2\pi}{\lambda},\tag{1.25}$$

where λ is the wavelength along the line being considered. It follows from Eq. (1.25) that the phase propagation constant has the units of rad/m (radians/metre).

1.3.1 Dispersion

The foregoing theory describes the propagation along a transmission line at a single frequency. But since all information-carrying signals contain more than one frequency, it is important to know how the propagation characteristics of a line change with frequency.

If all the frequencies contained in a signal travel at the same velocity, the transmission line is said to be dispersionless. If this is not so, and if the phase velocity,[1] v_P, is a function of frequency, the transmission line will exhibit dispersion. If dispersion is present, a signal containing a number of frequency components, such as a voltage pulse, will become distorted as it propagates along the line, with the degree of distortion increasing with the distance of propagation.

A useful concept in determining the degree of dispersion is group velocity, v_g. We can explain this concept by considering the transmission of a signal which consists of a number of sinusoids, each having a different frequency and amplitude. These frequency components will combine to form a composite pattern, with a particular envelope. The group velocity is the velocity with which this envelope propagates along the transmission line. It can be shown that the group velocity is given by

$$v_g = \frac{\delta\omega}{\delta\beta}.\tag{1.26}$$

If this velocity is independent of frequency, then the line will be dispersionless and the phase relationships between the frequency components of the signal will be maintained.

The reciprocal of the group velocity is known as the group delay, and is the slope of the β–ω response at a particular frequency. If β is a linear function of frequency, then the β–ω response will be a straight line and the group delay will be constant, and independent of frequency. To avoid distortion it is important that the group delay is constant over the full frequency range of the signal being transmitted.

1.3.2 Amplitude Distortion

Amplitude distortion will occur if the attention constant, α, is a varying function of frequency, thus causing the frequency components of a complex signal to suffer different amplitude changes as the signal propagates. For no attenuation distortion to occur, we require $\frac{\delta\alpha}{\delta f} = 0$. Normally attenuation distortion is not significant for RF and microwave circuit interconnections, since these interconnections are short and deliberately designed to be low loss.

1.4 Lossless Transmission Lines

The majority of transmission lines encountered in RF and microwave circuits are both short and deliberately manufactured to have low dissipative losses. Consequently, it is useful to consider how the foregoing theory is modified by considering transmission lines to be lossless.

A lossless line will have $R = G = 0$, and Eq. (1.10) will be modified such that the characteristic impedance of the line is real and given by

$$Z_O = \sqrt{\frac{L}{C}}.\tag{1.27}$$

Also, since there are zero losses in the line, the attenuation coefficient, α, will be zero and the propagation coefficient will only represent the phase behaviour of the wave on the line, i.e.

$$\gamma = j\beta = j\omega\sqrt{LC}\tag{1.28}$$

---

1 The phase velocity is the velocity with which phase of a sinusoid is transmitted down the line. It is represented by v_p, where $v_p = \omega/\beta$. Phase velocity is discussed in more detail in Appendix 1.C.5, in relation to propagation through waveguides.

and

$$\beta = \omega\sqrt{LC}. \tag{1.29}$$

With $\gamma = j\beta$, the expression for the input impedance of a transmission line is also modified, and Eq. (1.18) becomes

$$Z_{in} = \frac{Z_O(Z_L\cosh(j\beta l) + Z_O\sinh(j\beta l))}{(Z_O\cosh(j\beta l) + Z_L\sinh(j\beta l))}. \tag{1.30}$$

Recalling that $\cosh(jx) = \cos(x)$ and $\sinh(jx) = j\sin(x)$, Eq. (1.30) can be rewritten as

$$Z_{in} = \frac{Z_O(Z_L\cos(\beta l) + jZ_O\sin(\beta l))}{(Z_O\cos(\beta l) + jZ_L\sin(\beta l))} \tag{1.31}$$

or

$$Z_{in} = \frac{Z_O(Z_L + jZ_O\tan(\beta l))}{(Z_O + jZ_L\tan(\beta l))}. \tag{1.32}$$

This is an important expression giving the input impedance, Z_{in}, of a loss-free transmission line terminated in Z_L in terms of the phase propagation constant, β, characteristic impedance, Z_O, and line length, l.

Example 1.1 The following line constants apply to a lossless transmission line operating at 100 MHz: $L = 0.5\,\mu H/m$, $C = 180\,pF/m$.
Determine:

(i) The characteristic impedance of the line.
(ii) The phase propagation constant.
(iii) The velocity of propagation on the line.
(iv) The phase change over a 20 cm length of the line.

Solution

(i) $Z_O = \sqrt{\dfrac{L}{C}} = \sqrt{\dfrac{0.5 \times 10^{-6}}{180 \times 10^{-12}}}\Omega = 52.7\ \Omega,$

(ii) $\beta = \omega\sqrt{LC} = 2\pi \times 10^8 \times \sqrt{0.5 \times 10^{-6} \times 180 \times 10^{-12}}\ rad/m = 5.96\ rad/m,$

(iii) $v_P = \dfrac{1}{\sqrt{LC}} = \dfrac{1}{\sqrt{0.5 \times 10^{-6} \times 180 \times 10^{-12}}}\ m/s = 1.05 \times 10^8\ m/s,$

(iv) $\phi = \beta l = 5.96 \times 0.2\ rad = 1.192\ rad \equiv 68.29°.$

Example 1.2 A particular lossless transmission line has a characteristic impedance of 75 Ω, and a phase constant of 4 rad/m. Determine the input impedance of a 30 cm length of the transmission line when it is terminated by an impedance of $(100 - j50)\ \Omega$.

Solution

$$\phi = \beta l = 4 \times 0.3\ rad = 01.2\ rad \equiv 68.7°.$$

Using Eq. (1.32):

$$Z_{in} = \frac{Z_O(Z_L + jZ_O\tan(\beta l))}{(Z_O + jZ_L\tan(\beta l))}$$

$$= \frac{75 \times (100 - j50 + j75 \times \tan(68.7°))}{75 + j(100 - j50) \times \tan(68.7°)}\ \Omega$$

$$= 39.87\angle 3.26°\ \Omega$$

$$\equiv (39.81 + j2.27)\ \Omega.$$

1.5 Matched and Mismatched Transmission Lines

Using Eq. (1.32), we can establish the conditions for matching a lossless transmission line of characteristic impedance, Z_O, terminated in a load, Z_L:

(i) *Matched line.* If $Z_L = Z_O$, then $Z_{in} = Z_O$ and the line is described as being matched, and all of the energy travelling from the sending end will be absorbed by the load and there will be no reflected (reverse) wave.

(ii) *Totally mismatched line.* If the load impedance is replaced by either a short-circuit, or an open-circuit, the line is described as being totally mismatched and no energy will be dissipated in the termination. The input impedance is then entirely reactive.

With $Z_L = 0$ (i.e. a short-circuit) Eq. (1.32) becomes:

$$Z_{in} = Z_{S/C} = jZ_O \tan(\beta l). \tag{1.33}$$

With $Z_L = \infty$ (i.e. an open-circuit) Eq. (1.32) becomes:

$$Z_{in} = Z_{O/C} = \frac{Z_O}{j \tan(\beta l)} = -jZ_O \cot(\beta l). \tag{1.34}$$

(iii) *Partially mismatched line.* With an arbitrary value of load impedance, some of the incident energy will be reflected from the termination, giving rise to a standing wave as described in Section 1.6.

1.6 Waves on a Transmission Line

If $Z_L \neq Z_O$, a transmission line will be mismatched and some of the energy will be reflected from the load. Under these circumstances, the incident and reflected travelling waves will interact to form an interference pattern. Since the incident and reflected waves must be at the same frequency if the load is a passive impedance, the interference pattern will take the form of a standing wave, with the maxima and minima of the pattern in fixed positions. The distance between two adjacent maxima or minima must be $\lambda/2$, where λ is the wavelength on the line. A typical voltage standing wave pattern is shown in Figure 1.4.

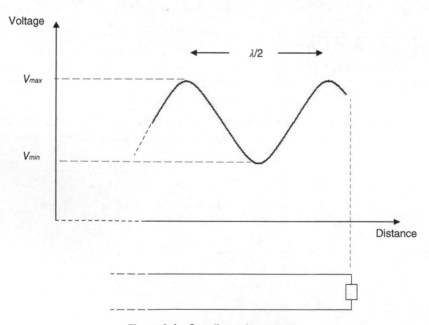

Figure 1.4 Standing voltage wave.

The voltage at the maximum point of the pattern will be $V_{max} = |V_F| + |V_R|$ and at the minimum point $V_{min} = |V_F| - |V_R|$, where V_F and V_R are the peak voltages of the forward and reflected waves, respectively. The degree of mismatch on a transmission can be specified by two parameters, namely the voltage standing wave ratio (VSWR) and the reflection coefficient, ρ. These two parameters are defined as follows.

VSWR is defined as

$$VSWR = \frac{V_{max}}{V_{min}} = \frac{|V_F| + |V_R|}{|V_F| - |V_R|}. \tag{1.35}$$

Since we know that $V_R = 0$ for a matched load, and $V_R = V_F$ for a total mismatch where all the energy is reflected, we can deduce the range of VSWR values as

$$1 \le (VSWR) \le \infty. \tag{1.36}$$

When $Z_L = Z_O$, the VSWR is unity.

Reflection coefficient, ρ, is defined as

$$\rho = \frac{V_R}{V_F}. \tag{1.37}$$

It follows that the range of the magnitude of ρ is given by

$$0 \le |\rho| \le 1, \tag{1.38}$$

with zero being the best value, and unity corresponding to total reflection from the load. It should be noted that ρ is a complex quantity, giving both magnitude and phase information, and this is an essential parameter in the design of matching networks, which will be discussed later in the chapter.

Clearly, there must be some relationship between VSWR and reflection coefficient, since both parameters provide information about the degree of reflection from a load, and this relationship is shown in Eq. (1.39)

$$VSWR = \frac{V_{max}}{V_{min}} = \frac{|V_F| + |V_R|}{|V_F| - |V_R|} = \frac{1 + \left|\dfrac{V_R}{V_F}\right|}{1 - \left|\dfrac{V_R}{V_F}\right|} = \frac{1 + |\rho|}{1 - |\rho|}. \tag{1.39}$$

Also, we can determine a relationship between reflection coefficient and impedance by first rewriting Eq. (1.11) in the form

$$V = V_F + V_R$$
$$I = \frac{V_F - V_R}{Z_O}. \tag{1.40}$$

Then

$$Z = \frac{V}{I} = Z_O \frac{V_F + V_R}{V_F - V_R} = Z_O \frac{1 + \dfrac{V_R}{V_F}}{1 - \dfrac{V_R}{V_F}} = Z_O \frac{1 + \rho}{1 - \rho}. \tag{1.41}$$

Rearranging Eq. (1.41) gives

$$\rho = \frac{Z - Z_O}{Z + Z_O}. \tag{1.42}$$

Example 1.3   What is the VSWR corresponding to a reflection coefficient of $0.4\angle -22°$?

Solution

$$VSWR = \frac{1 + |\rho|}{1 - |\rho|} = \frac{1 + 0.4}{1 - 0.4} = \frac{1.4}{0.6} = 2.33.$$

1.7 The Smith Chart

The Smith chart, developed by J.B. Smith in 1935, is a graphical tool used in the design of RF and microwave circuits. Whilst a technique that involves graphical manoeuvres is liable to significant reading and plotting errors, the Smith chart is still useful in giving a quick visual appreciation of a circuit problem, which can subsequently be reworked using CAD to give precise design information.

One of the main modern applications of the Smith chart is in measuring instrumentation to display circuit parameters as a function of frequency, and the chart forms an essential part of the display in modern network analyzers (see Chapter 7).

1.7.1 Derivation of the Smith Chart

Equation (1.42) can be written in terms of normalized impedances as

$$\rho = \frac{z-1}{z+1}, \tag{1.43}$$

where z represents a normalized impedance, defined as

$$z = \frac{Z}{Z_O}, \tag{1.44}$$

where Z_O is the characteristic impedance of the line.

(*It should be noted that we have used the normal convention whereby **normalized** values of impedance, resistance, and reactance are represented by lower-case letters z, r, and x, respectively.*)

Writing $z = r + jx$, Eq. (1.43) can be expanded as

$$\rho = \frac{(r+jx)-1}{(r+jx)+1} = \frac{(r-1)+jx}{(r+1)+x}. \tag{1.45}$$

Since we know the reflection coefficient is a complex quantity, representing magnitude and phase, we can write ρ as

$$\rho = U + jV. \tag{1.46}$$

Combining Eqs. (1.45) and (1.46) gives

$$U + jV = \frac{(r-1)+jx}{(r+1)+jx}. \tag{1.47}$$

We can solve Eq. (1.47) by equating the real and imaginary parts, and after some laborious, but routine maths, we obtain

$$\left(U - \frac{r}{r+1}\right)^2 + V^2 = \left(\frac{1}{r+1}\right)^2 \tag{1.48}$$

and

$$(U-1)^2 + \left(V - \frac{1}{x}\right)^2 = \left(\frac{1}{x}\right)^2. \tag{1.49}$$

It can be seen that Eqs. (1.48) and (1.49) are both in the form of equations representing circles in the $U - V$ plane.

For a particular value of r, Eq. (1.48) represents a circle of radius $\frac{1}{r+1}$ with a centre at $\left(\frac{r}{r+1}, 0\right)$. The value of r is therefore constant around any particular circle. These are termed constant normalized resistance circles and examples are drawn in Figure 1.5.

Similarly, for a particular value of x, Eq. (1.49) represents a circle of radius $\frac{1}{x}$ with centre at $\left(1, \frac{1}{x}\right)$. These are constant normalized reactance circles, and examples are drawn in Figure 1.6, where it should be noted that x can take both positive and negative values, since we can have both positive and negative reactance in a practical circuit. It can also be seen that the circle representing $x = 0$ has an infinite radius, and is therefore represented by a straight line coincident with the U-axis. For reference, the $r = 0$ circle has also been shown in Figure 1.6.

The sets of circles shown in Figures 1.5 and 1.6 can be combined onto a single $U - V$ plot, and this forms the Smith chart. A typical Smith chart is shown in Figure 1.7. Note that only reactance lines that lie within the $r = 0$ circle are shown, since normalized resistance values less than zero have no physical meaning. Note also that the resistance and reactance circles constituting the Smith chart are drawn in the polar diagram plane of the reflection coefficient, and concentric circles, centred on the $U = 0$ and $V = 0$ origin, represent values of constant reflection coefficient amplitude. These circles are not normally included on the Smith chart, but are drawn on by the user, as will be shown in subsequent examples.

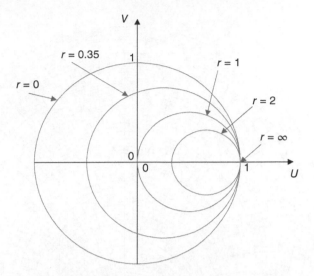

Figure 1.5 Normalized constant resistance circles.

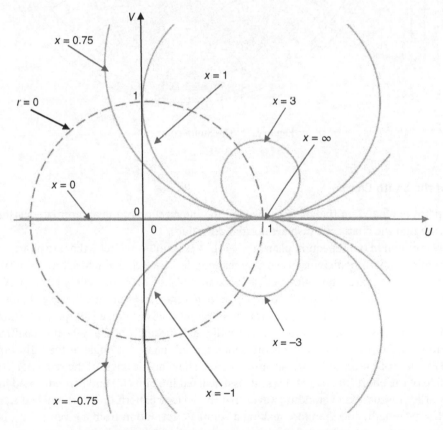

Figure 1.6 Normalized constant reactance circles.

The Smith chart shown in Figure 1.7 is an example of a commercially drawn chart, and it can be seen that in addition to the normalized resistance and reactance circles that have been discussed, a set of scales has been provided on the left-hand side. These scales are a useful aid for plotting radial distances on the chart, which correspond to particular values of VSWR and reflection coefficient. The use of these scales will be demonstrated in the solutions of worked examples later in the chapter.

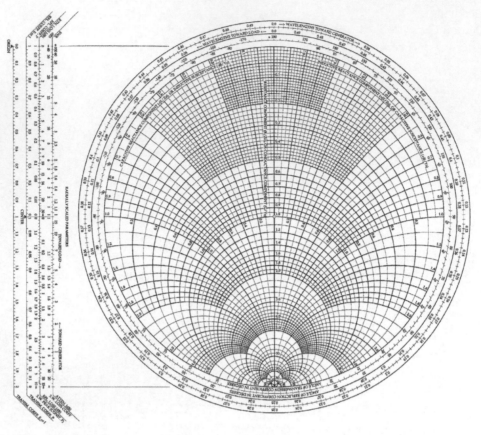

Figure 1.7 The Smith chart.

1.7.2 Properties of the Smith Chart

(i) An impedance point is plotted on the chart by locating the intersection of the appropriate resistance and reactance lines, remembering that the chart only displays normalized values.

(ii) Since the chart was drawn in the reflection plane, i.e. $\rho = U + jV$, points on circles that are plotted with their centre at the origin represent reflection coefficients of constant magnitude. These circles, which are not printed on the chart but need to be drawn by the user, are often referred to as constant VSWR circles (or usually just as VSWR circles). Moving around a constant VSWR circle corresponds to moving along a transmission line, thereby changing the angle of the reflection coefficient. However, the direction of rotation around a VSWR circle is important, since moving from the generator end of a transmission line towards the load will make the angle of the reflection coefficient more positive, and conversely moving from the load towards the generator will make the angle of the reflection coefficient more negative. To aid the user of the Smith chart, an annular scale showing the angle of the reflection coefficient is printed around the outside of the chart (the use of this scale is discussed in point (iii) and demonstrated in Example 1.4). We know from the earlier discussion of standing waves that the voltage pattern on a mismatched transmission line will repeat every half-wavelength, and therefore making a complete revolution from a given point on a VSWR circle, to return to the same point, must correspond to moving a distance $\lambda/2$ along the line. Appropriate wavelength scales are provided around the periphery of the chart. Note that there are two scales, denoting two different directions of movement. Distances on the Smith chart are always represented as electrical lengths, i.e. as fractions of a wavelength.

(iii) Reflection coefficients can be plotted directly on the chart. Radial distances correspond to the magnitude of the reflection coefficient on a linear scale, starting at 0 in the centre of the chart (the origin in the $U - V$ plane), and with a maximum of 1 at the maximum circumference. Some manufacturers of the Smith chart provide a reflection coefficient scale as an aid to plotting (Figure 1.7). Smith charts also contain circumferential scales corresponding to the angle of the reflection coefficient. So plotting a reflection coefficient point involves identifying the radial line through the appropriate angle, and then marking the required radial distance along this line.

(iv) A normalized impedance can be converted to a normalized admittance by rotating 180° around the chart. Once the normalized admittance has been plotted the resistance circles become conductance circles, and the reactance lines become susceptance lines.

The validity of this conversion from the impedance plane to the admittance plane can be established by first noting that a 180° rotation on the Smith chart corresponds to moving $\lambda/4$ along a transmission line; see the comment in point (ii). From Eq. (1.32) the impedance, $Z(l)$, at a distance l from the load is given by

$$\frac{Z(l)}{Z_O} = \frac{Z_L + jZ_O \tan(\beta l)}{Z_O + jZ_L \tan(\beta l)}. \tag{1.50}$$

Replacing l by $l + \lambda/4$, which corresponds to moving $\lambda/4$ along the transmission line we obtain

$$\frac{Z(l + \lambda/4)}{Z_O} = \frac{Z_L + jZ_O \tan(\beta(l + \lambda/4))}{Z_O + jZ_L \tan(\beta(l + \lambda/4))}$$

$$= \frac{Z_L + jZ_O \tan(\beta l + \beta\lambda/4)}{Z_O + jZ_L \tan(\beta l + \beta\lambda/4)},$$

i.e.

$$\frac{Z(l + \lambda/4)}{Z_O} = \frac{Z_L + jZ_O \tan(\beta l + \pi/2)}{Z_O + jZ_L \tan(\beta l + \pi/2)}. \tag{1.51}$$

Making use of the trigonometric relationship $\tan(x + \pi/2) = -\dfrac{1}{\tan(x)}$ we can rewrite Eq. (1.51) as

$$\frac{Z(l + \lambda/4)}{Z_O} = \frac{Z_L - jZ_O \dfrac{1}{\tan(\beta l)}}{Z_O - jZ_L \dfrac{1}{\tan(\beta l)}}. \tag{1.52}$$

Equation (1.52) can then be rearranged as

$$\frac{Z(l + \lambda/4)}{Z_O} = \frac{Z_O + jZ_L \tan(\beta l)}{Z_L + jZ_O \tan(\beta l)} = \frac{Z_O}{Z(l)}. \tag{1.53}$$

Thus, we see from Eq. (1.53) that the effect of moving a distance of $\lambda/4$ along the transmission line is to convert a normalized impedance into its reciprocal value, i.e. to convert a normalized impedance into a normalized admittance. Thus, any point which is plotted on the Smith chart as a normalized impedance can be converted directly to the equivalent normalized admittance by rotating 180° around the chart. This is a particularly useful technique when the chart is used in the analysis of circuits which involve a combination of series and parallel elements, as will be demonstrated in worked examples later in the chapter.

Some of the key features of the Smith chart are shown in Figures 1.8 through 1.13. Where points have been plotted on the chart to illustrate the principles involved, it should be appreciated that there will be plotting errors, as with any graphical technique. Consequently, where Smith charts have been used in this book to demonstrate RF design principles, readers should accept that precise data can only be obtained through use of appropriate CAD software.

Figure 1.8 shows examples of particular normalized constant resistance and reactance lines. The values of the resistance circles are shown on a vertical scale through the centre of the chart, and the values of the reactance line are shown on a scale around the periphery of the chart.

Impedance points are plotted on the chart by locating the intersection of the appropriate resistance and reactance lines. As an example, Figure 1.9 shows the position of the normalized impedance $0.3 + j0.6$, which is at the intersection of the 0.3 normalized resistance circle and the 0.6 normalized reactance line. Also shown in Figure 1.9 are impedance points of particular interest, namely the short-circuit, open-circuit, and matched impedance positions.

As mentioned earlier in the chapter, the Smith chart can also be used to plot and manipulate admittance data. In the admittance plane the 'real' circles printed on the chart become normalized conductance circles, and the 'imaginary' lines represent normalized susceptance. Figure 1.10 shows examples of admittances plotted on the chart. In the admittance plane the point $y = 0.3 + j0.6$ represents a normalized admittance with a normalized conductance of 0.3 and a normalized susceptance of 0.6.

VSWR circles were discussed in Section 1.7.2 (ii). These concentric circles can easily be plotted on the chart using the VSWR scale, which is one of the scales normally printed alongside the plotting area. The plot of a *VSWR = 4* circle is shown in Figure 1.11, where the radius of the circle has been obtained from the VSWR scale. Note that on the scales printed on most Smith charts the VSWR scale is identified simply as the SWR scale.

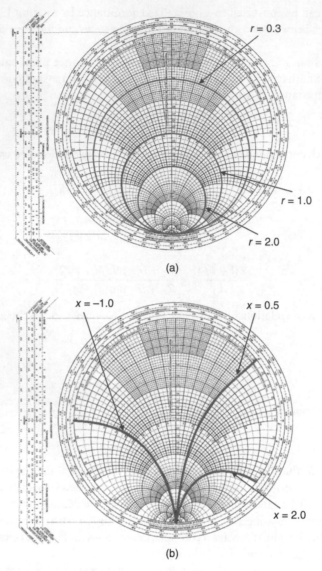

(a)

(b)

Figure 1.8 Examples of constant resistance lines (a) and reactance lines (b).

The procedure for plotting a reflection coefficient point on the chart is shown in Figure 1.12. In this case we show the plotting of point corresponding to a reflection coefficient $\rho = 0.7 \angle 60°$. Using the reflection coefficient scale a concentric circle of radius 0.7 is first plotted. A radial line is then drawn from the centre of the chart to pass through the required angle (60° in this example) on the reflection coefficient scale, which is printed around the periphery of the plotting area. Where the radial line intersects the drawn circle gives the location of the required reflection coefficient.

The impedance at any point on a loss-free transmission line terminated with a particular load lies on a VSWR circle. Using the Smith chart it is straightforward matter to find the impedance at a given distance from the load; the procedure is illustrated in Figure 1.13. The normalized impedance, z_L, of the load is first plotted, and a VSWR circle is drawn through z_L. A radial line drawn from the centre of the chart through z_L establishes the position, s_1, of the load on the wavelength scale printed around the outside of the chart. The impedance at an electrical distance d from the load is then found by moving clockwise (load-to-generator) around the wavelength scale to a new position s_2, where $d = s_2 - s_1$. A radial line is then drawn from s_2 to the centre of the chart. Where this radial line intersects, the VSWR circle gives the normalized impedance a distance d from the load. It is important to note that when using the Smith chart distances can only be represented as electrical distances, i.e. the distance expressed as a fraction of a wavelength at the frequency being used, since on the chart we only know that one revolution corresponds to half of one wavelength measured along the line.

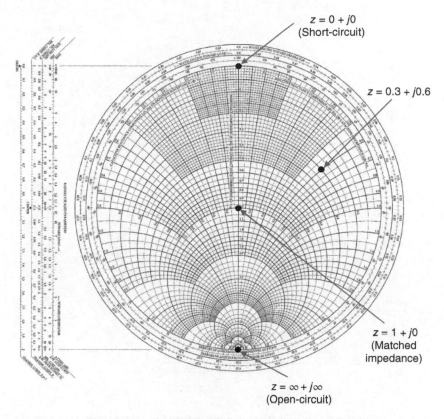

Figure 1.9 Examples of normalized impedance points.

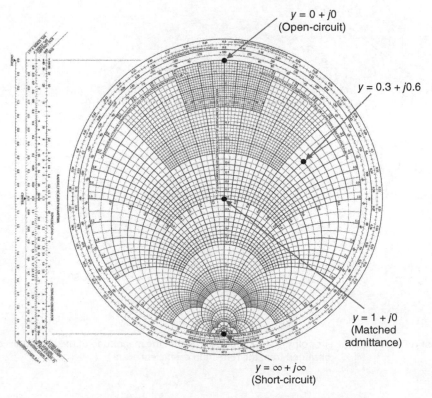

Figure 1.10 Examples of normalized admittance points.

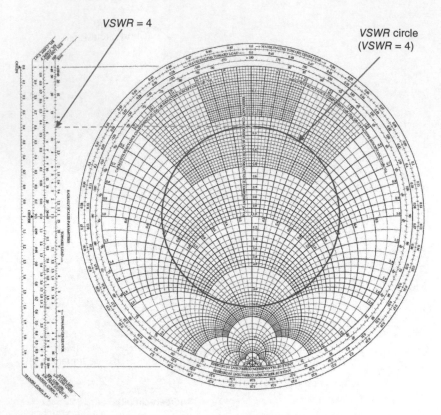

Figure 1.11 Plot of VSWR circle (*VSWR* = 4). (Note that the radius of the circle is obtained from the SWR scale.)

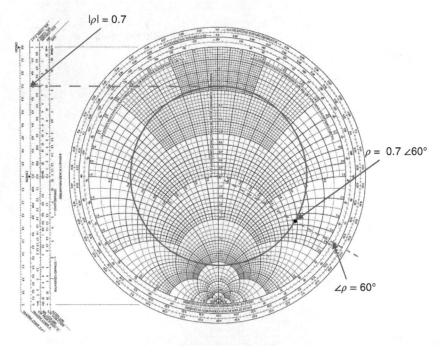

Figure 1.12 Plot of reflection coefficient point, $\rho = 0.7 \angle 60°$. (Note that the magnitude of the reflection coefficient point is obtained from the reflection coefficient scale.)

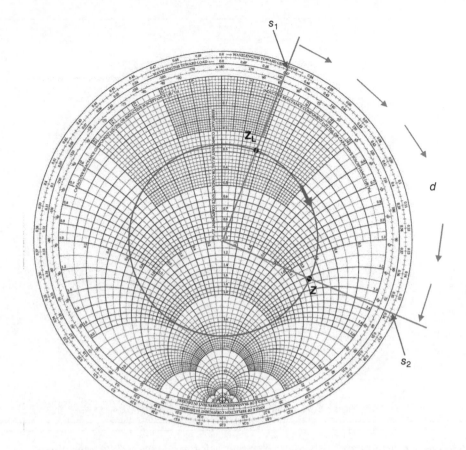

Figure 1.13 Impedance, Z, at a distance d in front of a load Z_L.

Example 1.4 A lossless transmission line having a characteristic impedance of 50 Ω is terminated by an impedance $(120 + j40)\,\Omega$.

Determine:

 (i) The normalized impedance of the load.
 (ii) The reflection coefficient of the load.
(iii) The VSWR on the line.

Solution

(i) $z_L = \dfrac{120 + j40}{50} = 2.4 + j0.8.$

(ii) and (iii)

Referring to the Smith chart shown in Figure 1.14: After plotting the normalized load impedance, and drawing the VSWR circle, we obtain

$$\rho = 0.46 \angle 17°.$$

$$VSWR = 2.7.$$

(Continued)

(Continued)

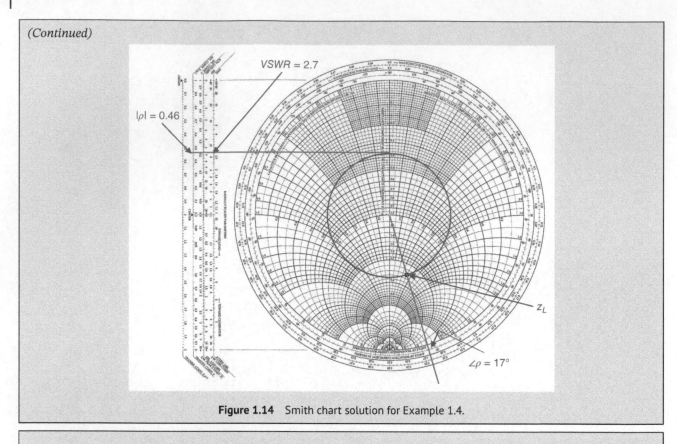

Figure 1.14 Smith chart solution for Example 1.4.

Example 1.5 A lossless transmission line having a characteristic impedance of 75 Ω is terminated by a load impedance $(18 - j30)\,\Omega$. Determine the impedance on the line at a distance of 0.175λ from the load.

Solution

Normalized load impedance, $z_L = \dfrac{18 - j30}{75} = 0.24 - j0.4$.

Moving 0.174λ clockwise around the VSWR circle from the plotted load impedance gives a point of intersection $z = 0.33 + j0.77$. Therefore, the required impedance is $Z = (0.33 + j0.77) \times 75\,\Omega = (24.75 + j57.75)\,\Omega$ (Figure 1.15).

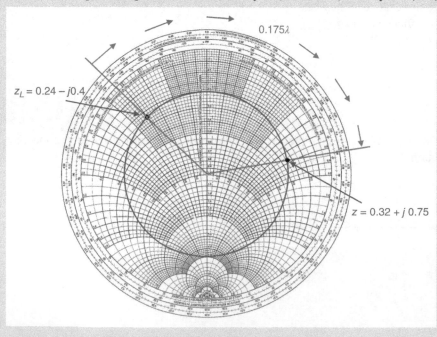

Figure 1.15 Smith chart solution for Example 1.5.

Example 1.6 A lossless transmission line having a characteristic impedance of 50 Ω is terminated by a load impedance $(18.5 + j25.0)$ Ω. The velocity of propagation on the line is 2×10^8 m/s.

Determine:

(i) The admittance of the load.

(ii) The admittance 35 mm from the load at a frequency of 700 MHz.

Solution (Referring to Figure 1.16)

(i) $z_L = \dfrac{18.5 + j25.0}{50} = 0.37 + j0.50.$

Rotating through 180° on the Smith chart gives y_L:

$$y_L = 0.90 - j1.37 \quad \Rightarrow \quad Y_L = (0.90 - j1.37) \times \frac{1}{50} \ \text{S} = (18.0 - j27.4) \ \text{mS}.$$

(ii) $\lambda = \dfrac{2 \times 10^8}{700 \times 10^6} \ \text{m} = 285.7 \ \text{mm} \quad \Rightarrow \quad 35 \ \text{mm} = \dfrac{35}{285.7}\lambda = 0.123\lambda.$

Rotating 0.123λ (clockwise) around VSWR circle gives y_1:

$$y_1 = 0.32 - j0.27 \quad \Rightarrow \quad Y_1 = (0.32 - j0.27) \times \frac{1}{50} \ \text{S} = (6.4 - j5.4) \ \text{mS}.$$

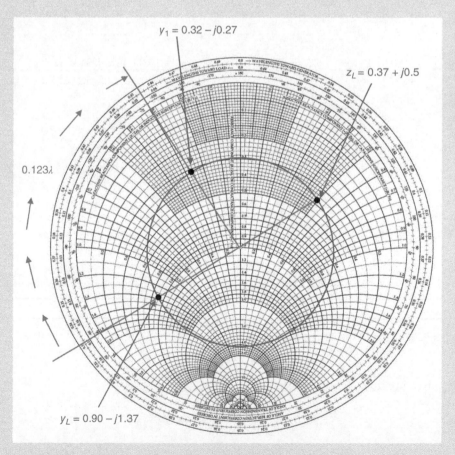

Figure 1.16 Smith chart solution for Example 1.6.

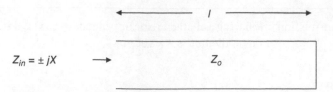

Figure 1.17 Schematic view of a transmission line stub.

1.8 Stubs

Stubs are short lengths of lossless transmission line that are terminated either in a short circuit or an open circuit. The input impedance of such stubs will be purely imaginary, as shown by Eqs. (1.33) and (1.34). Figure 1.17 shows a stub formed by a length of lossless transmission line of characteristic impedance, Z_O, terminated with a short circuit.

Thus, we can use a stub to create a positive or negative reactance (i.e. inductance or capacitance) of any magnitude, at a given frequency, purely by adjusting the length of the stub. Stubs provide very useful series or shunt matching elements at RF and microwave frequencies, where lumped reactive components suffer from unwanted parasitics, and where the physical geometry of lumped components may be incompatible with the circuit fabrication technology.

Example 1.7 A short-circuited stub, made from lossless 50 Ω transmission line, is to be used to create an inductance of 45 nH at a frequency of 820 MHz. If the velocity of propagation on the transmission line is 2.2×10^8 m/s, calculate the required length of the stub.

Solution
At 820 MHz:

$$450 \text{ nH} \quad \Rightarrow \quad jX_L = j(2\pi \times 820 \times 10^6 \times 45 \times 10^{-9}) \ \Omega = j231.9 \ \Omega,$$

$$\lambda = \frac{2.2 \times 10^8}{820 \times 10^6} \text{ m} = 268.30 \text{ mm}.$$

Using Eq. (1.33):

$$Z_{in} = jZ_O \tan(\beta l)$$

$$j231.9 = j50 \tan\left(\frac{2\pi}{268.30}l\right),$$

$$\tan\left(\frac{2\pi}{268.30}l\right) = \frac{231.9}{50} = 4.64,$$

$$\frac{2\pi}{268.30}l = 77.84 \times \frac{\pi}{180},$$

$$l = 58.00 \text{ mm}.$$

Comment: The length of the stub can also be found using the Smith chart, as shown through Example 1.8.

Example 1.8 Repeat Example 1.7 using the Smith chart.

Solution
Normalizing the required input reactance gives:

$$jx_{in} = \frac{j231.9}{50} = j4.64 \text{ (from Example 1.7)}.$$

Plotting $z = 0 + j4.64$ on the Smith chart and reading-off distance to $z_{S/C}$ gives $l = 0.216\lambda$ (Figure 1.18), i.e.

$$l = 0.216\lambda = 0.216 \times 268.3 \text{ mm} = 57.95 \text{ mm}.$$

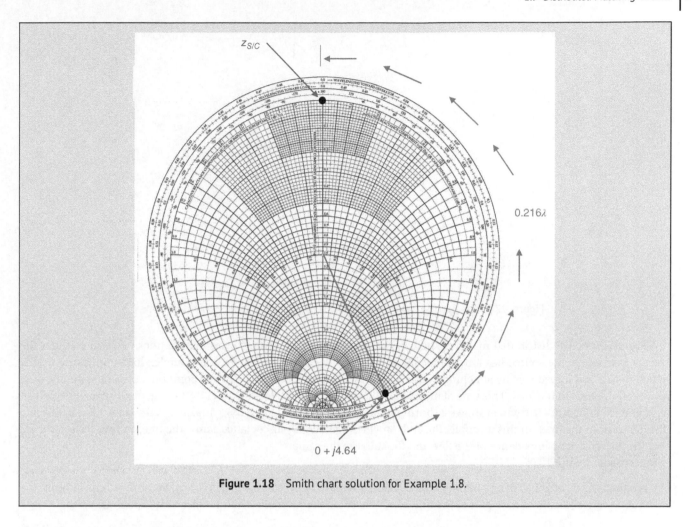

Figure 1.18 Smith chart solution for Example 1.8.

1.9 Distributed Matching Circuits

The primary function of a matching circuit is to create an impedance that will allow maximum power transfer from a source to a load. Matching networks are normally made from lossless distributed or lumped reactances. Distributed reactances are defined as those whose physical size is an appreciable fraction of a wavelength at the operating frequency, whereas the size of lumped reactances is small compared to a wavelength.

Figure 1.19 shows a load impedance, Z_L, connected to a source having an impedance Z_S, with an intervening matching network.

In order to ensure maximum power is transferred from the source to the load, the matching network must be designed so as to create an input impedance Z_{in}, which is the conjugate of the source impedance, i.e.

$$Z_{in} = Z_S^*. \tag{1.54}$$

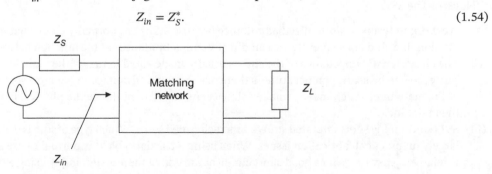

Figure 1.19 Matching network between source and load.

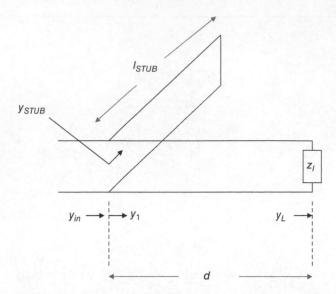

Figure 1.20 Single-stub tuner employing a short-circuited, shunt-connected stub.

A simple, effective distributed matching network can be made using lossless transmission lines. We will consider the special case where the source has a real impedance of Z_O at a specified frequency. Any complex load impedance can be matched to a real impedance by attaching a stub of the correct length at an appropriate point on a transmission line connecting the source to the load. This type of matching circuit is called a single-stub tuner (SST). The theory can be explained by reference to Figure 1.20, which shows a short-circuited shunt stub, of length l_{STUB} attached to the transmission line at a distance d from the load. In this example, the stub provides a matching susceptance across the line, and we have assumed that the characteristic impedance of the line and the stub are the same.

Referring to Figure 1.20:

(1) A shunt stub is attached at a point on the line where the normalized conductance is unity, because we can only use the stub to cancel the imaginary part of the line admittance. Let the normalized admittance at this point be y_1, so

$$y_1 = 1 \pm jb. \tag{1.55}$$

(2) The length of the stub is adjusted to create a normalized input susceptance that will cancel the residual susceptance on the main line, i.e. we need

$$y_{STUB} = \mp jb. \tag{1.56}$$

(3) Since the stub and the main transmission line are in parallel, the normalized input admittance at the input to the tuner will be the sum of the normalized stub admittance and the normalized admittance of the line

$$y_{in} = y_1 + y_{STUB} = (1 \pm jb) \mp jb = 1. \tag{1.57}$$

After de-normalizing we have $Y_{in} = Y_O$ and the line is therefore matched.

Notes on the SST:

(i) We work in terms of normalized admittance because stubs are normally connected in parallel with the main transmission line, and consequently we can directly sum admittances at the junction between the stub and the line.

(ii) The characteristic impedance of the stub is usually made equal to that of the main line, although this does not necessarily have to be so. In practical tuners there may be some advantage in having a stub of higher impedance than that of the main line, which means smaller dimensions, so as to minimize the physical discontinuity at the junction with the main line.

(iii) SSTs can employ short-circuited stubs or open-circuited stubs. Which type of stub is used in a particular design depends largely on practical fabrication issues. When using a coaxial cable it is normal to use short-circuited stubs because it is relatively easy to create a good short circuit at the end of the stub by simply using a metal disc to connect the centre conductor to the earthed sheath of the cable. If a coaxial line is left open at the end, the electromagnetic field will fringe into space, causing two problems. Firstly, the fringing field at the open end will cause the line to behave electrically as if it is longer than the actual length. Secondly, there will be some radiation from the open end, thereby introducing loss into the tuner.

When using planar circuits, such as a microstrip, the stubs are normally open-circuited, because of the physical problems in making connections through the substrate to the ground plane so as to create a short circuit. Substrates for RF and microwave circuits are often made of ceramic, which is a very hard material, and making connections through the substrate can be difficult and expensive, and usually involves laser drilling of the substrate.

(iv) The SST can theoretically be used to match any load impedance to a source, but it suffers from the disadvantage that the position of the stub must be changed to match different load impedances. This disadvantage can be overcome by using a double-stub tuner in which there are two matching stubs with a fixed distance between them. However, a double-stub tuner with a particular spacing between the stubs cannot match all possible values of load impedance. A triple-stub tuner, with the three stubs having a fixed spacing, is required to match any value of load impedance, although the practical use of this type of tuner can be difficult. An informative discussion of the theory and design of multi-stub tuners is given by Collin [1].

Procedure for designing an SST employing a short-circuited, shunt-connected stub, using the Smith chart:

(i) Plot the normalized load impedance, and convert this to the normalized load admittance.
(ii) Draw the VSWR circle through the load admittance.
(iii) Traverse the VSWR circle clockwise (i.e. load-to-generator) from the load admittance to intersect the unity conductance circle. The distance traversed (using the wavelength scale) gives the value of d, as an electrical length. The point of intersection with the unit circle is y_1 and equal to $1 \pm jb$.
(iv) Plot y_{STUB} ($=0 \mp jb$).
(v) Starting at y_{STUB} traverse the conductance circle counter-clockwise to $y_{S/C}$, the distance moved gives the electrical length of the stub, l_{STUB}. (The movement is counter-clockwise because as far as the stub is concerned we are moving from the generator end towards the short-circuited load.)
(vi) The design is completed by converting the electrical lengths of d and l_{STUB} into physical lengths using the appropriate guide wavelength.

Example 1.9   Design an SST that will match a load impedance $(80 - j65)\,\Omega$ to a $50\,\Omega$ source at a frequency of 1.3 GHz. The tuner is to employ a short-circuited stub connected in parallel with the main line. All of the cables used in the tuner have characteristic impedances of $50\,\Omega$, and a velocity of propagation of 2×10^8 m/s.

Solution

$$z_L = \frac{80 - j65}{50} = 1.6 - j1.3,$$

$$\lambda = \frac{2 \times 10^8}{1.3 \times 10^9}\ \text{m} = 153.85\ \text{mm}.$$

Referring to the Smith chart shown in Figure 1.21:

$$y_L = 0.38 + j0.31.$$

Rotating around the VSWR circle through y_L to intersect the unity conductance circle gives y_1:

$$y_1 = 1 + j1.14.$$

The distance moved gives d:

$$d = 0.166\lambda - 0.054\lambda = 0.112\lambda = 0.112 \times 153.85\ \text{mm} = 17.23\ \text{mm}.$$

For matching, we require $y_{STUB} = -j1.14$.

Rotating around the chart from y_{STUB} to $y_{S/C}$ gives the required length of the stub:

$$l_{STUB} = 0.250\lambda - 0.135\lambda = 0.115\lambda = 0.115 \times 153.85\ \text{mm} = 17.69\ \text{mm}.$$

(Continued)

(Continued)

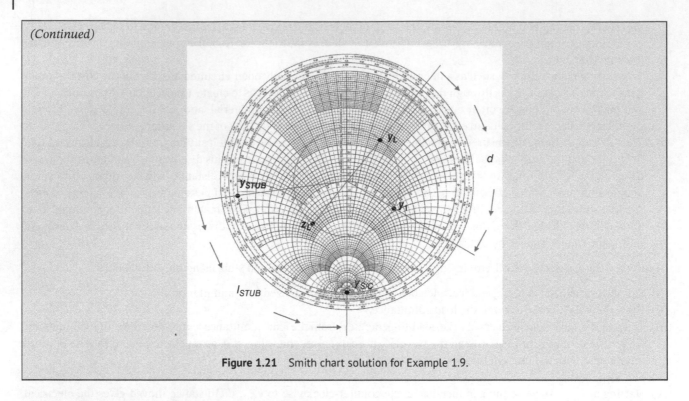

Figure 1.21 Smith chart solution for Example 1.9.

1.10 Manipulation of Lumped Impedances Using the Smith Chart

So far the discussion has focussed on the use of the Smith chart for situations involving transmission lines. Whilst the chart is often described as a transmission line calculator, it has many more general applications in electronics relating to the manipulation of impedance and admittance data.

Moving around the constant resistance line from any given impedance point on the chart corresponds to adding reactance to the impedance. Movements in the clockwise direction correspond to making the reactance more positive, and hence represent the addition of series inductance, whereas movements in the counter-clockwise direction correspond to making the reactance more negative, and therefore represent the addition of series capacitance. The effect of adding of inductance or capacitance to a given normalized impedance, z_1, is depicted in Figure 1.22.

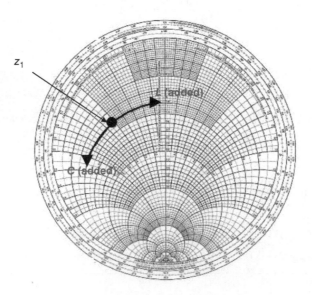

Figure 1.22 Effect of adding reactance to a normalized impedance, z_1.

Similarly, when working in the admittance plane, moving around the constant susceptance circles corresponds to adding the appropriate susceptance in shunt. In this case, clockwise movements represent the addition of capacitance, which has positive susceptance, and counter-clockwise movements the addition of inductance, which has negative susceptance.

Since we know that converting between impedance and admittance is very straightforward (a 180° rotation) on the Smith chart, it becomes very easy to analyze a network consisting of reactances connected in series and shunt. The technique for manipulating the data in such a network is demonstrated through Example 1.10.

Example 1.10   Using the Smith chart, determine the input impedance, Z_{in}, at 850 MHz of the network shown in Figure 1.23. The network consists of three lossless reactances, connected in a π-configuration, terminated by a load impedance $(20 - j15)\,\Omega$.

Firstly, the component values in the network must be normalized, so that data can be plotted on the Smith chart. For convenience, we will normalize with respect to 50 Ω.

Comment: When dealing with networks where there is no transmission line, and hence no characteristic impedance, we may normalize to any convenient value providing we de-normalize with respect to the same value. The normalization process in this case is merely a scaling operation to obtain a more convenient value to plot on the Smith chart.

$$z_L = \frac{20 - j15}{50} = 0.4 - j0.3.$$

Normalizing the reactance of the inductor:

$$5.15 \text{ nH}; j\frac{\omega L}{50} = j\frac{2\pi \times 850 \times 10^6 \times 5.15 \times 10^{-9}}{50} = j0.55.$$

Normalizing the susceptances of the capacitors:

$$6.93 \text{ pF}: j\frac{\omega C}{(1/50)} = j\omega C \times 50 = j2\pi \times 850 \times 10^6 \times 6.93 \times 10^{-12} \times 50 = j1.85,$$

$$4.87 \text{ pF}: j\frac{\omega C}{(1/50)} = j\omega C \times 50 = j2\pi \times 850 \times 10^6 \times 4.87 \times 10^{-12} \times 50 = j1.30.$$

The impedances and admittances at each of the junctions of the network are shown in Figure 1.24.

Working backwards from the load impedance, i.e. towards generator, we use the Smith chart to add the parallel components using admittance values, and add the series components using impedance values.

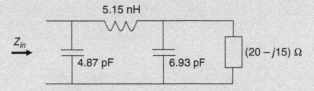

Figure 1.23   Circuit for Example 1.10.

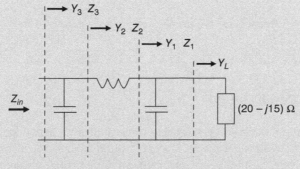

Figure 1.24   Nomenclature for solution for Example 1.10.

(Continued)

(Continued)

Step 1: Plot z_L

Step 2: Convert z_L to y_L

Step 3: Move 1.85 units clockwise around the susceptance line from y_L. This gives the position of y_1. *Note that this movement represents the addition of the 6.93 pF capacitor; we move clockwise because a capacitor has positive susceptance.*

Step 4: Convert y_1 to z_1

Step 5: Move 0.55 units clockwise around the reactance line from z_1. This gives the position of z_2. *Note that this movement represents the addition of the 5.15 nH inductor; we move clockwise because an inductor has positive reactance.*

Step 6: Convert z_2 to y_2

Step 7: Move 1.30 units clockwise around the susceptance line from y_2. This gives the position of y_3. *Note that this movement represents the addition of the 4.87 pF capacitor; we move clockwise because a capacitor has positive susceptance.*

Step 8: Convert y_3 to z_3

From the Smith chart (shown in Figure 1.25) we obtain:

$$z_3 = 0.36 + j0.37,$$

$$z_{in} \equiv z_3,$$

$$Z_{in} = (0.36 + j0.37) \times 50 \ \Omega = (18.0 + j18.5) \ \Omega.$$

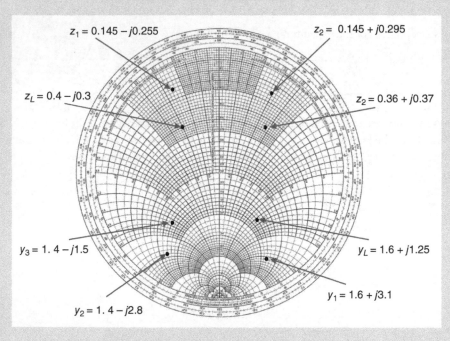

Figure 1.25 Smith chart solution for Example 1.10.

1.11 Lumped Impedance Matching

An alternative to the use of distributed transmission lines to create matching networks, as described in Section 1.9, is to use lumped elements in the form of lossless reactances. Normally, purely reactive elements are used in matching networks to avoid unnecessary dissipative losses. Although lumped elements suffer from unwanted parasitics at high frequencies, their use is essential in compact RF and microwave circuits (particularly integrated circuits) where size limitations prevent the use of transmission line elements, whose length may be a significant fraction of a wavelength. However, it must be remembered that the higher loss associated with lumped circuits means they tend to have lower Q-factors than distributed circuits.

1.11.1 Matching a Complex Load Impedance to a Real Source Impedance

A simple lumped-element matching network, consisting of two lossless reactances, one in series and the other in parallel, is shown in Figure 1.26, preceding a load whose impedance is Z_L. This network may be regarded as the lumped equivalent of the SST discussed in Section 1.9.

Theory: Let us consider that the matching network is to be used to match a load impedance, Z_L, to a source impedance of 50 Ω. The series reactance is used to create a normalized admittance, y_1, which has a real part of unity at the junction of the two reactances. The parallel reactance is then used to cancel out the normalized susceptance of y_1, thus making the normalized input admittance unity, and providing a match to the 50 Ω source.

The design procedure using the Smith chart can be deduced because we know three things concerning the impedance and admittance relationships:

(a) The real part of z_1 must be the same as the real part of z_L since these impedances differ only by the value of the series reactance, i.e.

$$z_1 = z_L \pm jx_S. \tag{1.58}$$

(b) The real part of y_1 must be the same as the real part of y_{in} since these admittances differ only by the value of the parallel susceptance, i.e.

$$y_{in} = y_1 \pm jb_P. \tag{1.59}$$

(c) For a match, the normalized input admittance must be of the form

$$y_{in} = 1 + j0. \tag{1.60}$$

Therefore, referring to the network configuration shown in Figure 1.26, the procedure using the Smith chart is:

(i) Plot the normalized load impedance, z_L.
(ii) Rotate the normalized unity resistance circle by 180°. *The reason for this construction is justified in step (iv).*
(iii) Traverse the constant resistance circle through z_L to intersect the rotated circle. The point of intersection is z_1, the movement from z_L to z_1 gives the value of the series component. *Note that there are two possible points of intersection on the rotated circle, giving rise to two possible vales of z_1, and hence two possible solutions.*
(iv) Move 180° from z_1 to find the position of y_1. *Note that the rotated circle was constructed to be the mirror-image of the normalized unity circle. Therefore, converting any normalized impedance point on this rotated circle to the equivalent normalized admittance will ensure that the normalized admittance lies on the unity conductance circle, thus satisfying the condition that the real part of y_1 must be unity.*
(v) Traverse the unity conductance circle from y_1 to the centre of the chart; this movement gives the value of the parallel component.

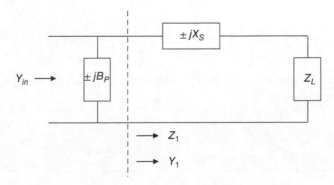

Figure 1.26 Lumped-element matching network.

Example 1.11   Design a lumped-element network that will match a load impedance $(20 - j40)\,\Omega$ to a 50 Ω source at a frequency of 2.4 GHz. The network is to be composed of two lossless reactances, with the configuration shown in Figure 1.26. Show that there are two possible solutions, and calculate the required reactive component values for each solution.

Solution

Normalizing the load impedance:
$$z_L = \frac{20 - j40}{50} = 0.4 - j0.8.$$

After rotating the unity resistance circle, and plotting the position of z_L, we see that the constant resistance circle through z_L will intersect the rotated circle at two points. These two points of intersection give rise to the two possible solutions.

First solution:

Consider the first intersection point, as shown in Figure 1.27.

We see that the addition of positive reactance is needed to move from z_L to z_1, the first point of intersection. Therefore, we need an inductance as the series component. After converting z_1 to the equivalent admittance, y_1, it can be seen that a negative susceptance is needed to move the admittance to the centre of the chart, which is the matched position. Thus, we need an inductor as the parallel component.

Using the data from the chart:
$$z_1 - z_L = (0.4 - j0.495) - (0.4 - j0.8) = j0.305$$

and
$$y_O - y_1 = 1.0 - (1.0 + j1.2) = -j1.2.$$

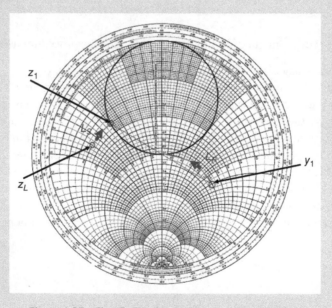

Figure 1.27   First Smith chart solution to Example 1.11.

Thus, we need a normalized series reactance of $j0.305$ and a normalized parallel susceptance of $-j1.2$,

i.e. $\quad j0.305 = \dfrac{j\omega L_S}{50} \quad \Rightarrow \quad L_S = \dfrac{50 \times 0.305}{\omega} = \dfrac{50 \times 0.305}{2\pi \times 2.4 \times 10^9}\ \mathrm{H} = 1.01\ \mathrm{nH},$

$\quad -j1.2 = -j\dfrac{1}{\omega L_P} \times 50 \quad \Rightarrow \quad L_P = \dfrac{50}{1.2 \times 2\pi \times 2.4 \times 10^9}\ \mathrm{H} = 2.76\ \mathrm{nH}.$

Second solution:

Looking at Figure 1.28, we see that by using a larger series inductor there is a valid intersection point at z_2, leading to a second solution.

Using data from the chart:

$$z_2 - z_L = (0.4 + j0.495) - (0.4 - j0.8) = j1.295,$$

$$y_O - y_2 = 1.0 - (1.0 - j1.2) = j1.2.$$

Thus, for the second solution we need a normalized series reactance of $j1.295$, and a normalized parallel susceptance of $j1.2$,

$$\text{i.e.}\quad j1.295 = j\frac{\omega L_S}{50} \quad \Rightarrow \quad L_S = \frac{50 \times 1.295}{\omega} = \frac{50 \times 1.295}{2\pi \times 2.4 \times 10^9}\ \text{H} = 4.29\ \text{nH},$$

$$j1.2 = j\omega C_P \times 50 \quad \Rightarrow \quad C_P = \frac{1.2}{2\pi \times 2.4 \times 10^9 \times 50} = 1.59\ \text{pF}.$$

Summary

The two possible matching networks that satisfy the requirements in Example 1.11 are shown in Figure 1.29.

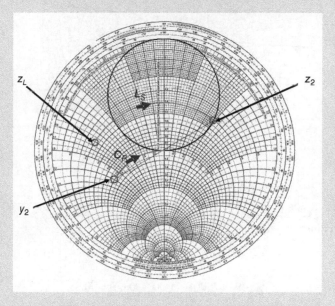

Figure 1.28   Second Smith chart solution to Example 1.11.

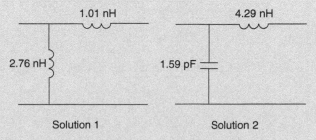

Figure 1.29   Matching networks for Example 1.11.

Example 1.12   The network of lossless reactances shown in Figure 1.26 is to be used to match a load impedance $(34 + j42f)\,\Omega$, where f is the frequency in GHz, to a 50 Ω source at a frequency of 2 GHz.

 (i) Calculate the required values of the reactive components. Show that there are two solutions, and calculate the values of the reactances in both cases.
 (ii) Choose one of the solutions found in part (i), and find the VSWR at the input to the network if the frequency is increased by 20%, and the component values are unchanged.

Solution
 (i) Normalizing the load impedance (Figure 1.30): $z_L = \dfrac{34 + j(42 \times 2)}{50} = 0.68 + j1.68$.

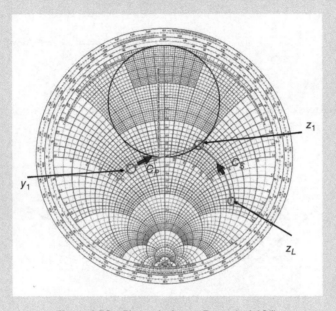

Figure 1.30   First solution to Example 1.12(i).

Using data from the Smith chart:

$$z_1 - z_L = (0.68 + j0.47) - (0.68 + j1.68) = -j1.21,$$

$$y_O - y_1 = 1.00 - (1.00 - j0.65) = j0.65.$$

Thus, we need a negative normalized series reactance of $-j1.21$ and a positive parallel normalized susceptance of $j0.65$,

i.e. $-j1.21 = -j\dfrac{1}{\omega C_S} \times \dfrac{1}{50} \;\Rightarrow\; C_S = \dfrac{1}{1.21 \times \omega \times 50} = \dfrac{1}{1.21 \times 2\pi \times 2 \times 10^9 \times 50}\,\mathrm{F} = 1.32\;\mathrm{pF},$

$$j0.65 = j\omega C_P \times 50 \;\Rightarrow\; C_P = \dfrac{0.65}{\omega \times 50} = \dfrac{0.65}{2\pi \times 2 \times 10^9 \times 50}\,\mathrm{F} = 1.03\;\mathrm{pF}.$$

Considering the second valid intersection point on the rotated circle gives the Smith chart solution shown in Figure 1.31.

Using data from the Smith chart:

$$z_2 - z_L = (0.68 - j0.47) - (0.68 + j1.68) = -j2.15,$$

$$y_O - y_2 = 1.00 - (1.00 + j0.65) = -j0.65.$$

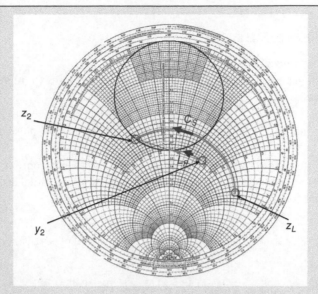

Figure 1.31 Second solution to Example 1.12(i).

Thus, we need negative series normalized reactance of $-j2.15$ and a negative parallel normalized susceptance of $-j0.65$,

i.e. $-j2.15 = -j\dfrac{1}{\omega C_S} \times \dfrac{1}{50}$ $\Rightarrow$ $C_S = \dfrac{1}{2.15 \times 50\omega} = \dfrac{1}{2.15 \times 2\pi \times 2 \times 10^9 \times 50}\,\text{F} = 0.74\ \text{pF},$

$-j0.65 = -j\dfrac{1}{\omega L_P} \times 50$ $\Rightarrow$ $L_P = \dfrac{50}{0.65 \times \omega} = \dfrac{50}{0.65 \times 2\pi \times 2 \times 10^9}\,\text{H} = 6.12\ \text{nH}.$

Summary

The two possible matching networks to satisfy the requirements in Example 1.12 are shown in Figure 1.32.

(ii) Choosing the *first* matching network from part (i):

New frequency: $f = 2.4$ GHz.

New value of load impedance: $Z_L^\dagger = (34 + j42 \times 2.4)\ \Omega = (34 + j100.8)\ \Omega.$

New value of normalized load impedance: $z_L^\dagger = \dfrac{34 + j100.8}{50} = 0.68 + j2.02.$

New value of normalized series reactance: $-j1.21 \times \dfrac{2}{2.4} = -j1.01.$

New value of normalized parallel susceptance: $j0.65 \times \dfrac{2.4}{2} = j0.78.$

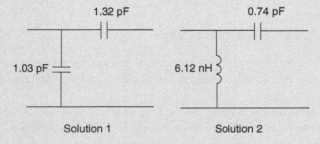

Solution 1   Solution 2

Figure 1.32 Matching networks for Example 1.12.

(Continued)

(Continued)

The procedure on the Smith chart is as follows:

(1) Plot the new normalized load impedance, $z_L^{\dagger}$ (see Figure 1.33).
(2) Traverse the constant resistance line through $z_L^{\dagger}$ counter-clockwise 2.02 units (to represent the new series reactance) to give a new value of $z_1^{\dagger}$.
(3) Convert $z_1^{\dagger}$ to $y_1^{\dagger}$.
(4) Traverse the constant conductance circle through $y_1^{\dagger}$ clockwise 0.78 units (to represent the new parallel susceptance) to give a value of y_{in}.
(5) Measure the radial distance from the centre of the chart to y_{in} and use the appropriate scale to find the VSWR at the input to the network. (Note that an alternative method to measuring the radial distance would be to plot the VSWR circle through y_{in} and find the VSWR as in Figure 1.11.)

Data from the Smith chart:

$$z_1^{\dagger} = z_L^{\dagger} + (-j1.01) = 0.68 + j2.02 - j1.01 = 0.68 + j1.01,$$

$$y_1^{\dagger} = 0.48 - j0.68,$$

$$y_{in} = y_1^{\dagger} + j0.78 = 0.48 - j0.68 + j0.78 = 0.48 + j0.10.$$

$$VSWR = 2.1.$$

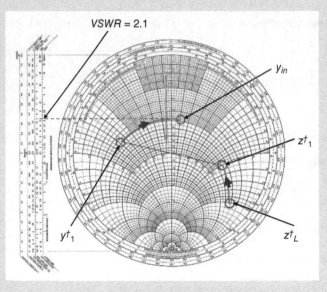

Figure 1.33 Solution to Example 1.12(ii).

Additional points to note about lumped-element matching:

(1) There are always two possible solutions for a particular network configuration (corresponding to the two possible points of intersection for Z_1). This gives the circuit designer an additional degree of design freedom in avoiding awkward-sized components.
(2) It will not be possible to achieve a match using the network configuration shown in Figure 1.26 if the value of the normalized load impedance lies within the unity resistance circle. In this case the configuration shown in Figure 1.34 must be used.

In the network shown in Figure 1.34, the parallel reactive element is used to create normalized impedance z_1 with a real part equal to unity. The imaginary part of z_1 is then cancelled using the series element of the matching network. The design procedure using the Smith chart is similar to that used for the circuit shown in Figure 1.26. The main difference is that we start by converting the load impedance into the equivalent normalized admittance, so that we can directly add the susceptance of the parallel element. As with the design of the earlier matching network, we need to draw a rotated

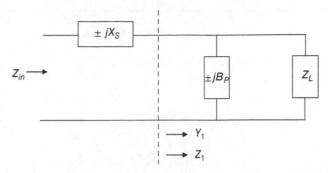

Figure 1.34 Lumped-element matching network 2.

unity circle on the Smith chart in order that we can convert y_1 to z_1 and ensure that the real part of z_1 is unity. The design procedure will be demonstrated through Example 1.13.

Example 1.13 Design a lumped-element network that will match a load impedance $(150 - j50)$ Ω to a 50 Ω source at a frequency of 4 GHz. The network is to be composed of two lossless reactances, with the configuration shown in Figure 1.34. Show that there are two possible solutions, and calculate the required reactance values for each solution.

Solution

Normalizing the load impedance: $z_L = \dfrac{150 - j50}{50} = 3 - j1$.

Having plotted z_L and converted to y_L, we see from Figure 1.35 that there are two possible directions of movement around the constant conductance circle to intersect the rotated circle, giving rise to two possible solutions.

First solution:

Moving clockwise from y_L to intersect the rotated circle we find y_1 as shown in Figure 1.35.

Using data from the chart:

$$y_1 - y_L = (0.3 + j0.455) - (0.3 + j0.1) = j0.355,$$

$$z_O - z_1 = 1 - (1 - j1.55) = j1.55.$$

Thus, we need a positive parallel normalized susceptance of $j0.355$ and a positive series normalized reactance of $j1.55$,

i.e. $j0.355 = j\omega C_P \times 50 \quad \Rightarrow \quad C_P = \dfrac{0.355}{50 \times \omega} = \dfrac{0.355}{50 \times 2\pi \times 4 \times 10^9} \mathrm{F} = 0.28 \ \mathrm{pF},$

$j1.55 = \dfrac{j\omega L_S}{50} \quad \rightarrow \quad L_S = \dfrac{1.55 \times 50}{\omega} = \dfrac{1.55 \times 50}{2\pi \times 4 \times 10^9} \mathrm{H} = 3.08 \ \mathrm{nH}.$

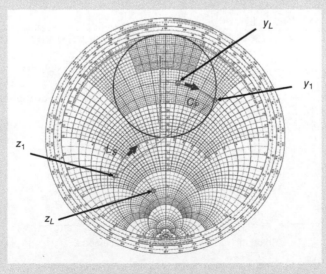

Figure 1.35 First solution to Example 1.13.

(Continued)

(Continued)

Second solution:

Moving counter-clockwise from y_L we obtain y_2 as shown in Figure 1.36.

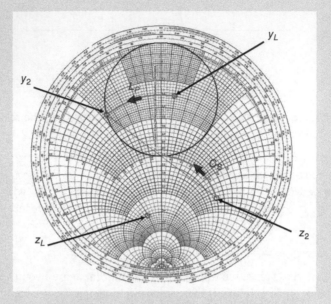

Figure 1.36 Second solution to Example 1.13.

Using data from the chart:

$$y_2 - y_L = (0.3 - j0.455) - (0.3 + j0.1) = -j0.555,$$

$$z_O - z_2 = 1 - (1 + j1.55) = -j1.55.$$

Thus, we need a negative parallel normalized susceptance of $-j0.555$ and a negative series normalized reactance of $-j1.55$,

i.e. $\quad -j0.555 = -\dfrac{j}{\omega L_P} \times 50 \quad \Rightarrow \quad L_P = \dfrac{50}{0.555 \times \omega} = \dfrac{50}{0.555 \times 2\pi \times 4 \times 10^9} \text{H} = 3.58 \text{ nH},$

$$-j1.55 = -\dfrac{j}{\omega C_S} \times \dfrac{1}{50} \quad \Rightarrow \quad C_S = \dfrac{1}{1.55 \times 2\pi \times 4 \times 10^9 \times 50} \text{F} = 0.51 \text{ pF}.$$

Summary

The two possible matching networks to satisfy the requirements in Example 1.13 are shown in Figure 1.37.

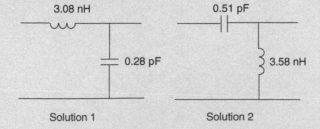

Figure 1.37 Matching networks for Example 1.13.

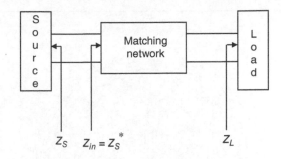

Figure 1.38 Matching complex impedances.

1.11.2 Matching a Complex Load Impedance to a Complex Source Impedance

In the previous section, we considered the matching of a complex load impedance to a real source. Whilst this is probably the most common situation encountered in practical designs, there may be situations, such as inter-stage matching in amplifiers, where we wish to match two complex impedances, i.e. match a complex load impedance to a complex source impedance. Such a match can be achieved at a single frequency by using two lossless reactances, as described in Section 1.11.1, but with a minor modification to the design technique using the Smith chart.

Suppose that we wish to match a complex source impedance, $Z_S = R_S + jX_S$, to a complex load impedance, $Z_L = R_L + jX_L$. The matching network must be designed to provide complex conjugate impedance matching between the source and the input to the network, i.e. the input impedance of the matching network must be equal to Z_S^*, as shown in Figure 1.38. (*Note that we have used the star notation to denote a complex conjugate quantity.*)

We will consider the matching network to have the same configuration as that shown in Figure 1.26. The design procedure on the Smith chart is similar to that previously discussed, with the exception that the series reactance must create a normalized admittance y_1 which has the same conductance as y_{in}. So the rotated circle must be formed by rotating the normalized conductance circle through z_L by 180°. The steps in the design are shown in Example 1.14. As with the previous examples of lumped element matching networks, there will be two valid solutions, corresponding to the two possible intersection points on the rotated circle.

Example 1.14 Design a lumped-element matching network to match a load impedance $(100 + j200)\,\Omega$ to a source whose impedance is $(25 + j130)\,\Omega$, at a frequency of 800 MHz. The matching network is to consist of two lossless reactances, connected in the configuration shown in Figure 1.26.

Solution

Normalizing the load impedance: $z_L = \dfrac{100 + j40}{50} = 2.0 + j0.8$.

Normalizing the source impedance: $z_S = \dfrac{25 + j130}{50} = 0.5 + j2.6$.

Note that the required normalized input impedance, z_{in}, of the matching network is given by

$$z_{in} = z_S^* = 0.5 - j2.6.$$

The initial steps in the design are to plot z_{in}, find the corresponding position of y_{in}, and rotate the constant conductance circle through z_L by 180°.

First solution:

It can be seen from Figure 1.39 that the normalized conductance line through y_{in} intersects the rotated circle at two points, leading to two possible solutions. The first solution will use the intersection point at y_1.

Using data from the Smith chart:

$$z_1 - z_L = (2.0 - j4.5) - (2.0 + 0.8) = -j5.3,$$

$$y_{in} - y_1 = (0.08 + j0.37) - (0.08 + j0.18) = j0.19.$$

(Continued)

(Continued)

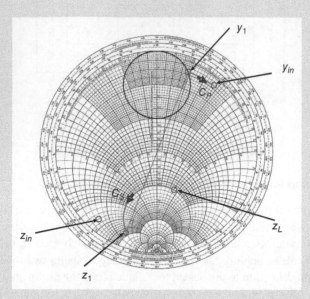

Figure 1.39 First solution to Example 1.14.

Thus, we need a negative series normalized reactance of $-j5.3$ and a positive parallel normalized susceptance of $j0.19$,

$$\text{i.e.} \quad -j5.3 = -j\frac{1}{\omega C_S} \times \frac{1}{50} \quad \Rightarrow \quad C_S = \frac{1}{5.3 \times \omega \times 50} = \frac{1}{5.3 \times 2\pi \times 800 \times 10^6 \times 50} \text{F} = 0.75 \ \text{pF},$$

$$j0.19 = j\omega C_P \times 50 \quad \Rightarrow \quad C_P = \frac{0.19}{\omega \times 50} = \frac{0.19}{2\pi \times 800 \times 10^6 \times 50} \text{F} = 0.76 \ \text{pF}.$$

Second solution:

Choosing the second intersection point, y_2, gives the solution shown in Figure 1.40.

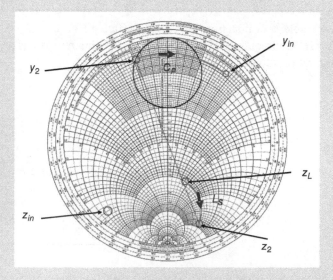

Figure 1.40 Second solution to Example 1.14.

From the Smith chart we obtain:

$$z_2 - z_L = (2.0 + j4.5) - (2.0 + j0.8) = j3.7,$$

$$y_{in} - y_2 = (0.08 + j0.37) - (0.08 - j0.18) = j0.55.$$

Thus, we need a positive series normalized reactance of $j3.7$ and a positive normalized parallel susceptance of $j0.55$,

$$\text{i.e.} \quad j3.7 = j\omega L_S \times \frac{1}{50} \quad \Rightarrow \quad L_S = \frac{3.7 \times 50}{\omega} = \frac{3.7 \times 50}{2\pi \times 800 \times 10^6} \text{H} = 36.80 \ \text{nH},$$

$$j0.55 = j\omega C_p \times 50 \quad \Rightarrow \quad C_P = \frac{0.55}{\omega \times 50} = \frac{0.55}{2\pi \times 800 \times 10^6 \times 50} F = 2.19 \text{ pF}.$$

Summary

The two possible matching networks to satisfy the requirements in Example 1.14 are shown in Figure 1.41.

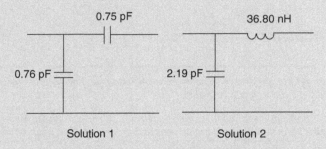

Figure 1.41 Matching networks for Example 1.14

1.12 Equivalent Lumped Circuit of a Lossless Transmission Line

One of the problems frequently encountered in RF and microwave circuits, whether in monolithic or hybrid formats, is the lack of space to implement distributed designs in which the dimensions of the circuit are appreciable fractions of a wavelength. A useful stratagem for overcoming this problem is to replace transmission line sections with their equivalent lumped circuits. It can be shown (Appendix 1.E) that a simple π-network of reactances exhibits the same electrical behaviour as a matched section of lossless transmission line. Figure 1.42 shows a transmission line and its equivalent circuit.

The values of the reactances shown in the equivalent circuit of Figure 1.42 can be calculated using the following expressions:

$$L = \frac{Z_O}{2\pi f} \sin(\theta) = \frac{Z_O}{2\pi f} \sin(\beta l), \tag{1.61}$$

$$C = \frac{1}{2\pi f Z_O} \tan\left(\frac{\theta}{2}\right) = \frac{1}{2\pi f Z_O} \tan\left(\frac{\beta l}{2}\right), \tag{1.62}$$

where Z_O is the characteristic impedance of the transmission line, f is the frequency of operation, and l is the length of transmission line.

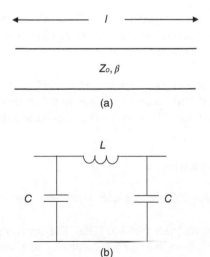

Figure 1.42 Equivalent lumped network of a transmission line: (a) length of lossless transmission line and (b) equivalent π-network.

For the particular case of a quarter-wavelength line, $\beta l = 90°$, and Eqs. (1.61) and (1.62) become

$$L = \frac{Z_O}{2\pi f},$$

(1.63)

$$C = \frac{1}{2\pi f Z_O}.$$

(1.64)

1.13 Supplementary Problems

Note: (i) *All of the transmission lines specified in these problems are assumed to be lossless. Similarly, where reactances are specified in a question they are also assumed to be lossless.*

   (ii) *It is intended that the problems be solved using the Smith chart. The numerical answers presented at the end of the book were obtained using Smith charts and it should be appreciated that these answers may contain small plotting errors.*

Q1.1   A 50 Ω transmission line is terminated by an impedance $(70 - j20)$ Ω. Determine the impedance at a distance of 0.35λ from the termination.

Q1.2   A 75 Ω transmission line is terminated by a $(150 + j40)$ Ω load. Determine the electrical length from the load to the nearest point where the voltage on the line is a maximum.

Q1.3   A 50 Ω transmission line is terminated by a load whose impedance is $(62 - j120)$ Ω. Determine the reflection coefficient of the load, and the VSWR on the line.

Q1.4   A 500 MHz generator having an output impedance of 50 Ω is to be connected to a 50 Ω load using a 2.5 m length of 50 Ω cable. If a 75 Ω cable is used by mistake, determine the reflection coefficient at the output of the generator. The velocity of propagation along the cable is 2.2×10^8 m/s.

Q1.5   An impedance terminating a 75 Ω transmission line causes a VSWR of 4.5 on the line when the frequency is 750 MHz. If the distance between the termination and the nearest voltage minimum on the line is 6 cm, determine the impedance of the termination, given that the velocity of propagation on the line is 2×10^8 m/s.

Q1.6   A transmission line that has a characteristic impedance of 100 Ω, is terminated by an impedance $(220 + j80)$ Ω. Determine the admittance of the termination and the electrical length (i.e. the length expressed as a fraction of a wavelength) between the load and the nearest point where the normalized conductance is unity.

Q1.7   A transmission line has a characteristic impedance of 50 Ω and a velocity of propagation of 2.2×10^8 m/s. The line is terminated by an impedance, Z_L. Determine the value of Z_L if the impedance on the line 0.75 m from the load is $(30 - j85)$ Ω when the frequency is 100 MHz.

Q1.8   A 75 Ω transmission line is terminated by an impedance $(60 - j95)$ Ω. Design an SST, using a short-circuited stub, which will match the terminating impedance to the line at a frequency of 500 MHz, given that the velocity of propagation on the line is 2.0×10^8 m/s. The stub is to have a characteristic impedance of 75 Ω and be connected in parallel with the main line.

Q1.9   Repeat Q1.8 using an open-circuited stub.

Q1.10   Repeat Q1.8 if both the transmission line and the stub have a characteristic impedance of 50 Ω.

Q1.11   A 4 nH inductor is required at a frequency of 800 MHz. The inductor is to be constructed using a length of short-circuited, 50 Ω transmission line on which the velocity of propagation is 2.2×10^8 m/s. Determine the required length of the transmission line.

Q1.12 A 10 pF capacitor at 1.5 GHz is to be made from a length of short-circuited, 75 Ω transmission line on which the velocity of propagation is 2.2×10^8 m/s. Find the required length of the transmission line.

Q1.13 Find the value of Z_{in} shown in Figure 1.43, at a frequency of 750 MHz.

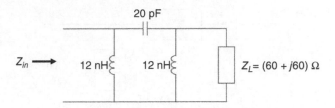

Figure 1.43 Circuit for Q1.13.

Q1.14 Find the value of Z_{in} shown in Figure 1.44, at a frequency of 2 GHz.

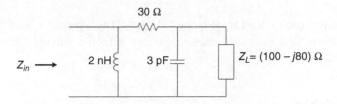

Figure 1.44 Circuit for Q1.14.

Q1.15 Find the value of Z_{in} shown in Figure 1.45, at a frequency of 400 MHz.

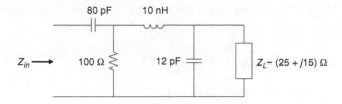

Figure 1.45 Circuit for Q1.15.

Q1.16 A matching network consisting of a lossless series reactance and a lossless shunt susceptance is shown in Figure 1.46. Determine the component values needed to match a load impedance $(10 - j15)$ Ω to a 50 Ω source at a frequency of 550 MHz. Show that there are two possible solutions and calculate the component values in both cases.

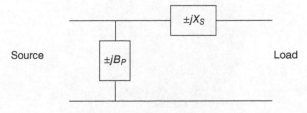

Figure 1.46 Circuit for Q1.16.

Q1.17 (i) A matching network consisting of a lossless series reactance and a lossless shunt susceptance is shown in Figure 1.47. The load impedance is $(120 - j10f)$ Ω, where f is the frequency in GHz. Find the component values needed to match the load impedance to a 50 Ω source at a frequency of 3 GHz. Show there are two possible solutions and find the component values in both cases.

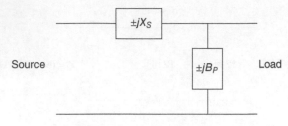

Figure 1.47   Circuit for Q1.17.

(ii) Suppose that the frequency decreases by 15%. Find the reflection coefficient at the input of each of the matching networks calculated in part (i).

Q1.18 A network is required to match a load impedance $(20 - j15)$ Ω to a source whose impedance is $(25 - j35)$ Ω, at frequency of 1.5 GHz. Show that a network consisting of two inductors will provide a suitable match, and find the values of the inductors.

Appendix 1.A Coaxial Cable

1.A.1   Electromagnetic Field Patterns in Coaxial Cable

The electric and magnetic fields within a coaxial cable are shown in Figure 1.48. The electric field forms radial lines between the conductors, with the magnetic field forming closed loops around the centre conductor.

It can be seen that the electric and magnetic fields are orthogonal, and lie in a plane which is at 90° to the direction of propagation, i.e. in a plane that is at 90° to the axis of the cable. This is described as the transverse electromagnetic (TEM) mode of propagation.

1.A.2   Essential Properties of Coaxial Cables

The characteristic impedance, Z_O, of coaxial cable is given by [2]

$$Z_O = \frac{1}{2\pi} \sqrt{\frac{\mu}{\varepsilon}} \ln \left(\frac{r_O}{r_i} \right),$$ (1.65)

where μ and ε are the permeability and permittivity respectively of the dielectric filling the cable, and r_O and r_i are the cable dimensions, as shown in Figure 1.49.

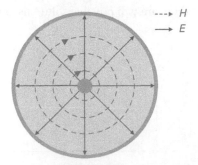

Figure 1.48   Electromagnetic field within a coaxial cable.

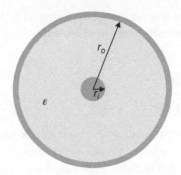

Figure 1.49 Nomenclature for a coaxial cable.

The propagation of a wave along coaxial cable depends on the value of the propagation constant (γ), which was defined in Section 1.3, and which provides information on the attenuation and the phase change per unit length in the cable. The attenuation is determined by the sum of the losses of the conductors and dielectric in the cable. The conductor loss at a frequency, f, for coaxial cable using copper conductors is given by [2] as

$$\alpha_C = \frac{9.5 \times 10^{-5} \times (r_O + r_i)\sqrt{f\varepsilon_r}}{r_O r_i \ln\left(\frac{r_O}{r_i}\right)} \quad \text{dB/unit length,} \tag{1.66}$$

where ε_r is the relative permittivity (dielectric constant) of the material filling the cable. The dielectric loss is given by [2] as

$$\alpha_d = 27.3\sqrt{\varepsilon_r}\frac{\tan\delta}{\lambda_O} \quad \text{dB/unit length,} \tag{1.67}$$

where $\tan\delta$ is the loss tangent of the cable, and λ_O is the free-space wavelength at the frequency of interest. Note that the loss tangent is a parameter used to specify loss in an insulating material, and this will be discussed in more detail in Chapter 3.

The velocity of propagation along a coaxial cable depends only on the permittivity of the dielectric filling the cable, and since the propagation is TEM the velocity is given simply by

$$v_p = \frac{c}{\sqrt{\varepsilon_r}}, \tag{1.68}$$

where ε_r is the relative permittivity (dielectric constant) of the material filling the cable, and c is the velocity of light.

Thus, at a given frequency, f, the propagation constant for coaxial cable can be written as

$$\gamma = \alpha + j\beta = (\alpha_c + \alpha_d) + j\frac{2\pi f}{c}\sqrt{\varepsilon_r}. \tag{1.69}$$

Another important parameter relating to coaxial cable is the cut-off wavelength, λ_c. This gives the frequency above which higher-order modes can propagate in a given size of coaxial cable. The cut-off wavelength is related to the physical parameters of the cable by [2]

$$\lambda_C = \pi\sqrt{\varepsilon_r}(r_O + r_i). \tag{1.70}$$

It is important to ensure that coaxial cable used for a particular application is working below the cut-off wavelengths of higher-order modes, so that a single unique mode is propagating, for two reasons:

(i) In order to calculate the precise phase change along a given length of coaxial cable, it is necessary to know that only one mode, with known propagation characteristics, is travelling along the line.
(ii) Where transitions are made from coaxial cable to another transmission medium, such as waveguide or microstrip, it is necessary to know the precise electromagnetic field pattern within the cable.

As the frequency of operation increases, the size of the cable must decrease. Typical coaxial cable used for RF work has a diameter around 6 mm, with a centre conductor having a diameter around 1 mm. Flexible cable with these dimensions is usually filled with a plastic dielectric, which has a relative permittivity of 2.3. So, using Eq. (1.70), the cut-off wavelength is approximately 17 mm, which corresponds to a cut-off frequency of 17.6 GHz, and this is the maximum frequency at which the cable should be used. If there was a requirement to use coaxial cable at millimetre-wave frequencies, say 50 GHz, the overall diameter of the cable would need to be less than around 2.5 mm. At 100 GHz, the coaxial cable and connectors used in millimetre-wave network analyzers are normally 1 mm in diameter.

Appendix 1.B Coplanar Waveguide

1.B.1 Structure of Coplanar Waveguide (CPW)

Two examples of coplanar transmission lines are illustrated in Figure 1.50. The principal feature of this type of line is that both the signal and grounded conductors are on the same side of the substrate. Conventional CPW lines have the ground plane extending to the edge of the substrate, but in many practical situations this is not possible and the ground plane is restricted to two wide strips on either side of the signal line. This latter type of structure is known as finite ground plane CPW.

1.B.2 Electromagnetic Field Distribution on a CPW Line

The *E*-field distribution of a coplanar line is depicted in Figure 1.51.

There are three practical points that should be noted about the distribution:

(i) There is a higher concentration of electric field within the substrate than above the substrate due to the substrate having a higher dielectric constant than air.
(ii) A significant proportion of the field will fringe into the air above the substrate, so that if the line is enclosed within a metallic package there must be adequate space above the line to avoid the fringing field coupling to the metal of the package.
(iii) If the substrate is thin there may also be fringing of the electromagnetic field into the air beneath the substrate, so again some caution would be needed in packaging the coplanar line.

The magnetic field on a coplanar line, not shown in Figure 1.51, forms closed loops around the signal tracks, and at frequencies up to the low microwave region, typically below 20 GHz, propagation can be regarded as quasi-TEM, which means

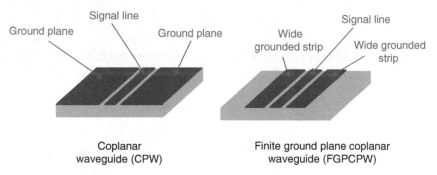

Figure 1.50 Coplanar waveguides.

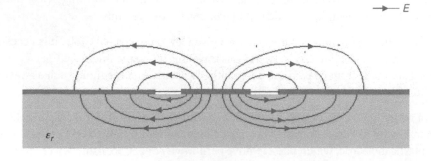

Figure 1.51 *E*-field distribution on a coplanar line.

that the electric and magnetic fields are directed entirely in a transverse plane perpendicular to the direction of propagation. At high microwave frequencies the propagation becomes non-TEM, since longitudinal components of the magnetic field exist. Collin [1] suggests that simple design formulae (as presented in Section A.1.2.3) based on quasi-TEM propagation may be used up to 50 GHz, if the CPW line dimensions are small compared to the wavelength, which is the case for monolithic microwave integrated circuits (MMIC) circuits. Where CPW lines are used at higher millimetre-wave frequencies, a full-wave analysis is required to accurately determine the propagation characteristics. Many CPW circuits operate in the millimetre-wave band above the cut-off frequency for surface-wave modes. However, Riaziat and colleagues [3] showed that for most MMIC applications, there was minimal interaction between CPW propagation and surface-wave modes.

1.B.3 Essential Properties of Coplanar (CPW) Lines

The velocity of propagation, v_p, along a *CPW* line is given by

$$v_p = \frac{c}{\sqrt{\varepsilon_{r,eff}^{CPW}}}, \tag{1.71}$$

where c is the velocity of light, and $\varepsilon_{r,eff}^{CPW}$ is the effective dielectric constant of the medium, and takes into account the dielectric constant of the substrate and the proportion of the electromagnetic field that fringes into the air above and below the substrate. The value of the effective dielectric constant is usually taken to be [3, 4]

$$\varepsilon_{r,eff}^{CPW} = \frac{\varepsilon_r + 1}{2}, \tag{1.72}$$

where ε_r is the dielectric constant of the substrate supporting the CPW line.

The characteristic impedance, Z_O, of a CPW line is given by [4]

$$Z_O = \frac{30\pi}{\sqrt{\varepsilon_{r,eff}^{CPW}}} \left[\frac{1}{\pi} \ln \left(\frac{2(1 + \sqrt{k'})}{(1 - \sqrt{k'})} \right) \right] \quad 0 \leq k \leq 0.71, \tag{1.73}$$

$$Z_O = \frac{30\pi}{\sqrt{\varepsilon_{r,eff}^{CPW}}} \left[\frac{1}{\pi} \ln \left(\frac{2(1 + \sqrt{k})}{(1 - \sqrt{k})} \right) \right]^{-1} \quad 0.71 \leq k \leq 1, \tag{1.74}$$

where

$$k = \frac{s}{s + 2w}, \tag{1.75}$$

$$k' = \sqrt{1 - k^2}, \tag{1.76}$$

and s and w are defined in Figure 1.52.

The loss in a CPW line is the sum of the conductor and dielectric losses. Collin [1] derives an expression for the total conductor loss in dB/unit length as

$$\alpha_c = \frac{A}{2Z_O} \left[\pi + \ln \left(\frac{4\pi s}{t} \right) - kB \right] + \frac{kA}{2Z_O} \left[\pi + \ln \left(\frac{4\pi(s + 2w)}{t} \right) - \frac{B}{k} \right], \tag{1.77}$$

where

$$A = \frac{R_m}{4s(1 - k^2)[K(k)]^2}, \tag{1.78}$$

$$B = \ln \left(\frac{1 + k}{1 - k} \right), \tag{1.79}$$

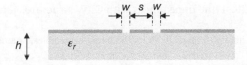

Figure 1.52 Nomenclature for a coplanar waveguide.

and

t = conductor thickness

R_m = conductor surface resistance

$K(k)$ = Complete elliptic integral[2] of first kind with a modulus k.

The dielectric loss in a CPW line in dB/unit length is given by [4]

$$\alpha_d = 2.73 \frac{\varepsilon_r}{\sqrt{\varepsilon_{r,eff}^{CPW}}} \frac{(\varepsilon_{r,eff}^{CPW} - 1)}{(\varepsilon_r - 1)} \frac{\tan\delta}{\lambda_O}, \tag{1.80}$$

where

ε_r = dielectric constant of substrate

$\tan\delta$ = loss tangent of substrate.

The propagation constant for CPW at a frequency f is then given by

$$\gamma = \alpha_c + \alpha_d + j \frac{2\pi f}{c} \sqrt{\varepsilon_{r,eff}^{CPW}}, \tag{1.81}$$

where α_c and α_d are found by substituting in Eqs. (1.77) through (1.80).

1.B.4 Summary of Key Points Relating to CPW Lines

(1) It is usually easier to mount packaged active components in hybrid RF circuits using CPW, compared to microstrip, because the components are often in fabricated in flat beam-lead packages which require one of more of the leads to be earthed. In microstrip, this is an awkward situation that requires the use of vertical connections (VIAs) to the ground plane.

(2) CPW are preferred for MMICs because the provision of ground planes on the same side of the substrate as the signal tracks gives good isolation between closely spaced signal lines.

(3) The symmetrical geometry of CPW permits easy implementation of transitions from microstrip or slotline; this allows flexibility in design, particularly for multilayer circuits.

(4) It is easy to convert from coaxial line to CPW, since both structures have symmetrical geometries with the signal line between ground planes. Thus, it is easy to mount coaxial connectors on packaged planar circuits that use CPW. Connectors mounted on RF and microwave packages are often a weak point in the design, and this will be discussed in more detail in Chapter 3, which deals with fabrication issues.

(5) Coplanar lines exhibit low dispersion, particularly for small line geometries. Jackson [5] showed that at 60 GHz the dimensions of CPW could be chosen to give better results than microstrip, in terms of both dispersion and conductor loss.

(6) The separation of the lines in coplanar structures is usually small, and this imposes limitations on the power handling capability, which is normally rather low.

(7) CPW lines generally need larger packaging enclosures than microstrip because of the greater fringing field; in CPW there may be fringing field from both the upper and lower surfaces of the structure.

(8) Electrical parameters such as characteristic impedance and velocity of propagation depend on a number of parameters, so it is less easy to use graphical design curves as can be done for microstrip (see Appendix 1.D and Chapter 2). The only practical design technique for CPW is to use appropriate CAD packages.

(9) In general, coplanar lines require relatively small line spacing to give useful values of characteristic impedance. For example, using 25 mil ($h = 0.025'' \equiv 0.635$ mm) thick alumina, which is a very common substrate for RF and microwave circuits, a line of characteristic impedance 50 Ω requires a signal track width (w) of 100 μm and a line spacing (s) of 90 μm. However, the range of practical line impedances with CPW structures tends to be greater than with microstrip.

---

2 The value of the complete elliptic integral with a given modulus k can be found from tables, or as an approximation [1] by summing the first three terms in the following series:

$$K(k) = \frac{\pi}{2} \left(1 + \frac{k^2}{4} + \frac{9k^4}{64} + \dots \right) \qquad k \le 0.4.$$

Typical range of impedances:

Microstrip	$15\,\Omega \rightarrow 110\,\Omega$
CPW	$20\,\Omega \rightarrow 250\,\Omega$

(10) The small geometries of CPW or FGPCPW lines make them the preferred format for higher millimetre-wave frequency applications.

Appendix 1.C Metal Waveguide

1.C.1 Waveguide Principles

Waveguides are hollow metal tubes through which the microwave field patterns are propagated. The tubes are usually rectangular or circular in cross-section. Waveguides provide low loss transmission and are capable of handling high powers.

The walls of a rectangular waveguide are termed the broad and narrow walls, with their dimensions represented by a and b, as shown in Figure 1.53.

Previous mention was made of TEM waves; these are waves in which the electric and magnetic fields are entirely directed in a plane perpendicular to the direction of propagation. The transverse plane is considered to extend to infinity which means that the electric and magnetic fields have uniform density. A TEM wave is also described as plane wave. Such a wave cannot exist within a medium bounded by a metal conductor, because the boundary conditions would be violated at the metal surfaces. So a different pattern of electric and magnetic fields must exist within a metal waveguide; various different patterns can exist, and these are known as the waveguide modes.

1.C.2 Waveguide Propagation

The pattern of the electric and magnetic fields within a rectangular waveguide is often illustrated by considering the interference pattern due to the intersection of two plane waves in free space. Figure 1.54a shows two plane waves travelling at angles θ to the horizontal, with the electric field in a direction perpendicular to the paper. The solid lines represent lines of peak magnetic field, and the dotted lines represent lines of zero magnetic field. The peaks are separated by half the free-space wavelength. The electric field is shown using the normal convention that a circle represents a field line directed into the paper, and a solid dot a field line directed out of the paper.

The interference pattern resulting from the intersection of the two plane waves is shown in Figure 1.54b. From this pattern it can be seen that the magnetic field has formed closed loops, and that there are maxima and minima points in the electric field. Also, it can be seen that the electric fields are zero along the lines xx' and yy'. Thus conducting sheets, perpendicular to the plane of the diagram, could be placed along these two lines without disrupting the field pattern. This indicates the electromagnetic field pattern that would exist between the narrow walls of the rectangular waveguide. Metal plates forming the broad walls of the waveguide would be in an orthogonal plane, perpendicular to the electric field and parallel to the magnetic field, and would thus have no effect on the field pattern. So, as the wave propagates, the pattern of magnetic loops and the associated electric field travel inside the waveguide.

1.C.3 Rectangular Waveguide Modes

Many different patterns of electric and magnetic field can exist within a rectangular waveguide and these are called the waveguide modes. The modes are categorized into two types: *TE* modes in which the electric field lies entirely in the transverse plane, and *TM* modes in which the magnetic field lies entirely within the transverse plane. Subscripts are used to

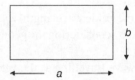

Figure 1.53 Cross-section of a rectangular waveguide.

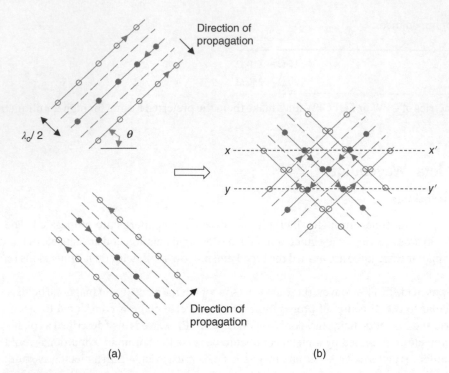

Figure 1.54 Intersection of two plane waves: (a) the individual waves; (b) the resulting interference pattern.

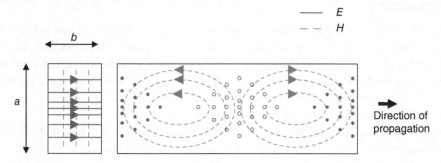

Figure 1.55 Sketch of electric and magnetic field patterns for TE_{10} mode in a rectangular waveguide.

identify particular modes. Thus, we have TE_{mn} and TM_{mn} modes. The first subscript, m, specifies the number of loops of variation of the field across the broad dimension, and the second subscript, n, gives the number of loops of variation of the field across the narrow dimension. In this context, a loop of variation refers to the magnitude of the particular field varying through zero–maximum–zero.

Thus, the electromagnetic field pattern we deduce from the intersection of plane waves, and shown in Figure 1.55, would be described as the TE_{10} mode, since there is one loop of variation of the E-field across the broad dimension, and the E-field is constant across the narrow dimension.

1.C.4   The Waveguide Equation

The waveguide equation is an expression relating the guide wavelength, λ_g, along the guide, the free-space wavelength, and the guide dimensions. The equation can be derived by considering the geometry of a single loop of magnetic field, as shown in Figure 1.56.

The line LN in Figure 1.56 represents a plane wave travelling at the velocity of light, c, at an angle θ to the axis of the waveguide. From the earlier consideration of the intersection of plane waves we know that the axial length of one magnetic loop will be $\lambda_g/2$, and it follows from the interference pattern of two plane waves that the distance MP must be $\lambda_O/2$, where

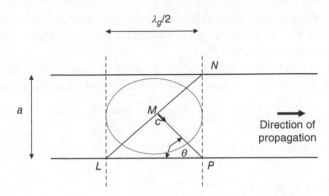

Figure 1.56 Single loop of magnetic field in a rectangular waveguide.

λ_O is the free-space wavelength. Considering the right-angled triangles LMP and MNP, we have

$$\cos\theta = \frac{\lambda_O/2}{\lambda_g/2} = \frac{\lambda_O}{\lambda_g} \quad\text{and}\quad \sin\theta = \frac{\lambda_O/2}{a} = \frac{\lambda_O}{2a}.$$

Now

$$\cos^2\theta + \sin^2\theta = 1.$$

Therefore,

$$\left(\frac{\lambda_O}{\lambda_g}\right)^2 + \left(\frac{\lambda_O}{2a}\right)^2 = 1,$$

i.e.

$$\frac{1}{\lambda_g^2} + \frac{1}{(2a)^2} = \frac{1}{\lambda_O^2}$$

or

$$\frac{1}{\lambda_g^2} + \frac{1}{\lambda_c^2} = \frac{1}{\lambda_O^2}, \tag{1.82}$$

where $\lambda_c = 2a$ and is defined as the cut-off wavelength. The cut-off wavelength is the longest wavelength for normal propagation of the TE$_{10}$ mode along a guide whose broad dimension is a; the corresponding frequency is the cut-off frequency, f_c. At frequencies less than f_c energy will only penetrate into the waveguide in the form of an evanescent mode, and will suffer very high attenuation. Thus, propagation is only possible in the TE$_{10}$ mode at frequencies where the free-space wavelength is greater than $2a$.

1.C.5 Phase and Group Velocities

It is clear from Figure 1.56 that the time taken for point L to travel to point P must be the same as that for point M to travel to point P, i.e.

$$\frac{\lambda_g/2}{v_P} = \frac{\lambda_O/2}{c} \quad\Rightarrow\quad v_P = \frac{\lambda_g}{\lambda_O}c = \frac{c}{\cos\theta}, \tag{1.83}$$

where v_P is the velocity with which point L travels along the waveguide, and is termed the phase velocity. It is obvious from Eq. (1.83) that the phase velocity is greater than the velocity of light. But this is the velocity with which a particular point (or phase) of the electromagnetic pattern will travel, and not the velocity at which energy will travel. The velocity with which energy travels along the guide is the group velocity, v_g, and it can be seen from Figure 1.56 that this can be found by resolving c in the axial direction, i.e.

$$v_g = c \times \cos\theta. \tag{1.84}$$

Combining Eqs. (1.83) and (1.84) gives

$$v_P \times v_g = c^2. \tag{1.85}$$

1.C.6 Field Theory Analysis of Rectangular Waveguides

Whilst the explanation of waveguide propagation based on the intersection of two plane waves, as described in Section A.1.3.2, provides a useful pictorial view, particularly as an introduction to the topic, a more rigorous and comprehensive analysis can be obtained using classical field theory. In this approach, Maxwell's equations are solved using the boundary conditions set by the walls of the rectangular waveguide. A complete development of the theory is given in Collin [1], but it is useful here to state the principal results.

Based on field theory analysis [1], the components of the electric and magnetic fields for the TE modes in rectangular waveguide are given in Eq. (1.86) below, and are based on the coordinate system shown in Figure 1.57.

$$E_x = \frac{j\omega\mu n\pi}{bk_c^2} H_O \cos\left(\frac{m\pi x}{a}\right) \sin\left(\frac{n\pi y}{b}\right) e^{j(\omega t - \beta z)},$$

$$E_y = -\frac{j\omega\mu m\pi}{ak_c^2} H_O \sin\left(\frac{m\pi x}{a}\right) \cos\left(\frac{n\pi y}{b}\right) e^{j(\omega t - \beta z)},$$

$$E_z = 0,$$

$$H_x = \frac{j\beta m\pi}{ak_c^2} H_O \sin\left(\frac{m\pi x}{a}\right) \cos\left(\frac{n\pi y}{b}\right) e^{j(\omega t - \beta z)},$$

$$H_y = \frac{j\beta n\pi}{bk_c^2} H_O \cos\left(\frac{m\pi x}{a}\right) \sin\left(\frac{n\pi y}{b}\right) e^{j(\omega t - \beta z)},$$

$$H_z = H_O \cos\left(\frac{m\pi x}{a}\right) \cos\left(\frac{n\pi y}{b}\right) e^{j(\omega t - \beta z)}, \tag{1.86}$$

where

$$k_c = \sqrt{\left(\frac{m\pi}{a}\right)^2 + \left(\frac{n\pi}{b}\right)^2}, \tag{1.87}$$

and H_O is the magnetic field strength at the input to the waveguide.

Using Eq. (1.86) the field components for the TE_{10} mode are shown in Eq. (1.88).

$$E_x = 0,$$

$$E_y = -\frac{j\omega\mu\pi}{ak_c^2} H_O \sin\left(\frac{\pi x}{a}\right) e^{j(\omega t - \beta z)},$$

$$E_z = 0,$$

$$H_x = \frac{j\beta\pi}{ak_c^2} H_O \sin\left(\frac{\pi x}{a}\right) e^{j(\omega t - \beta z)},$$

$$H_y = 0,$$

$$H_z = H_O \cos\left(\frac{\pi x}{a}\right) e^{j(\omega t - \beta z)}. \tag{1.88}$$

Through inspection of Eq. (1.88), we see that the field distribution agrees with that in Figure 1.55, which was deduced from the intersection of two plane waves. The E-field only has a component in the y-direction, and is zero at the waveguide walls ($x = 0$ and $x = a$) and is a maximum in the centre of the guide ($x = a/2$).

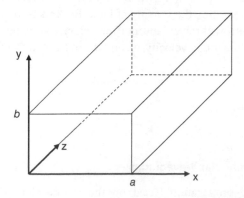

Figure 1.57 Coordinate system for electromagnetic field components.

For completeness the field components for the TM_{mn} modes are stated in Eq. (1.89), but these have very limited application in modern RF work.

$$E_x = -\frac{j\beta m\pi}{ak_c^2} E_O \cos\left(\frac{m\pi x}{a}\right) \sin\left(\frac{n\pi y}{b}\right) e^{j(\omega t - \beta z)},$$

$$E_y = -\frac{j\beta n\pi}{bk_c^2} E_O \sin\left(\frac{m\pi x}{a}\right) \cos\left(\frac{n\pi y}{b}\right) e^{j(\omega t - \beta z)},$$

$$E_z = E_O \sin\left(\frac{m\pi x}{a}\right) \sin\left(\frac{n\pi y}{b}\right) e^{j(\omega t - \beta z)},$$

$$H_x = \frac{j\omega\varepsilon n\pi}{bk_c^2} E_O \sin\left(\frac{m\pi x}{a}\right) \cos\left(\frac{n\pi y}{b}\right) e^{j(\omega t - \beta z)},$$

$$H_y = -\frac{j\omega\varepsilon m\pi}{ak_c^2} E_O \cos\left(\frac{m\pi x}{a}\right) \sin\left(\frac{n\pi y}{b}\right) e^{j(\omega t - \beta z)},$$

$$H_z = 0, \tag{1.89}$$

where E_O is the electric field strength at the input to the waveguide.

1.C.7 Waveguide Impedance

The impedance, Z_O, of an electromagnetic wave propagating in free space is given by

$$Z_O = \left|\frac{E}{H}\right| = \sqrt{\frac{\mu_O}{\varepsilon_O}} = 120\pi = 377 \ \Omega. \tag{1.90}$$

In a waveguide, the free-space impedance is modified by the presence of the waveguide walls. For the TE_{10} mode the waveguide impedance will be

$$Z_g = \left|\frac{E_y}{H_x}\right|. \tag{1.91}$$

Substituting from Eq. (1.88) into Eq. (1.91) gives

$$Z_g = \frac{\omega\mu\pi H_O / ak_c^2}{\beta\pi H_O / ak_c^2} = \frac{\omega\mu}{\beta} = \frac{2\pi f\mu}{2\pi / \lambda_G} = f\mu\lambda_g$$

$$= \frac{c}{\lambda_O}\mu\lambda_g = \frac{1}{\sqrt{\mu\varepsilon}\lambda_O}\mu\lambda_g = \sqrt{\frac{\mu}{\varepsilon}}\frac{\lambda_g}{\lambda_O},$$

i.e.

$$Z_g = Z_O \frac{\lambda_G}{\lambda_O}. \tag{1.92}$$

1.C.8 Higher-Order Rectangular Waveguide Modes

The previous theory has focused on one particular mode, namely the TE_{10} mode. This mode is often described as the dominant mode; it is the mode which propagates at the lowest frequency for a given size of waveguide, and there will be a frequency range over which it is the only mode that can propagate. Thus, the TE_{10} mode is the most useful for practical waveguide transmission and devices. The cut-off wavelength for a particular mode can be found using Eq. (1.93) below:

$$\lambda_c = \frac{1}{\sqrt{\left(\frac{m}{2a}\right)^2 + \left(\frac{n}{2b}\right)^2}}. \tag{1.93}$$

The corresponding cut-off frequencies are found from Eq. (1.94) below:

$$f_c = c\sqrt{\left(\frac{m}{2a}\right)^2 + \left(\frac{n}{2b}\right)^2}. \tag{1.94}$$

Table 1.1 Designations for a rectangular waveguide.

WG designation	Recommended operating range (GHz)	$f_{c\,(TE10)}$ (GHz)
WG90	8.20–12.40	6.56
WG28	26.50–40.00	21.08
WG10	75.00–110.00	59.01
WG5	140.00–220.00	115.750

1.C.9 Waveguide Attenuation

Typical variations of attenuation (α) as a function of frequency for modes in rectangular waveguide are sketched in Figure 1.58. Two particular modes are sketched, namely the TE_{10} mode and the TE_{11} mode, which is the higher-order mode with the closest cut-off frequency to that of the dominant mode.

It can be seen from Figure 1.58 that the attenuation increases sharply as the frequency approaches the cut-off value. Above cut-off, the attenuation slowly increases with frequency due to dissipative losses in the waveguide walls; the dissipative losses increase with increasing frequency due to the skin effect, which will be discussed in more detail in Chapter 2. It is also apparent from Figure 1.58 that there is a frequency region where only the TE_{10} mode will propagate, and within this region Δf indicates those frequencies where the attenuation is minimum. Thus, we can regard a waveguide of a particular size as having an operational bandwidth indicated by Δf.

1.C.10 Sizes of Rectangular Waveguide and Waveguide Designation

It is clear from the previous discussion that waveguide size is a crucial issue, and that for operation using the dominant mode the required size will decrease as the frequency increases. The sizes of rectangular waveguide are designated using the WG notation, and some examples of commonly used waveguides are given in Table 1.1. In the table the recommended operating range for the TE_{10} mode corresponds to Δf in Figure 1.58.

1.C.11 Circular Waveguide

The propagation of high-frequency electromagnetic waves along waveguides having a circular cross-section involves similar principles to those already discussed for rectangular guides. The main differences are that the analysis uses spherical

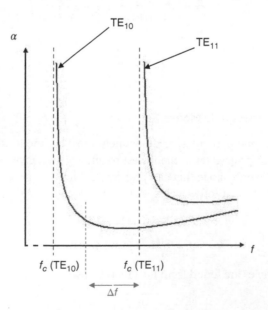

Figure 1.58 Variation of attenuation for rectangular waveguide modes.

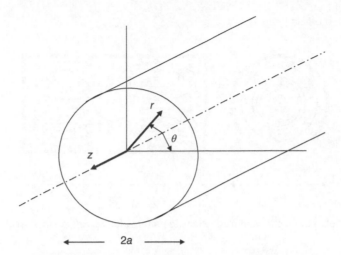

Figure 1.59 Coordinate system for a circular waveguide.

polar coordinates, to be compatible with the geometry of the guide, and that solutions to Maxwell's equations are found in terms of Bessel functions. The modes of propagation in circular waveguide can be classified as TE_{nm} and TM_{nm}, where the subscript m refers to the loops of variation of the field across the radius of the guide, and the subscript n to the double loops (full sinusoid) of variation in the circumferential direction. The coordinate system for circular waveguides is shown in Figure 1.59.

The fields for the TE_{nm} modes in circular waveguide are stated in Eq. (1.95) below:

$$E_r = -\frac{\omega\mu n}{rk_c^2}H_O J_n(k_c r)\cos(n\theta)e^{j(\omega t - \beta z)},$$

$$E_\theta = \frac{j\omega\mu}{k_c}H_O J_n'(k_c r)\cos(n\theta)e^{j(\omega t - \beta z)},$$

$$E_z = 0,$$

$$H_r = -\frac{j\beta}{k_c}H_O J_n'(k_c r)\cos(n\theta)e^{j(\omega t - \beta z)},$$

$$H_\theta = -\frac{\beta n}{rk_c^2}H_O J_n(k_c r)\cos(n\theta)e^{j(\omega t - \beta z)},$$

$$H_z = H_O J_n(k_c r)\cos(n\theta)e^{j(\omega t - \beta z)}. \tag{1.95}$$

Note that a function of the form $J_n(x)$ is called a Bessel function of the first kind, of order m, and argument x; the values of the function are normally found from tables [6]. The amplitude of the Bessel function oscillates between positive and negative values, and decays as the argument increases. There will be certain values of the argument which make the function zero, and these are called the roots of the function. The differential of a Bessel function of the first kind is written as $J'_n(x)$.

The mode of particular interest in circular waveguide is TE_{01} because it exhibits the unusual property of decreasing attenuation with increasing frequency. The field pattern for the TE_{01} mode is sketched in Figure 1.60, in which it can be seen that the E-field forms closed loops around the axis of the guide, with the magnetic field forming closed loops in axial planes through the centre of the guide. This means that the current will flow in the circumferential direction in the waveguide walls.

The symmetry of the field and current about the central axis means that the mode is also useful for forming rotating waveguide joints, since a joint in circular TE_{01} waveguide will not significantly disturb the current flow. But it is the low loss property of the TE_{01} mode which is its most attractive feature, and this mode can be used for low-loss feeder connections to antennas and for constructing very high-Q cavities, and this latter application will be discussed in more detail in Chapter 3 in the context of the measurement of dielectrics. It should be noted that whilst the TE_{01} mode has useful practical properties, it is not the dominant mode in circular waveguide and caution must be taken when exciting the TE_{01} mode not to generate unwanted modes. The dominant mode in circular waveguide is TE_{11}, and the attenuation characteristic for this mode is shown in Figure 1.61.

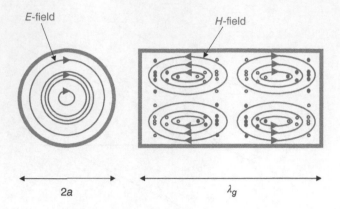

Figure 1.60 Sketch of electric and magnetic field patterns for TE_{01} mode in a circular waveguide.

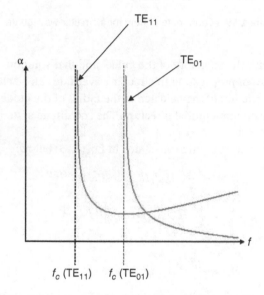

Figure 1.61 Attenuation curves of typical modes in a circular waveguide.

We know that for any mode in metallic waveguide the tangential component of electric field must be zero at the waveguide walls, and considering the expression for E_θ in Eq. (1.95) this means $J'_n(k_c r) = 0$ when $r = a$. Therefore, for the TE_{01} mode, where $n = 0$, we require $J'_0(k_c a) = 0$. From tables of Bessel functions the first root of $J'_0(x)$ occurs when $x = 3.83$.

Karbowiak [7] showed that the cut-off wavelength for the TE_{01} mode in circular waveguide of radius a is given by

$$\lambda_c = \frac{2\pi a}{x}.$$

(1.96)

It then follows that the cut-off frequency for the TE_{01} mode is

$$f_c = \frac{c}{\lambda_c}.$$

(1.97)

As an example, for a 50 mm diameter circular waveguide

$$\lambda_c = \frac{2\pi \times 25}{3.83} \text{ mm} = 41.02 \text{ mm},$$

$$f_c = \frac{3 \times 10^8}{41.02 \times 10^{-3}} \text{ Hz} = 7.31 \text{ GHz}.$$

Karbowiak [7] also showed that the attenuation of the TE_{01} mode (in dB/m) for a frequency, f_O, which is well above the cut-off value is given by

$$\alpha = 17.37 \frac{F^2 R_m}{2a},$$

(1.98)

where

$F = f_c/f_o$.

R_m = surface resistance of the metal waveguide.

It is clear from Eq. (1.98) that the attenuation of the TE_{01} mode decreases with increasing frequency, and also that increasing the diameter of the waveguide will decrease the attenuation at any given frequency.

Appendix 1.D Microstrip

The cross-section of a microstrip line is shown in Figure 1.62.

The line consists of a low-loss insulating substrate, having a relative permittivity ε_r, with conductor covering the lower surface to form a ground plane, and a signal track on the upper surface. The key parameters of a microstrip line are the substrate relative permittivity (ε_r), the substrate thickness (h), and the track width (w). It is these three parameters that determine the characteristic impedance (Z_O) of the line and the velocity of propagation (v_p) along the line. The track thickness need only be considered if the track is exceptionally thick, typically more than 10 μm.

The characteristic impedance of a microstrip line is related to the geometric parameters of the line by the following expressions [8].

For $\frac{w}{h} < 3.3$:

$$Z_O = \frac{119.9}{\sqrt{2(\varepsilon_r + 1)}} \ln\left(4\frac{h}{w} + \sqrt{16\left(\frac{h}{w}\right)^2 + 2}\right). \tag{1.99}$$

For $\frac{w}{h} > 3.3$:

$$Z_O = \frac{119.9\pi}{2\sqrt{\varepsilon_r}} \chi, \tag{1.100}$$

where

$$\chi = \frac{w}{2h} + \frac{\ln 4}{\pi} + \frac{\ln(e\pi^2/16)}{2\pi}\left(\frac{\varepsilon_r - 1}{\varepsilon_r^2}\right) + \frac{\varepsilon_r + 1}{2\pi\varepsilon_r}\left[\ln\left(\frac{\pi e}{2}\right) + \ln\left(\frac{w}{2h} + 0.94\right)\right] \tag{1.101}$$

and $e = 2.718$.

Microstrip is an open structure and some of the electromagnetic field fringes into the air above the microstrip line. Since the propagating wave is not entirely within the substrate we need to specify an effective relative permittivity $(\varepsilon_{r,eff}^{MSTRIP})$ for the propagation medium, where

$$1 \leq \varepsilon_{r,eff}^{MSTRIP} \leq \varepsilon_r. \tag{1.102}$$

The effective relative permittivity of a microstrip line having a characteristic impedance Z_O can be found from the following expression [8]

$$\varepsilon_{r,eff}^{MSTRIP} = \frac{\varepsilon_r + 1}{2}\left[1 + \frac{29.98}{Z_O}\left(\frac{2}{\varepsilon_r + 1}\right)\left(\frac{\varepsilon_r - 1}{\varepsilon_r + 1}\right)\left(\ln\left[\frac{\pi}{2}\right] + \frac{1}{\varepsilon_r}\ln\left[\frac{4}{\pi}\right]\right)\right]^{-2}, \tag{1.103}$$

where ε_r is the relative permittivity of substrate.

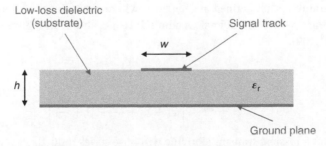

Figure 1.62 Cross-section of a microstrip line.

Once the effective relative permittivity of the substrate is known the velocity of propagation along the microstrip line can be found from

$$v_p = \frac{c}{\sqrt{\varepsilon_{r,eff}^{MSTRIP}}}, \tag{1.104}$$

where c is the velocity of light (3×10^8 m/s).

The wavelength along the microstrip line (also known as the substrate wavelength) at a given frequency (f) will be

$$\lambda_s = \frac{v_p}{f} = \frac{c}{f\sqrt{\varepsilon_{r,eff}^{MSTRIP}}}. \tag{1.105}$$

Hammerstad and Bekkadal [9] give the following widely used expression for the conductor loss, α_c, in microstrip

$$\alpha_c = \frac{0.072\sqrt{f}}{wZ_O}\left(1 + \frac{2}{\pi}\tan^{-1}[1.4(\Delta R_m \sigma)^2]\right) \quad \text{dB/unit length,} \tag{1.106}$$

where

f = working frequency
w = microstrip track width
Z_O = characteristic impedance of microstrip line
Δ = RMS surface roughness of metal
R_m = surface resistance of metal
σ = conductivity of metal

The dielectric loss in a microstrip line can be calculated from the expression given by Gupta [4] as

$$\alpha_d = 27.3\frac{\varepsilon_r(\varepsilon_{r,eff}^{MSTRIP} - 1)\tan\delta}{\sqrt{\varepsilon_{r,eff}^{MSTRIP}}(\varepsilon_r - 1)\lambda_O} \quad \text{dB/unit length,} \tag{1.107}$$

where

ε_r = relative permittivity of substrate
$\varepsilon_{r,eff}^{MSTRIP}$ = effective relative permittivity of substrate
λ_O = free-space wavelength at the working frequency
$\tan\delta$ = loss tangent of substrate (see Chapter 3).

The propagation constant, γ, for a particular microstrip line can then be found from

$$\gamma = \alpha_c + \alpha_d + j\frac{2\pi f}{c}\sqrt{\varepsilon_{r,eff}^{MSTRIP}}, \tag{1.108}$$

by substituting the appropriate values of α_c, α_d, and $\varepsilon_{r,eff}^{MSTRIP}$ for the given line.

Appendix 1.E Equivalent Lumped Circuit Representation of a Transmission Line

The equivalence between a two-port network and a length of lossless transmission line can be deduced using transmission, or *ABCD*, parameters. These parameters are defined in Chapter 5, where network parameters are discussed in more detail.

A two-port network comprising a π-network of lossless admittances, as shown in Figure 1.63, can be represented [8] by the *ABCD* matrix given in Eq. (1.109).

$$\begin{bmatrix} A & B \\ C & D \end{bmatrix}_{\pi\text{-network}} = \begin{bmatrix} 1 + \dfrac{Y_2}{Y_3} & \dfrac{1}{Y_3} \\ Y_1 + Y_2 + \dfrac{Y_1 Y_2}{Y_3} & 1 + \dfrac{Y_1}{Y_3} \end{bmatrix}. \tag{1.109}$$

We know from Section 1.2 that a lossless transmission line will have series inductance and shunt capacitance, and so a representative equivalent π-network for an electrically short length is shown in Figure 1.64.

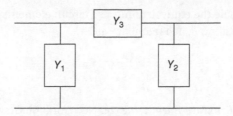

Figure 1.63 π-Network of admittances.

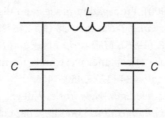

Figure 1.64 π-Network representing an electrically short transmission line.

Comparing Figures 1.63 and 1.64 we have:

$$Y_1 = Y_2 = j\omega C,$$
$$Y_3 = \frac{1}{j\omega L} = -j\frac{1}{\omega L}. \tag{1.110}$$

The *ABCD* matrix representing the circuit shown in Figure 1.64 will then be as shown in Eq. (1.111).

$$\begin{bmatrix} A & B \\ C & D \end{bmatrix}_{equivalent\ \pi\text{-}network} = \begin{bmatrix} 1 - \omega^2 LC & j\omega L \\ j(2\omega C - \omega L[\omega C]^2) & 1 - \omega^2 LC \end{bmatrix}. \tag{1.111}$$

A length of lossless transmission line, having a characteristic impedance Z_O and a propagation constant β, and shown in Figure 1.65, can also be represented by an *ABCD* matrix [10], and this is given in Eq. (1.112).

$$\begin{bmatrix} A & B \\ C & D \end{bmatrix}_{Tx\ line} = \begin{bmatrix} \cos\beta l & jZ_O\sin\beta l \\ jY_O\sin\beta l & \cos\beta l \end{bmatrix}. \tag{1.112}$$

If the electrical performance of the π-network shown in Figure 1.64 is to be equivalent to that of a short transmission line, the *ABCD* matrix in Eq. (1.111) must be identical to that in Eq. (1.112). Equating the matrix elements in these two equations gives

$$\cos\beta l = 1 - \omega^2 LC,$$
$$jZ_O\sin\beta l = j\omega L,$$
$$jY_O\sin\beta l = j(2\omega C - \omega L[\omega C]^2). \tag{1.113}$$

Rearranging the terms in Eq. (1.113) gives

$$L = \frac{Z_O}{\omega}\sin\beta l,$$
$$C = \frac{1}{\omega Z_O}\tan\left(\frac{\beta l}{2}\right). \tag{1.114}$$

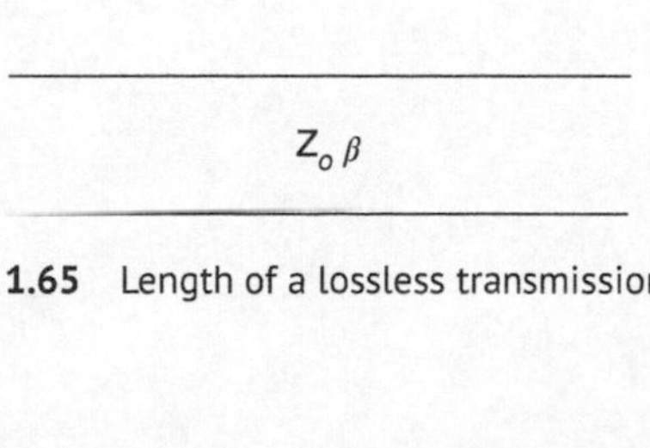

Figure 1.65 Length of a lossless transmission line.

Equation (1.114) can be used to calculate the equivalent lumped circuit elements of a lossless transmission line of characteristic impedance, Z_O, and phase constant, β, at a frequency $\omega/2\pi$.

References

1 Collin, R.E. (1992). *Foundations for Microwave Engineering*, 2e. New York: McGraw Hill.

2 Chang, K., Bahl, I., and Nair, V. (2002). *RF and Microwave and Component Design for Wireless Systems*. New York: Wiley.

3 Riaziat, M., Majidi-Ahy, R., and Feng, I.J. (1990). Propagation modes and dispersion characteristics of coplanar waveguides. *IEEE Transactions on Microwave Theory and Techniques* 38 (3): 245–251.

4 Gupta, K.C., Garg, R., Bahl, I., and Bhartia, P. (1996). *Microstrip Lines and Slotlines*, 2e. Boston, MA: Artech House.

5 Jackson, R.W. (1996). Considerations in the use of coplanar waveguide for millimeter-wave integrated circuits. *IEEE Transactions on Microwave Theory and Techniques* 34 (12): 1450–1456.

6 Kreyszig, E. (1993). *Advanced Engineering Mathematics*. New York: Wiley.

7 Karbowiak, A.E. (1965). *Trunk Waveguide Communication*. London: Chapman and Hall.

8 Edwards, T.C. and Steer, M.B. (2000). *Foundations of Interconnect and Microstrip Design*. London: Wiley.

9 Hammerstad, E.O. and Bekkadal, F.A. (1975). *Microstrip Handbook*, ELAB Report STF 44A74169. Norway: University of Trondheim.

10 Pozar, D.M. (2001). *Microwave and RF Design of Wireless Systems*. New York: Wiley.

2

Planar Circuit Design I

Designing Using Microstrip

2.1 Introduction

Microstrip is the most common form of interconnection used in RF and microwave planar circuits. The concept of microstrip, together with a discussion of its advantages, was introduced in Chapter 1; for convenience, a diagram showing the key dimensions of a microstrip line is reproduced in Figure 2.1.

Also presented in Chapter 1 were design equations that enable the characteristic impedance and propagation constant of a microstrip line of given dimensions to be calculated. In this chapter, the practical issues associated with microstrip design will be discussed in more detail, and examples given of the design of some common microstrip components.

2.2 Electromagnetic Field Distribution Across a Microstrip Line

Propagation along a microstrip line is quasi-Transverse Electromagnetic (quasi-TEM). TEM refers to propagation where the electric and magnetic fields are entirely directed in a plane that is perpendicular to the direction of propagation. It should be noted that microstrip cannot support a pure TEM wave since the wave is travelling in a mixed dielectric medium, i.e. substrate and air. It is important in microstrip design to choose line parameters that minimize unwanted modes; this issue is discussed in more detail in Section 2.5.

A sketch of the electric field surrounding a microstrip line is shown in Figure 2.2.

There will also be a magnetic field, not shown in Figure 2.2, where the magnetic field lines form closed loops around the microstrip track, and intersect the electric field line at right-angles. It can be seen from Figure 2.2 that the greatest electromagnetic field density occurs beneath the microstrip line, with the field fringing sideways into the substrate, and also into the air above the substrate.

2.3 Effective Relative Permittivity, $\varepsilon_{r,eff}^{MSTRIP}$

The fringing fields associated with a microstrip line mean that a propagating wave will travel in a mixed dielectric medium, i.e. partly in the substrate, and partly in the air above the microstrip line. Thus, we need to specify an effective relative permittivity, $\varepsilon_{r,eff}^{MSTRIP}$, where

$$1 \leq \varepsilon_{r,eff}^{MSTRIP} \leq \varepsilon_r, \tag{2.1}$$

where ε_r is the relative permittivity (dielectric constant) of the substrate. Using the concept of effective relative permittivity, we can specify the wavelength along a microstrip line (usually known simply as the substrate wavelength) at a particular frequency, f, as

$$\lambda_s = \frac{v_p}{f} = \frac{c}{f\sqrt{\varepsilon_{r,eff}^{MSTRIP}}}, \tag{2.2}$$

where

$\lambda_s =$ substrate wavelength
$v_p =$ velocity of propagation along microstrip line
$c =$ velocity of light (3×10^8 m/s).

RF and Microwave Circuit Design: Theory and Applications, First Edition. Charles E. Free and Colin S. Aitchison.
© 2022 John Wiley & Sons Ltd. Published 2022 by John Wiley & Sons Ltd.
Companion website: www.wiley.com/go/free/rfandmicrowave

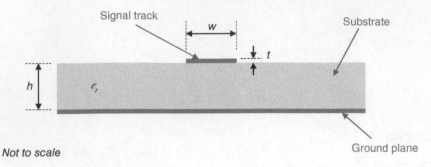

Figure 2.1 Cross-section of a microstrip line showing key dimensions.

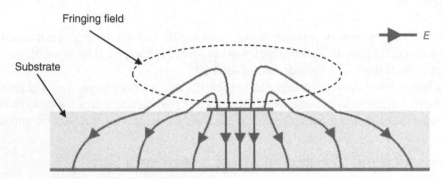

Figure 2.2 Sketch of electric field surrounding a microstrip line.

2.4 Microstrip Design Graphs and CAD Software

Knowing the dimensions of a microstrip line, precise values of the characteristic impedance of the line and the velocity of propagation can be found from various computer-aided-design (CAD) software packages. In addition to a range of commercial software packages there are a number of free basic packages available on the Internet that enable users to relate the physical characteristics of a microstrip line to the key electrical parameters.

However, in order to demonstrate basic microstrip design principles in this chapter we will make use of design graphs to obtain the essential circuit parameters. But it must be emphasized that the reading errors associated with these graphs mean that they are not accurate enough for practical design work. The relevant design graphs are shown in Appendix 2.A. There are two graphs, the first giving the characteristic impedance, Z_o, as a function of w/h, and the second giving $\varepsilon_{r,eff}^{MSTRIP}$ as a function of w/h. The two graphs shown were drawn for a substrate having a relative permittivity of 9.8 (different graphs are required for different substrate permittivities).

It is useful to note from the graphs that the characteristic impedance of a microstrip line on a substrate of given thickness decreases as the width of the line increases. Following from this, the graphs indicate that there is a practical limit to the range of characteristic impedance that can be obtained from a microstrip line on a substrate of given thickness and relative permittivity. Very low characteristic impedances require lines that are too wide to be practical, and conversely high impedances require lines that are too narrow to be conveniently fabricated. In the case of lines on a substrate of relative permittivity 9.8, the graphs in Appendix 2.A suggest that the impedance range is around 25–90 Ω.

Using the design graphs is straightforward, usually following the sequence

$$Z_o \Rightarrow \frac{w}{h} \Rightarrow \varepsilon_{r,eff}^{MSTRIP}, \tag{2.3}$$

and this is demonstrated through Example 2.1.

Example 2.1   A microstrip line having a characteristic impedance of 52 Ω is to be fabricated on a substrate that has a thickness of 0.5 mm and a relative permittivity of 9.8. Determine:

 (i) The required line width
 (ii) The effective relative permittivity of the line
(iii) The wavelength on the line at a frequency of 2 GHz.

Solution:
 (i) Using the microstrip impedance design graph in Appendix 2.A:

$$Z_o = 52 \ \Omega \quad \Rightarrow \quad \frac{w}{h} = 0.82,$$

$$w = 0.82 \times 0.5 \ \text{mm} = 0.41 \ \text{mm}.$$

 (ii) Using the microstrip relative permittivity design graph in Appendix 2.A:

$$\frac{w}{h} = 0.82 \quad \Rightarrow \quad \varepsilon_{r,eff}^{MSTRIP} = 6.55.$$

(iii) $\lambda_s = \dfrac{c}{f\sqrt{\varepsilon_{r,eff}^{MSTRIP}}} = \dfrac{3 \times 10^8}{2 \times 10^9 \times \sqrt{6.55}} \ \text{m} = 0.0586 \ \text{m} = 58.6 \ \text{mm}.$

2.5   Operating Frequency Limitations

The operating frequencies for microstrip are limited so as to avoid generating TM modes, and also to avoid transverse microstrip resonances. (*TM stands for transverse magnetic, and refers to modes in which the magnetic field lies entirely within a plane that is orthogonal to the direction of propagation.*)

 (i) Strong coupling exists between the normal quasi-TEM mode of propagation and the lowest-order TM mode when the phase velocities of these two modes are close. This occurs when [1, 2]

$$f_{TEM} = \frac{c \tan^{-1}(\varepsilon_r)}{\pi h\sqrt{2(\varepsilon_r - 1)}}. \tag{2.4}$$

This leads to a restriction on the maximum substrate thickness, h_{max}, given by [1]

$$h_{max} = \frac{0.354\lambda_o}{\sqrt{\varepsilon_r - 1}}, \tag{2.5}$$

where λ_o is the free-space wavelength at the frequency of operation.

(ii) If the microstrip track is sufficiently wide, then a transverse resonance can occur. The effective resonant length is $(w + 2x)$ where w is the width of the line and x is a length representing the fringing at the track edges. Usually, x is taken to be $0.2h$. It follows that the cut-off wavelength, λ_{cf}, for transverse resonance is given by

$$\frac{\lambda_{cf}}{2} = w + 2h = w + 0.4h, \tag{2.6}$$

with a corresponding cut-off frequency, f_{cf}, given by

$$f_{cf} = \frac{c}{(2w + 0.8h)\sqrt{\varepsilon_r}}. \tag{2.7}$$

2.6 Skin Depth

At microwave frequencies the current flowing in a conductor is not uniformly distributed through the cross-section of the conductor, but is concentrated on the surface of the conductor. This phenomenon is known as the skin effect, and is a very important concept that has practical implications for the design of microstrip circuits.

The skin effect can be explained by considering a cylindrical conductor carrying alternating current, as follows:

(a) The alternating current produces a changing magnetic field (magnetic flux) within the conductor, thus causing an induced voltage within the conductor.
(b) The direction of the induced voltage opposes the flow of current (Lenz's law).
(c) The value of the induced voltage will depend on the rate of change of the magnetic flux (Faraday's law).
(d) The maximum induced voltage will occur where the change of magnetic flux is greatest, i.e. at the centre of the conductor.
(e) The current at the centre of the conductor is therefore suppressed, and the current is forced towards the surface of the conductor.
(f) Since the magnitude of the effect depends on the rate of change of magnetic flux, i.e. on the frequency of the alternating current, it will become very significant at high RF and microwave frequencies.

The typical distribution of high-frequency current across the diameter of a cylindrical conductor is shown in Figure 2.3. The skin depth (δ_s) is the parameter used to quantify the skin effect. It is defined as the depth into the conductor at which the magnitude of the current has decreased to $1/e$ of the surface value ($e = 2.718$), and is indicated in Figure 2.3 where the surface current is denoted by I_s. The value of δ_s can be calculated from

$$\delta_s = \frac{1}{\sqrt{\pi \mu f \sigma}}, \tag{2.8}$$

where μ is the permeability of the conductor, f is the frequency of operation, and σ is the conductivity of the conductor. (Usually for a conductor the permeability is the permeability of free space, i.e. $\mu = \mu_0 = 4\pi \times 10^{-7}$ H/m.)

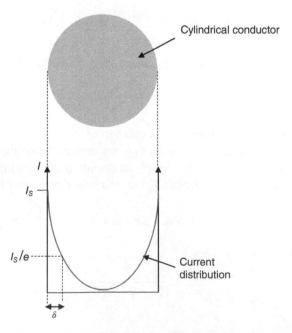

Figure 2.3 Sketch of the magnitude of the RF current distribution across a cylindrical conductor.

Example 2.2 Determine the skin depth in a gold conductor at the following frequencies: (i) 1 GHz; (ii) 10 GHz; (iii) 100 GHz.

Solution:
Note that the conductivity of gold is 4.1×10^7 S/m.

(i) $\delta_s = \dfrac{1}{\sqrt{\pi \times 4 \times \pi \times 10^{-7} \times 10^9 \times 4.1 \times 10^7}}$ m $= 2.485$ μm

(ii) $\delta_s = \dfrac{1}{\sqrt{\pi \times 4 \times \pi \times 10^{-7} \times 10^{10} \times 4.1 \times 10^7}}$ m $= 0.786$ μm

(iii) $\delta_s = \dfrac{1}{\sqrt{\pi \times 4 \times \pi \times 10^{-7} \times 10^{11} \times 4.1 \times 10^7}}$ m $= 0.249$ μm

The data obtained in Example 2.2 show the very small values of δ_s that apply in the microwave frequency range. At 100 GHz, the value of δ_s is comparable with the RMS surface roughness of many conductors, and indicates that very close attention should be given to the surface quality of materials used at high frequencies (this issue is considered in more detail in Chapter 3, where the materials and processes used in the manufacture of microstrip circuits are discussed).

Also, it is useful to consider that in a conductor around 65% of the current will be flowing in one skin depth below the surface. This leads to a practical value for the required thickness of conductors in microstrip circuits, and this is commonly taken to be $5\delta_s$. Thinner conductors will exhibit higher resistance, whilst conductors that are thicker than the $5\delta_s$ value offer no advantage in terms of lower resistance, since very little current will flow at the centre of the conductor.

2.7 Examples of Microstrip Components

The theory and design of some common microstrip components are introduced in this section.

2.7.1 Branch-Line Coupler

The branch-line coupler is a four-port component that can be used as a power splitter, or to combine signals from separate sources. The layout of a microstrip branch-line coupler is shown in Figure 2.4.

The coupler consists of four quarter-wavelength sections, which connect the four ports of the components; the characteristic impedances of the ports are Z_o. A signal applied at port 1 will split between ports 3 and 4, leaving port 2 isolated. Two ports of a microwave circuit or network are said to be isolated when a signal applied at one of the ports does not appear at the other. In the case of the branch-line coupler, we can explain the isolation of port 2 from port 1 by considering the path lengths around the ring of quarter-wavelength sections. A signal applied at port 1 will split at the junction with the ring;

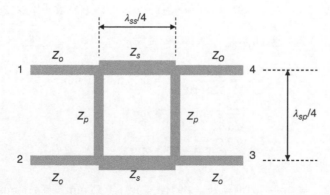

Figure 2.4 Microstrip branch-line coupler.

with half of the signal travelling counter-clockwise through a single $\lambda/4$ section to reach port 2, and with the other half of the signal travelling clockwise through three $\lambda/4$ sections to reach port 2. Thus, the distances travelled by the clockwise and counter-clockwise signals will differ by $\lambda/2$ at port 2, i.e. they will be out-of-phase, and they will cancel, leaving port 2 isolated.

The characteristic impedances of the series and parallel sections, Z_s and Z_p, respectively, determine the power distribution. For a 3 dB division of power we require $Z_p = Z_o$ and $Z_s = Z_o/\sqrt{2}$.

Additional points to note:

(a) For a 3 dB branch-line coupler, when a signal is applied at port 1, the equal amplitude signals at ports 3 and 4 will be 90° out of phase, i.e. they will be in phase quadrature, because the path lengths around the ring differ by $\lambda/4$.

(b) Since port 2 is isolated from port 1, two generators of different frequencies could be connected to ports 1 and 2, and their outputs combined at ports 3 and 4. The two generators would be isolated from each other. If the generators were of different frequencies the normal practice would be to design the quarter-wavelength sections at the mean of the two frequencies.

(c) The branch-line coupler is an example of a microstrip design in which there are significant discontinuities caused by sudden changes in the physical geometry of the microstrip lines, in this case at the corners of the ring. For accurate design the lengths of the quarter-wavelength sections should be specified to the geometric centres of the corners, and appropriate compensation applied to account for the discontinuities. The issue of microstrip discontinuities is considered in more detail in Chapter 4.

(d) The branch-line coupler is a relatively narrow-band component, since its correct operation depends on the lengths of the arms being a quarter-wavelength. Nevertheless, it still has many uses in hybrid microstrip circuits, particularly for signal combining.

(e) The operation of the coupler requires each of the arms to be a quarter-wavelength long. But the arms have different characteristic impedances which means different widths, and since the substrate wavelength is a function of track width, the physical lengths of the sides of the coupler will not be equal. For this reason the physical appearance of the coupler is in the form of a rectangle rather than a square.

Example 2.3 Design a microstrip branch-line coupler to work at 12 GHz. The coupler is to provide a 3 dB power split, and have port impedances of 50 Ω. It is to be fabricated on a substrate that has a relative permittivity of 9.8, and a thickness of 0.25 mm.

Solution:

For a 3 dB power split we require:

Impedance of parallel arm, $Z_p = 50 \, \Omega$.

Impedance of series arm, $Z_s = \dfrac{50}{\sqrt{2}} \, \Omega = 35.4 \, \Omega$.

Using the microstrip design graphs in Appendix 2.A, we obtain:

$$50 \, \Omega \quad \Rightarrow \quad \frac{w}{h} = 0.9 \quad \Rightarrow \quad w = 0.9 \times h = 0.9 \times 0.25 \text{ mm} = 0.225 \text{ mm},$$

$$\frac{w}{h} = 0.9 \quad \Rightarrow \quad \varepsilon_{r,eff}^{MSTRIP} = 6.6 \quad \Rightarrow \quad \lambda_{sp} = \frac{\lambda_o}{\sqrt{6.6}},$$

$$\lambda_o = \frac{c}{f} = \frac{3 \times 10^8}{12 \times 10^9} \text{ m} = 0.25 \text{ m} = 25 \text{ mm},$$

$$\lambda_{sp} = \frac{25}{\sqrt{6.6}} \text{ mm} = 9.73 \text{ mm},$$

$$\frac{\lambda_{sp}}{4} = \frac{9.73}{4} \text{ mm} = 2.43 \text{ mm},$$

$$35.4 \, \Omega \quad \Rightarrow \quad \frac{w}{h} = 1.9 \Rightarrow w = 1.9 \times h = 1.9 \times 0.25 \text{ mm} = 0.475 \text{ mm},$$

$$\frac{w}{h} = 1.9 \quad \Rightarrow \quad \varepsilon_{r,eff}^{MSTRIP} = 7.05 \quad \Rightarrow \quad \lambda_{sp} = \frac{\lambda_o}{\sqrt{7.05}},$$

$$\frac{\lambda_{ss}}{4} = \frac{25}{4 \times \sqrt{7.05}} \text{ mm} = 2.35 \text{ mm}.$$

The dimensions of the completed design are shown in Figure 2.5.

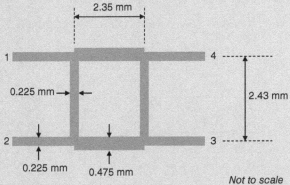

Figure 2.5 Completed design for Example 2.3.

2.7.2 Quarter-Wave Transformer

The quarter-wave transformer ($\lambda/4$ transformer) is a particularly useful microstrip component that enables two resistances (or two lines of different characteristic impedance) to be matched at a given frequency. The transformer is shown in Figure 2.6 connecting a line of characteristic impedance Z_o to a load Z_L; it consists of a quarter-wavelength section of line having characteristic impedance, Z_{oT}, such that the input impedance, Z_{in}, is given by

$$Z_{in} = \frac{Z_{oT}^2}{Z_L}, \tag{2.9}$$

and thus Z_{oT} can be selected so that $Z_{in} = Z_o$.

One of the reasons why the quarter-wave transformer is particularly suitable for implementation in microstrip is that any value of Z_{oT} (within the typical microstrip impedance range 25–90 Ω) can be obtained simply by adjusting the width of the line. In other transmission mediums, such as coaxial cable, only a few standard values of line characteristic impedance are available.

For convenience, we may also write Eq. (2.9) as

$$Z_{oT} = \sqrt{Z_1 Z_2}, \tag{2.10}$$

where Z_1 and Z_2 are the resistive impedances that are to be matched.

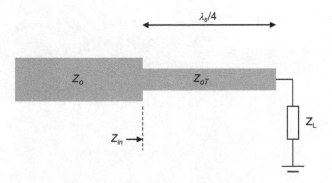

Figure 2.6 Quarter-wave transformer.

Example 2.4   Design a microstrip $\lambda/4$ transformer to match a generator with an output impedance of 50 Ω to a planar antenna that has an input impedance of 75 Ω, at a frequency of 2.5 GHz. The transformer is to be implemented on a substrate that has a relative permittivity of 9.8, and a thickness of 1 mm.

Solution:
Characteristic impedance of the transformer, $Z_{oT} = \sqrt{50 \times 75}\ \Omega = 61.2\ \Omega$.
   Using microstrip design graphs in Appendix 2.A, we obtain:

$$61.2\ \Omega \quad \Rightarrow \quad \frac{w}{h} = 0.55 \quad \Rightarrow \quad w = 0.55h = 0.55 \times 1\ \text{mm} = 0.55\ \text{mm},$$

$$\frac{w}{h} = 0.55 \quad \Rightarrow \quad \varepsilon_{r,eff}^{MSTRIP} = 6.38 \quad \Rightarrow \quad \lambda_s = \frac{\lambda_o}{\sqrt{6.38}},$$

$$\lambda_o = \frac{c}{f} = \frac{3 \times 10^8}{2.4 \times 10^9}\ \text{m} = 0.125\ \text{m} = 125\ \text{mm},$$

$$\frac{\lambda_s}{4} = \frac{125}{4 \times \sqrt{6.38}}\ \text{mm} = 12.37\ \text{mm}.$$

The dimensions of the completed design are shown in Figure 2.7.

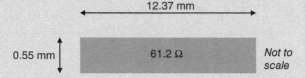

Figure 2.7   Completed design for Example 2.4.

The single-section $\lambda/4$-transformer is relatively narrow band and will only give a perfect match at the frequency that makes the length of the matching section a quarter-wavelength. However, the component can be broad-banded by having a number of $\lambda/4$ sections in cascade, as shown in Figure 2.8. It should be noted that the electrical length of each section is the same, namely $\lambda/4$ or 90°, but the physical lengths of the sections will be different since the substrate wavelengths vary with the widths of the sections.

A broad-band, maximally flat response (Butterworth response) will be obtained [3] when the input reflection coefficient, Γ, is of the form

$$\Gamma = A(1 + e^{-j2\theta}), \tag{2.11}$$

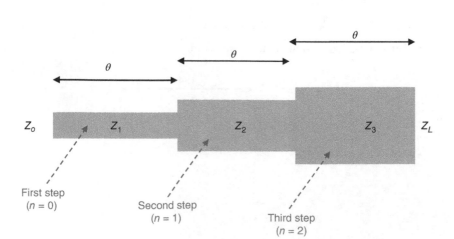

Figure 2.8   Microstrip three-step $\lambda/4$ transformer.

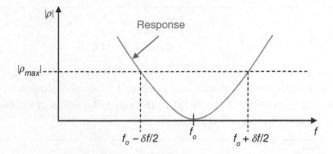

Figure 2.9 Sketch of the reflection coefficient magnitude of a multi-step $\lambda/4$ transformer.

where A is a constant related to the impedances to be matched, and θ is the electrical length of each section of the transformer. The characteristic impedances of adjacent sections must satisfy the expression

$$\ln\left(\frac{Z_{n+1}}{Z_n}\right) = 2^{-N}C_n^N \ln\left(\frac{Z_L}{Z_o}\right), \tag{2.12}$$

where

N = number of sections (steps) in the transformer

Z_n = impedance of the nth step

C_n^N = binomial coefficient[1]

Z_L = terminating impedance

Z_o = source impedance

The expression in Eq. (2.12) is based on the theory of multiple reflections and a comprehensive derivation is presented in Collins [3], who also showed that the fractional bandwidth of a multi-step transformer is given by

$$\frac{\delta f}{f_o} = 2 - \frac{4}{\pi}\cos^{-1}\left|\frac{2\rho_{max}}{\ln\left(\frac{Z_L}{Z_o}\right)}\right|^{\frac{1}{N}}, \tag{2.13}$$

where δf is the bandwidth of the transformer, defined as the frequency range over which the magnitude of the reflection coefficient at the input is less than some specified maximum value, ρ_{max}. The centre frequency of the transformer is f_o. For clarity, the reflection coefficient response of a typical multi-step $\lambda/4$ transformer is shown in Figure 2.9.

Example 2.5 Design a three-step $\lambda/4$ transformer to match a 25 Ω load to a 50 Ω microstrip line at a frequency of 5 GHz. The transformer is to have a maximally flat response, and is to be fabricated on a substrate that has a relative permittivity of 9.8 and a thickness of 0.8 mm.

Solution:
The impedances of the three steps can be found using Eq. (2.12), with $N = 3$, $Z_L = 25\ \Omega$, and $Z_o = 50\ \Omega$.

Step 1: $n = 0$ $\ln\left[\dfrac{Z_1}{50}\right] = 2^{-3}C_0^3 \ln\left[\dfrac{25}{50}\right] \Rightarrow Z_1 = 45.85\ \Omega$,

Step 2: $n = 1$ $\ln\left[\dfrac{Z_2}{45.85}\right] = 2^{-3}C_1^3 \ln\left[\dfrac{25}{50}\right] \Rightarrow Z_2 = 35.36\ \Omega$,

Step 3: $n = 2$ $\ln\left[\dfrac{Z_3}{35.36}\right] = 2^{-3}C_2^3 \ln\left[\dfrac{25}{50}\right] \Rightarrow Z_3 = 27.27\ \Omega$.

Note that a useful check on the calculation is to find the impedance of a fictitious fourth step, which should correspond to the value of Z_L, i.e.

(Continued)

---

1 The binomial coefficient can be evaluated using $C_n^N = \frac{N!}{(N-n)!n!}$.

(Continued)

$$n = 3 \qquad \ln\left[\frac{Z_4}{27.27}\right] = 2^{-3}C_3^3 \ln\left[\frac{25}{50}\right] \quad \Rightarrow \quad Z_4 = 25.00 \ \Omega \equiv Z_L.$$

To complete the design, we need to use the microstrip design graphs in Appendix 2.A (or appropriate CAD software) to find the widths and lengths of each step. Note that each step will have an electrical length of 90°, but the physical lengths will differ slightly because the substrate wavelength varies with the width of the track.

Using the design graphs the following data are obtained:

Z_n (Ω)	w/h	w (mm)	$\varepsilon_{r,eff}^{MSTRIP}$	λ_s (mm)	$\lambda_s/4$ (mm)
45.85	1.10	0.88	6.70	23.18	5.80
35.36	1.85	1.48	7.04	22.61	5.65
27.27	2.90	2.32	7.37	22.10	5.53

The dimensions of the completed design are shown in Figure 2.10.

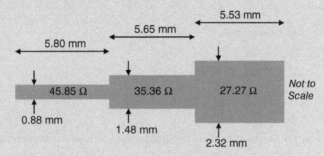

Figure 2.10 Completed design for Example 2.5.

Example 2.6 Find the bandwidth of the transformer designed in Example 2.5 if the maximum VSWR that can be tolerated at the input is 1.5.

Solution:
Firstly we note that

$$(VSWR)_{max} = \frac{1 + |\rho_{max}|}{1 - |\rho_{max}|},$$

i.e.

$$1.5 = \frac{1 + |\rho_{max}|}{1 - |\rho_{max}|} \quad \Rightarrow \quad |\rho_{max}| = 0.2.$$

To find the bandwidth we need to substitute in Eq. (2.13), i.e.

$$\frac{\delta f}{5 \times 10^9} = 2 - \frac{4}{\pi}\cos^{-1}\left|\frac{2 \times 0.2}{\ln\left(\frac{25}{50}\right)}\right|^{\frac{1}{3}} = 1.254,$$

$$\delta f = 1.254 \times 5 \times 10^9 \ \text{Hz} = 6.27 \ \text{GHz},$$

i.e.

$$\text{Bandwidth} = 6.27 \ \text{GHz}.$$

Comment: This is equivalent to a fractional bandwidth of

$$\frac{\delta f}{f_o} = \frac{6.27}{5} = 1.254 \equiv 125.4\%$$

and shows the very large bandwidth that can be obtained from using comparatively few steps in a multi-step transformer.

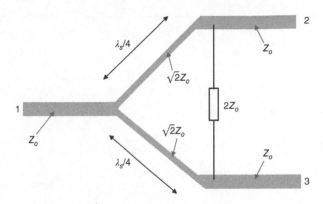

Figure 2.11 Wilkinson power divider.

2.7.3 Wilkinson Power Divider

The Wilkinson power divider is a three-port microstrip component that can be used for power dividing or combining. It has many uses in planar microwave circuitry, particularly for corporate feeds for planar antennas, and also for combining the inputs and outputs in multi-stage microwave power amplifiers.

The layout of a Wilkinson power divider with characteristic impedance, Z_o, is shown in Figure 2.11.

The transmission line from port 1 is split, and connected to ports 2 and 3 through quarter-wave transformers. A resistor is connected between ports 2 and 3, with a value equal to twice that of the characteristic impedance, and thus the Wilkinson power divider behaves as a four-port branch-line coupler with one port internally terminated.

Additional points to note:

(i) Function of the matching resistor: this effectively creates an inaccessible fourth port, so the device can be simultaneously matched at all three accessible ports.

   Also, the use of the resistor improves the isolation when there is a mismatch on an output port.

(ii) The form of component shown in Figure 2.11 is too large for monolithic microwave integrated circuit (MMIC) implementation, particularly below 10 GHz. Three methods can be used to reduce the size:

   (a) Meander the $\lambda/4$ sections – but the problem is avoiding unwanted coupling and inductance in the meandered sections.

   (b) Replace the $\lambda/4$ sections with lumped equivalent components – but a possible problem is excess loss in the inductor.

   (c) Replace the $\lambda/4$ sections with a lumped-distributed equivalent circuit – no major problems, except a limitation on the size reduction possible.

Example 2.7   Design a Wilkinson power splitter, including the input and output lines, to provide a 3 dB power split at a frequency of 2.5 GHz, with port impedances of 50 Ω. The design is to be implemented in microstrip on a substrate that has a relative permittivity of 9.8, and a thickness of 1 mm.

Solution:

Impedance of lines connecting each port = 50 Ω.

   Impedances of the $\lambda/4$ sections = $\sqrt{2} \times 50$ Ω = 70.7 Ω.

   The dimensions of the circuit are found using the microstrip design graphs in Appendix 2.A.

   Input and output lines:

$$50 \ \Omega \quad \Rightarrow \quad \frac{w}{h} = 0.9 \quad \Rightarrow \quad w = 0.9 \times h = 0.9 \times 1 \text{ mm} = 0.9 \text{ mm}.$$

$\lambda/4$ transformers:

$$70.7 \ \Omega \quad \Rightarrow \quad \frac{w}{h} = 0.35 \Rightarrow w = 0.35 \times h = 0.35 \times 1 \text{ mm} = 0.35 \text{ mm},$$

$$\frac{w}{h} = 0.35 \quad \Rightarrow \quad \varepsilon_{r,eff}^{MSTRIP} = 6.25 \Rightarrow \lambda_s = \frac{\lambda_o}{\sqrt{6.25}},$$

(Continued)

(Continued)

$$\lambda_o = \frac{c}{f} = \frac{3 \times 10^8}{2.5 \times 10^9} \text{ m} = 0.12 \text{ m} = 120 \text{ mm,}$$

$$\lambda_s = \frac{120}{\sqrt{6.25}} \text{ mm} = 48.0 \text{ mm} \quad \Rightarrow \quad \frac{\lambda_s}{4} = 12.0 \text{ mm.}$$

The dimensions of the completed design are shown in Figure 2.12.

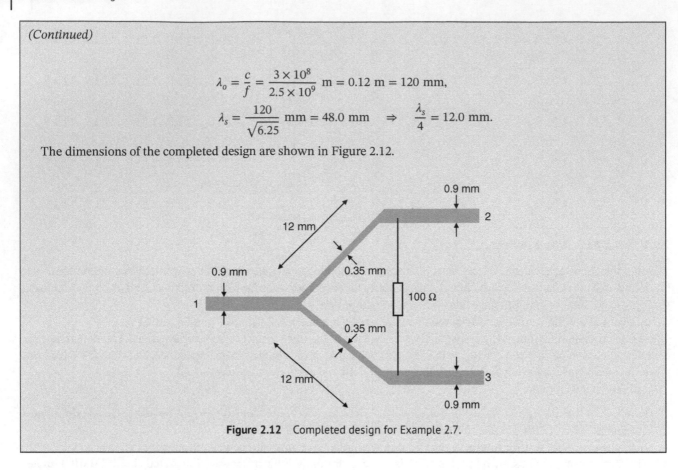

Figure 2.12 Completed design for Example 2.7.

2.8 Microstrip Coupled-Line Structures

A number of four-port microstrip passive components rely for their operation upon the coupling between closely spaced microstrip transmission lines. A length of two such coupled lines is illustrated in Figure 2.13.

Current flowing in one of the microstrip lines will induce a signal in the other and we then describe the two lines as being electromagnetically coupled. The magnitude of the coupling depends primarily on the spacing between the lines, with the greatest coupling for smallest spacing.

2.8.1 Analysis of Microstrip Coupled Lines

A complex electromagnetic field exists between any pair of coupled microstrip lines. However, the analysis can be simplified by noting that at any instant of time the voltages on the two coupled lines must either be at the same potential, or at opposite potential. This leads to the concept of odd and even modes: the even mode refers to the field pattern that exists between the

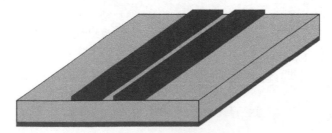

Figure 2.13 Coupled microstrip lines.

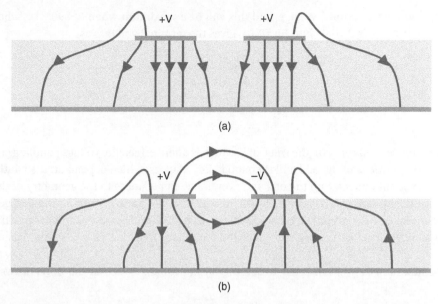

Figure 2.14 Sketch of electric field distributions for even and odd modes: (a) even mode and (b) odd mode.

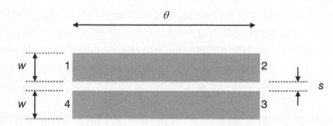

Figure 2.15 Nomenclature for microstrip coupled lines.

lines when they are at the same potential, and the odd mode to the pattern that exists when they are at opposite potentials. The electric field distribution corresponding to each mode is sketched in Figure 2.14.

The terminal voltages on a coupled line structure may be found independently for the odd and even modes, using straightforward transmission line theory, and the total voltage through summation of the terminal voltages for each mode. We will present the principle results of the coupled-line analysis, the details of the solution are routine and given in many textbooks [1].

The conventional notation for identifying the ports of two coupled lines of electrical length, θ, is shown in Figure 2.15.

If it is assumed that the odd and even modes have the same velocity, and therefore the same transmission phase, θ, then for an input of amplitude V at port 1, the terminal voltages will be

$$V_1 = V,$$

$$V_2 = \frac{\sqrt{1-k^2}}{\sqrt{1-k^2}\cos\theta + j\sin\theta}V,$$

$$V_3 = 0,$$

$$V_4 = \frac{jk\sin\theta}{\sqrt{1-k^2}\cos\theta + j\sin\theta}V, \tag{2.14}$$

where

$$k = \frac{Z_{oe} - Z_{oo}}{Z_{oe} + Z_{oo}}, \tag{2.15}$$

and Z_{oe} and Z_{oo} are the impedances of the even and odd modes, respectively.

In Eq. (2.14), V_4 represents the coupled voltage, and this will be a maximum when $\theta = 90°$, i.e. when the length of the coupled section is $\lambda/4$. Substituting $\theta = 90°$ into Eq. (2.14) gives the terminal voltages as

$$
\begin{aligned}
V_1 &= V, \\
V_2 &= -j\sqrt{1 - k^2}V, \\
V_3 &= 0, \\
V_4 &= kV.
\end{aligned}
\tag{2.16}
$$

The coupled voltage then depends only on the value of k, which is often termed the voltage coupling coefficient. The value of k is a function of the magnitudes of the odd and even mode impedances, which depend on the width, w, and spacing, s, of the coupled lines. The equations relating the odd and even mode impedances to the geometry of the coupled lines are given in [1]. A graphical technique was developed by Akhtarzad et al. [4] for determining the line geometry required to give particular impedances. But the graphical technique tends to be rather inaccurate, and for practical designs the line geometry required to realize particular values for the odd and even mode impedances is best found using appropriate CAD software.

It is shown in appendix A of [1] that in order to achieve an impedance match at the ports of two coupled lines it is necessary that

$$
Z_o = \sqrt{Z_{oe}Z_{oo}},
\tag{2.17}
$$

where Z_o is the port impedance.

It should also be noted that since the odd and even modes have different electromagnetic field patterns, the two modes will have different phase velocities, and hence different wavelengths. Thus, in calculating the axial length of a section of coupled microstrip, normal practice is to use the average of the odd and even mode wavelengths. So for $\theta = 90°$ the required length, L, is given by

$$
L = \frac{\lambda_{av}}{4} = \frac{\lambda_{s,even} + \lambda_{s,odd}}{8},
\tag{2.18}
$$

where $\lambda_{s,even}$ and $\lambda_{s,odd}$ are the substrate wavelengths for the even and odd modes, respectively.

The issue of differing wavelengths for the even and odd modes is further addressed in Section 2.8.2.3.

2.8.2 Microstrip Directional Couplers

The layout of a microstrip directional coupler is shown in Figure 2.16. Microstrip directional couplers consist of two microstrip lines, which are coupled together over a length L. The connections to the four ports are splayed out, so as to identify a definite coupling length, L. There is no specific angle for the amount of splay, i.e. the angle φ, but as in the case of the Wilkinson power splitter, there is a compromise between avoiding unwanted coupling and creating a significant discontinuity (microstrip discontinuities are considered in detail in Chapter 4). A splay angle around $\varphi = 90°$ is normally used.

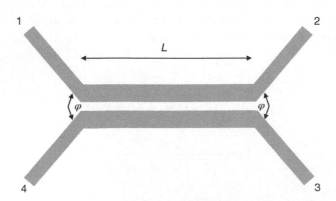

Figure 2.16 Layout of a microstrip directional coupler.

Microstrip directional couplers are examples of backward wave couplers. If we consider a coupler with correctly matched ports, and a signal is applied to port 1, some of the energy travelling along the main line from port 1 to port 2 will be coupled to the secondary line and will appear at port 4. No signal will appear at port 3, which is isolated. Thus, the coupler has directional properties: signals travelling from port 1 to port 2 will only couple to port 4, and signals travelling from 2 to 1 will only couple to port 3.

There are four parameters used to specify the performance of a directional coupler: Coupling (C); Directivity (D); Isolation (I); and Transmission (T). These parameters are defined in terms of power ratios. If a power P_1 is incident at port 1, and the power coupled to port 4 is P_4, then the coupling (C) in dB is defined as

$$C = 10 \log \left(\frac{P_1}{P_4} \right) \text{ dB.} \tag{2.19}$$

Theoretically, port 3 should be isolated, and the effectiveness of the coupler in directing coupled energy specifically to port 4 is specified in terms of the directivity (D), where

$$D = 10 \log \left(\frac{P_4}{P_3} \right) \text{ dB,} \tag{2.20}$$

and so for an ideal directional coupler $D = \infty$. The degree of isolation of port 3 from port 1 is specified as

$$I = 10 \log \left(\frac{P_1}{P_3} \right) \text{ dB,} \tag{2.21}$$

and for an ideal directional coupler, with $P_3 = 0$, the isolation is infinite. Power which is not coupled from port 1 to port 4 will travel directly from port 1 to port 2, and thus the transmission, T, is defined as

$$T = 10 \log \left(\frac{P_1}{P_2} \right) \text{ dB.} \tag{2.22}$$

Two points should be noted:

(i) Directional couplers are specified primarily in terms of the coupling parameter, C. Thus, a 20 dB directional coupler is one in which the signal power at the coupled port is 20 dB below the signal power at the input port.
(ii) It should always be remembered that the dB is defined as 10 times $\log_{10}$ of the ratio of two powers, i.e.

$$\text{dB} \equiv 10 \log \left(\frac{P_1}{P_2} \right). \tag{2.23}$$

The corresponding voltage relationship is derived from

$$\text{dB} = 10 \log \left(\frac{V_1^2 / Z_1}{V_2^2 / Z_2} \right). \tag{2.24}$$

If $Z_1 = Z_2$, then we may define the dB in terms of a simple voltage ratio, i.e.

$$\text{dB} = 10 \log \left(\frac{V_1^2}{V_2^2} \right) = 20 \log \left(\frac{V_1}{V_2} \right). \tag{2.25}$$

Thus, if the ports are all matched we may write the four directional coupler parameters as

$$C = 20 \log \left(\frac{V_1}{V_4} \right) \text{ dB.}$$

$$D = 20 \log \left(\frac{V_4}{V_3} \right) \text{ dB.}$$

$$I = 20 \log \left(\frac{V_1}{V_3} \right) \text{ dB.}$$

$$T = 20 \log \left(\frac{V_1}{V_2} \right) \text{ dB.} \tag{2.26}$$

Example 2.8   A matched 12 dB microstrip directional coupler has port impedances of 50 Ω. Determine:

(i) The value of the coupling coefficient, k
(ii) The values of the odd and even mode impedances.

Solution:
(i) Using Eq. (2.26):

$$12 = 20 \log \left| \frac{V_1}{V_4} \right| = 20 \log \left(\frac{1}{k} \right),$$

$$k = 0.251.$$

(ii) Using Eq. (2.17):

$$50 = \sqrt{Z_{oe} Z_{oo}}. \tag{a}$$

Using Eq. (2.15):

$$0.251 = \frac{Z_{oe} - Z_{oo}}{Z_{oe} + Z_{oo}}. \tag{b}$$

From (a) we have $Z_{oo} = \dfrac{2500}{Z_{oe}}$ and substituting this into (b) and solving for Z_{oe} gives $Z_{oe} = 64.62 \ \Omega$ and $Z_{oo} = 38.69 \ \Omega$.

Example 2.9   Port 1 of an ideal 20 dB microstrip coupled-line directional coupler shown in Figure 2.16 is supplied with 10 mW at the working frequency. The port impedances are matched to 50 Ω. Determine:

(i) The power leaving ports 2, 3, and 4 of the coupler
(ii) The corresponding RMS voltages at ports 2, 3, and 4.

Solution:
(i) $C = 20 \ \text{dB} \equiv 100$ (power ratio)

Power leaving port 4: $P_4 = \dfrac{10}{100} \ \text{mW} = 0.1 \ \text{mW},$

Power leaving port 3: $P_3 = 0,$

Power leaving port 2: $P_2 = (10 - 0.1) \ \text{mW} = 9.9 \ \text{mW}.$

(ii) Noting that $P = \dfrac{V_{RMS}^2}{R} \ \Rightarrow \ V_{RMS} = \sqrt{PR}$ we have

RMS voltage at port 4: $V_4 = \sqrt{0.1 \times 10^{-3} \times 50} \ \text{V} = 70.71 \ \text{mV},$

RMS voltage at port 3: $V_3 = 0,$

RMS voltage at port 2: $V_2 = \sqrt{9.9 \times 10^{-3} \times 50} \ \text{V} = 703.56 \ \text{mV}.$

Example 2.10   A 12 dB directional coupler is designed to work at 10 GHz. Determine the coupling at a frequency of 9 GHz. Comment upon the result.

Solution:
At the design frequency of 10 GHz

$$12 = 20 \log \left(\frac{1}{k} \right) \ \Rightarrow \ k = 0.251.$$

The electrical length of the coupler is given by

$$\theta = \beta l = \frac{2\pi}{\lambda} l = \frac{2\pi}{v_p} fl,$$

i.e.

$$\theta \propto f.$$

Therefore, if f is changed to 9 GHz (i.e. a decrease of 10%), θ will also reduce by 10%. At the design frequency of 10 GHz, $\theta = 90°$, and so at

9 GHz, $\theta = 90 - 9° = 81°$.

Using Eq. (2.14):

$$\frac{V_4}{V_1} = \frac{jk \sin \theta}{\sqrt{1 - k^2} \cos \theta + j \sin \theta} = \frac{jk \tan \theta}{\sqrt{1 - k^2} + j \tan \theta}.$$

Let the voltage coupling coefficient at 9 GHz be k_1, then

$$k_1 = \left| \frac{j0.251 \tan 81°}{\sqrt{1 - (0.251)^2} + j \tan 81°} \right| = \left| \frac{j0.251 \times 6.314}{0.968 + j6.314} \right| = 0.248,$$

i.e.

$$k_1 = 0.248 \equiv 20 \log \left(\frac{1}{0.248} \right) \text{ dB} = 12.11 \text{ dB}.$$

Comment: This calculation shows that directional couplers are relatively wideband components, with a change of 10% in frequency causing less than 1% change in the dB value of coupling.

2.8.2.1 Design of Microstrip Directional Couplers

There are five main steps in the design of a microstrip coupled-line directional coupler:

(1) Determine the values of C, Z_0, and f_o from the specification.
(2) Evaluate Z_{oe} and Z_{oo}.
(3) Find w/h and s/h using appropriate CAD software.
(4) Find $\lambda_{s,even}$ and $\lambda_{s,odd}$ using appropriate CAD software.
(5) Calculate the length, L, of the coupled region using Eq. (2.18), i.e.

$$L = \frac{\lambda_{av}}{4} = \frac{\lambda_{s,even} + \lambda_{s,odd}}{8}.$$

Note that this design procedure, in common with those of all simple microstrip coupled-line structures, involves an approximation. This arises from the odd and even modes having different phase velocities, which means that the coupled length must be calculated as the average of the wavelengths of the two modes.

Example 2.11 Design a microstrip directional coupler that will provide 15 dB coupling at a frequency of 8 GHz. The coupler is to have port impedances of 50 Ω, and is to be fabricated on a 0.6 mm thick substrate that has a relative permittivity of 9.8.

Solution:

$$15 \text{ dB} \quad \Rightarrow \quad 15 = 20 \log \left(\frac{1}{k} \right) \quad \Rightarrow \quad k = 0.178,$$

From Eq. (2.15) :

$$k = 0.178 = \frac{Z_{oe} - Z_{oo}}{Z_{oe} + Z_{oo}}. \tag{a}$$

From Eq. (2.17) :

$$Z_0 = 50 = \sqrt{Z_{oe} Z_{oo}}. \tag{b}$$

Solving Eqs. (a) and (b) gives:

$$Z_{oe} = 59.85 \ \Omega,$$

$$Z_{oo} = 41.77 \ \Omega.$$

(Continued)

(Continued)

Using appropriate CAD software the line geometry to give the above values of even and odd mode impedances is:

$$w = 585\ \mu m \quad \text{and} \quad s = 275\ \mu m.$$

The line geometry also determines the wavelengths of the two modes, and again from CAD software we have:

$$\lambda_{s,even} = 14.13\ \text{mm} \quad \text{and} \quad \lambda_{s,odd} = 16.72\ \text{mm}.$$

The required length of the coupled region is then found from Eq. (2.18) as:

$$L = \frac{14.13 + 16.72}{8}\ \text{mm} = 3.86\ \text{mm}.$$

The single microstrip lines connecting the coupled region to the four ports must each have a characteristic impedance of 50 Ω, and we find the widths using the microstrip design graphs in Appendix 2.A:

$$50\ \Omega \quad \Rightarrow \quad \frac{w}{h} = 0.9 \quad \Rightarrow \quad w = 0.9 \times h = 0.9 \times 0.6\ \text{mm} = 450\ \mu m.$$

The dimensions of the completed design are shown in Figure 2.17.

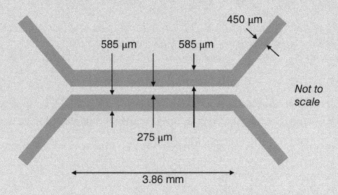

Figure 2.17 Completed design for Example 2.11.

2.8.2.2 Directivity of Microstrip Directional Couplers

The theoretical directivity of microstrip directional couplers is infinite if the simplifying assumption is made that the transmission phase changes of the even and odd modes are equal, i.e. $\theta_e = \theta_o$. In practice, this will not be the case, since the velocity of the odd mode is around 12% higher than the velocity of the even mode for typical microstrip geometries. A rigorous analysis, in terms of the individual phase changes of the even and odd modes, gives the following expression for the directivity [1]

$$D = \frac{V_4}{V_3} = \left| \frac{(aZ_{oe} - bZ_{oo})Z_0}{a(Z_0 Z_{oe} \cos \theta_e + jZ_{oe}^2 \sin \theta_e) - b(Z_0 Z_{oo} \cos \theta_o + jZ_{oo}^2 \sin \theta_o)} \right|, \tag{2.27}$$

where

$$a = 2Z_0 Z_{oo} \cos \theta_o + j(Z_{oo}^2 + Z_0^2) \sin \theta_o,$$

$$b = 2Z_0 Z_{oe} \cos \theta_e + j(Z_{oe}^2 + Z_0^2) \sin \theta_e, \tag{2.28}$$

and θ_e and θ_o are the transmission phases of the even and odd modes, respectively, through the coupled section.

The directivity of practical microstrip directional couplers is limited to around 20 dB at the design frequency, due to the unequal velocities of the even and odd modes. However, the directivity can be improved significantly by employing the modifications described in the following section.

2.8.2.3 Improvements to Microstrip Directional Couplers

Simple microstrip directional couplers suffer a loss of directivity because of the need to average the odd and even mode wavelengths when calculating the length of the coupled region. Two techniques can be used to equalize the transmission phases of the odd and even modes over the length of the coupled region.

(1) Use of a dielectric overlay

A thick layer of dielectric, having the same relative permittivity as the substrate, is placed over the coupled region, as shown in Figure 2.18.

With the overlay present the electromagnetic field originally fringing into the air will now be in the same medium as the field in the substrate, and so the velocities of the two modes will be the same, as will the transmission phases, i.e.

$$v_{p,even} = v_{p,odd} \tag{2.29}$$

and

$$\theta_e = \theta_o. \tag{2.30}$$

Whilst the overlay technique is very simple in concept, it has a number of practical disadvantages. It requires some form of epoxy adhesive, with material characteristics matching those of the substrate, to fix the overlay in place. Moreover, in a production environment it necessitates a non-standard surface-mount operation which could be very costly to implement.

(2) Use of the Podell technique

Podell [5] developed a technique to overcome the problem of the two modes having different velocities that was based on a simple change to the geometry of the coupled lines. In the Podell technique the coupling gap is made in the form of a zig-zag, as shown in Figure 2.19, so as to constrain the faster odd mode to travel further than the even mode.

The overall width of the coupler is unchanged from the straight gap version, so that the propagation of the even mode is essentially unaffected by the shaping of the gap since the electromagnetic field associated with the even mode is essentially independent of the position of the gap.

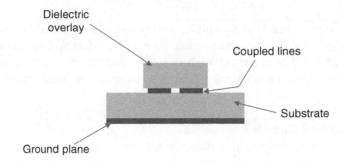

Figure 2.18   Cross-section of microstrip coupled lines with dielectric overlay.

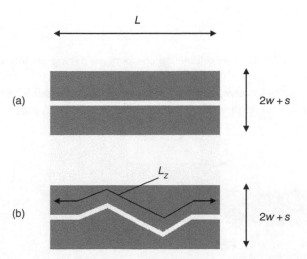

Figure 2.19   Coupled microstrip lines, showing the structure of a zig-zag coupling gap: (a) linear gap and (b) zig-zag gap.

Since the odd mode is travelling faster than the even mode, the transmission phase of each mode through the coupler can be equalized by making the ratio of the zig-zag length to that of the straight-through length equal to the ratio of the velocities of the two modes, i.e.

$$\frac{L_z}{L} = \frac{v_{p,odd}}{v_{p,even}}. \tag{2.31}$$

The practical advantage of the Podell technique is that it only involves a simple modification to the circuit layout at the photo-lithographic stage, and does not therefore add significantly to the cost of production. In terms of technical performance, the use of the Podell method can, typically, improve the directivity of the coupler by around 15 dB. Thus, the technique increases the directivity of microstrip couplers from around 20 dB to well over 30 dB. The only minor drawback to the technique is that the zig-zag path does introduce some significant discontinuities into the gap, which may become a problem at high microwave frequencies.

2.8.3 Examples of Other Common Microstrip Coupled-Line Structures

2.8.3.1 Microstrip DC Break

The microstrip DC break is shown in Figure 2.20, and consists of two closely coupled microstrip fingers, each of length $\lambda_s/4$, integrated into a microstrip line.

This type of component is a useful alternative to a surface-mount capacitor in situations where it is necessary to provide DC isolation on a microstrip line, for example to permit independent DC biasing of adjacent active devices. The microstrip DC break does not suffer from the package parasitics often associated with surface-mount capacitors at microwave frequencies. Moreover, it can provide low insertion loss, typically <0.2 dB, with a relatively wide bandwidth. The one drawback to the component is the requirement for a very small coupling gap. However, this requirement can be overcome by using a multi-layered format, which will be described in Chapter 3.

The condition for matching the coupled-line section to the main line is

$$Z_o = \sqrt{Z_{oe}Z_{oo}}, \tag{2.32}$$

where Z_o is the impedance of the main line, and Z_{oo} and Z_{oe} are the odd and even mode impedances, respectively, of the coupled lines. However, meeting this matching condition usually requires finger widths that are quite narrow compared to the width of the main line, and it is necessary to apply compensation for microstrip discontinuities. Also, Free and Aitchison [6] found that this type of circuit configuration can produce significant excess phase, which has to be considered in a practical circuit design.

2.8.3.2 Edge-Coupled Microstrip Band-Pass Filter

The layout of a third-order microstrip band-pass filter is shown in Figure 2.21.

The filter comprises three $\lambda_s/2$ resonant sections, with each section coupled to the adjacent resonator though a $\lambda_s/4$ coupled-line. The detailed design of this type of filter is presented in Chapter 5, which discusses the design and implementation of various types of RF and microwave filters.

2.8.3.3 Lange Coupler

The Lange coupler [7] is a wide-band four-port microstrip component, having an interdigitated structure, as shown in Figure 2.22.

The coupler shown consists of four interleaved microstrip fingers, with an overall length of $\lambda_s/4$. Wire bonds are used to connect alternate fingers, and serve to maximize the power transfer from port 1 to port 4.

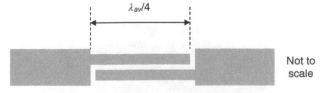

Figure 2.20 Microstrip DC break.

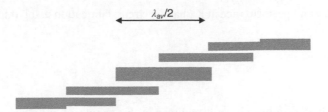

Figure 2.21 Edge-coupled microstrip band-pass filter.

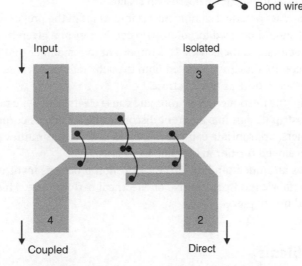

Figure 2.22 Lange coupler, showing the port designations.

The use of an interdigitated structure means that it is possible to achieve tight coupling between the input port and the coupled port, without the need for very small gaps between microstrip tracks. The interdigitated structure originally reported by Lange [7] provided 3 dB coupling between the input and coupled ports. To achieve the same degree of coupling with a single edge-coupled microstrip structure, as shown in Figure 2.16, would require a coupling gap, s, significantly less than 10 μm.

Lange proposed the original idea of the interdigitated coupler, although he did not provide any practical design formation. Various authors have addressed the practical design of the Lange coupler, and the most useful and widely quoted design equations are those developed by Osmani [8]. The expressions given by Osmani enable the even and odd mode impedances of adjacent lines to be found in terms of the voltage coupling coefficient (k), the even number of fingers (N), and the terminating impedance (Z_o), and are:

$$Z_{oo} = Z_o \left[\frac{1-k}{1+k} \right]^{0.5} \frac{(N-1)(1+q)}{(k+q) + (N-1)(1-k)}, \tag{2.33}$$

$$Z_{oe} = Z_{oo} \frac{k+q}{(N-1)(1-k)}, \tag{2.34}$$

where

$$q = [k^2 + (1-k^2)(N-1)^2]. \tag{2.35}$$

Once the even and odd mode impedances are found from the above expressions, the width (w) and spacing (s) of the microstrip fingers can be determined, as described earlier in the chapter. The length of the fingers is found using Eq. (2.18).

The Lange coupler requires the use of wire bonds, but care must be taken to ensure these do not contribute significant inductance; this is normally done by using flat ribbon bonds and making sure the lengths of the bonds are $<\lambda_o/8$. In recent work, Bikiny et al. [9] eliminated the need for wire bonds by fabricating the Lange in a multilayer format. In their arrangement, alternate fingers were fabricated on different layers of the structure, separated by a thin dielectric layer. Since alternate fingers were on different layers, there was no need for bonding wires and the required connections between alternate fingers were made using narrow strips of conductor. Bikiny used a thick-film technique to fabricate the multilayer structure and reported [9] excellent wide-band performance at Ka-band[2]; circuit fabrication using thick-film techniques is discussed in more detail in Chapter 3.

2.9 Summary

The basic concepts of microstrip design have been introduced in this chapter. A number of key issues that underpin successful high-frequency circuit design using microstrip have been identified:

(a) Precise line geometries are essential for accurate microstrip design.
(b) An appreciation of skin depth and its practical significance is important in the preparation of microstrip circuits. Skin depth affects the thickness that should be used for microstrip conductors at a given frequency, and also the quality of the surface finish of both conductors and dielectrics. It is important to remember that the surface finish of substrates may be very significant where conductors are deposited onto the substrate, and where consequently the underside of the conductor has the same surface profile as the substrate.
(c) Discontinuities in microstrip circuits introduce reflections and cause electromagnetic radiation, and should be avoided where possible. Where a microstrip design has inherent discontinuities, such as corners in branch-line couplers, or steps in width in $\lambda/4$ transformers, appropriate compensations for the discontinuities must be applied. The theme of microstrip discontinuities is developed further in Chapter 4.
(d) Good microstrip design requires an understanding of the fabrication processes involved in the production of the circuit, and the influence that the fabrication process has on electrical performance. Modern techniques for producing microstrip circuits are discussed in Chapter 3.

2.10 Supplementary Problems

Q2.1 Why is it necessary to specify the *effective* relative permittivity when dealing with microstrip circuits?

Q2.2 Would you expect the value of effective relative permittivity to increase or decrease with an increase in the width of the microstrip track?

Q2.3 A microstrip line has an effective relative permittivity of 3.8. What is the substrate wavelength at a frequency of 5 GHz?

Q2.4 Complete the following table, which refers to the design parameters for microstrip lines fabricated on a substrate that has a relative permittivity of 9.8, and a thickness of 635 μm, for use at 10 GHz.

Z_o (Ω)	w (μm)	λ_s (mm)	V_p (m/s)
25			
50			
75			
100			

Q2.5 A 50 Ω microstrip line is fabricated on a 1 mm thick alumina substrate ($\varepsilon_r = 9.8$). What would be the approximate percentage change in the substrate wavelength at 10 GHz if a thick dielectric, also having $\varepsilon_r = 9.8$, were placed on top of the track?

---

2 Ka-band is the designated frequency band from 26.5 to 40 GHz.

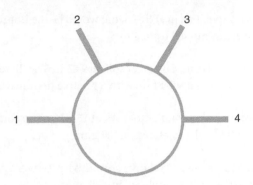

Figure 2.23 Hybrid ring for use with Q2.8.

Q2.6 If microstrip circuits are to be fabricated using copper conductors, find the minimum track thicknesses required at the following frequencies: 1, 10, 100 GHz. (Note that the resistivity of copper is 17.5×10^{-9} Ω-m.)

Q2.7 A microstrip circuit is to be fabricated on a substrate that has a relative permittivity of 9.8 and a thickness of 0.635 mm. The substrate is coated with copper that has a thickness of 32 μm. A wet etching process is to be used to fabricate microstrip lines of characteristic impedance 50 and 70 Ω. Find the percentage errors in the line impedances if no allowance is made for undercutting during the etching process.

Q2.8 Figure 2.23 shows the schematic layout of a microstrip hybrid ring. This consists of a circular ring of microstrip having a characteristic impedance of Z_R. The ring has four ports with the relative spacing as shown. The characteristic impedance of the line feeding each port is Z_o, and for matching at each port we must have $Z_R = Z_o/\sqrt{2}$.

   (a) If a signal of unit voltage amplitude is applied to port 1, find the amplitude and phases of the signals leaving the other three ports, when the ports are correctly terminated.

   (b) Explain why signal sources can be connected to ports 1 and 3 without mutual interference.

   (c) Design a microstrip hybrid ring to operate at a frequency of 10 GHz. The ring is to have port impedances of 50 Ω and is to be fabricated on a substrate that has a relative permittivity of 9.8 and a thickness of 0.5 mm.

   (d) Suppose that the hybrid ring designed in part (c) were covered with a thick layer of dielectric, also having a relative permittivity of 9.8. Estimate the new optimum frequency of operation.

Q2.9 Design a single-section microstrip quarter-wave transformer to match a load impedance of 80 Ω to a 50 Ω microstrip line at a frequency of 6 GHz. The transformer is to be fabricated on a substrate that has a relative permittivity of 9.8 and a thickness of 0.5 mm.

Q2.10 Repeat Q2.9, but consider the transformer to be fabricated in coaxial cable that is filled with a dielectric having a relative permittivity of 2.3. Comment upon the result.

Q2.11 Design a four-section microstrip $\lambda/4$-transformer to match a load of 25 Ω to a 50 Ω microstrip line at a frequency of 12 GHz. The transformer is to have a maximally flat response and is to be fabricated on a substrate that has a relative permittivity of 9.8 and a thickness of 0.635 mm.

Q2.12 Design a three-section microstrip $\lambda/4$-transformer to match a load of 80 Ω to a 50 Ω microstrip line at a frequency of 15 GHz. The transformer is to have a maximally flat response and is to be fabricated on a substrate that has a relative permittivity of 9.8 and a thickness of 0.5 mm.

Q2.13 Assuming that the transformer designed in Q2.11 is to be used in a hybrid microstrip circuit where the maximum VSWR at the input to the transformer is 1.3, determine the bandwidth over which the transformer will work satisfactorily.

Q2.14   Considering the VSWR conditions specified in Q2.13, what would be the improvement in bandwidth if the number of sections in the transformer were increased from 4 to 6?

Q2.15   Design a Wilkinson power divider to give a 3 dB power split at 9 GHz. The divider is to have port impedances of 50 Ω and be implemented in microstrip using a substrate with a relative permittivity of 9.8 and a thickness of 1.2 mm.

Q2.16   Design a 50 Ω 3 dB microstrip branch-line coupler to work at 15 GHz. The coupler is to be fabricated on a substrate that has a relative permittivity of 9.8 and a thickness of 0.4 mm.

Q2.17   Suppose that a thick layer of dielectric, also of relative permittivity 9.8 were placed on top of the coupler designed Q2.16. Explain how the performance of the coupler would change.

Q2.18   A 50 Ω microstrip edge-coupled directional coupler is to be designed to provide 14 dB of coupling at 15 GHz. Find the even and odd mode impedances needed for the coupled section.

Q2.19   Determine the frequency range over which the coupler designed in Q2.18 will maintain coupling to within 0.5 dB of the designed value.

Appendix 2.A Microstrip Design Graphs

The microstrip design graphs given in this appendix apply to a substrate having a relative permittivity of 9.8. Whilst these graphs are convenient for demonstrating design procedures in microstrip, they should not be used for practical design work since there may be significant reading errors, and accurate line geometry is essential if microstrip components are to yield high, predictable performance.

For practical design work, CAD software such as ADS® should be used.

Accurate microstrip calculators can also be found as Freeware on the Internet, although these calculators tend to be very limited in their functionality.

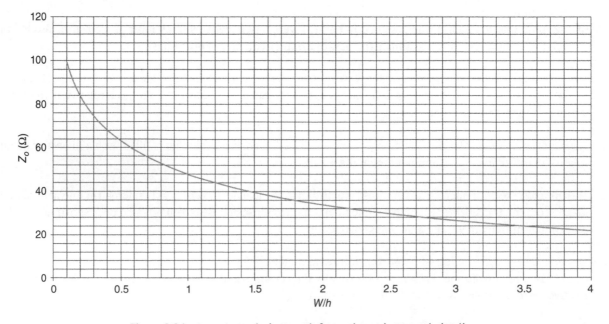

Figure 2.24   Impedance design graph for a microstrip transmission line.

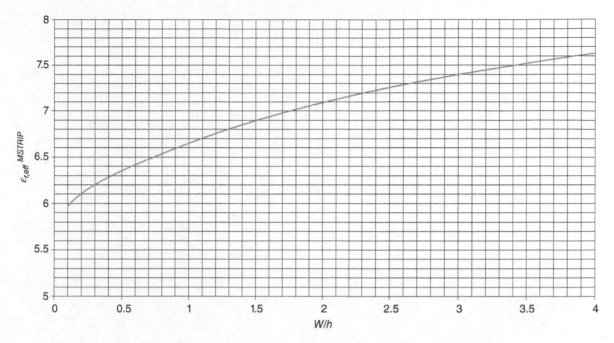

Figure 2.25 Effective permittivity design graph for a microstrip transmission line.

References

1 Edwards, T.C. and Steer, M.B. (2000). *Foundations of Interconnection and Microstrip Design*. Chichester, UK: Wiley.

2 Glover, I.A., Pennock, S.R., and Shepherd, P.R. (2005). *Microwave Devices, Circuits and Subsystems*. Chichester, UK: Wiley.

3 Collin, R.E. (1992). *Foundations for Microwave Engineering*, 2e. New York: McGraw Hill.

4 Akhtarzad, S., Rowbottom, T.R., and Jones, P.B. (1975). The design of coupled microstrip lines. *IEEE Transactions on Microwave Theory and Techniques* 23: 486–492.

5 Podell, A. (1970). A high directivity microstrip coupler technique. *Proceedings of IEEE International Microwave Symposium* (May 1970), pp. 33–36

6 Free, C.E. and Aitchison, C.S. (1984). Excess phase in microstrip DC blocks. *Electronics Letters* 20 (21): 892–893.

7 Lange, J. (1969). Interdigitated stripline quadrature hybrid. *IEEE Transactions on Microwave Theory and Techniques* 17: 1150 1151.

8 Osmani, R.M. (1981). Synthesis of Lange couplers. *IEEE Transactions on Microwave Theory and Techniques* 29 (2): 168–170.

9 Bikiny, A., Quendo, C., Rius, E. et al. (2009). Ka-band Lange coupler in multilayer thick-film technology. *Proceedings of IEEE International Microwave Symposium*, Boston USA (June 2009), pp. 1001–1004

3

Fabrication Processes for RF and Microwave Circuits

3.1 Introduction

As frequencies increase into the high RF and microwave regions, the materials and fabrication technologies used to manufacture planar circuits assume greater significance. Good circuit design at these high frequencies is very much dependent on the selection of appropriate materials and fabrication processes, and it is essential that the circuit designer should have an appreciation of the properties and limitations of the available circuit materials and associated manufacturing processes. This chapter introduces the main fabrication processes and materials that are available, and provides an indication of their relative merits.

The chapter commences with a review of the parameters used to describe the performance of high-frequency circuit materials, and then expands the discussion to list the essential requirements for materials for RF and microwave circuit applications.

The choice of a fabrication process for a particular circuit application is of critical importance, and has a significant effect on the electrical circuit design. Recent developments in materials for high-frequency applications offer the circuit designer a wide choice of circuit structures. One particularly useful development, which provides the designer with significantly greater flexibility, is that of relatively low-cost multilayer circuit processes. Once the sole province of the monolithic circuit designer, multilayer techniques using thick-film or low-temperature co-fired ceramic (LTCC) techniques are now a cost-effective option for the hybrid designer. The use of multilayer structures permit strong, effective coupling between planar conductors, and leads to improved performance for filters and antenna feeds. LTCC multilayer structures in particular have led to the practical implementation of the system-in-package (SiP) concept, in which circuit interconnections and lumped passives are integrated within a single ceramic package. SiP structures are discussed later in the chapter.

In the area of thick-film technology, the availability of photo-sensitive materials enables low-cost, high-resolution circuits to be manufactured in a semi-automated process, and this technology has attracted significant interest in both the industrial and research environments. More recently, the use of inkjet techniques to directly print a single layer, and 3D high-frequency structures has provided additional flexibility in circuit manufacture. For completeness, the chapter includes a review of the various manufacturing technologies, together with an analysis of their high-frequency performance.

The characterization of circuit materials is important both for the materials manufactures and the users, and the chapter concludes with a discussion of the various measurement techniques that are available. The discussion focuses primarily on microwave and millimetre-wave techniques, since it is in these frequency regions where material properties are particularly significant.

3.2 Review of Essential Material Parameters

3.2.1 Dielectrics

Normally, the dielectrics used for RF applications such as coaxial transmission lines and planar substrates are extremely good insulators and so exhibit negligible conductivity. However, when a dielectric is subjected to a varying electric field, various other physical mechanisms occur within the material, which cause a damping effect and lead to dielectric loss. For example, the dielectric in a parallel plate capacitor can be represented by a simple GC parallel circuit, as shown in Figure 3.1, where the resistor, represented here by its conductance G, is used to account for the dielectric loss.

RF and Microwave Circuit Design: Theory and Applications, First Edition. Charles E. Free and Colin S. Aitchison.
© 2022 John Wiley & Sons Ltd. Published 2022 by John Wiley & Sons Ltd.
Companion website: www.wiley.com/go/free/rfandmicrowave

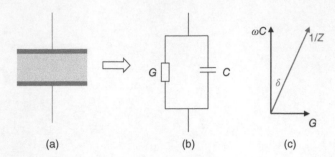

Figure 3.1 Parallel-plate capacitor (a), with its equivalent circuit (b) and vector admittance diagram (c).

Also shown in Figure 3.1c is the admittance vector diagram for the GC combination, from which it can be seen that

$$\tan \delta = \frac{1}{\omega CR}. \tag{3.1}$$

$\tan \delta$ is referred to as the loss tangent of the dielectric. From simple circuit theory Eq. (3.1) can also be written as

$$\tan \delta = \frac{1}{Q}, \tag{3.2}$$

where Q is the Q-factor[1] of the dielectric.

An alternative to using a conductance G to represent the dielectric loss is simply to write the relative permittivity of the dielectric in a complex format as

$$\varepsilon_r = \varepsilon' - j\varepsilon'', \tag{3.3}$$

where ε' is the value of permittivity used at DC and low frequencies, and ε'' is a quantity representing the loss in the dielectric. It can be shown [1] that the loss tangent and the components of the complex permittivity are related by

$$\tan \delta = \frac{\varepsilon''}{\varepsilon'}. \tag{3.4}$$

The values of ε' and $\tan \delta$ remain reasonably constant with frequency over the lower part of the RF spectrum, but at microwave (and particularly millimetre-wave) frequencies, significant changes can occur. Consequently, dielectric characterization techniques are of particular importance when selecting or developing dielectric materials for microwave applications.

Propagation of electromagnetic waves through dielectric material may be described using the concepts introduced in Chapter 1, namely that the propagation constant, γ, is of the form

$$\gamma = \alpha + j\beta, \tag{3.5}$$

where α is the attenuation constant in the units of Np/m, and β is the phase propagation constant in the units of rad/m.

In the particular case of microstrip the dielectric loss in dB per line wavelength (λ_s) is given by the widely referenced formula [2]

$$\alpha_d = 27.3 \frac{\varepsilon_r(\varepsilon_{r,eff}^{MSTRIP} - 1) \tan \delta}{\varepsilon_{r,eff}^{MSTRIP}(\varepsilon_r - 1)} \ \mathrm{dB}/\lambda_s, \tag{3.6}$$

where the symbols have their usual meanings. Note that the effective relative permittivity is used to take account of fringing, since some of the energy travels in the air above the substrate.

Example 3.1 A low-loss dielectric has a complex relative permittivity given by

$$\varepsilon_r = 2.6 - j0.018.$$

Determine:

(i) The loss tangent of the dielectric
(ii) The Q-factor of the dielectric.

---

1 Q-factor is defined later in the chapter in Eq. (3.16)

Solution

(i) Using Eq. (3.4):

$$\tan \delta = \frac{0.018}{2.6} = 6.92 \times 10^{-3}.$$

(ii) Using Eq. (3.2):

$$Q = \frac{1}{\tan \delta} = \frac{1}{6.92 \times 10^{-3}} = 144.51.$$

Example 3.2   A dielectric has a real value of relative permittivity of 2.6, a Q of 210, and a propagation constant given by

$$\gamma = \frac{2\pi}{\lambda_d}(0.5 \tan \delta + j),$$

where the symbols have their usual meanings.

Determine (at a frequency of 10 GHz):

(i)   The wavelength in the material
(ii) The ratio of the velocity of propagation in the material to that in air
(iii) The phase change through a 10 mm length of the material
(iv) The attenuation constant in dB/m
(v)   The loss in dB through 30 mm of the material.

Solution

(i) $\lambda_d = \dfrac{\lambda_o}{\sqrt{2.6}} = \dfrac{30}{\sqrt{2.6}}$ mm = 18.61 mm.

(ii) $\dfrac{v_p(\text{material})}{v_p(\text{air})} = \dfrac{1}{\sqrt{2.6}} = 0.62.$

(iii) $\phi = \dfrac{2\pi}{\lambda_d} \times l = \dfrac{2\pi}{18.61} \times 10 \text{ rad} = 1.07\pi \text{ rad} \equiv 192.60°.$

(iv) $Q = 210 \quad \Rightarrow \quad \tan \delta = \dfrac{1}{210} = 4.46 \times 10^{-3},$

$$\alpha = \frac{2\pi}{\lambda_d} \times 0.5 \tan \delta = \frac{2\pi}{18.61} \times 0.5 \times 4.76 \times 10^{-3} \text{ Np/mm}$$

$$= 8.04 \times 10^{-4} \text{ Np/mm} = 8.04 \times 10^{-1} \text{ Np/m}$$

$$= 8.04 \times 8.686 \times 10^{-1} \text{ dB/m} = 6.98 \text{ dB/m}.$$

(v) Loss $= 6.98 \times 0.030 \text{ dB} = 0.21 \text{ dB}.$

Example 3.3   Determine the dielectric loss in dB of a 100 mm length of 50 Ω microstrip line at 15 GHz, given that the line is fabricated on a substrate that has a Q of 195, a relative permittivity of 9.8, and a thickness of 0.4 mm.

Solution

$$Q = 195 \quad \Rightarrow \quad \tan \delta = \frac{1}{195} = 5.13 \times 10^{-3}.$$

Using microstrip design graphs (or CAD): $Z_o = 50 \, \Omega \quad \Rightarrow \quad \varepsilon_{r,eff}^{MSTRIP} = 6.6$.

$$15 \text{ GHz} \quad \Rightarrow \quad \lambda_o = 20 \text{ mm},$$

$$\lambda_s = \frac{20}{\sqrt{6.6}} \text{ mm} = 7.78 \text{ mm}.$$

Using Eq. (3.6):

$$\alpha_d = 27.3 \times \frac{9.8 \times (6.6 - 1) \times 5.13 \times 10^{-3}}{6.6 \times (9.8 - 1)} \text{ dB per 7.78 mm}$$

$$= 0.13 \text{ dB per 7.78 mm}.$$

Therefore, the loss in a 100 mm length of line is:

$$\text{Loss} = 0.13 \times \frac{100}{7.78} \text{ dB} = 1.67 \text{ dB}.$$

3.2.2 Conductors

The attenuation due to the finite conductivity of conductors is one of the main contributing factors to loss in RF and microwave circuits. Conductor losses may be divided into those due to the bulk resistivity of the material and those due to surface roughness. As the frequency increases surface losses tend to dominate due to the skin effect, which was discussed in Chapter 2.

The conductor loss in a microstrip line is normally evaluated using an expression developed by Hammerstad and Bekkadal [2], namely

$$\alpha_c = 0.072 \frac{\sqrt{f}}{w Z_o} \lambda_s \left(1 + \frac{2}{\pi} \tan^{-1} \left[\frac{1.4 \Delta^2}{\delta_s^2} \right] \right) \text{ dB}/\lambda_s, \tag{3.7}$$

where

f = frequency of operation in GHz
w = width of microstrip line
Z_o = characteristic impedance of microstrip line
Δ = RMS surface roughness
δ_s = skin depth.

Example 3.4   Determine the dielectric and conductor losses (in dB/m) at 5 GHz for a 70 Ω microstrip line given the following material parameters:

Dielectric:	Relative permittivity = 9.8
	Thickness = 0.6 mm
	Loss tangent = 0.0004
Conductor:	Copper ($\sigma = 56 \times 10^6$ S/m)
	RMS surface roughness = 0.63 µm

Solution

Using the microstrip design curves (or CAD):

$$70\ \Omega \quad \Rightarrow \quad \frac{w}{h} = 0.35 \quad \Rightarrow \quad \varepsilon_{r,eff}^{MSTRIP} = 6.25.$$

$$\lambda_s = \frac{\lambda_o}{\sqrt{6.25}} = \frac{60}{\sqrt{6.25}}\ \text{mm} = 24\ \text{mm}.$$

$$w = 0.35 \times h = 0.35 \times 0.6\ \text{mm} = 0.21\ \text{mm}.$$

Using Eq. (3.6):

$$\alpha_d = 27.3 \times \frac{9.8 \times (6.25 - 1) \times 0.0004}{6.25 \times (9.8 - 1)}\ \text{dB}/\lambda_s = 0.01\ \text{dB}/\lambda_s.$$

$$\alpha_d = 0.01 \times \frac{1000}{24}\ \text{dB/m} = 0.42\ \text{dB/m}.$$

The skin depth, δ_s, is found using Eq. (2.8) in Chapter 2:

$$\delta_s = \sqrt{\frac{2}{\omega\mu_o\sigma}} = \sqrt{\frac{2}{2\pi \times 5 \times 10^9 \times 4\pi \times 10^{-7} \times 56 \times 10^6}}\ \text{m}$$

$$= 0.951\ \mu\text{m},$$

where we have assumed the relative permeability of the conductor to be unity.

Using Eq. (3.7):

$$\alpha_c = 0.072 \times \frac{\sqrt{5}}{0.21 \times 70} \times 24 \times \left(1 + \frac{2}{\pi}\tan^{-1}\left[\frac{1.4 \times (0.63)^2}{(0.951)^2}\right]\right)\ \text{dB}/\lambda_s$$

$$= 0.355\ \text{dB}/\lambda_s$$

$$= 0.355 \times \frac{1000}{24}\ \text{dB/m} = 14.79\ \text{dB/m}.$$

Example 3.5   In Example 3.4, what percentage of the conductor loss is due to surface roughness?

Solution

Putting $\Delta = 0$ in Eq. (3.7) gives the bulk conductor loss, i.e.

$$\alpha_c(\text{bulk}) = 0.072 \times \frac{\sqrt{5}}{0.21 \times 70} \times 24\ \text{dB}/\lambda_s$$

$$= 0.263\ \text{dB}/\lambda_s.$$

Then

$$\alpha_c(\text{surface}) = (0.355 - 0.263)\ \text{dB}/\lambda_s$$

$$= 0.092\ \text{dB}/\lambda_s.$$

Percentage of loss due to surface roughness is

$$\frac{0.092}{0.355} \times 100\% = 25.92\%.$$

The results from Example 3.4 show that the conductor loss is significantly greater than the dielectric loss, and this is normally the case for modern low-loss substrate materials. However, the expression given in Eq. (3.7) becomes less accurate as the frequency increases, and the skin depth becomes comparable with the RMS surface roughness. Figure 3.2 shows how the skin depth varies with frequency for three common materials used in RF and microwave circuits. It can be seen that

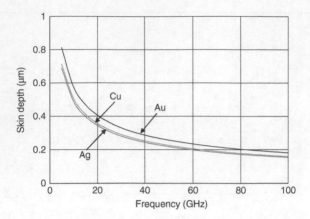

Figure 3.2 Variation of skin depth with frequency for gold (Au), copper (Cu), and silver (Ag).

the skin depth decreases dramatically above 10 GHz, and consequently a number of authors suggest that this should be the upper frequency limit for the use of Eq. (3.7).

Although the expression for conductor loss given in Eq. (3.7) is still widely used, it was developed some time ago. More recent research using full 3D simulation, and more complex models to represent the surface of conductors, has led to improved techniques to determine surface losses at higher microwave frequencies. Using a high-frequency structure simulator, Sain and Melde [3] have shown good agreement between simulated and measured conductor loss on a conductor-backed coplanar line at frequencies up to 40 GHz. In a similar work, Iwai and Mizatani [4] proposed a modification to Eq. (3.7) to better represent surface losses at millimetre-wave frequencies. Their work led to a modified expression for conductor loss

$$\alpha_c = 0.072 \frac{\sqrt{f}}{w Z_o} \lambda_s \left(1 + \frac{32}{\pi} \tan^{-1} \left[\frac{0.24 \Delta^2}{\delta_s^2} \right] \right) \ dB/\lambda_s, \tag{3.8}$$

which gives a higher predicted loss at millimetre-wave frequencies. Data published in [4] shows reasonably good agreement between measured and predicted loss for Δ/δ ratios up to 2, using the modified expression given in Eq. (3.8).

Another parameter that is often used when dealing with the performance of a conductor at high frequencies is that of surface impedance. This is essentially the impedance between two electrodes placed along opposite edges of a square on the surface of the conductor, and so has the units of Ohms per square, which is normally written in short-hand form as $Z_\square$. For smooth conductors the surface impedance is

$$Z_{\square,smooth} = \frac{1+j}{\sigma \delta_s}, \tag{3.9}$$

where the symbols have their usual meanings. Gold and Helmreich [5] showed how the concept of surface impedance could be used for modelling conductor roughness, by modifying Eq. (3.9) as

$$Z_{\square,rough} = \frac{1}{\sigma_{eff} \delta(\sigma_{eff})} + j \frac{1}{\sigma_{bulk} \delta(\mu_{r,eff})}, \tag{3.10}$$

where σ_{eff} is an effective conductivity that represents the increased loss due to rough surfaces, and $\mu_{r,eff}$ is an effective permeability that represents the effect that surface roughness has on the inductance of the surface and hence the propagation delay. Using this concept, Gold and Helmreich showed good agreement between simulation and measurement of transmission loss for a typical transmission line at millimetre-wave frequencies.

3.3 Requirements for RF Circuit Materials

This section summarizes the principal requirements for conductors and dielectrics used in RF and microwave circuits.

Conductors:

(1) Low bulk resistivity: *Low resistivity means high conductivity and low transmission loss. This is important for planar line structures, where the conductor loss is the main source of loss.*

(2) Good surface finish: *A good surface finish means low surface roughness, and consequently low surface losses. This becomes more important as the frequency increases and the skin depth diminishes, so that most of the current may be flowing within the rough surface.*

(3) Stable surface: *It is important that the quality of the surface should not deteriorate with time, or with exposure to the atmosphere as this will increase the surface losses. Copper and silver, for example, both oxidize when exposed to the atmosphere, causing a high resistance surface layer, which increases their surface losses. Gold, although expensive, provides a very stable surface and is often used to plate[2] RF and microwave circuits and components. Silver is normally used for thick-film circuits (discussed later in the chapter) and a thin layer of low-loss dielectric such as glass is often printed over the conducting tracks to prevent oxidization.*

(4) Ability to provide good line/space resolution: *Many planar components, such as couplers and filters, rely on having coupling between closely spaced conductors, and so RF and microwave conductors must be capable of being processed to provide small coupling gaps, often less than 10 μm.*

(5) Well-defined track geometry: *Poorly defined lines lead to errors in line impedance and transmission phase.*

(6) Low variation of properties with temperature: *As with all circuits, it is important that the conductors should retain their physical and electrical properties over a reasonable range of temperature.*

Dielectrics:

(1) Choice of ε_r: *The choice of dielectric constant is essential to achieve optimum RF performance. Substrates for planar circuits such as a microstrip normally require a high dielectric constant (typically around 10) to provide compact circuits, whereas substrates used for planar antennas need a low permittivity to give efficient radiation. For dielectric resonators (discussed in Chapter 11), low-loss dielectrics with ε_r values in the range 15–50 are needed to provide high Q values.*

(2) Low-loss tangent: *Normally the loss tangent should be less than 0.001 for acceptable dielectric loss, for both planar lines and antennas.*

(3) Isotropic permittivity: *The permittivity should be the same in all directions, so that a planar component such as a filter fabricated on a microstrip will exhibit the same performance irrespective of its orientation in the circuit. Some soft substrates are rolled under pressure during manufacture, and consequently exhibit different permittivity in the direction of rolling relative to other directions.*

(4) Precisely defined ε_r: *The electrical behaviour of RF and microwave circuits is critically dependent on the value of ε_r of the substrate. This should be known precisely and should be stable with frequency.*

(5) Low variation of a dielectric constant coefficient with temperature: *This is normally abbreviated to T_f and expressed in parts per million per °C (ppm/°C).*

(6) Consistent dimensions: *Substrate thickness has a significant effect on electrical performance in microstrip and multi-layer circuits.*

(7) Dimensional stability with temperature: *This is particularly important for the substrates used in thick-film fabrication (covered later in the chapter) where the substrate may be subjected to temperatures up to 1000 °C during the firing of the metal conductors.*

(8) Low surface roughness: *When a conductor is deposited on a dielectric substrate the underside of the conductor will assume the same roughness as the dielectric, and so a rough substrate will lead to high surface losses in the conductor. This is particularly important for a microstrip where most of the track current flows in the underside of the metal.*

(9) High thermal conductivity: *This is desirable to assist in dissipating heat from planar circuits.*

Dielectrics for use in planar circuits such as a microstrip can be classified loosely as 'hard' and 'soft' substrates. As the name implies 'hard' substrates are those which are physically hard, and frequently brittle, and which require the use of special mechanical processing techniques. Ceramic substrates, such as alumina, are examples of hard substrates. Vertical holes in these substrates need to be laser-drilled. These holes, when filled with metal, provide conducting interconnections between different layers of a circuit. 'Soft' substrates are flexible, physically soft and easy to process mechanically, although usually their electrical performance is inferior to that of hard substrates. Alumina has become the preferred material for most high-frequency applications. As can be seen from Table 3.1, high-purity alumina (99.5%) has excellent properties at 10 GHz, with very low dielectric loss and a very low surface roughness, which is achieved through surface polishing. But

---

2 Gold plating refers to the deposition of a very thin layer of gold on the surface of a conductor. A gold plating layer, thinner than the skin depth, can be used on top of a silver conductor so the composite structure has the high conductivity properties of silver, with the gold preventing surface oxidization of the silver.

Table 3.1 Typical data for common 'hard' substrates at 10 GHz.

Material	ε_r	tan δ (at 10 GHz)	Δ (μm)
96% Alumina	9.6	0.0006	10
99.5% Alumina	10.2	0.0001	0.1
Quartz	3.8	0.0001	0.01
LTCC (Dupont 9 K7™)	7.1	0.001	—

Table 3.2 Typical data for common 'soft' substrates at 10 GHz.

Material	ε_r	tan δ (at 10 GHz)	Δ (μm)
RT/duroid™ 6011	10.4	0.001	6
RT/duroid 5880	2.3	0.001	6
FR4 Woven board	4.4	0.01	6

there is a very significant cost penalty when selecting high-purity alumina, and for most applications 96% alumina gives adequate performance.

Tables 3.1 and 3.2 provide a summary of the properties of some common dielectric materials used for high-frequency planar circuits, but these are only a few of a very wide range of materials currently available, and a more comprehensive discussion can be found by Cruickshank [6].

3.4 Fabrication of Planar High-Frequency Circuits

An extensive range of fabrication techniques is now available for the manufacture of high-frequency circuits, and the principal methods are described in the following sections.

3.4.1 Etched Circuits

This is a straightforward application of the traditional method for manufacturing printed circuit board (PCB) circuits. The only difference between etching circuits for RF applications, and those for lower frequency use, is that for RF work care must be taken to maintain specific track widths and spacing. In Chapter 2, we saw how the width of a microstrip line directly affected the characteristic impedance of the line, and also the velocity of propagation. The basic steps in the process of etching a microstrip circuit are shown in Figure 3.3.

The track width of the final circuit shown in Figure 3.3 is w. However, this will be less than the width of track on the photo-mask, since at the etching stage the etchant will tend to eat sideways as well as downwards thereby undercutting the resist layer. The amount of undercutting is determined by the etch ratio, which is the ratio of the downward etch rate to the sideways etch rate. So if we assume an etch ratio of unity, and a track thickness of t, the final circuit will have a width, w, given by

$$w = w' - 2t, \tag{3.11}$$

where w' is the track width on the photo-mask. Failure to compensate for undercutting when designing the photo-mask can have serious consequences for high-frequency circuits, as illustrated in Example 3.6. In order to compensate for undercutting it is necessary to know the thickness of the metal covering the substrate. Manufacturers of copper-coated substrate material tend to quote weight of metal (copper) per unit area, from which the user must calculate the track thickness, t. For material manufactured in the United States, the weight of metal will be quoted in imperial units, i.e. oz/ft^2 (ounces per square foot), and appropriate dimensional conversions must be made. For example, if the manufacturer quotes 1 oz/ft^2 for the copper deposited on a particular substrate this is equal to 305.1 g/m^2. Then using the specific density of copper (8930 kg/m^3) the thickness of the copper layer can be calculated as 34.2 μm.

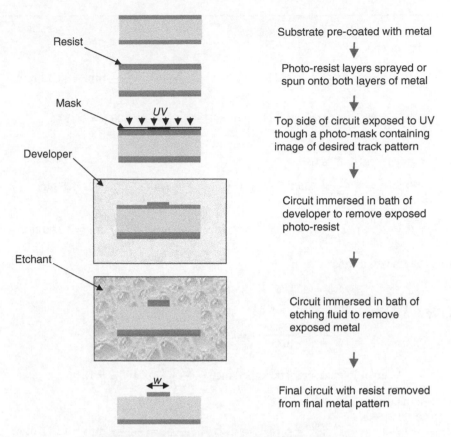

Figure 3.3 Basic steps in etching a microstrip circuit.

Example 3.6   An etching process is used to fabricate a 10 GHz microstrip circuit on a substrate that has a relative permittivity of 9.8 and a thickness of 0.635 mm. The substrate is covered with copper that has a weight of 1 oz/ft². Calculate the percentage error in the characteristic impedances and substrate wavelengths for 50 and 70 Ω lines if no allowance is made for undercutting.

Solution

Using data in text for copper: 1 oz/ft² $\Rightarrow$ $t = 34.2\,\mu m$.

  50 Ω lines

  Using microstrip design curves from Chapter 2 (or CAD):

$$50\,\Omega \quad \Rightarrow \quad \frac{w}{h} = 0.9 \quad \Rightarrow \quad w = 0.9h = 0.9 \times 635\,\mu m = 571.5\,\mu m,$$

$$\frac{w}{h} = 0.9 \quad \Rightarrow \quad \varepsilon_{r,eff}^{MSTRIP} = 6.6 \quad \Rightarrow \quad \lambda_s = \frac{\lambda_o}{\sqrt{6.6}} = \frac{30}{\sqrt{6.6}}\,mm = 11.68\,mm.$$

After undercutting the track width is:

$$w' = (571.5 - [2 \times 34.2])\,\mu m = 503.1\,\mu m.$$

Using microstrip design curves:

$$\frac{w'}{h} = \frac{503.1}{635} = 0.79 \quad \Rightarrow \quad Z_o = 53\,\Omega.$$

$$\% \text{ error in characteristic impedance} = \frac{53 - 50}{50} \times 100\% = 6\%.$$

(Continued)

(Continued)

Using microstrip design curves:

$$\frac{w'}{h} = 0.70 \quad \Rightarrow \quad \varepsilon_{r,eff}^{MSTRIP} = 6.54 \quad \Rightarrow \quad \lambda_s = \frac{\lambda_o}{\sqrt{6.54}} = \frac{30}{\sqrt{6.54}} \text{ mm} = 11.73 \text{ mm}.$$

$$\% \text{ error in substrate wavelength} = \frac{11.73 - 11.68}{11.68} \times 100\% = 0.43\%.$$

70 Ω lines

Using microstrip design curves (or CAD):

$$70 \, \Omega \quad \Rightarrow \quad \frac{w}{h} = 0.35 \quad \Rightarrow \quad w = 0.35h = 0.35 \times 635 \, \mu m = 222.25 \, \mu m,$$

$$\frac{w}{h} = 0.35 \quad \Rightarrow \quad \varepsilon_{r,eff}^{MSTRIP} = 6.25 \quad \Rightarrow \quad \lambda_s = \frac{\lambda_o}{\sqrt{6.25}} = \frac{30}{\sqrt{6.25}} \text{ mm} = 12.00 \text{ mm}.$$

After undercutting the track width is:

$$w' = (222.25 - [2 \times 34.2]) \, \mu m = 153.85 \, \mu m.$$

Using microstrip design curves:

$$\frac{w'}{h} = \frac{153.85}{635} = 0.24 \quad \Rightarrow \quad Z_o = 79 \, \Omega.$$

$$\% \text{ error in characteristic impedance} = \frac{79 - 70}{70} \times 100\% = 12.86\%.$$

Using microstrip design curves:

$$\frac{w'}{h} = 0.24 \quad \Rightarrow \quad \varepsilon_{r,eff}^{MSTRIP} = 6.14 \quad \Rightarrow \quad \lambda_s = \frac{\lambda_o}{\sqrt{6.14}} = \frac{30}{\sqrt{6.14}} \text{ mm} = 12.11 \text{ mm}.$$

$$\% \text{ error in substrate wavelength} = \frac{12.11 - 12.00}{12.00} \times 100\% = 0.92\%.$$

As was shown in Example 3.6, the effects of neglecting undercutting will depend on the substrate thickness. Manufacturers of copper-coated soft substrate material offer coatings with weights down to $1/8 \, oz/ft^2$, which is equivalent to a track thickness of approximately 4.2 μm. Although this is thin copper it is quite adequate for many microwave circuits where the skin depth is less than 1 μm. The use of thin copper layer enables quite precise circuits to be simply etched, with very little compensation for undercutting. As an example, Figure 3.4 shows an enlargement of part of a 10 GHz DC finger break, where the gap between the fingers is 30 μm. The circuit was etched on a soft-substrate material coated with 1/8 oz copper, and it can be seen that a simple etching process leads to a well-defined gap, and the etched dimensions of the gap were within 2% of the designed values.

3.4.2 Thick-Film Circuits (Direct Screen Printed)

Thick-film processing of low-frequency electronic circuits has been well-established over many decades, and it has now also become one of the viable techniques for manufacturing circuits for applications well into the millimetre-wave region. This technology has a number of advantages over simple etched circuits; it is cheap, suitable for mass production, and most notably offers the potential for producing multilayer structures. Conductors and dielectrics for use in a thick-film process are supplied in the form of thick pastes. The basic thick-film process involves these pastes through a metal screen and then firing the result in furnace. Both conductor and dielectric patterns can be processed in this way, but we will describe the basic steps in terms of a conductor pattern:

(1) The process requires the use of a screen, formed from a mesh of fine steel wires typically with 300 or 400 wires per inch (i.e. ~100 or 130 wires per cm).
(2) The mesh is covered with a photo-sensitive emulsion, and a negative image of the required conductor pattern is developed in the emulsion, leaving clear areas corresponding to the conductor pattern.

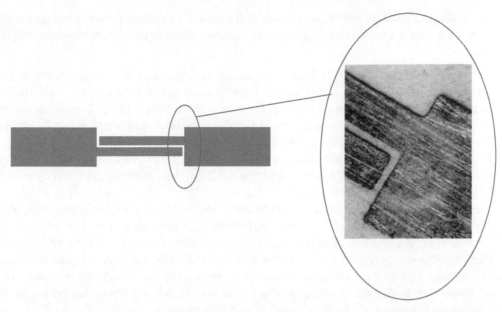

Figure 3.4 Magnified view of an etched 30 µm gap in a 10 GHz microstrip DC break.

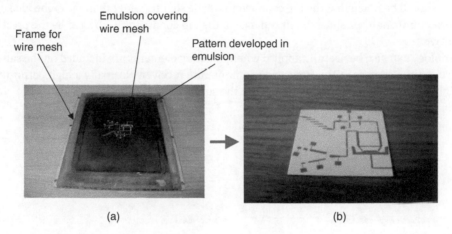

Figure 3.5 Example of a thick-film screen: (a) thick-film wire mesh screen and (b) final circuit printed on alumina substrate.

(3) The conductor, which is in the form of a paste, is then squeezed through the screen onto a ceramic substrate, using a specially designed thick-film printer, so as to form the required conductor pattern.

(4) The substrate, now coated with the wet conductor pattern, is dried at a relatively low temperature (<100 °C) to remove volatile components. It is then passed through a furnace programmed with a particular firing cycle to convert the conductor paste to solid metal. All of the common conductor metals (copper, gold, and silver) can be processed in this way, although the precious metals gold and silver are most common for high-frequency circuits.

The steps described in the previous section are very much an outline, and reference [7] is recommended for those readers requiring more detailed information on thick-film materials and the associated fabrication processes. An example of a typical thick-film screen is shown in Figure 3.5. The screen shown contains the pattern of a 10 GHz microstrip circuit.

Whilst thick-film technology provides a simple, low cost method of manufacturing high-frequency circuits, the need to print patterns though a wire mesh screen has some limitations in terms of the smallest feature size that can be achieved, and also because the edges of the conductors are serrated due to the shape of the apertures in the screen. These limitations are overcome by the thick-film developments described in Section 3.4.3.

However, the opportunity to produce multilayer circuits with the basic thick-film process provides the high-frequency circuit designer with some useful additional design techniques. Three examples where multilayer techniques can be used to practical advantage are:

(1) Microstrip lines with a high characteristic impedance are often required in filters and matching networks. High impedance lines require narrow tracks, and producing these lines on typical substrate materials, such as alumina, is limited by practical problems involved in fabricating very narrow lines. Tian et al. [8] showed that by printing a layer of low permittivity dielectric beneath a conductor on an alumina substrate (as shown in Figure 3.6) could significantly increase the characteristic impedance of the line without decreasing its width.

Tian et al. [8] considered an alumina substrate with $\epsilon_r = 9.9$ and $h = 635\ \mu m$ and showed that the characteristic impedance of a 50 μm wide microstrip line increased from 110 Ω (without underlay) to 156 Ω by including a 100 μm thick underlay whose relative permittivity was 3.9.

(2) Many planar components, such as filters, directional couplers, and DC breaks, rely on edge coupling between parallel microstrip lines to achieve a particular electrical performance. In some cases the required spacing between two microstrip lines in the same plane can be of the order of 10 μm, which can be difficult to fabricate. Using multilayer techniques, edge coupling can be replaced by limited broad-face coupling by simply printing the two conductors on different planes separated by a thin dielectric layer. Figure 3.7 shows the layout of a thick-film microstrip DC break where a multilayer approach has been used to interpose a dielectric layer between the two coupled fingers. This structure enables the two fingers to overlap, giving strong partial broad-face coupling.

Tian et al. [9] used both simulation and practical measurement to show that the configuration illustrated in Figure 3.7 was capable of giving very wide bandwidth performance. Their results showed an insertion loss <0.2 dB over a frequency range 2.5–10 GHz. An enlarged view of the finger coupling is given in Figure 3.8, showing details of the conductor overlap. It should be noted that the top conductor is made slightly wider than the embedded conductor; this is to maintain the correct impedances, since the impedance of the top conductor is affected by the presence of the printed, low impedance layer.

(3) Multilayer technology can also be usefully applied when radiating elements are included on the same package as other high-frequency circuitry, such as in the front-end of a miniature receiver. Normally a high permittivity substrate such as alumina is chosen for microwave circuitry because the substrate wavelength is small and the circuits can be made

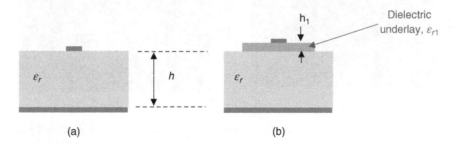

Figure 3.6 Multilayer microstrip line: (a) microstrip and (b) microstrip with underlay.

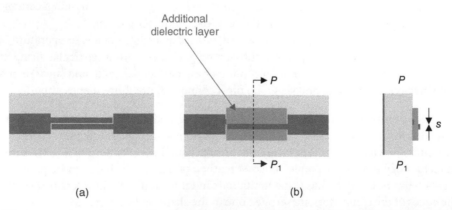

Figure 3.7 Layout of a multilayer DC break: (a) single-layer DC break and (b) tick-film multilayer DC break.

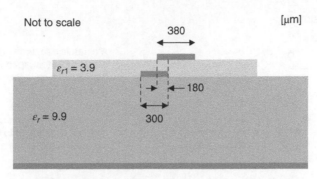

Figure 3.8 Details of coupling in a multilayer DC break [9].

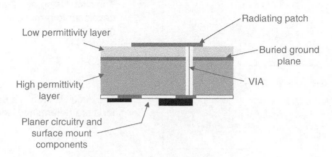

Figure 3.9 Multilayer package.

very compact. But for radiating elements, such as patch antennas, a low permittivity substrate is desirable, since this enables energy to be radiated more efficiently (see Chapter 12). A multilayer package that satisfies the requirements for both permittivies is illustrated schematically in Figure 3.9.

An alternative to the arrangement shown in Figure 3.9 is to have the antenna also on the high permittivity substrate, and to print low permittivity material only beneath the radiating edges of the antenna. This technique is discussed in more detail in Chapter 12.

The discussion so far has focussed on printing conductors and dielectrics using thick-film. However, the range of materials that can be printed using this technology is far more extensive, and provides the RF and microwave circuit designer with useful opportunities, particularly in reducing unwanted parasitics associated with surface-mount packages. Printing of resistive thick-film pastes is a well-established technology [7] and can avoid the need to include chip resistors for biasing purposes in a high-frequency circuit. Capacitors are very easily created by printing high permittivity dielectric between two flat electrodes. Both printed resistors and capacitors can easily be buried within a multilayer structure. Another useful material for high-frequency applications is Barium Strontium Titanate (BST), which has ferroelectric properties whereby its dielectric constant can be changed by varying the electric field strength across the material. BST is available in the form of a thick-film paste, and so can easily be incorporated into multilayer thick-film circuits. The drawback to BST material is that it exhibits a very high-loss tangent (∼0.01), but this is not a significant problem if only a small amount of the material is used, for example in a thin layer within a multilayer structure. Osman and Free [10] showed that the measured dielectric constant of a thick-film BST sample could be changed by around 15% over the frequency range 1–8 GHz, through the application of an electric field of strength 2.5 V/μm. They subsequently used BST to form a capacitor within a miniature ring-resonator filter, to provide tuning of the filter's centre frequency.

3.4.3 Thick Film Circuits (Using Photoimageable Materials)

The potential of thick-film technology for manufacturing high-quality microwave and millimetre-wave circuits has been significantly enhanced through recent developments in photoimageable dielectric and conductor pastes. These pastes are sensitive to UV light and obviate the need to print patterns through a screen. This enables the materials to be patterned directly using photo-lithographic techniques, which leads to the ability to realize high-resolution patterns of conductor and dielectric, with feature dimensions down to the order of 10 μm.

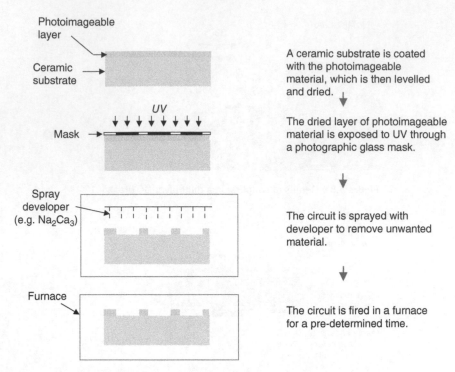

Photoimageable layer

Ceramic substrate

A ceramic substrate is coated with the photoimageable material, which is then levelled and dried.

Mask

UV

The dried layer of photoimageable material is exposed to UV through a photographic glass mask.

Spray developer (e.g. Na$_2$Ca$_3$)

The circuit is sprayed with developer to remove unwanted material.

Furnace

The circuit is fired in a furnace for a pre-determined time.

Figure 3.10   Basic photoimageable thick-film fabrication process.

The process of making thick-film circuits using photoimageable materials is outlined in Figure 3.10, and involves four basic steps:

(i) A layer of photoimageable material (conductor or dielectric) is printed so as to cover the surface of a supporting substrate. The printed layer is left a short while to level, and is then dried.
(ii) The photoimageable layer is then exposed to UV through a suitable (negative) mask.
(iii) The areas of the layer that have been developed through exposure to UV are washed away, normally using a special unit that rotates the substrate while applying the developer through a fine spray.
(iv) The circuit with the desired pattern of conductor or dielectric is fired in a furnace to achieve the final thick-film pattern.

Steps 1–4 can be successively repeated to build up a multilayer structure of conductor and dielectric material. If desired, dielectrics having different permittivities can be included. Also, materials forming passive components can be used, as previously described.

One particularly useful feature of the photoimageable process is the ability to easily create vertical interconnections (VIAs), by simply imaging holes in the dielectric layers and subsequently filling them with metal through a suitable mask. The VIAs can be used as simple electrical connections between conductors in different levels, or as thermal VIAs to conduct excess heat from within the multilayer package. Thermal VIAs are particularly useful for removing heat from high-power components such as power amplifiers, which tend to be rather inefficient at high frequencies.

Whilst photoimageable thick-film technology can be used to fabricate a wide range of variety of electronic circuits, one particular feature that has attracted a lot of attention is the ability to fabricate surface integrated waveguides (SIWs). This is a waveguide that is integrated within the printed dielectric layers, with the top and bottom broad faces of the waveguide formed from printed layers of conductor, and the vertical side walls formed from VIA fences. A VIA fence is simply a closely spaced linear array of conducting VIAs. If the spacing between adjacent VIAs is small compared to the wavelength in the dielectric, the VIA fence will act like a solid conductor. Figure 3.11 illustrates the basic SIW structure.

Figure 3.11a shows the integrated rectangular waveguide with the dimensions identified using the conventional notations, a and b. The required waveguide height is achieved through the use of a number of printed dielectric layers. One such layer is shown in Figure 3.11b. The method of fabricating the SIW is to first print and fire a conducting layer on the supporting substrate to form the bottom broad wall of the waveguide. A layer of photoimageable dielectric is then printed on the ground plane, and photo-imaged with the array of holes forming the two VIA fences. The circuit is then developed and fired, so as to leave a layer of dielectric with an array of holes. The holes are then filled with metal, through a suitable

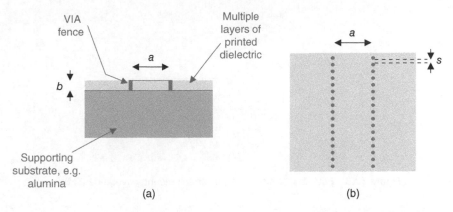

(a)   (b)

Figure 3.11   Structure of surface integrated waveguide (SIW): (a) cross-section of SIW and (b) one layer of printed dielectric showing filled VIAs.

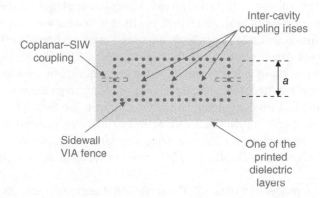

Figure 3.12   Schematic view of VIA arrays in a four-cavity SIW filter.

mask, and the circuit fired again. This process of printing and firing the dielectric layers is repeated until the desired waveguide height, b, is achieved. Typically, each layer of dielectric is of the order of 10 μm thick, and between 10 and 20 firings are required to achieve a suitable waveguide height.

Two particular uses of SIW have attracted a lot of attention in the literature, namely for filters and antennas. Figure 3.12 illustrates one of the printed dielectric layers in a four-cavity filter, and shows how the VIA arrays are arranged. The input and output connections to the SIW filter are normally provided through coplanar coupling probes on the top surface of the structure.

A particularly good introduction to SIW filters is provided through a series of articles in IEEE *Microwave Magazine* [11–13], which provide a wealth of theoretical and practical data, together with an extensive list of references.

Whilst SIW cavity filters have the advantage of being fully integrated within the ceramic package, they suffer from the disadvantage that the cavities are necessarily filled with dielectric, and the losses associated with the dielectric have detrimental effects on the insertion loss and roll-off of the filter. One technique that overcomes these losses is to couple the integrated SIW line to a surface mount waveguide (SMW) which contains air-filled metallic cavities that form the filter, as shown in Figure 3.13.

The combination of SIW and SMW is particularly useful at millimetre-wave frequencies where the dielectric losses become very significant and where the wavelength is small enough to make the size of the SMW cavities a realistic proposition. Consequently, at these frequencies the absence of dielectric filling the filter cavities gives them a much higher Q-value, and at the same time the short wavelength makes the cavities relatively small. Schorer et al. [14] compared the performance of SIW filters with those using SMW cavities at K-band. They found significant benefits accrued from using SMW filters, and showed that the insertion loss and roll-off improved by a factor of around 3.7 using SMW cavities that were only a few millimetres high.

The other area in which SIW technology has received significant attention is that of slot antennas. A very compact antenna can be made by cutting slots in the upper surface of an SIW; this type of antenna is discussed in more detail in Section 12.15 of Chapter 12.

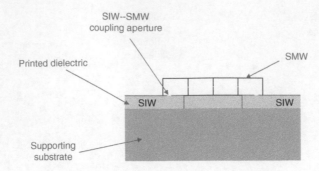

Figure 3.13 Four-cavity filter using a surface mount waveguide (SMW)

3.4.4 Low-Temperature Co-Fired Ceramic Circuits

The major drawback to the use of photoimageable thick-film technology is that a large number of sequential firing cycles are required, since each printed layer must be imaged, developed, and fired before the next layer is deposited. Even using a belt furnace for the firing, where the firing cycle may only take 20 minutes, the production of a relatively simple structure is very time consuming and is certainly not attractive in a production environment. LTCC technology can be used to manufacture very similar multilayer structures to those using photoimageable techniques, but enables the layers to be processed in parallel. Only a single firing cycle is required, thus making it relatively quick to produce a complex multilayer package.

LTCC dielectric material is supplied from the material manufacturer in the form of flexible sheets; these sheets are normally available to the user rolled onto a cylindrical former and so are often referred to as LTCC tapes. Various sheet thicknesses are available and can be specified by the user (circuit designer); typical values are 5 mil (127 µm) and 10 mil (254 µm). The dielectric constant of commonly available LTCC tapes is in the range 5–10, with loss tangents varying between 0.001 and 0.003.

There are five main stages in the production of an LTCC circuit, and these are shown schematically in Figure 3.14:

Stage 1:	The LTCC tape is cut to size. Unfired LTCC tape is often referred to as green tape, and the process of cutting to size as blanking.
Stage 2:	The layers of green tape that are to form the multilayer package are punched or laser cut to provides holes for VIAs and cavities.
Stage 3:	The layers of green tape are screen-printed with compatible conductive pastes to fill the VIAs and deposit required conductor patterns.
Stage 4:	The layers of green tape are stacked, with appropriate mechanisms to achieve correct alignment (alignment marks are normally included on each layer, and a conventional optical mask aligner can be used). The stack of tapes is then vacuum-sealed in a plastic bag, and laminated or bonded under pressure in an isostatic chamber at low temperature (a few tens of °C).
Stage 5:	The plastic bag is removed and the laminated stack of tapes is fired in either a belt or chamber furnace, following a particular time–temperature profile to finally produce a solid ceramic component. Using a belt furnace is particularly advantageous for this work, as the work-piece can be moved through various firing zones in which the environmental conditions can be precisely controlled, thereby providing an optimum firing profile with a relatively fast throughput. For LTCC the maximum temperature in the firing profile is of the order of 850 °C.

During the firing cycle of the LTCC fabrication process shrinkage will occur. Shrinkage in the $x - y$ plane can have a significant adverse effect on RF and microwave circuits where the relative positions of the conductors are of critical importance, as in components such as directional couplers and filters, which exploit the properties of coupled lines, and where the spacing between lines is an essential design parameter. This problem can be overcome by using self-constrained LTCC tapes, such as Heralock™, where the tape has been designed to shrink primarily in the z-direction, thus maintaining good precision in the critical $x - y$ plane.

An idealized view of an LTCC module is given in Figure 3.15, and shows the potential of the technology for integrating a number of circuit functions and components within a single ceramic package. This type of circuit assembly is often referred to as SiP. An in-depth discussion of the materials and processes involved in SiP technology is given by Gaynor [15].

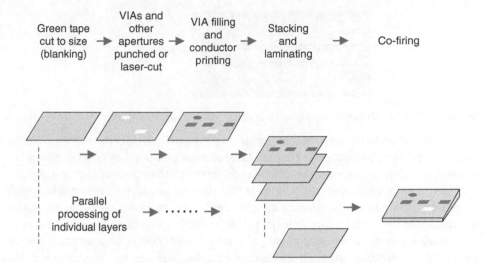

Figure 3.14 LTCC production process.

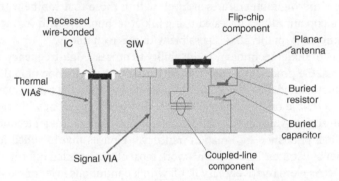

Figure 3.15 Features of a typical LTCC (SiP) module.

SIWs can be easily created within LTCC structures, but as with SIWs in multilayer photoimageable thick-film circuits they suffer from the disadvantage of being filled with dielectric. However, Henry et al. [16] demonstrated that air-filled SIW channels can be formed using LTCC technology, and these yielded a relatively low measured insertion loss of ~0.14 dB/mm over a frequency range 140–190 GHz.

3.5 Use of Ink Jet Technology

The use of inkjet printing to manufacture high-frequency circuits is a fast emerging technology that offers significant advantages over the more traditional techniques that have been discussed in the previous sections. The inkjet process is well known and involves the computer controlled deposition of inks through a small diameter nozzle. In the present context 'ink' refers to the conductor and dielectric pastes that are to be used to build up the electronic circuit. There are a number of obvious advantages to this technology:

 (i) Production times are short, since there is no lithographic process involved.
 (ii) The process is economic, since the required conductor or dielectric material is added to a substrate, rather than with the subtractive fabrication methods where the unwanted material is washed away as in the etched and photoimageable thick-film processes.
(iii) It is easy to print onto soft substrates, such as paper or thin plastics, which is important for applications such as RF tags and conformal antennas.
(iv) The sintering temperatures required by inkjet printed material are relatively low (<300 °C), and so a wide range of substrate materials can be used.

The inkjet process requires the use of a special printer in which computer control allows both the lateral position of the inkjet head, and the amount of ink deposition, to be precisely controlled. Using this technique, conductor and dielectric

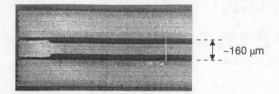

Figure 3.16 CPW line fabricated using inkjet technology. *Source*: Reproduced from [17] with permission.

circuit features with sizes less than 50 μm are possible. Conductive inks are normally made from colloidal solutions of silver nanoparticles, and after printing are sintered at temperatures of a few hundred degrees centigrade. For conductors, the percentage of conducting metal in an ink, and the sintering temperature are important in producing high-quality circuit interconnections. In microwave applications it is important that sintered conductor lines should have low-loss (i.e. low resistivity) and well-defined geometries, which means straight edges, rectangular cross-sections, and smooth surfaces. Belhaj et al. [17] have investigated the performance of inkjet printed conductors on flexible Kaplon substrates at frequencies up to 67 GHz. In their work a commercial ink composed of 40% (by weight) silver nanoparticles was used. They investigated the dependence of the resitivity of printed lines with sintering temperature, and measured a resistivity of ~80 Ω-cm at 100 °C, which dropped to ~10 Ω-cm at 200 °C, and then levelled off. These results show the significance of sintering temperature, and also that the sintering temperatures are well within those that can be withstood by a variety of plastic substrates. Figure 3.16 shows a coplanar line fabricated using inkjet technology, with Ag nanoparticle ink, and indicates the high resolution and good line edge quality that is possible with this technology.

Whilst one of the key benefits of inkjet technology is the ability to process high-frequency circuits on thin, flexible substrates at moderate temperatures, the process is also very attractive for the production of multilayer circuits on thick, rigid substrates. Inkjet printed conductors and dielectrics offer the potential to build the same system-on-package modules that were described in earlier sections based on photimageable thick-film technology, but with significant advantages in terms of reduced cost and production times. Amongst the significant current work on inkjet technology and its high-frequency applications, Kim et al. [18] have published information multilayer circuits that included fully inkjet-printed VIAs and SIW structures on thick polymer substrates. Prior to Kim's work most of the reported information on inkjet-filled VIAs was for thin substrates, where the VIAs were relatively easy to fill with a continuous layer of metal. Filling straight cylindrical VIAs in thick substrates proved difficult, even with several multiple prints over the holes, due to the shrinkage of the metal during firing. A novel feature of the work in [18] was the use of stepped VIAs to introduce a gradual transition between the two sides of the substrate, and it was found that this permitted more efficient metallization. The concept of the stepped VIA was subsequently justified through the construction of a SIW cavity resonator in which the vertical walls of the resonator were formed from arrays of stepped VIAs; good agreement between measured and simulated S-parameter data at ~5.8 GHz has been published [18].

The use of inkjet technology for RF and microwave applications is not limited to printing conductors and dielectrics. Nikfalazar et al. [19] demonstrated the capability of the inkjet process for printing BST material by fabricating a tuneable S-band phase shifter. They produced a multilayer circuit containing ferroelectric (BST) varactors in a metal–insulator–metal (MIM) configuration as the tuning elements, and demonstrated good agreement between measured and simulated phase shift values over the frequency band 1–4 GHz. Another example of the versatility of inkjet technology for high-frequency work was provided by Chen and Wu [20] who showed that carbon nanotube solutions can successfully be printed as part of an all-inkjet printed phased array antenna operating at 5 GHz.

3.6 Characterization of Materials for RF and Microwave Circuits

Precise knowledge of the properties of materials used in high-frequency circuits is essential for good design. Of particular importance are the loss and relative permittivity (dielectric constant) of the dielectric materials used in planar circuits, and the main focus of this section will be on the measurement of these parameters at microwave and millimetre-wave frequencies.

3.6.1 Measurement of Dielectric Loss and Dielectric Constant

The traditional method of determining the properties of a dielectric material is to measure the changes in Q-factor and resonant frequency when a specimen of the material under test is inserted into a cavity resonator. Because of the importance

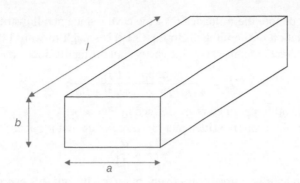

Figure 3.17 Dimensions of a rectangular waveguide cavity.

of the cavity resonator in these measurements, the following section is devoted to a discussion of the properties of these resonators.

3.6.1.1 Cavity Resonators

A cavity resonator can be formed from a section of hollow metal rectangular waveguide closed at both ends, with the dimensions shown in Figure 3.17.

If an electromagnetic wave is introduced into the cavity, a standing wave pattern will be set up if the length (l) of the cavity is some multiple of half a guide wavelength, i.e.

$$l = p \times \frac{\lambda_g}{2}, \tag{3.12}$$

where p is an integer. In general, for transverse electric modes, the mode within a resonant cavity is denoted by TE_{mnp} where m and n have been defined in Appendix 1.C.

We can find an expression for the resonant frequency of the cavity by substituting from Eqs. (3.12) and (1.83)–(1.72) to give

$$\left(\frac{p}{2l}\right)^2 + \left(\frac{m}{2a}\right)^2 + \left(\frac{n}{2b}\right)^2 = \left(\frac{f}{c}\right)^2. \tag{3.13}$$

The resonant frequency is then

$$f = c\sqrt{\left(\frac{p}{2l}\right)^2 + \left(\frac{m}{2a}\right)^2 + \left(\frac{n}{2b}\right)^2}. \tag{3.14}$$

The dominant cavity mode is represented by TE_{101}, and from Eq. (3.14) its resonant frequency will be

$$f_o = c\sqrt{\left(\frac{1}{2l}\right)^2 + \left(\frac{1}{2a}\right)^2}. \tag{3.15}$$

The Q-factor of a resonant cavity is defined in the normal manner as

$$Q = \frac{\omega_o \times \text{average energy stored}}{\text{power dissipated}}, \tag{3.16}$$

where $\omega_o = 2\pi f_o$, and f_o is the resonant frequency. The energy stored in the cavity is given by the volume integral

$$W = \frac{1}{2}\mu_o \int_v |H|^2 dv, \tag{3.17}$$

where $|H|$ is the peak magnetic field strength within the cavity. The power dissipated will be due to the finite conductivity of the walls which allows wall currents to flow, and leads to a power loss given by the surface integral

$$P = \frac{1}{2}R_s \int_s |H_t|^2 ds, \tag{3.18}$$

where H_t is the magnetic field tangential to the walls and R_s is the surface resistance. Substituting from Eqs. (3.17) and (3.18) into (3.16) gives the cavity Q-factor as

$$Q_u = \frac{\omega_o \mu_o \int_v |H|^2 dv}{R_s \int_s |H_t|^2 ds}. \tag{3.19}$$

The suffix 'u' has been used to denote the unloaded Q of the cavity, since no allowance has been made for a coupling aperture in the cavity wall. Evaluation of the integrals in Eq. (3.19) is straightforward [21, 22] and leads to the following expression for the unloaded Q of a rectangular waveguide cavity supporting the dominant TE_{101} mode

$$Q_u = \frac{abl(a^2 + l^2)}{\delta_s(2a^3b + 2l^3b + a^3l + l^3a)},$$

(3.20)

where δ_s is the skin depth, and the other parameters are defined in Figure 3.17.

The unloaded Q of a resonant cavity is often written in a more approximate format as

$$Q_u = \frac{2}{\delta_s} \times \frac{V}{A},$$

(3.21)

where V and A are the volume and surface area, respectively, of the cavity, but this expression is only correct for cavities that have a cubic shape. However, Eq. (3.21) is useful in that it indicates that in general the highest Q values are obtained with simple cavity shapes that have a high ratio of volume to surface area.

Example 3.7 A copper rectangular waveguide resonant cavity supporting the TE_{101} mode has the following dimensions: $a = 22.86$ mm, $b = 10.16$ mm, $l = 20$ mm. Determine the resonant frequency and the unloaded Q of the cavity. [σ(copper) $= 5.8 \times 10^7$ S/m]

Solution

To find the resonant frequency

At resonance $\quad l = \frac{\lambda_g}{2} \quad \Rightarrow \quad \lambda_g = 2l = 2 \times 20 \text{ mm} = 40 \text{ mm},$

$$\lambda_c = 2a = 2 \times 22.86 \text{ mm} = 45.72 \text{ mm}.$$

Using Eq. (1.72): $\quad \dfrac{1}{\lambda_o^2} = \dfrac{1}{(45.72)^2} + \dfrac{1}{(40)^2} \quad \Rightarrow \quad \lambda_o = 30.10 \text{ mm},$

$$\lambda_o = 30.10 \text{ mm} \quad \Rightarrow \quad f_o = 9.97 \text{ GHz}.$$

Resonant frequency $= 9.97$ GHz.

To find the Q of the cavity

The skin depth is given by Eq. (2.8):

$$\delta_s = \frac{1}{\sqrt{\pi \mu_o f \sigma}}$$

$$= \frac{1}{\sqrt{\pi \times 4\pi \times 10^{-7} \times 9.97 \times 10^9 \times 5.8 \times 10^7}} \text{ m}$$

$$= 0.66 \text{ μm}.$$

Using Eq. (3.20):

$$Q_u = \frac{1}{\delta_s} \times \frac{x}{y},$$

where

$$x = 22.86 \times 10.16 \times 20 \times [22.86^2 + 20^2] \times 10^{-3} = 4285.52,$$

$$y = [2 \times 22.86^3 \times 10.6] + [2 \times 20^3 \times 10.16] + [22.86^3 \times 20] + [20^3 \times 22.86]$$

$$= 827\ 109.5619$$

giving

$$Q_u = \frac{1}{0.66 \times 10^{-6}} \times \frac{4285.52}{827\ 109.56} = 7850.49.$$

Unloaded Q of cavity $= 7850.49$.

So far we have considered unloaded cavities, i.e. cavities without any means of exciting the signal within the cavity. The Q of the cavity is then denoted by Q_u. In order to excite a signal within a cavity, there must, in practice, be either a coupling hole in one of the walls of the cavity, or a coupling loop of wire projecting into the cavity. In either case the coupling mechanism will have the effect of loading the cavity.

The unloaded cavity can be represented by a series *RLC* resonant circuit, and the *Q* is then given by

$$Q_u = \frac{\omega L}{R}.$$

Coupling the cavity to the outside world will provide additional loss, but no additional reactance, since any additional reactance from the coupling will be compensated for by alteration of the internal reactance of the cavity to maintain the original resonant frequency. However, the external load resistance will be coupled into the cavity, and will add to the resistance (R) of the resonant circuit representing the cavity, thus lowering the *Q* of the cavity. If the external (load) resistance coupled into the cavity is R_L, the loaded *Q*, Q_L, of the cavity will be given by

$$Q_L = \frac{\omega L}{R + R_L}.$$

Inverting each side of the last expression gives

$$\frac{1}{Q_L} = \frac{R + R_L}{\omega L} = \frac{R}{\omega L} + \frac{R_L}{\omega L},$$

i.e.

$$\frac{1}{Q_L} = \frac{1}{Q_u} + \frac{1}{Q_e}, \tag{3.22}$$

where Q_e is defined as the external *Q* of the cavity; Q_e thus represents the ratio of the same total stored energy as the unloaded cavity to only the loss coupled from the outside world.

One of the common methods of making a cavity resonator is to use a short-circuited length of rectangular waveguide, with a transverse plate containing a coupling iris, positioned at some distance from the short circuit, as shown in Figure 3.18. The length *l* is thus the length of the resonant cavity.

In Figure 3.18, a small circular coupling iris is shown at the centre of the transverse plate. This iris behaves as a shunt inductive susceptance with a normalized value given by

$$b_L = -j\frac{3ab}{8\beta r^3}, \tag{3.23}$$

where $\beta = \frac{2\pi}{\lambda_g}$ and λ_g is the guide wavelength. Thus, in the plane PP′ the equivalent circuit of the resonator can be represented as a normalized susceptive inductance b_L in parallel with a normalized admittance y_{in}, which represents the input admittance of a length of waveguide terminated with a short circuit. From Eq. (1.33) we know

$$y_{in} = -j\cot\beta l. \tag{3.24}$$

Therefore, for resonance we require

$$b_L + y_{in} = 0, \tag{3.25}$$

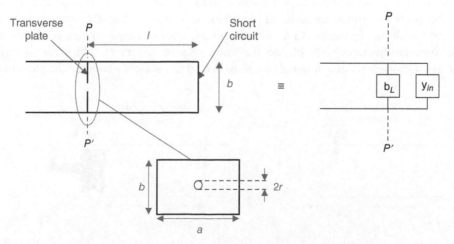

Figure 3.18 Waveguide cavity resonator and its equivalent circuit.

i.e.

$$\frac{3ab}{8\beta r^3} + \cot \beta l = 0. \tag{3.26}$$

Since the iris presents a shunt inductance the length of the cavity must be slightly less than $\lambda_g/2$ in order that y_{in} provides a capacitance to cancel the inductance of the iris. However, the value of shunt inductance provided by a typical iris is very small, and only a very slight reduction in the cavity length is needed, as shown by Example 3.8.

Example 3.8   A 6 GHz cavity resonator having the form shown in Figure 3.18 is to be made using a rectangular waveguide having the following internal dimensions: $a = 40.39$ mm, $b = 20.19$ mm. If the circular coupling iris has a radius of 3 mm, determine the required length (l) of the cavity.

Solution

At 6 GHz: $\lambda_o = \dfrac{3 \times 10^8}{6 \times 10^9}$ m $= 50$ mm.

Using Eq. (1.72) to find the guide wavelength:

$$\frac{1}{(50)^2} = \frac{1}{(2 \times 40.39)^2} + \frac{1}{\lambda_g^2} \;\Rightarrow\; \lambda_g = 63.66 \text{ mm}.$$

Then,

$$\beta = \frac{2\pi}{63.66} \text{ rad/m} = 0.0987 \text{ rad/m}.$$

Substituting in Eq. (3.26):

$$\frac{3 \times 40.39 \times 20.19}{8 \times 0.0987 \times 3^3} + \cot \beta l = 0,$$

$$114.74 + \cot \beta l = 0,$$

$$\beta l = 0.9972\pi,$$

$$l = 0.9972\pi \times \frac{63.66}{2\pi} \text{ mm} = 31.74 \text{ mm}.$$

Comment: The calculated length is only slightly less than $\lambda_g/2$ ($=63.66/2$ mm $= 31.83$ mm), which is the resonant length of the cavity when ignoring the effect of the iris.

The cavity resonators described so far are slightly limited in their application because the resonant frequency is fixed. It is possible to achieve some limited manual tuning of resonant cavities, usually at the expense of a slight decrease in the cavity Q-factor. The two most common tuning techniques are illustrated in Figure 3.19.

In the case of screw tuning shown in Figure 3.19a, a metal post is screwed into the waveguide parallel to the E-field. This has the effect of introducing an admittance in parallel with that of the cavity. The diameter of the post and the depth of penetration into the guide determine the value of the admittance, and thereby the amount of frequency tuning. An alternative arrangement is shown in Figure 3.19b, and this simply uses a plunger to change the resonant length of the cavity, and hence the resonant frequency. The plunger is normally driven by a micrometre screw to provide fine frequency tuning. Also, the plunger is non-contacting to avoid variations in contact resistance between the plunger and the waveguide

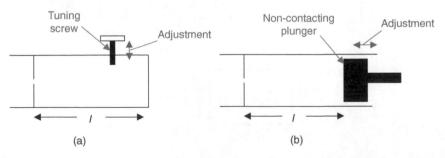

(a)   (b)

Figure 3.19   Manual tuning of resonant cavities: (a) use of tuning screw and (b) use of non-contacting plunger.

walls; this is achieved by machining slots in the sides of the plunger which reflect short circuits between the face of the plunger and the sides of the waveguide.

Cavity resonators formed from rectangular waveguide are attractive because of their relative simplicity and the high Q-values that they provide. However, their use is problematic as the frequency increases into the millimetre-wave part of the spectrum because the internal dimensions of the waveguide decrease and become inconveniently small. A good solution at millimetre-wave frequencies is to make use of the properties of over-moded circular waveguide supporting the TE_{01} mode. Over-moded means that there is the potential for more than one mode to exist within the waveguide. As was seen in Appendix 1.C of Chapter 1 the TE_{01} mode in circular waveguide exhibits the property of decreasing attenuation with increasing frequency, which means that it has the potential for providing very high Q values for cavity resonators operating at millimetre-wave frequencies. An expression for the Q of a circular cavity resonator can be determined by evaluating Eq. (3.19) for the appropriate electromagnetic field distribution. Collin [22] showed that for TE_{mnp} modes in a cylindrical cavity resonator the Q is given by

$$Q = \frac{\lambda_o}{\delta_s} \frac{\left[1 - \left(\frac{m}{x}\right)^2\right]\left[x^2 + \left(\frac{p\pi a}{l}\right)^2\right]^{1.5}}{2\pi \left[x^2 + \frac{2a}{l}\left(\frac{p\pi a}{l}\right)^2 + \left(1 - \frac{2a}{l}\right)\left(\frac{mp\pi a}{xl}\right)^2\right]}, \tag{3.27}$$

where

p = number of guide half-wavelengths in axial length of cavity
a = radius of cavity
l = length of cavity
δ_s = skin depth
λ_o = free-space wavelength
m and x are defined in Appendix 1.C.

Example 3.9 Determine the Q at 50 GHz of a cylindrical cavity resonator supporting the TE_{019} mode if the cavity is made of copper and has a radius of 25 mm.
  ($\sigma_{copper} = 5.8 \times 10^7$ S/m)

Solution

$$f_o = 50 \text{ GHz} \quad \Rightarrow \quad \lambda_o = \frac{3 \times 10^8}{50 \times 10^9} \text{ m} = 6 \text{ mm}.$$

Using Eq. (1.86) to obtain λ_c: $\lambda_c = \frac{2\pi a}{x} = \frac{2\pi \times 25}{3.83} \text{ mm} = 41.02 \text{ mm}$.

Using Eq. (1.72): $\frac{1}{(6)^2} = \frac{1}{(41.02)^2} + \frac{1}{\lambda_g^2} \quad \Rightarrow \quad \lambda_g = 6.07 \text{ mm}$.

Length of cavity at resonance: $l = 9 \times \frac{\lambda_g}{2} = 9 \times \frac{6.07}{2} \text{ mm} = 27.32 \text{ mm}$.

Skin depth: $\delta_s = \frac{1}{\sqrt{\pi \mu_o f \sigma}} = \frac{1}{\sqrt{\pi \times 4\pi \times 10^{-7} \times 50 \times 10^9 \times 5.8 \times 10^7}} \text{ m} = 0.30 \text{ μm}$.

Substituting in Eq. (3.27) and noting that $m = 0$:

$$Q = \frac{6 \times 10^{-3}}{0.3 \times 10^{-6}} \times \frac{\left[(3.83)^2 + \left(\frac{9 \times \pi \times 25}{27.32}\right)^2\right]^{1.5}}{2\pi \left[(3.83)^2 + \frac{2 \times 25}{27.32}\left(\frac{9 \times \pi \times 25}{27.32}\right)^2\right]}$$

$$= 20 \times 10^3 \times \frac{17\,899.51}{7793.10}$$

$$= 45\,736.82.$$

In order to achieve the very high Q value found in Example 3.9, the cavity must be over-moded. The number of modes, N, that can be supported simultaneously at a given frequency in a circular waveguide of radius a is given by

$$N = 10.20\left(\frac{a}{\lambda_o}\right)^2, \tag{3.28}$$

where λ_o is the free-space wavelength.

Example 3.10 Determine the number of possible modes that can exist within the cavity resonator specified in Example 3.9.

Solution
Using Eq. (3.28):

$$N = 10.2 \times \left(\frac{25}{6}\right)^2 = 177.$$

If a waveguide is heavily over-moded, i.e. it has a large value of N, it is difficult to avoid exciting unwanted modes. Also, when measuring the Q-value of an over-moded cavity it is difficult to identify the correct resonance curve from amongst the resonances of the unwanted modes. These problems can be overcome through the use of a mode filter. A mode filter is a section of waveguide that inhibits the propagation of all TE and TM modes, except those in the TE_{0n} family. The physical realization of the required mode filter is possible because all TE and TM modes, except those in the TE_{0n} family, have axial wall current components. Thus, the required mode filter should have anisotropic wall impedance, with low impedance in the circumferential direction to allow the low-loss propagation of the TE_{01} mode, and high axial impedance to attenuate all the unwanted modes. These requirements can be met by making the mode filter from a section of helical waveguide, whose structure is shown schematically in Figure 3.20. The normal method of manufacturing this type of filter is to start by winding enamelled-coated copper wire on a precisely dimensioned steel mandrel having the same diameter as the desired internal diameter of the waveguide. The wire is normally of the order of 40 swg (standard wire gauge), which has a diameter ~130 μm. The wire is wound in a tight helix and secured in place with resin, the outside of which is coated with absorbing material to prevent wave propagation along the outside of the helix. Finally, the structure is encased in a protective cylinder, and the mandrel withdrawn. Because a small diameter wire has been used, the circumferential impedance is very low, but the axial impedance is very high since axial currents would have to travel through the enamel surrounding the wire core. The difference between the transmission loss of the TE_{01} mode travelling through wire-wound helical waveguide and through solid copper tube of the same diameter depends on the amount of enamelling, i.e. on the ratio d/D as defined in Figure 3.20. But it can be shown [23] that if $d/D > 0.8$, the difference in loss is less than 9%.

Whilst the helical mode filter will inhibit the propagation of all modes other than those in the TE_{0n} family, it does mean that the TE_{01} will only exist on its own at frequencies below the cut-off value of the TE_{02} mode, as shown in Example 3.11.

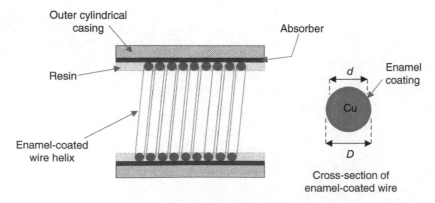

Figure 3.20 Typical structure of a helical waveguide.

Example 3.11   Determine the frequency range over which only the TE_{01} mode will propagate through a helical waveguide whose internal diameter is 50 mm.

Comment: The first root of $J'_0(x)$ occurs when $x = 3.8317$, and the second root when $x = 7.0156$.

Solution

Using Eq. (1.86):

Cut-off wavelength of TE_{01} mode: $\lambda_{c,01} = \dfrac{2\pi \times 25}{3.8317}$ mm $= 41.00$ mm.

Cut-off wavelength of TE_{02} mode: $\lambda_{c,02} = \dfrac{2\pi \times 25}{7.0156}$ mm $= 22.39$ mm.

Then,

$$f_{c,01} = \frac{c}{\lambda_{c,01}} = \frac{3 \times 10^8}{41.00} \text{Hz} = 7.32 \text{ GHz},$$

$$f_{c,02} = \frac{c}{\lambda_{c,01}} = \frac{3 \times 10^8}{22.39} \text{ Hz} = 13.40 \text{ GHz}.$$

Frequency range: $(7.32–13.40)$ GHz.

3.6.1.2 Dielectric Characterization by Cavity Perturbation

The traditional method of measuring the dielectric properties of a material at microwave frequencies is to insert a small sample of the material into a resonant cavity, so as to cause a small perturbation in the resonant frequency and Q of the cavity. From the change in resonant frequency the dielectric constant can be found, and from the change in Q the loss tangent can be determined. Figure 3.21 shows a typical arrangement, where a narrow specimen of the material under test is inserted through a small aperture into a rectangular resonant cavity in a position of maximum electric field.

The sample shown in Figure 3.21 has a rectangular cross-section, but the shape is not critical. In practice, the shape of the sample is determined by the ease with which it can be manufactured, and the need to know the volume of the sample with reasonable precision. For convenience, the sample is usually inserted through a small aperture, as shown in Figure 3.21b, although the presence of the aperture will disturb the wall currents. Theoretically, a better solution is to have a fully enclosed cavity which can be disassembled/assembled to insert and remove the sample, but with cavities whose Q-values are in the region of 10 000 it is very difficult to mechanically assemble cavities to give consistent value of Q. It is important that the presence of the sample should not significantly disturb the field pattern within the cavity and so the sample should have a small volume, low loss, and be of relatively low dielectric constant. Thus, the method is suitable for most materials used as substrates in RF and microwave circuits, but for materials with high dielectric constants an alternative method should be used. Using the procedure outlined above the dielectric properties of a sample can be determined from:

$$\varepsilon' = \varepsilon_r = \left(\frac{f_o - f_1}{2f_o} \times \frac{V_c}{V_s} \right) + 1 \tag{3.29}$$

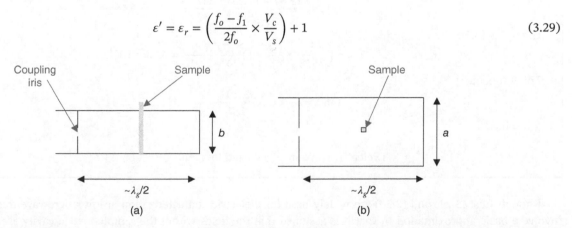

Figure 3.21   Sample mounted in a rectangular waveguide cavity: (a) side view of resonant cavity and (b) top view of resonant cavity.

and

$$\varepsilon'' = \left(\frac{1}{Q_1} - \frac{1}{Q_o} \right) \times \frac{V_c}{4V_o}, \tag{3.30}$$

where

f_o = resonant frequency of cavity without sample
f_1 = resonant frequency of cavity with sample
Q_o = Q of cavity without sample
Q_1 = Q of cavity with sample
V_s = volume of sample within cavity
V_c = volume of empty cavity

Example 3.12　The dielectric properties of a cylindrical rod of insulating material, diameter = 1 mm, were measured using a cavity perturbation technique, with the sample positioned as shown in Figure 3.21 within a rectangular waveguide cavity supporting the TE_{101} mode. The cavity cross-sectional data was: $a = 22.86$ mm and $b = 10.19$ mm. The following measurement data were obtained:

Cavity empty:	Resonant frequency = 12 GHz
	Q = 8704
Cavity + rod:	Resonant frequency = 11.23 GHz
	Q = 7538

Determine the dielectric constant and loss tangent of the dielectric rod.

Solution
At 12 GHz the length of the cavity is $\lambda_g/2$ for the TE_{101} mode where:

$$\frac{1}{\lambda_o^2} = \frac{1}{(2a)^2} + \frac{1}{\lambda_g^2} \quad \Rightarrow \quad \frac{1}{(25)^2} = \frac{1}{(2 \times 22.86)^2} + \frac{1}{\lambda_g^2} \quad \Rightarrow \quad \lambda_g = 29.86 \text{ mm},$$

$$l = \frac{29.86}{2} \text{ mm} = 14.93 \text{ mm}.$$

Volume of rod within cavity = $V_s = \pi \times 1 \times 10.19$ mm = 32.02 mm.
Volume of empty cavity = $V_c = (22.86 \times 10.19 \times 14.93)$ mm^3 = 3477.84 mm^3.
Using Eq. (3.29):

$$\varepsilon' = \varepsilon_r = \frac{12.00 - 11.23}{2 \times 12} \times \frac{3477.84}{32.02} + 1 = 4.48.$$

Using Eq. (3.30):

$$\varepsilon'' = \left(\frac{1}{7538} - \frac{1}{8704} \right) \times \frac{3477.84}{4 \times 32.02} = 4.83 \times 10^{-4}.$$

Loss tangent:

$$\tan \delta = \frac{\varepsilon''}{\varepsilon'} = \frac{4.83 \times 10^{-4}}{4.48} = 1.08 \times 10^{-4}.$$

Summary:

Dielectric constant = 4.48 and loss tangent = 1.08×10^{-4}.

Although Eqs. (3.29) and (3.30) are widely used for dielectric characterization at low microwave frequencies they do involve a basic approximation in that it is assumed that the fields within the sample-loaded cavity are uniform. Recent work by Orloff et al. [24] showed significant improvements in measurement accuracy by making some corrections to the basic method to account for the non-uniform fields created by the presence of the sample within the cavity. The details of the corrections proposed by Orloff are outside the scope of this textbook, but the details are well-explained in [24], which also presents data on measured samples of a known quartz dielectric to verify the new technique.

One of the difficulties that are encountered with the cavity perturbation technique is determining the precise volume of the dielectric specimen, which is necessarily kept small so as to not significantly disturb the fields within the cavity. Equations (3.29) and (3.30) can be rearranged to give

$$\tan \delta = \frac{\varepsilon_r - 1}{2\varepsilon_r} \frac{f_o}{f_o - f_1} \left(\frac{1}{Q_1} - \frac{1}{Q_o} \right), \tag{3.31}$$

which means that in situations where the dielectric constant is already known the loss tangent can be determined without the need to know volume of the sample within the cavity. Since the dielectric constant of materials does not change substantially with frequency up to the low microwave region this is a useful expression for situations where the dielectric constant is often known from low-frequency measurements, and where the primary interest is in determining the loss in material.

An alternative to using a thin rod of the dielectric material under test is to use a thin membrane of the material and insert it through a slit in the waveguide cavity wall. Providing that the slit does not significantly interrupt the current flow there will be negligible radiation. Figure 3.22 shows how non-radiating slits may be cut in a rectangular waveguide cavity (supporting TE_{10n} modes) and an over-moded circular waveguide cavity (supporting TE_{01n} modes). In both cases the long dimension of the slit is parallel to the wall currents, and for narrow slits there is very little interruption to the current paths.

In the case of the circular waveguide cavity shown in Figure 3.22b, a section of the cavity is made from helical waveguide to act as a mode filter and suppress the unwanted modes. It should also be noted that the circular cavity is fed through an off-centre coupling iris to excite the TE_{01} mode; the position of the coupling iris corresponds to the maximum of the electric field.

The cavity perturbation techniques discussed so far have used homogeneous dielectric samples. But there is a need at RF and microwave frequencies to be able to characterize layers of dielectric that are not self-supporting but need to be printed or deposited on a rigid base. An example is thick-film dielectric which has to be printed on top of a supporting substrate, normally alumina. Consequently, there is a need to characterize dielectric materials that form two-layer samples. One technique to characterize a thick-film layer is to partially coat a thin low-loss substrate with the material under test, as shown in Figure 3.23. Typically the substrate would be 250 μm thick high purity alumina.

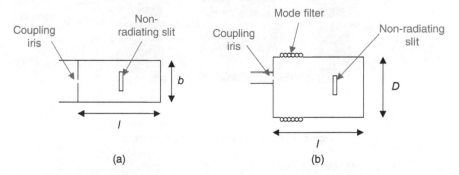

Figure 3.22 Non-radiating slits in waveguide cavities: (a) rectangular waveguide cavity and (b) circular waveguide cavity.

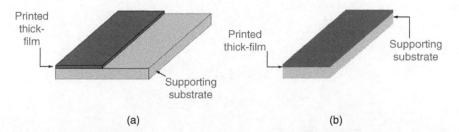

Figure 3.23 Two-layer thick-film test samples: (a) substrate partially coated with thick-film layer under test and (b) substrate sawn to produce a two-layer test specimen and a reference sample.

The partially coated substrate is divided using a ceramic saw to provide the two-layer test sample and a reference piece having the same dimensions and properties as that supporting the thick-film layer. The resonant frequency and Q of the cavity are then measured with the uncoated tile in place to form the reference values, f_o and Q_o. The reference tile is then replaced with the two-layer test specimen and the measurements repeated to enable the dielectric constant and loss tangent of the thick-film layer to be calculated using the previous theory. The method assumes that there is no variation in dielectric properties across the original alumina tile, and also that the reference piece is located precisely in the same position as the tile supporting the thick-film layer under test.

3.6.1.3 Use of the Split Post Dielectric Resonator (SPDR)

The split post dielectric resonator (SPDR) is widely used as a method of measuring the properties of low-loss dielectric laminate samples in the frequency range 1–35 GHz. The development of the SPDR is largely due to the work of Krupta [25, 26], who has published extensively on the theory and application of this measurement technique. The principal features of an SPDR are shown in Figure 3.24.

The dielectric resonator consists of a short vertical cylinder of dielectric, which is split into two parts. The sample-under-test is inserted between the two halves of the dielectric. The dielectric resonator is mounted on insulating supports within a metal enclosure. Two coaxial feeds provide RF connections, with horizontal coupling loops to energize and detect electromagnetic fields within the metal cavity. The SPDR is designed to operate with the $TE_{01\delta}$ mode within the dielectric resonator; the 0 and 1 suffixes have the same meaning as for the TE_{01} mode in circular metal waveguide, and the δ indicates that the axial length of the resonator is less than half a wavelength due to the fringing fields at each end. The electric field lines of the $TE_{01\delta}$ mode form closed loops in the plane of the sample as shown in Figure 3.25.

The measurement samples used in SPDR instruments are normally made larger than the diameter of the dielectric resonator to ensure that the dielectric-under-test completely fills the area between the two halves of the resonator. It

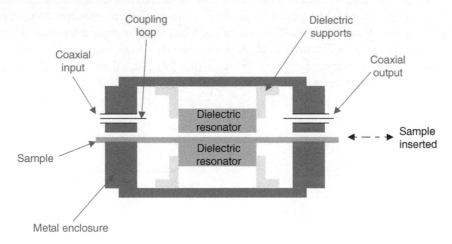

Figure 3.24 Split post dielectric resonator.

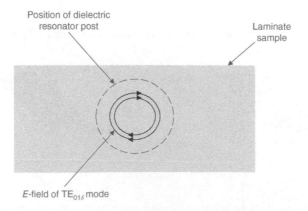

Figure 3.25 Planar view of sample in SPDR.

is important that the laminar samples used in SPDR measurements have consistent thickness, so that there are no non-uniform air gaps between the two halves of the resonator. In general terms, the higher the dielectric constant of the material the thinner should be the sample, and manufacturers of SPDR instruments normally provide guidance on the optimum sample thickness for particular ranges of dielectric constant. Measurements to determine the complex permittivity of a material using an SPDR are similar to those for perturbed cavities discussed in the previous section, in that the resonant frequency and Q value are measured with and without the sample. The complex permittivity can then be determined from the following equations [26]:

$$\varepsilon' = \varepsilon_r = 1 + \frac{f_o - f_1}{h f_o K_g} \tag{3.32}$$

and

$$\tan \delta = \frac{\varepsilon''}{\varepsilon'} = \left(\frac{1}{Q} - \frac{1}{Q_{DR}} - \frac{1}{Q_c} \right) \times \frac{1}{P_{es}}, \tag{3.33}$$

where

f_o = resonant frequency of empty fixture
f_1 = resonant frequency of fixture with sample in place
Q = Q value of the empty metal fixture (ignoring losses)
Q_{DR} = Q value of fixture including the dielectric parts
Q_c = Q value of fixture including effects of metal losses
K_ε = function of the sample thickness and relative permittivity
P_{es} = energy filling factor of the sample.

For a particular SPDR instrument, the value of K_ε can be evaluated [26] for various combinations of ε_r and h (sample thickness) and an iterative procedure used to find ε_r from Eq. (3.32). The value of P_{es} is found from the ratio of the volume integrals of the sample and the empty fixture [26]. Krupta et al. [26] identified the uncertainty in the thickness of the sample as being the major source of uncertainty in SPDR measurements. Based on their measurements using a 4 GHz SPDR it was concluded that the dielectric constant could be measured with an uncertainty of 0.3%, and a resolution of 2×10^{-5} in the value of the loss tangent. These values make the instrument very attractive for measuring the properties of substrates for RF and microwave applications, since the measurement errors quoted for the SPDR technique are small compared to those encountered in standard circuit production.

In addition to measuring the dielectric properties of homogeneous samples, the SPDR is very useful for measuring the properties of thick-film layers, which were discussed in the previous section. By making three sets of measurements, (i) with the cavity empty, (ii) with the cavity containing the supporting substrate, and (iii) with the cavity containing the substrate coated with the thick-film under test, the properties of the thick- film layer can be deduced. Dziurdzia and colleagues [27] used this technique to determine the properties of samples of photoimageable thick-film dielectric at frequencies in the vicinity of 20 GHz, and concluded that the uncertainty in the measurement technique was less than 2% for samples prepared on sapphire substrates.

3.6.1.4 Open Resonator

The use of closed-cavity measurement techniques becomes more problematic as the frequency of operation increases into the millimetre-wave region, mainly because dominant mode cavities become inconveniently small. An alternative technique for making complex permittivity measurements at millimetre-wave frequencies is to use an open resonator. The principle of the open resonator is well established in the literature [28, 29] and provides a precise measurement tool for dielectric analysis at high frequencies. The key features of a hemispherical open resonator are shown in Figure 3.26; the resonator shown is described as a hemispherical resonator because it consists of one reflector with a spherical profile and one flat reflector, as opposed to a resonator with two spherical reflectors.

In Figure 3.26, the millimetre-wave signal is injected into a resonant region though a waveguide feed at the centre of spherical mirror. The signal resonates between the concave surface of the spherical mirror and the top surface of a flat mirror. The mirrors are normally made of brass with precisely ground reflecting surfaces, which are normally gold-plated to reduce surface losses and increase the Q of the resonator. The sample-under-test is placed on top of the plane mirror, and the Q and resonant frequency of the instrument are measured with and without the sample in place. From the changes in resonant frequency and Q the complex permittivity can be evaluated.

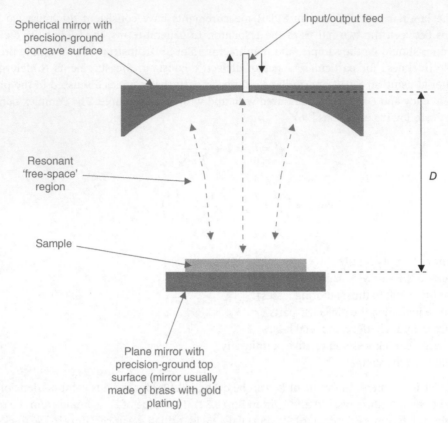

Figure 3.26 Hemispherical open resonator.

The theory of the open resonator is well documented in the literature, and a concise and informative discussion is provided by Komiyama et al. [30], who made complex permittivity measurements on a range of substrate materials at 100 GHz. They provide details of the somewhat lengthy procedure for extracting the value of complex permittivity from measurements of resonant frequency and resonator Q. In a similar work, Hirvonen and colleagues [31] also measured the dielectric constant and loss tangent of materials at 100 GHz using an open resonator, but included a correction to allow for the thickness of the sample-under-test. Since practical test samples are normally planar, with a flat upper surface, the contour of the upper surface does not match the phase front of the beam reflected from the spherical mirror. This is not of any great consequence for thin samples, but will clearly introduce phase errors as the sample thickness increases. In general, the measurement data obtained using open resonators is considered to be very accurate, with quoted [31] uncertainty in the measurement of dielectric constant being 0.02–0.04% for $\varepsilon_r \geq 2$ and 6–40×10^{-6} in the value of the loss tangent for $10^{-4} \leq \tan \delta \leq 10^{-3}$ at 100 GHz.

There are essentially two methods for extracting measurement data from an open resonator. The most common method is to connect the open resonator to a millimetre-wave network analyzer, and with the length (D) of the resonator kept constant, to display the resonance curves as a function of frequency. This is termed the frequency-variation method, and was the method used to obtain the data in [31]. It is a fast measurement scheme and offers the potential of using an averaging technique to reduce the effects of noise. The alternative method, termed the cavity-length-variation method, is to mount the flat mirror on a translation stage and to vary the length of the cavity in small precise steps, and so build up the profile of the resonances. This method was reported by Afsar et al. [32] who used 20 nm steps to vary the cavity length. It was proposed that this technique was both simpler and less costly than the frequency-variation method, primarily because it did not require the use of an expensive millimetre-wave network analyzer. Also, it was postulated that the new technique was potentially more accurate than the frequency-variation method because it required only fixed frequency sources, and therefore did not suffer from the noise associated with swept sources. Subsequently, Afsar et al. [33] published an in-depth study of the accuracy of the two methods. In this study they made 60 GHz measurements on a range of polymer and ceramic samples with dielectric constants varying from 2.1 to 9.6. Their general conclusion was that the frequency-variation method provided the better overall accuracy, but at the cost of a more complex system.

3.6.1.5 Free-Space Transmission Measurements

The dielectric measurement techniques discussed so far in this chapter have employed either open or closed resonant cavities. Whilst these techniques are known to provide extremely accurate data, they are essentially fixed-frequency (narrowband) methods, and costly to implement at millimetre-wave frequencies. An alternative, well-established technique is to deduce the dielectric properties of a material from S-parameter measurements of a sample located between two antennas in free space. This method has the advantage of being potentially broadband, and is attractive at millimetre-wave frequencies where the size of the antennas is small, thus making a relatively compact test set-up. The principal features of free-space measurement system are shown schematically in Figure 3.27.

The measurement system shown in Figure 3.27 consists of 2 mm-wave horn antennas, with the sample-under-test positioned midway between them with the plane of the sample perpendicular to the axes of the antennas. The horn antennas are loaded with dielectric lenses (see Chapter 12) to focus the electromagnetic beams onto the sample, i.e. the sample must be in the focal plane of the antennas. A millimetre-wave vector network analyzer (VNA) is connected between the inputs to the two antennas. By using a free-space TRL calibration procedure (see Chapter 13) the S-parameters of the sample can be measured by the VNA, and displayed as a function of frequency. A comprehensive discussion of the free-space measurement system is provided by Ghodgaonkar et al. [34], in respect of a system working at 16 GHz. Although the system described in [34] is not in the mm-band, the methodology is easily applied to higher frequencies. For example, Osman et al. [35] used the same approach to obtain data on LTCC samples at G-band (145–155 GHz).

A variant on the free-space method using two antennas, is to use a single antenna and measure the reflections from a sample backed by a conducting plate, and positioned in the focal plane of the antenna. Figure 3.28 shows the simple arrangement of a sample with a conducting back plate.

Using the same approach as [34], we can view the sample with a conducting backing plate as a transmission line of characteristic impedance Z_s terminated with a short circuit. Then from Eq. (1.33)

$$Z_{in} = jZ_s \tan \beta_s d \tag{3.34}$$

and using Eq. (1.42)

$$\rho_{in} = \frac{jZ_s \tan \beta_s d - Z_o}{jZ_s \tan \beta_s d + Z_o}. \tag{3.35}$$

From the definition of S-parameters given in Eq. (5.3)

$$\rho_{in} \equiv S_{11}$$

and so

$$S_{11} = \frac{jZ_s \tan \beta_s d - Z_o}{jZ_s \tan \beta_s d + Z_o}. \tag{3.36}$$

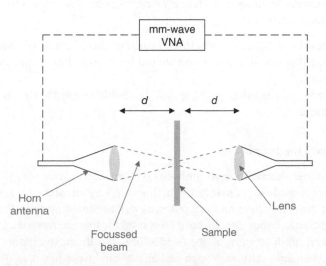

Figure 3.27 Free-space millimetre-wave measurement system.

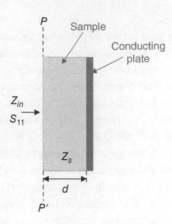

Figure 3.28 Sample-under-test with a conductor backing plate.

For a non-magnetic sample

$$Z_s = \sqrt{\frac{\mu_0}{\varepsilon_0 \varepsilon^*}}, \tag{3.37}$$

where μ_0 and ε_0 are the permeability and permittivity of free space, respectively, and ε^* is the complex relative permittivity of the sample. Also,

$$Z_0 = \sqrt{\frac{\mu_0}{\varepsilon_0}}. \tag{3.38}$$

Combining Eqs. (3.37) and (3.38) we have

$$\frac{Z_s}{Z_0} = \frac{1}{\sqrt{\varepsilon^*}} = (\varepsilon^*)^{-0.5}. \tag{3.39}$$

Substituting from Eq. (3.39) into Eq. (3.36) gives

$$S_{11} = \frac{j(\varepsilon^*)^{-0.5} \tan \beta_s d - 1}{j(\varepsilon^*)^{-0.5} \tan \beta_s d + 1}, \tag{3.40}$$

where

$$\beta_s = \frac{2\pi}{\lambda_s} = \frac{2\pi}{\lambda_0} \times (\varepsilon^*)^{-0.5}, \tag{3.41}$$

and λ_0 is the free-space wavelength at a given measurement frequency. Having measured S_{11} at a particular frequency the value of ε^* can be found through the solution of Eq. (3.40).

Whilst the free-space measurement technique is relatively straightforward at millimetre-wave frequencies, two precautions are necessary to achieve accurate results:

(i) The area of the sample should be sufficiently large to avoid diffraction effects at the sample edges. For a square sample it is normally recommended that the side dimension should be at least three times the maximum dimension of the illuminated area on the sample.

(ii) In the case of the two-antenna system a time gating technique should be used to eliminate the effect of multiple reflections between the two antennas.

3.6.2 Measurement of Planar Line Properties

Whilst the measurement techniques described in the previous sections can provide very precise information on the properties of dielectrics, measurements made on planar test structures are very attractive for the circuit designer because they embrace all of the properties of the conductors and substrates, as well as providing an indication of the effectiveness of the fabrication process. Various resonant planar circuits have been used for line characterization, including linear resonators in the form of stubs or half-wavelength sections of open-ended line. But linear resonators necessarily involve significant discontinuities in the form of open ends. Although open-end effects can be predicted, as discussed in Chapter 4, the most widely used circuit for planar measurements is the resonant ring, which consists of a closed loop of transmission line,

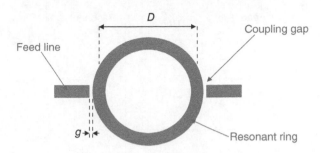

Figure 3.29 Microstrip resonant ring.

thereby avoiding open ends. The microstrip resonant ring was originally proposed by Troughton [36] in 1968, and it is still extensively used for line characterization, primarily for dispersion measurements. Both microstrip and coplanar waveguide (CPW) formats have been used for resonant ring measurements at frequencies well into the millimetre-wave part of the frequency spectrum. In addition to resonant circuit configurations, simple arrangements of non-resonant linear lines can also provide useful measurement data, and these are discussed briefly in Section 3.6.2.2.

3.6.2.1 The Microstrip Resonant Ring

The conventional layout of a microstrip resonant ring is shown in Figure 3.29, and consists of a closed loop of microstrip with two linear feeds, which are coupled to the ring through small gaps. Normally, data are obtained from resonant rings using transmission measurements, which require two feeds, although reflection measurements using a single feed can provide the same information.

The ring will exhibit resonance when the circumference, specified around the mean diameter, D, is equal to an integer number of substrate wavelengths, i.e.

$$\pi D = n\lambda_s, \tag{3.42}$$

where n is an integer. It follows from Eq. (3.42) that the resonant frequency (f_o) of the ring will be

$$f_o = \frac{nc}{\pi D \sqrt{\varepsilon_{r,eff}^{MSTRIP}}} \tag{3.43}$$

and

$$\varepsilon_{r,eff}^{MSTRIP} = \left(\frac{nc}{f_o \pi D}\right)^2, \tag{3.44}$$

where $\varepsilon_{r,eff}^{MSTRIP}$ is the effective relative permittivity of the ring.

Using a VNA[3] connected to the ends of the feed lines to measure the transmission loss, a set of discrete resonance curves can be displayed against frequency, each corresponding to a different value of n. By measuring the resonant frequency of a given resonance response the effective relative permittivity can be found through use of Eq. (3.44). Also, by measuring the Q of the resonance curve the line loss can be found from

$$\alpha_T = \frac{n}{Q_L D}, \tag{3.45}$$

where α_T is the total line loss in Np/m, and Q_L is the Q-factor of the resonant ring. Whilst the resonant ring is a simple, convenient measurement tool, there are a number of points that need to be considered if accurate data are to be obtained:

(i) There will necessarily be gaps between the feed lines and the ring, and these will create both reactive and resistive loading on the ring. The choice of gap size is therefore important for accurate measurements; the gap must be small enough to couple a measurable signal into the ring, but not so small that loading effects become significant. Various authors have addressed the issue of the equivalent circuit of the gap in order that compensation can be applied to measured data. Yu and Chang [37] used a transmission-line analysis to examine the effect of the gap on the performance of a microstrip ring resonator at frequencies around 3.5 GHz. They modelled the gap as an L-shaped combination of capacitors, as shown in Figure 3.30.

---

3 VNA is an abbreviation of vector network analyzer; this instrument is discussed in detail in Chapter 7.

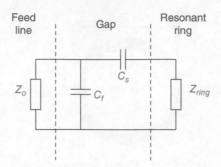

Figure 3.30 Simple model of a resonant ring coupling gap.

The series capacitance (C_s) represents the coupling between the end of the feed line and the ring, and the parallel capacitance (C_f) represents the fringing at the end of the feed line; the characteristic impedance of the feed line is represented by Z_o and the ring impedance by Z_{ring}. The capacitance values were related to the dimensions of the gap by modifying the standard equations for a gap in a linear microstrip line (see Chapter 4). Very good agreement between theoretical and measured resonant frequencies was obtained using their model [37]. They also concluded that if the ratio of the gap size (g) to the width of the feed line was greater than 0.4, the effect of the gap on the ring resonant frequencies was negligible. In a subsequent study, Bray and Roy [38] developed an improved lumped-element model to represent the gap, based on a T-gap discontinuity. This made the reasonable assumption that the curvature of a large diameter ring would appear to be linear in the vicinity of the coupling gap. They extracted the capacitance values for their model using EM simulation. This model was able to predict the resonant frequencies of the ring to within 0.11% over a very wide frequency range, from 5 to 40 GHz.

(ii) Since the ring is circular there will be curvature effects, and appropriate compensation must be applied to the measured data if these are to be representative of straight lines used in microstrip designs. The effects of curvature were originally investigated by Owens [39], who published compensation data for microstrip resonant rings used over the frequency range 2–12 GHz. More recently Faria [40] presented updated information on the effects of curvature and concluded that for small radii of curvature, Owens compensation data was an underestimate. As would be expected, the effects of curvature are more significant for rings formed from wide tracks with small radii of curvature. This suggests that for measurement purposes it is better to have large diameter rings with tracks of moderate width. Using tracks that have a characteristic impedance of 50 Ω is sensible, since the measured data can be more readily applied to practical designs where lines having a characteristic impedance of 50 Ω are most common. However, using very wide diameter rings can be the source of another problem, in that there may be discontinuities within the ring due to variations in substrate or conductor properties when wide areas are considered. Discontinuities within a microstrip ring can lead to mode splitting and the appearance of resonance curves with a double peak.

(iii) It is important to ensure that resonant rings are large enough to avoid electromagnetic field interactions across the ring, whereby the ring would act as a disc resonator. However, this is not usually a problem since the need to avoid significant curvature effects dictates that the ring diameter should be large.

(iv) Since the losses measured with a microstrip resonant ring should be representative of the losses in a linear microstrip line of the same width, it is important to ensure that the ring losses do not include a significant radiation component, as there will not be radiation from an ideal straight microstrip line. This can usually be achieved through sensible choice of the test ring dimensions. It is well known that conductor discontinuities on thick, low permittivity substrates will radiate quite effectively, and this is the basis of microstrip antenna theory (see Chapter 12). Therefore, ensuring that resonant rings used for measurement purposes have a small curvature, i.e. a large radius, and are fabricated on thin substrates will effectively eliminate the radiation loss, and the measured loss will be that due to the conductor and dielectric alone. A useful rule-of-thumb is to ensure that the mean radius of the ring is at least 10 times the width of the track.

Whilst the microstrip ring resonator is a simple, efficient, and widely used circuit for materials characterization, its use at millimetre-wave frequencies becomes problematic, because of the need to have very thin substrates to avoid generating higher-order modes, as discussed in Chapter 2. However, the concept of the ring resonator can be employed at very high frequencies by using planar geometries, such as slotline and CPW, where the conductor is only deposited on one side of the substrate. Examples of typical slotline and CPW ring resonator geometries are shown in Figure 3.31.

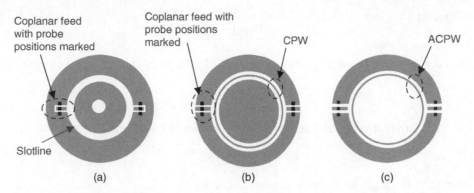

Figure 3.31 Alternative ring resonator formats: (a) Slotline, (b) CPW, and (c) asymmetric CPW.

Hopkins and Free [41] demonstrated the effectiveness of the slotline ring resonator by using the configuration shown in Figure 3.31a to make ultra-wideband dispersion measurements over the frequency range 2–220 GHz. (*It should be noted that the two outer connections of the CPW probe will earth the outer ring, and so no earth connection is required to the inner ring.*) Measurement data in [41] were taken from resonances displayed on a VNA over two frequency ranges 5–100 GHz and 140–220 GHz, and results for the extracted values of $\varepsilon_{r,eff}$ agreed with theoretical predictions to within 3.5%, thus validating the effectiveness of the measurement structure. An alternative to the slotline configuration is to use CPW as shown in Figure 3.31b. However, CPW presents a practical problem in that the isolated central disc of conductor must be earthed to ensure the correct CPW mode is generated. Earthing the central conductor can be done by using one or more VIAs to connect the disc to a ground plane on the other side of the substrate, but this detracts from the main benefit of CPW in that conductors are no longer confined to a single plane. Benarabi et al. [42] addressed this problem by using an asymmetrical CPW ring resonator, as shown in Figure 3.31c. Although the primary focus of the work reported in [42] was on the microwave performance of new silver composite conductors, it also provides useful information on the design of ACPW structures.

3.6.2.2 Non-resonant Lines

The measurement techniques described in the previous sections can provide high measurement accuracy primarily because they make use of resonant circuits in one form or another. These circuits utilize voltage magnification, which increases the sensitivity of the measured parameters. A much simpler technique is to make use of non-resonant lines of different lengths. Whilst this will not have same degree of accuracy as the more complex resonant structures, with modern measurement instrumentation it can provide useful information on the properties of planar lines. Also, using non-resonant lines enables swept measurements over a continuous range of frequencies to be obtained, whereas resonant structures can only provide data at a number of discrete frequencies. The basic non-resonant circuit configuration is shown in Figure 3.32, and consists of two microstrip lines of equal width, but unequal length.

The meandered line shown in Figure 3.32 simply provides an additional path length compared to the straight through line. The spacing between the meandered sections must be large enough to avoid edge coupling, and the corners should be chamfered to minimize the corner discontinuity. If the same voltage, V_1, is applied to each line the loss per unit length of

Figure 3.32 Non-resonant test circuit.

the lines at a given frequency is given by

$$\alpha = \frac{|V_{2M}| - |V_{2S}|}{\Delta L},$$

(3.46)

where ΔL is the difference in length between the two lines. Similarly, the phase change per unit length at a given frequency is given by

$$\phi = \frac{\angle(V_{2M}) - \angle(V_{2S})}{\Delta L}.$$

(3.47)

The substrate wavelength and effective relative permittivity can then be found from

$$\phi = \frac{2\pi}{\lambda_s} = \frac{2\pi}{\lambda_o}\sqrt{\varepsilon_{r,eff}^{MSTRIP}},$$

(3.48)

where λ_o is the free-space wavelength at the measurement frequency.

The use of the non-resonant line technique assumes that the measurement conditions are the same for both the straight line and the meandered line, and in particular that the characteristics of the launchers are identical. For microstrip lines a high degree of launcher repeatability is possible by mounting the test circuit in a 'Universal Test Fixture' as shown in Figure 3.33a. With this type of fixture the test circuit is clamped between two sprung jaws, one of which is fixed in position and the other movable in the $x - y$ plane through the use of sliders. RF signals are connected to the jig through miniature coaxial connectors and contact made through the use of flat tabs, which can be precisely positioned over the ends of the microstrip lines. Repeatability of circuit mounting with this type of jig is extremely good, typically being better than ± 0.1 dB up to 20 GHz, with phase repeatability better than $\pm 1°$.

Non-resonant lines with a CPW format can also be used for testing at higher millimetre-wave frequencies, with the test circuit mounted in a conventional probe station as shown in Figure 3.33b.

It should also be noted that when designing CPW lines with a finite ground plane, correct choice of the widths of the ground strips is important in minimizing line losses. The cross-section of a FGCPW line is shown in Figure 3.34.

Ghione and Goano [43] showed that the following relationship must be satisfied if the influence of the finite ground on line loss is to be negligible (<10% of the ideal case of $c = \infty$).

$$c > 2b.$$

(3.49)

The use of the swept frequency technique can also provide informative displays showing the significance of the various sources of loss as a function of frequency.

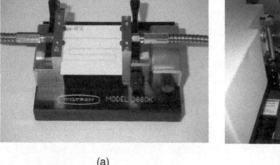
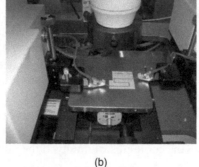

(a)   (b)

Figure 3.33 Mounting of planar test circuits: (a) microstrip test circuit mounted in a universal test fixture and (b) CPW test circuit mounted on a conventional probe station.

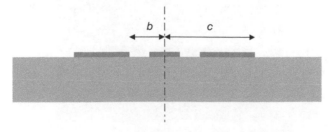

Figure 3.34 Cross-section of a symmetrical FGCPW line.

Example 3.13   The total line loss in dB/mm obtained from a broadband sweep of a 50 Ω microstrip line is shown in Figure 3.35.

The line data are:	Substrate:	Relative permittivity = 9.8
		Thickness = 0.25 mm
	Conductor:	Copper ($\sigma = 5.87 \times 10^7$ S/m)
		RMS surface roughness = 370 nm

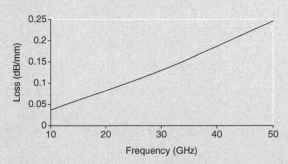

Figure 3.35   Microstrip line loss.

Neglecting radiation loss, determine:

(i) The loss tangent of the dielectric at 10, 30, and 50 GHz

(ii) The percentage contributions of bulk conductor loss, conductor surface loss, and dielectric loss at 10, 30, and 50 GHz, and show the results graphically. Comment upon the results.

Solution

(i) Using microstrip graphs:

$$50\,\Omega \quad \Rightarrow \quad \frac{w}{h} = 0.9 \quad \Rightarrow \quad w = 0.9 \times 0.25 \text{ mm} = 0.225 \text{ mm}.$$

$$\frac{w}{h} = 0.9 \quad \Rightarrow \quad \varepsilon_{r,eff}^{MSTRIP} = 6.6.$$

10 GHz: $\lambda_o = \dfrac{c}{f} = \dfrac{3 \times 10^8}{10 \times 10^9}$ m = 30 mm $\lambda_s = \dfrac{30}{\sqrt{6.6}}$ mm = 11.68 mm.

30 GHz: $\lambda_o = \dfrac{c}{f} = \dfrac{3 \times 10^8}{30 \times 10^9}$ m = 10 mm $\lambda_s = \dfrac{10}{\sqrt{6.6}}$ mm = 3.89 mm.

50 GHz: $\lambda_o = \dfrac{c}{f} = \dfrac{3 \times 10^8}{50 \times 10^9}$ m = 6 mm $\lambda_s = \dfrac{6}{\sqrt{6.6}}$ mm = 2.34 mm.

Calculating the skin depth: $\delta_s = (\pi \mu_o f \sigma)^{-0.5}$

$$\delta_s(10 \text{ GHz}) = (\pi \times 4\pi \times 10^{-7} \times 10^{10} \times 5.87 \times 10^7)^{-0.5} \text{ m} = 0.66 \text{ μm},$$

$$\delta_s(30 \text{ GHz}) = 0.38 \text{ μm},$$

$$\delta_s(50 \text{ GHz}) = 0.30 \text{ μm}.$$

Reading total loss from graph: $\alpha_t = 0.038$ dB/mm at 10 GHz,

$\alpha_t = 0.130$ dB/mm at 30 GHz,

$\alpha_t = 0.246$ dB/mm at 50 GHz.

Now $\alpha_t = \alpha_d + \alpha_c$,

(Continued)

(Continued)

where

α_d = dielectric loss

α_c = total conductor loss (sum of bulk and surface loss).

The value of α_c at the three frequencies being considered can be found through substitution in Eq. (3.8):

$$\alpha_c(10 \text{ GHz}) = \frac{0.072 \times \sqrt{10}}{0.225 \times 50} \left(1 + \frac{32}{\pi} \tan^{-1}\left[\frac{0.24 \times 0.370^2}{0.66^2}\right]\right) \text{ dB/mm}$$

$$= 0.035 \text{ dB/mm}.$$

Similarly,

$$\alpha_c(30 \text{ GHz}) = 0.117 \text{ dB/mm},$$

$$\alpha_c(50 \text{ GHz}) = 0.207 \text{ dB/mm}.$$

The dielectric loss can now be found by subtracting the calculated conductor loss from the total loss read from the swept frequency response:

$$\alpha_d(10 \text{ GHz}) = \alpha_t(10 \text{ GHz}) - \alpha_c(10 \text{ GHz}) = (0.038 - 0.035) \text{ dB/mm} = 0.003 \text{ dB/mm},$$

$$\alpha_d(30 \text{ GHz}) = \alpha_t(30 \text{ GHz}) - \alpha_c(30 \text{ GHz}) = (0.130 - 0.117) \text{ dB/mm} = 0.013 \text{ dB/mm},$$

$$\alpha_d(50 \text{ GHz}) = \alpha_t(50 \text{ GHz}) - \alpha_c(50 \text{ GHz}) = (0.246 - 0.207) \text{ dB/mm} = 0.039 \text{ dB/mm}.$$

Using Eq. (3.6):

$$10 \text{ GHz}: 0.003 = 27.3 \times \frac{9.8 \times (6.6 - 1) \times \tan\delta}{6.6 \times (9.8 - 1)} \times \frac{1}{11.68} \quad \Rightarrow \quad \tan\delta = 0.0014,$$

$$30 \text{ GHz}: 0.013 = 27.3 \times \frac{9.8 \times (6.6 - 1) \times \tan\delta}{6.6 \times (9.8 - 1)} \times \frac{1}{3.89} \quad \Rightarrow \quad \tan\delta = 0.0020,$$

$$50 \text{ GHz}: 0.039 = 27.3 \times \frac{9.8 \times (6.6 - 1) \times \tan\delta}{6.6 \times (9.8 - 1)} \times \frac{1}{2.34} \quad \Rightarrow \quad \tan\delta = 0.0035.$$

(ii) We can find the bulk conductor loss by putting $\Delta = 0$ in Eq. (3.8) giving:

$$\alpha_{c,bulk}(10 \text{ GHz}) = \frac{0.072 \times \sqrt{10}}{0.225 \times 50} \text{ dB/mm} = 0.020 \text{ dB/mm},$$

$$\alpha_{c,bulk}(30 \text{ GHz}) = \frac{0.072 \times \sqrt{30}}{0.225 \times 50} \text{ dB/mm} = 0.035 \text{ dB/mm},$$

$$\alpha_{c,bulk}(50 \text{ GHz}) = \frac{0.072 \times \sqrt{50}}{0.225 \times 50} \text{ dB/mm} = 0.045 \text{ dB/mm}.$$

The surface loss is given by $\alpha_c, surface = \alpha_c - \alpha_{c, bulk}$, i.e.

$$\alpha_{c,surface}(10 \text{ GHz}) = (0.035 - 0.020) \text{ dB/mm} = 0.015 \text{ dB/mm},$$

$$\alpha_{c,surface}(30 \text{ GHz}) = (0.117 - 0.035) \text{ dB/mm} = 0.082 \text{ dB/mm},$$

$$\alpha_{c,surface}(50 \text{ GHz}) = (0.207 - 0.045) \text{ dB/mm} = 0.162 \text{ dB/mm}.$$

Summary:

f (GHz)	α_t (dB/mm)	α_d (dB/mm)	$\alpha_{c,bulk}$ (dB/mm)	$\alpha_{c,surface}$ (dB/mm)
10	0.038	0.003 ≡ **7.9%**	0.020 ≡ **52.6%**	0.015 ≡ **39.5%**
30	0.130	0.013 ≡ **10.0%**	0.035 ≡ **26.9%**	0.082 ≡ **63.1%**
50	0.246	0.039 ≡ **15.9%**	0.045 ≡ **18.2%**	0.162 ≡ **65.9%**

Just for illustrative purposes the data presented in the table can also be shown using Pie-charts, as in Figure 3.36. *Comment: As the frequency increases the losses due to the dielectric, and particularly those due to the roughness of the conductor surface, assume much greater importance.*

It is also useful to add the calculated conductor losses to the swept frequency response to emphasize the relative significance of the sources of loss, as shown in Figure 3.37.

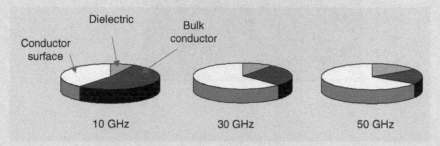

Figure 3.36 Constituent components of line loss.

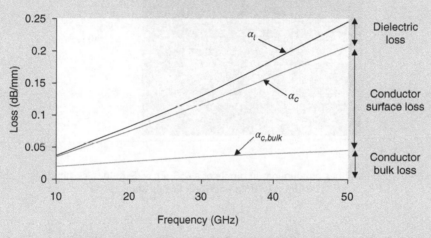

Figure 3.37 Influence of various sources of line loss.

3.6.3 Physical Properties of Microstrip Lines

The significance of surface roughness in determining the loss in planar interconnections has already been discussed. It is therefore important to be able to examine and measure the surface roughness of conductors and dielectrics, and also the profile of deposited conductors. Mechanical profilers (sometimes referred to as Talysurf profilers) are normally used for this purpose. In these instruments a sharply pointed stylus is drawn over the surface. Vertical deflections are sensed by a transducer, and the resulting electrical signals are displayed as a function of the lateral distance travelled by the stylus. The tip of the stylus is normally made of diamond, and its precise shape is critical in obtaining accurate surface profiles. The tip is normally curved, with a radius of curvature of the order of 2–3 μm, and clearly this limits the maximum curvature than can be recorded. Figure 3.38 shows examples of surface profiles of etched gold thick-film lines on alumina, obtained using a standard laboratory profiler. The responses give an indication of the surface roughness, both of the conductor and dielectric, as well as the cross-sectional geometry of the lines. Ideally, the lines should have a rectangular cross-section so that the characteristic impedance is precisely defined. In both cases shown the lines exhibit very good roll-off at the line edges, i.e. sharp roll-off, although for the 78 μm line it appears that the roll-off is rather poor, but this is due to the expanded horizontal scale.

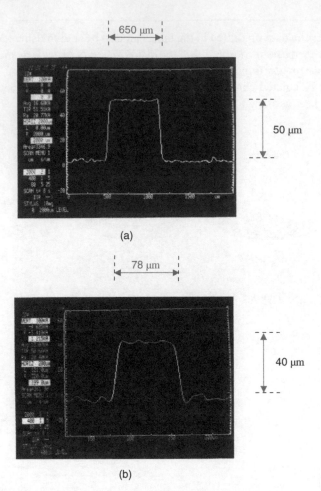

(a)

(b)

Figure 3.38 Profiles of gold thick-film lines on alumina: (a) profile of a 650-μm wide gold thick-film line on alumina and (b) profile of a 78-μm wide gold thick-film line on alumina.

Surface profilers of the type described above rely upon measuring the physical movement of the stylus, and converting this to a measurable electrical signal. Whilst this provides good profile information, and an indication of RMS surface roughness, it has limitations for detailed surface analysis. At high millimetre-wave frequencies, where the skin depth is less than 200 nm, it is often necessary to study the surface of materials in more detail, for example to examine the effects of surface polishing. A convenient instrument for this type of surface examination is the atomic force microscope (AFM). The AFM, employs a fine metal stylus mounted on the end of a cantilever. As the tip of the stylus is scanned over the surface of the material under test the instrument measures the forces on the tip. With this arrangement the AFM is capable of measuring depth changes of the order of 10 p.m. with lateral resolutions of the order of 100 p.m. Examples of AFM scans of a gold conductor deposited on 99.5% alumina are shown in Figure 3.39. This figure also shows the effect of chemically polishing the surface of the gold. In this instance the chemical polishing was achieved by immersing the gold conductor for a short period in a strong etching solution made from iodine and potassium iodide.

It can be seen in Figure 3.39 that the etching solution has preferentially removed the peaks from the surface of the gold, and this will reduce the surface roughness. Some caution is needed when using this type of chemical polishing to avoid creating pin-holes in the troughs of the surface. However, at high millimetre-wave frequencies where the skin depth is comparable (or even less) than the RMS surface roughness, careful chemical polishing of the surface of a planar conductor can have a significant beneficial effect in reducing conductor surface loss.

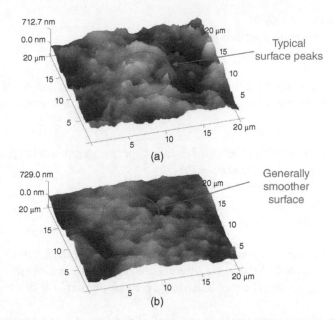

Figure 3.39 AFM scans of surface of gold conductor: (a) untreated gold surface and (b) gold surface after chemical polishing.

3.7 Supplementary Problems

Q3.1 The complex relative permittivity of a particular dielectric material is 2.32−j0.007. Determine the loss tangent and Q of the material.

Q3.2 The phase change through a 50 mm length of dielectric at 950 MHz is 144°. If the dielectric material has a Q of 125, what is the complex relative permittivity of the material?

Q3.3 Measurements on a 3 cm thick block of dielectric material at a particular frequency show the transmission loss through the block is 0.35 dB, and the transmission phase change is 0.3 rad. What is the propagation constant of the material?

Q3.4 A microstrip line having a characteristic impedance of 70 Ω is fabricated on a 0.635 mm thick alumina substrate ($\varepsilon_r = 9.8$). Determine the dielectric loss (in dB/m) in the line at 8.5 GHz, given that the loss tangent of the alumina is 0.0008.

Q3.5 The following data apply to a 50 Ω microstrip line fabricated on an LTCC substrate that has a relative permittivity of 9.8:
Substrate: thickness = 1.5 mm
Conductor: copper ($\sigma = 5.76 \times 10^7$ S/m)
         RMS surface roughness = 0.31 µm
If the loss of a 50 mm length of the line at 2.4 GHz is 0.120 dB, determine the loss tangent of the dielectric.

Q3.6 A 50 Ω microstrip line designed for use at 94 GHz has the following parameters:
Substrate: $\varepsilon_r = 9.8$
         $\tan \delta = 0.0005$
         $h = 0.2$ mm

Conductor: Gold ($\sigma = 4.45 \times 10^7$ S/m)

RMS surface roughness = 0.605 μm

Determine the total line loss in dB/mm.

Q3.7 A 12 GHz resonant cavity supporting the TE_{101} mode is made from a length of rectangular copper waveguide having internal dimensions of $a = 22.86$ mm and $b = 10.16$ mm. Determine the unloaded Q of the cavity ($\sigma_{copper} = 5.76 \times 10^7$ S/m).

Q3.8 A 40 GHz resonant cavity is made from a length of circular copper waveguide having an internal diameter of 50 mm. The cavity supports the TE_{013} mode. Determine:
  (i) The required length of the unloaded cavity
  (ii) The unloaded Q of the cavity.
      ($\sigma_{copper} = 5.76 \times 10^7$ S/m).

Q3.9 An aperture-coupled resonant cavity is to be constructed from a length of short-circuited X-band rectangular waveguide (as shown in Figure 3.18), that has internal dimensions of $a = 22.86$ mm and $b = 10.16$ mm, and is to support the dominant TE_{101} mode. The coupling iris is a centred circular hole in a transverse plate.
Determine:
  (i) The required radius of the aperture if the cavity is 19 mm long, and is to resonate at 10 GHz
  (ii) The required length of the cavity if the radius of the aperture is 3 mm, and the resonance is to occur at 12 GHz.
(Use a Smith chart to confirm the answers.)

Q3.10 The following data were obtained when the loss tangent of a low-loss dielectric having a relative permittivity of 6 was measured using a resonant cavity perturbation method:
Resonant frequency of unloaded cavity = 9.6 GHz
Q-factor of the unloaded cavity = 3500
Change in resonant frequency when specimen loaded = 3%
Change in Q-factor when specimen loaded = 4.2%
Determine:
  (i) The resonant frequency of the loaded cavity
  (ii) The Q-factor of the loaded cavity
  (iii) The loss tangent and Q of the dielectric

Q3.11 A 60 GHz high-Q cavity is to be made from a length of circular waveguide supporting the TE_{011} mode. The waveguide has a diameter of 25 mm. One end of the cavity is terminated by a short circuit and the other by a transverse plate with a small coupling iris. Determine the required length of the cavity if the iris has a normalized susceptance of $-j5.2$.

Q3.12 The following data apply to a 50 Ω microstrip line used in an antenna feed system:

Substrate:	$\varepsilon_r = 9.8$
	$h = 0.3$ mm
	$\tan \delta = 0.0015$
Conductor:	Copper ($\sigma = 5.87 \times 10^7$ S/m)

If the system specification requires that the loss in the microstrip feed line should not exceed 0.08 dB/mm at 25 GHz, determine the maximum acceptable RMS surface roughness of the conductor.

Q3.13 Draw a graph showing how the total loss (in dB/mm) of the microstrip line specified in Q3.12 varies as a function of Δ/δ for $0.5 \leq \Delta/\delta \leq 2$.

Q3.14 Repeat the graph in Q3.13, but for a line of 70 Ω characteristic impedance. Comment upon the result.

References

1 Harrop, P.J. (1972). *Dielectrics*. London: Butterworth.

2 Hammerstad, E.O. and Bekkadal, F. (1975). *A Microstrip Handbook*, ELAB Report, STF44 A74169. University of Trondheim Norway.

3 Sain, A. and Melde, K.L. (2013). Broadband characterization of coplanar waveguide interconnects with rough conductor surfaces. *IEEE Transactions on Components, Packaging and Manufacturing Technology* 3 (6): 1038–1046.

4 Iwai, T. and Mizatani, D. (2015). Motoaki Tani measurement of high-frequency conductivity affected by conductor surface roughness using dielectric rod resonator method. *Proceedings of IEEE International Symposium on Electromagnetic Compatibility*, Dresden, Germany (Auguest 2015), pp. 634–639.

5 Gold, G. and Helmreich, K. (2015). Surface impedance concept for modelling conductor roughness. *Proceedings of IEEE International Microwave Symposium*, Phoenix, AZ (May 2015).

6 Cruickshank, D.B. (2011). *Microwave Materials for Wireless Applications*. Norwood, MA: Artech House.

7 Pitt, K.E.G. (ed.) (2005). *Handbook of Thick Film Technology*. Port Erin, Isle of Man, UK: Electrochemical Publications Ltd.

8 Tian, Z., Free, C.E., Aitchison, C., Barnwell, P., and Wood, J. (2002). Multilayer thick-film microwave components and measurements. *Proceedings of 35th International Symposium on Microelectronics*, Denver, CO (4–6 September 2002).

9 Tian, Z., Free, C.E., Barnwell, P., Wood, J., and Aitchison, C. (2001). Design of novel multilayer microwave coupled line structures using thick-film technology. *Proceedings of 31st European Microwave Conference*, London (24–26 September 2001).

10 Osman, N. and Free, C.E. (2014). Miniature rectangular ring band-pass filter with embedded barium strontium titanate capacitors. *Proceedings of Asia Pacific Microwave Conference*, Sendai, Japan (November 2014), pp. 306–308.

11 Chen, X.-P., Wu, K. et al. (2014). *IEEE Microwave Magazine* 15 (5): 108–116.

12 Chen, X.-P. and Wu, K. (2014). Substrate integrated waveguide filters: design techniques and structure innovations. *IEEE Microwave Magazine* 15 (6): 121–133.

13 Chen, X.-P. and Wu, K. (2014). Substrate integrated waveguide filters: practical aspects and design considerations. *IEEE Microwave Magazine* 15 (7): 75–83.

14 Schorer, J., Bornemann, J., and Rosenberg, U. (2014). Comparison of surface mounted high quality filters for combination of substrate integrated and waveguide technology. *Proceedings of 2014 Asia Pacific Microwave Conference*, Sendai, Japan (4–7 November 2014), pp. 929–931.

15 Gaynor, M.P. (2007). *System-in-Package: RF Design and Applications*. Norwood, MA: Artech House.

16 Henry, M., Osman, N., Tick, T., and Free, C.E. (2008). Integrated air-filled waveguide Antennas in LTCC for G-band operation. *Proceedings of Asia Pacific Microwave Conference*, Hong Kong (December 2008).

17 Belhaj, M.M., Wei, W., Palleecchi, E., Mismer, C., Roch-jeune, I., and Happy, H. (2014). Inkjet printed flexible transmission lines for high frequency applications up to 67 GHz. *Proceedings of 9th European Microwave Integrated Circuit Conference*, Rome, Italy (6–7 October 2014), pp. 584–587.

18 Kim, S., Shamim, A., Georgiadis, A. et al. (2016). Fabrication of fully inkjet-printed Vias and SIW structures on thick polymer substrates. *IEEE Transactions on Components, Packaging and Manufacturing Technology* 6 (3): 486–496.

19 Nikfalazar, M., Zheng, Y., Wiens, A., Jakoby, R., Friederich, A., Kohler, C., and Binder, J.R. (2014). *Proceedings of 44th European Microwave Conference*, Rome, Italy (6–9 October 2014), pp. 504–507.

20 Chen, M.Y., Pham, D., Subbaraman, H. et al. (2012). Conformal ink-jet printed C-band phased-Array antenna incorporating carbon nanotube field-effect transistor based reconfigurable true-time delay lines. *IEEE Transactions on Microwave Theory and Techniques* 60 (1): 179–184.

21 Waldron, R.A. (1969). *Theory of Guided Electromagnetic Waves*. London: Van Nostrand Reinhold.

22 Collin, R.E. (1992). *Foundations for Microwave Engineering*. New York: McGraw-Hill.

23 Karbowiak, A.E. (1965). *Trunk Waveguide Communication*. London: Chapman and Hall.

24 Orloff, N.D., Obrzut, J., Long, C.J. et al. (2014). Dielectric characterization by microwave cavity perturbation corrected for nonuniform fields. *IEEE Transactions on Microwave Theory and Techniques* 62 (9): 2149–2159.

25 Krupta, J.A., Geyer, R.G., Baker-Jarvis, J., and Ceremuga, J. (1996). Measurements of the complex permittivity of microwave circuit board substrates using split dielectric resonator and re-entrant cavity techniques. *Proceedings of 7th International Conference on Dielectric Materials, Measurements and Applications*, Bath, UK (23–26 September 1996), pp. 21–24.

26 Krupta, J., Clarke, R.N., Rochard, O.C., and Gregory, A.P. (2000). Split post dielectric resonator technique for precise measurements of laminar dielectric specimens – measurement uncertainties. *Proceedings of 13th International Conference on Microwaves, Radar and Wireless Communications*, Wroclaw, Poland (22–24 May 2000), pp. 305–308.

27 Dziurdzia, B., Krupta, J., and Gregorczyk, W. (2006). Characterization of thick-film dielectric at microwave frequencies. *Proceedings of 16th International Conference on Microwaves, Radar and Wireless Communications*, Krakow, Poland (22–24 May 2006), pp. 361–364.

28 Cullen, A.L. and Yu, P.K. (1971). The accurate measurement of permittivity by means of an open resonator. *Proceedings of the Royal Society of London* A325: 493–509.

29 Cullen, A.L., Nagenthiram, P., and Williams, A.D. (1972). Improvement in open resonator permittivity measurement. *Electronics Letters* 8 (23): 577–579.

30 Komiyama, B., Kiyokawa, M., and Matsui, T. (1991). Open resonator for precision measurements in the 100 GHz band. *IEEE Transactions on Microwave Theory and Techniques* 39 (10): 1792–1796.

31 Hirvonen, T.M., Vainikainen, P., Lozowski, A., and Raisanen, A.V. (1996). Measurement of dielectrics at 100 GHz with an open resonator connected to a network analyzer. *IEEE Transactions on Instrumentation and Measurement* 45 (4): 780–786.

32 Afsar, M.N., Ding, H., and Tourshan, K. (1999). A new open resonator technique at 60 GHz for permittivity and loss tangent measurement of low-loss materials. *Proceedings of 1999 IEEE International Microwave Symposium*, Anaheim, CA (13–19 June 1999), pp. 1755–1758.

33 Afsar, M.N., Moonshiram, A., and Wang, Y. (2004). Assessment of random and systematic errors in millimeter-wave dielectric measurement using open resonator and Fourier transform spectroscopy systems. *IEEE Transactions on Instrumentation and Measurement* 53 (4): 899–906.

34 Ghodgaonkar, D.K., Varadan, V.V., and Varadan, V.K. (1989). A free-space method for measurement of dielectric constants and loss tangents at microwave frequencies. *IEEE Transactions on Instrumentation and Measurement* 37 (3): 789–793.

35 Osman, N., Leigh, R., and Free, C.E. (2009). Characterization of LTCC material at G-band. *Proceedings of 42nd International Symposium of Microelectronics*, San Jose, CA (1–5 November 2009), pp. 260–267.

36 Troughton, P. (1968). High Q-factor resonators in microstrip. *Electronics Letters* 4 (24): 520–522.

37 Yu, C.-C. and Chang, K. (1997). Transmission-line analysis of a capacitively coupled microstrip ring resonator. *IEEE Transactions on Microwave Theory and Techniques* 45 (11): 2018–2024.

38 Bray, J.R. and Roy, L. (2003). Microwave characterization of a microstrip line using a two-port ring resonator with an improved lumped-element model. *IEEE Transactions on Microwave Theory and Techniques* 51 (5): 1540–1547.

39 Owens, R.P. (1976). Curvature effect in microstrip ring resonators. *Electronics Letters* 12 (14): 356–357.

40 Faria, J.A.B. (2009). A novel approach to ring resonator theory involving even and odd mode analysis. *IEEE Transactions on Microwave Theory and Techniques* 57 (4): 856–862.

41 Hopkins, R. and Free, C.E. (2008). Ultra-wideband slotline dispersion measurements using ring resonator. *Electronics Letters* 44 (21): 1262–1264.

42 Benarabi, B., Bayard, B., Kahlouche, F. et al. (2017). Asymmetric coplanar ring resonator (ACPW) for microwave characterization of silver composite conductors. *IEEE Transactions on Microwave Theory and Techniques* 65 (6): 2139–2144.

43 Ghione, G. and Goano, M. (1997). The influence of ground plane width on the ohmic losses of coplanar waveguides with finite lateral ground planes. *IEEE Transactions on Microwave Theory and Techniques* 45 (9): 1640–1642.

4

Planar Circuit Design II

Refinements to Basic Designs

4.1 Introduction

At RF and microwave frequencies there are many circuit components and interconnections that exhibit significant non-ideal electrical performance. In particular, discontinuities in conducting tracks cause radiation and unwanted reactance. Similarly, the packaging of active and passive components usually introduces unwanted resistance and reactance at high frequencies. It is important for the circuit designer to appreciate how high-frequency effects may influence the performance of circuit designs, and how these designs may need to be refined in order to produce good performance. Whilst there is a wealth of CAD software, including electromagnetic simulation, now available to aid the high-frequency circuit designer, it is useful for the designer to be able to identify potential sources of non-ideal performance at the prototype design stage. Thus, the purpose of this chapter is to introduce some of the key areas where non-ideal performance occurs, and where possible quantify the effects through the use of worked examples.

Initially, the occurrence and RF consequences of some common microstrip discontinuities will be discussed, and information presented on typical compensation techniques. Since most practical RF circuits are encased in some form of protective package the effect of metallic enclosures on circuit performance will be discussed, and recommendations given on the optimum size of enclosure that should be used. The chapter concludes with a discussion of the non-ideal behaviour of some common passive components at RF and microwave frequencies.

4.2 Discontinuities in Microstrip

Whenever there is a sudden change in the geometry of a microstrip line, there will be fringing electric and magnetic fields that will influence the behaviour of the circuit. Microstrip discontinuities have been the subject of extensive research, and there is a wealth of literature providing design equations and guidance on the use and accuracy of these equations over different frequency ranges. Most RF and microwave CAD software contains a library of typical microstrip discontinuities that can be applied to practical designs, but it is useful for the circuit designer to have an appreciation of the significance of the discontinuities, and so we will discuss some of the significant discontinuities and provide supporting performance and dimensional data. For readers requiring more information, there are two excellent specialist textbooks [1, 2] that provide in-depth discussions on the theoretical analysis of microstrip discontinuities, with detailed information on the accuracy of commonly used design equations.

4.2.1 Open-End Effect

One of the most common discontinuities is the microstrip open-end. The typical electric field distribution around the end of an open microstrip line is depicted in Figure 4.1.

The fringing field at the open end of the microstrip line can be represented as a capacitance, C_f, connected between the line and the ground plane, as shown in the transmission line model of an open-end microstrip in Figure 4.2a.

This capacitance will have the effect of making the electrical length of the line appear longer than the physical length, and can be represented by an additional length, l_{eo}, as shown in Figure 4.2b. This additional length is normally small, of

RF and Microwave Circuit Design: Theory and Applications, First Edition. Charles E. Free and Colin S. Aitchison.
© 2022 John Wiley & Sons Ltd. Published 2022 by John Wiley & Sons Ltd.
Companion website: www.wiley.com/go/free/rfandmicrowave

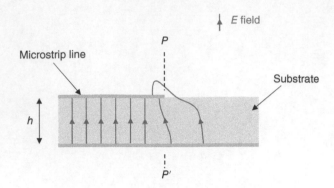

Figure 4.1 Electric field surrounding a microstrip open-end discontinuity.

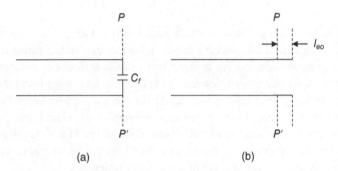

Figure 4.2 Transmission line representation of open-end effect.

the order of a few hundred micrometres, but it can have a very significant effect in some situations, such as in open-ended stubs in matching networks. The most common expression quoted in the literature for calculating l_{eo} is given by

$$l_{eo} = 0.412h \frac{\varepsilon_{r,eff}^{MSTRIP} + 0.3}{\varepsilon_{r,eff}^{MSTRIP} - 0.258} \left[\frac{w/h + 0.262}{w/h + 0.813} \right], \tag{4.1}$$

where $\varepsilon_{r,eff}^{MSTRIP}$ is the effective relative permittivity of the microstrip line whose width is w, and h is the thickness of the substrate. This expression was originally developed by Hammerstad and Bekkadal [3] and has been found to give an accurate representation of microstrip open-end effects at frequencies up to 20 GHz. Thus, in practical microstrip designs, any line with an open end, such as a stub or resonating patch, should be shortened by an amount given by l_{eo} to compensate for the fringing effect. In the case of a resonating patch, as will be seen in Chapter 12, there are two open ends, and so the patch length should be shortened by $2 \times l_{eo}$.

Example 4.1 An 8 GHz open-circuited microstrip stub is to be designed to give an input reactance of $-j16\,\Omega$. The stub is to have a characteristic impedance of 50 Ω and is to be fabricated on a substrate that has a relative permittivity of 9.8 and a thickness of 0.6 mm. Determine the required length of the stub, applying the necessary compensation for the open-end effect.

Solution
The input reactance of an open-circuited stub is given by Eq. (1.34) as

$$Z_{in} = -jZ_o \cot \beta l,$$

where Z_o is the characteristic impedance of the stub, l is the length of the stub, $\beta = \dfrac{2\pi}{\lambda_S}$ and λ_S is the substrate wavelength.

Using CAD or the microstrip design graphs given in Chapter 2:

$$Z_o = 50\,\Omega \quad \Rightarrow \quad \frac{w}{h} = 0.9 \quad \Rightarrow \quad \varepsilon_{r,eff} = 6.6.$$

Now

$$f = 8\,\text{GHz} \quad \Rightarrow \quad \lambda_o = 37.5\,\text{mm}.$$

Thus,

$$\lambda_S = \frac{37.5}{\sqrt{6.6}} \text{ mm} = 14.60 \text{ mm}.$$

Substituting data into the expression for the input reactance of the stub gives

$$-j16 = -j50 \cot\left(\frac{2\pi}{14.60} \times l\right)$$

and

$$l = 2.93 \text{ mm}.$$

We must shorten the stub to compensate for the open-end effect. Using Eq. (4.1) we have:

$$l_{eo} = 0.412 \times 0.6 \times \frac{6.6 + 0.3}{6.6 - 0.258} \times \frac{0.9 + 0.262}{0.9 + 0.813} \text{ mm}$$

$$= 0.182 \text{ mm}.$$

Thus, the required length of the stub is $(2.93 - 0.182)$ mm $= 2.748$ mm.

Example 4.2 Determine the percentage error in the magnitude of the input reactance of the stub designed in Example 4.1 if no compensation is applied for the open-end effect. Comment upon the result.

Solution
Without open-end compensation the effective length of the stub will be $(2.930 + 0.182)$ mm $= 3.112$ mm. The input reactance is then

$$Z_{in} = -j\,50 \cot\left(\frac{2\pi}{14.597} \times 3.112\right)\Omega$$

$$= -j\,50 \cot(76.75°)\,\Omega$$

$$= -j11.773\,\Omega.$$

Percentage error in the magnitude of the input reactance is then

$$\%\text{error} = \left|\frac{-16 - (-11.773)}{-16}\right| \times 100\% = 26.42\%.$$

Comment: Although the value of l_{eo} is small (i.e. 0.182 mm) it can have a significant effect on the input reactance of a stub if the nominal stub length is close to a quarter wavelength, since the argument of the cotangent will be close to 90° where small changes in the argument have significant effect on the value of the function.

4.2.2 Step-Width

Symmetrical steps in microstrip often occur in practical designs, notably in filters and quarter-wave transformers. A typical symmetrical microstrip step is shown in Figure 4.3, at the junction of a low-impedance line (Z_{o1}) with high impedance line (Z_{o2}).

A number of authors have investigated the behaviour of the electromagnetic field in the vicinity of the step, and a good discussion is given in Fooks and Zakarevicius [4]. The development of circuit models to represent the behaviour of the step are based on the physical arguments that the narrowing of the conductor in the vicinity of the step will introduce excess series inductance, and fringing at the step discontinuity will introduce shunt capacitance. Based on these arguments, a

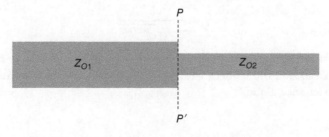

Figure 4.3 Symmetrical step in microstrip.

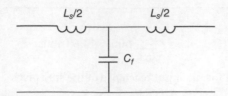

Figure 4.4 T-network representing a microstrip step discontinuity.

T-network consisting of two series inductors and a shunt capacitor, as shown in Figure 4.4, is usually used to represent the step discontinuity, where C_f represents the fringing capacitance and L_S represents the excess inductance.

In order to compensate for the fringing capacitance, Hammerstad and Bekkadal [3] proposed that the wide line should be shortened by δl_C as defined in Eq. (4.2)

$$\delta l_C = \left(1 - \frac{w_2}{w_1}\right) l_{eo}, \tag{4.2}$$

where w_2 is the width of the narrow line, w_1 is the width of the wide line, and l_{eo} represents the open-end effect of the wide line. This is a reasonable assumption, since as w_2 approaches zero δl_C will approach the value for an open-circuited line, and as w_2 approaches w_1 the fringing capacitance will become insignificant and δl_C will approach zero. Garg et al. [2] developed the following expression to represent the excess inductance, L_S, at the discontinuity

$$\frac{L_S}{h} = 40.5\left(\frac{w_1}{w_2} - 1\right) - 75 \log\left(\frac{w_1}{w_2}\right) + 0.2\left(\frac{w_1}{w_2} - 1\right)^2 \text{ nH/m}, \tag{4.3}$$

where h is the thickness of the substrate in metres. Normal practice is then to shorten the length of the narrow line by δl_L to compensate for the series inductance. The length δl_L is related the excess inductance by

$$L_S = \delta l_L \times L, \tag{4.4}$$

where L is the inductance per unit length of the narrow line in nH/m. The value of L can be determined from the properties of the narrow microstrip lines using Eqs. (1.27) and (1.29). Rewriting Eq. (1.27) we have

$$Z_{o2} = \sqrt{\frac{L}{C}}, \tag{4.5}$$

where Z_{o2} is the characteristic impedance of the narrow microstrip line whose width is w_2, and L and C are the inductance and capacitance per unit length respectively, of this line. Equation (1.29) gives

$$\beta = \frac{2\pi}{\lambda_{S2}} = \frac{2\pi f}{v_{p2}} = 2\pi f \sqrt{LC}, \tag{4.6}$$

where λ_{S2} and v_{p2} are the wavelength and velocity of propagation, respectively, on the narrow line. Equation (4.6) can be rewritten as

$$v_{p2} = \frac{c}{\sqrt{\varepsilon_{r,eff2}^{MSTRIP}}} = \frac{1}{\sqrt{LC}}, \tag{4.7}$$

where $\varepsilon_{r,eff2}^{MSTRIP}$ is the effective relative permittivity of the narrow line.

Combining Eqs. (4.5) and (4.7) gives

$$L = \frac{Z_{o2}\sqrt{\varepsilon_{r,eff2}^{MSTRIP}}}{c}. \tag{4.8}$$

Substituting from Eq. (4.8) into Eq. (4.4) gives

$$\delta l_L = \frac{cL_S}{Z_{o2}\sqrt{\varepsilon_{r,eff2}^{MSTRIP}}}. \tag{4.9}$$

It is worth noting that for most of the microstrip steps encountered in practical designs the inductance associated with the step discontinuity is very much a second-order effect, with $\delta l_L \ll \delta l_C$, and so the step discontinuity can often be adequately represented by the fringing capacitance alone.

Example 4.3 Determine the compensation needed to account for the discontinuity at the junction between a 50 Ω microstrip line and a 70 Ω microstrip line, when the lines are fabricated on a substrate that has a relative permittivity of 9.8 and a thickness of 0.8 mm. Assume that the junction is symmetrical.

Solution
Using the microstrip design graphs given in Chapter 2 (or using CAD):

50 Ω line:

$$\frac{w_1}{h} = 0.9 \quad \Rightarrow \quad w_1 = 0.9 \times 0.8 \, \text{mm} = 0.72 \, \text{mm},$$

$$\frac{w_1}{h} = 0.9 \quad \Rightarrow \quad \varepsilon_{r,eff1} = 6.6.$$

70 Ω line:

$$\frac{w_2}{h} = 0.35 \quad \Rightarrow \quad w_2 = 0.35 \times 0.8 \, \text{mm} = 0.280 \, \text{mm},$$

$$\frac{w_2}{h} = 0.35 \quad \Rightarrow \quad \varepsilon_{r,eff1} = 6.25.$$

(1) Compensation for fringing capacitance
Using Eqs. (4.2) and (4.1):

$$\delta l_C = \left(1 - \frac{w_2}{w_1}\right) 0.412 h \frac{\varepsilon_{r,eff1} + 0.3}{\varepsilon_{r,eff1} - 0.258} \left[\frac{w_1/h + 0.262}{w_1/h + 0.813}\right]$$

$$= \left(1 - \frac{0.280}{0.720}\right) \times 0.412 \times 0.8 \times 10^{-3} \times \frac{6.6 + 0.3}{66 - 0.258} \times \frac{0.9 + 0.262}{0.9 + 0.813} \, \text{m}$$

$$= 1.487 \times 10^{-4} \, \text{m} = 148.70 \, \mu\text{m}.$$

(2) Compensation for series inductance
Using Eq. (4.3):

$$\frac{L_S}{h} = 40.5 \left(\frac{w_1}{w_2} - 1.0\right) - 75 \log \left(\frac{w_1}{w_2}\right) + 0.2 \left(\frac{w_1}{w_2} - 1.0\right)^2 \text{nH/m}$$

$$= 40.5 \times \left(\frac{0.720}{0.280} - 1\right) - 75 \log \left(\frac{0.720}{0.280}\right) + 0.2 \times \left(\frac{0.720}{0.280} - 1\right)^2 \text{nH/m}$$

$$= 33.374 \, \text{nH/m},$$

i.e.

$$L_S = 33.374 h \, \text{nH} = 33.374 \times 0.8 \times 10^{-3} \, \text{nH} = 0.027 \, \text{nH}.$$

Using Eq. (4.9):

$$\delta l_L = \frac{c L_S}{Z_{o2} \sqrt{\varepsilon_{r,eff2}}} = \frac{3 \times 10^8 \times 0.027 \times 10^{-9}}{70 \times \sqrt{6.35}} \, \text{m} = 45.92 \, \mu\text{m}.$$

Example 4.4 Apply microstrip step compensation to the 3-step $\lambda/4$ transformer designed in Example 2.5.

Solution
Figure 4.5 shows the design from Example 2.5, with the three discontinuities identified. Note that a step discontinuity has been included between the first step and the 50 Ω feed line, as this will affect the length of the fist step.

(Continued)

(Continued)

Summary of line data from Example 2.5:

$Z_n(\Omega)$	w/h	w(mm)	$\epsilon_{r,eff}^{MSTRIP}$	λ_s(mm)	$\lambda_s/4$(mm)
45.85	1.10	0.88	6.70	23.18	5.80
35.36	1.85	1.48	7.04	22.61	5.65
27.27	2.90	2.32	7.37	22.10	5.53

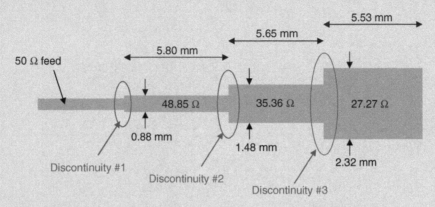

Figure 4.5 Design taken from Example 2.5.

In order to determine the compensation for the first discontinuity we also need the width of the 50 Ω feed line:

$$Z_o = 50\,\Omega \quad \Rightarrow \quad \frac{w}{h} = 0.9 \quad \Rightarrow \quad w = 0.9h = 0.9 \times 0.8\,\text{mm} = 0.72\,\text{mm}$$

Discontinuity 1:
Using eqns. (4.2) and (4.1):

$$\delta l_{c1} = \left(1 - \frac{0.72}{0.88}\right) \times 0.412 \times 0.8 \times \left(\frac{6.70 + 0.3}{6.70 - 0.258}\right) \times \left(\frac{1.10 + 0.262}{1.10 + 0.813}\right)\,\text{mm}$$
$$= 0.046\,\text{mm} \simeq 0.05\,\text{mm}$$

Note that we do not need to calculate the series inductance for discontinuity #1, since this will be absorbed into the 50 Ω feed line.

Discontinuity 2:
Using eqns. (4.2) and (4.1):

$$\delta l_{c2} = \left(1 - \frac{0.88}{1.48}\right) \times 0.412 \times 0.8 \times \left(\frac{7.04 + 0.3}{7.04 - 0.258}\right) \times \left(\frac{1.85 + 0.262}{1.85 + 0.813}\right)\,\text{mm}$$
$$= 0.115\,\text{mm} \simeq 0.12\,\text{mm}$$

Using Eq. (4.3):

$$\frac{L_{S2}}{h} = 40.5 \times \left(\frac{1.48}{0.88} - 1.0\right) - 75\log\left(\frac{1.48}{0.88}\right) + 0.2 \times \left(\frac{1.48}{0.88} - 1.0\right)^2\,\text{nH/m}$$
$$= 10.77\,\text{nH/m},$$

i.e.

$$L_{S2} = 10.77h \text{ nH/m} = 10.77 \times 0.8 \times 10^{-3} \text{ nH} = 0.0086 \text{ nH}.$$

Using Eq. (4.9):

$$\delta l_{L2} = \frac{3 \times 10^8 \times 0.0086 \times 10^{-9}}{35.36 \times \sqrt{7.04}} \text{ m} = 27.50 \text{ μm}.$$

Discontinuity 3:

Using Eqs. (4.2) and (4.1):

$$\delta l_{C3} = \left(1 - \frac{1.48}{2.32}\right) \times 0.412 \times 0.8 \times \left(\frac{7.37 + 0.3}{7.37 - 0.258}\right) \times \left(\frac{2.90 + 0.262}{2.90 + 0.813}\right) \text{ mm}$$
$$= 0.110 \text{ mm}.$$

Using Eq. (4.3):

$$\frac{L_{S3}}{h} = 40.5 \times \left(\frac{2.32}{1.48} - 1.0\right) - 75 \log \left(\frac{2.32}{1.48}\right) + 0.2 \times \left(\frac{2.32}{1.48} - 1.0\right)^2 \text{ nH/m}$$
$$= 8.41 \text{ nH/m},$$

i.e.

$$L_{S3} = 8.41h \text{ nH} = 8.41 \times 0.8 \times 10^{-3} \text{ nH} = 0.00673 \text{ nH}.$$

Using Eq. (4.9):

$$\delta l_{L3} = \frac{3 \times 10^8 \times 0.00673 \times 10^{-9}}{27.27 \times \sqrt{7.37}} \text{ m} = 27.27 \text{ μm}.$$

Modified step lengths:

$$5.80 \quad \Rightarrow \quad 5.80 - \delta l_{C1} - \delta l_{L2} = (5.80 - 0.046 - 0.028) \text{ mm} = 5.73 \text{ mm},$$
$$5.65 \quad \Rightarrow \quad 5.65 - \delta l_{C2} - \delta l_{L3} = (5.65 - 0.12 - 0.027) \text{ mm} = 5.50 \text{ mm},$$
$$5.53 \quad \Rightarrow \quad 5.53 - \delta l_{C3} = (5.53 - 0.11) \text{ mm} = 5.42 \text{ mm}.$$

Final design (Figure 4.6):

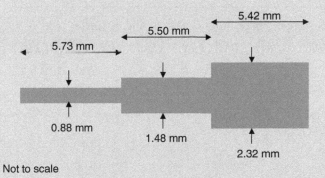

Not to scale

Figure 4.6 Final design for Example 4.3.

4.2.3 Corners

Corners will occur in most practical microstrip circuit designs. A right-angled bend in microstrip will cause significant fringing of the electromagnetic field, leading to excess shunt capacitance at the corner. Forming the corner from a gradual

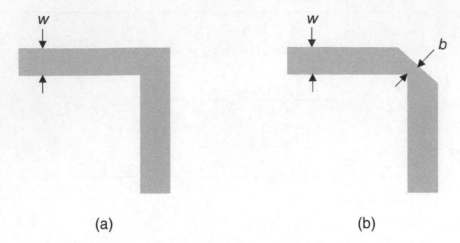

Figure 4.7 Right-angled bends in microstrip: (a) 90° corner and (b) compensated 90° corner.

bend in the microstrip line will avoid the fringing, but will occupy significant substrate space. For compact designs it is better to use right-angled corners and to apply the appropriate compensation.

Various methods have been investigated for compensating for the excess capacitance at a microstrip corner. The simplest and most popular method is to chamfer the corner as shown in Figure 4.7. Chamfering has the effect of decreasing the excess shunt capacitance and increasing the series inductance, thus maintaining the required L/C ratio, and hence the match, around the corner. The amount of chamfering has been the subject of investigations by a number of researchers, and Easter and colleagues [5] suggest an optimum degree of chamfering given by

$$1 - \frac{b}{\sqrt{2w}} = 0.6, \tag{4.10}$$

where the dimensional parameters are shown in Figure 4.7.

Edwards and Steer [1] have presented the results of several electromagnetic simulations that support the use of Eq. (4.10), although they caution that the expression should be used only where the permittivity of the substrate approximates that of alumina, i.e. $\varepsilon_r \approx 9.8$.

Example 4.5 Determine the amount of chamfering that should be applied to a right-angled bend in a 50 Ω microstrip line that is fabricated on a substrate that has a thickness of 0.8 mm and a relative permittivity of 9.8.

Solution
Using microstrip design graphs (or CAD):

$$50\,\Omega \quad \Rightarrow \quad \frac{w}{h} = 0.9 \quad \Rightarrow \quad w = 0.9h = 0.9 \times 0.8 \text{ mm} = 0.72 \text{ mm}.$$

Using Eq. (4.10):

$$1 - \frac{b}{\sqrt{2w}} = 0.6 \quad \Rightarrow \quad b = 0.41 \text{ mm}.$$

Final design (Figure 4.8):

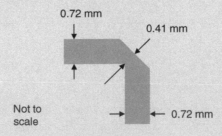

Figure 4.8 Final design for Example 4.4.

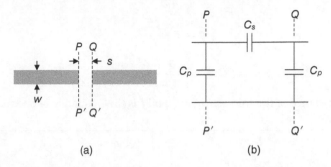

Figure 4.9 Microstrip gap and its equivalent circuit: (a) microstrip gap and (b) equivalent circuit.

4.2.4 Gaps

High-frequency active and passive chip components are easily mounted across gaps in a microstrip line. However, it is important to understand the behaviour of the gaps in order that they do not have an adverse effect on the performance of the surface-mounted component. Microstrip gaps are also important where they provide coupling to resonant structures such as the resonant ring, which was discussed in Chapter 3.

A microstrip gap is normally modelled as a π-network consisting of three capacitors, as shown in Figure 4.9.

The values of the capacitors can be expressed in terms of the odd and even mode capacitances, C_{odd} and C_{even}, such that

$$C_{odd} = 2C_s + C_p \tag{4.11}$$

and

$$C_{even} = 2C_p. \tag{4.12}$$

Garg and Bahl [6] developed approximate closed-form expressions to evaluate the values of C_{odd} and C_{even} for a gap spacing of s in a microstrip line of width w on a substrate of thickness h as follows:

$$\frac{C_{odd}}{w} = \left[\frac{s}{w}\right]^{m_o} e^{k_o} \text{ pF/m}, \tag{4.13}$$

$$\frac{C_{even}}{w} = 12\left[\frac{s}{w}\right]^{m_e} e^{k_e} \text{ pF/m}, \tag{4.14}$$

where k and m are given below for both odd and even modes.

For $\quad 0.1 \le \dfrac{s}{w} \le 1.0 \qquad m_o = \dfrac{w}{h}\left[0.619\log\left(\dfrac{w}{h}\right) - 0.3853\right]$

$$k_o = 4.26 - 1.453\log\left(\frac{w}{h}\right). \tag{4.15}$$

For $\quad 0.1 \le \dfrac{s}{w} \le 0.3 \qquad m_e = 0.8675,$

$$k_o = 2.043\left(\frac{w}{h}\right)^{0.12}. \tag{4.16}$$

For $\quad 0.3 \le \dfrac{s}{w} \le 1.0 \qquad m_e = 1.565\left(\dfrac{w}{h}\right)^{-0.16} - 1,$

$$k_o = 1.97 - 0.03\left(\frac{h}{w}\right). \tag{4.17}$$

The expressions given above for the odd and even mode capacitances were originally developed for $\varepsilon_r = 9.6$, but Garg and Bahl [6] provided scaling factors for other values of ε_r in the range $2.5 \le \varepsilon_r \le 15$ in the form

$$C_{odd}(\varepsilon_r) = C_{odd}(9.6) \times \left(\frac{\varepsilon_r}{9.6}\right)^{0.8}. \tag{4.18}$$

$$C_{even}(\varepsilon_r) = C_{even}(9.6) \times \left(\frac{\varepsilon_r}{9.6}\right)^{0.9}. \tag{4.19}$$

Once the odd and even mode capacitances are found for a particular gap, the insertion loss of the gap, IL_{dB}, can be determined from straightforward network analysis, as shown in Appendix 4.A, and is given by

$$IL_{dB} = 20\log|0.5(a + jb)|, \tag{4.20}$$

where

$$a = 2 + \frac{2C_p}{C_s}, \tag{4.21}$$

$$b = 4\pi f\, C_p Z_0 + \frac{2\pi f\, C_p^2 Z_0}{C_s} - \frac{1}{2\pi f\, C_s Z_0}, \tag{4.22}$$

and Z_0 is the characteristic impedance of the microstrip line, and f is the frequency of operation.

Example 4.6

(i) Determine the odd and even mode capacitances for a 600 μm wide gap in a 50 Ω microstrip line fabricated on a substrate that has a relative permittivity of 9.8, and a thickness of 0.8 mm.
(ii) Find the insertion loss of the gap at 10 GHz.

Solution

(i) Using microstrip design graphs (or CAD):

$$50\,\Omega \quad \Rightarrow \quad \frac{w}{h} = 0.9 \quad \Rightarrow \quad w = 0.9h = 0.9 \times 0.8 \text{ mm} = 0.72 \text{ mm}.$$

Relative gap size:

$$\frac{s}{w} = \frac{0.6}{0.72} = 0.833.$$

Using Eq. (4.15):

$$m_o = 0.9 \times (0.619 \times \log(0.9) - 0.3853) = -0.372.$$

$$k_o = 4.26 - 1.453 \times \log(0.9) = 4.326.$$

Using Eq. (4.17), and noting that s/w is within the range 0.3–1.0:

$$m_e = 1.565 \times (0.9)^{-0.16} - 1 = 0.592,$$

$$k_e = 1.97 - 0.03 \times (0.9)^{-1} = 1.937.$$

Substituting into Eqs. (4.13) and (4.14) to find the odd and even capacitances:

$$\frac{C_{odd}}{w} = (0.833)^{-0.372} e^{4.326} \text{ pF/m} = 80.96 \text{ pF/m}.$$

$$C_{odd} = 80.96 \times 0.72 \times 10^{-3} \text{ pF} = 58.29 \text{ fF}.$$

$$\frac{C_{even}}{w} = 12 \times (0.833)^{0.592} e^{1.937} \text{ pF/m} = 74.719 \text{ pF/m},$$

$$C_{even} = 74.719 \times 0.72 \times 10^{-3} \text{ pF} = 53.798 \text{ fF}.$$

We must now correct the values of odd and even mode capacitance for the value of substrate relative permittivity given in the problem, using Eqs. (4.18) and (4.19):

$$C_{odd}(9.8) = C_{odd}(9.6) \times \left(\frac{9.8}{9.6}\right)^{0.8} = 58.29 \times \left(\frac{9.8}{9.6}\right)^{0.8} \text{ fF} = 59.26 \text{ fF}.$$

$$C_{even}(9.8) = C_{even}(9.6) \times \left(\frac{9.8}{9.6}\right)^{0.9} = 53.798 \times \left(\frac{9.8}{9.6}\right)^{0.9} \text{ fF} = 54.806 \text{ fF}.$$

(ii) We can find the insertion loss using Eq. (4.20), but we must first find the values of C_p and C_s.
Using Eq. (4.12):

$$C_p = \frac{C_{even}}{2} = \frac{54.81}{2} \text{ fF} = 27.41 \text{ fF}.$$

Using Eq. (4.11):

$$C_s = \frac{C_{odd} - C_p}{2} = \frac{59.26 - 27.41}{2} \text{ fF} = 15.93 \text{ fF}.$$

Using Eq. (4.21):

$$a = 2 + \frac{2 \times 27.41}{15.93} = 5.44.$$

Using Eq. (4.22):

$$b = 2 \times 2\pi \times 10^{10} \times 27.41 \times 10^{-15} \times 50$$
$$+ \frac{2\pi \times 10^{10} \times (27.41 \times 10^{-15})^2 \times 50}{15.93 \times 10^{-15}}$$
$$- \frac{1}{2\pi \times 10^{10} \times 15.93 \times 10^{-15} \times 50}$$
$$= -19.66.$$

Substituting in Eq. (4.20) gives

$$(IL)_{\mathrm{dB}} = 20 \log |0.5(5.44 - j19.66)| \ \mathrm{dB} = 20.17 \ \mathrm{dB}.$$

Using the same procedure as that in Example 4.6, the insertion losses for various gap sizes were determined, and the results are plotted in Figure 4.10.

As would be expected, Figure 4.10 shows the insertion loss increasing as the gap size increases. However, it is useful to note that the insertion loss is only of the order of 15 dB when the gap is 300 μm, which is the order of gap size that is often used to surface mount beam-lead PIN diodes. The insertion loss of 15 dB indicates that in this case a significant fraction of the RF energy would bypass the diode. This could be quite serious for an on–off switching diode, where there would be significant leakage in the off state. The consequences of this are discussed in more detail in Chapter 10, which deals with the use of switching diodes in microstrip phase shifters.

The use of the three-capacitor model to represent a microstrip gap is well established in the literature, and provides reasonably accurate design data. However, more recent work by Alexopoulos and Wu [7] suggests that this model slightly underestimates the insertion loss of typical gaps at X-band by around 2 dB, when compared with a more complex model that includes resistors to represent surface and radiation losses.

4.2.5 T-Junctions

Symmetrical T-junctions occur in many practical microstrip circuits, for example, in corporate power dividers and combiners, in connections to branch line couplers, and in stub filters. The general form of a symmetrical microstrip T-junction is shown in Figure 4.11a, with its lumped-element microwave equivalent circuit in Figure 4.11b.

For the special case of a 50 Ω impedance main line (i.e. $Z_{o1} = 50 \ \Omega$) on a substrate having a relative permittivity of 9.9, Garg and colleagues [2] quoted the following closed-form approximate expressions to enable the reactances in the equivalent circuit to be calculated for w_1/h and w_2/h within the range 0.5–2.0:

$$\frac{C_j}{w_1} = \frac{100}{\tanh(0.0072Z_{o2})} + 0.64Z_{o2} - 261 \ \mathrm{pF/m}. \tag{4.23}$$

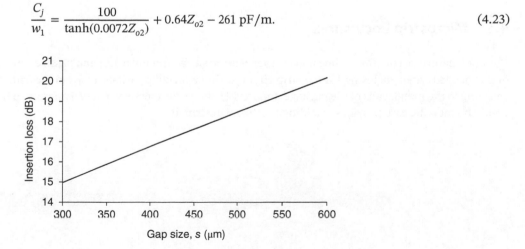

Figure 4.10 Variation of the insertion loss of a gap in a 50 Ω microstrip line as a function of the gap width, s ($\varepsilon_r = 9.8$, $h = 0.8$ mm).

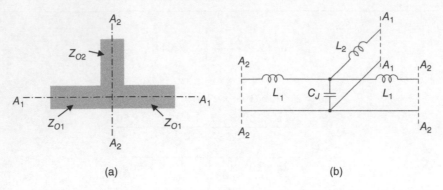

(a)   (b)

Figure 4.11 Symmetrical microstrip T-junction and its equivalent circuit: (a) microstrip T-junction and (b) equivalent circuit.

$$\frac{L_1}{h} = -\frac{w_2}{h}\left(\frac{w_2}{h}\left[-0.016\frac{w_1}{h} + 0.064\right] + \frac{0.016h}{w_1}\right)L_{w1}\ \text{nH/m.} \tag{4.24}$$

$$\frac{L_2}{h} = \left(\frac{xw_2}{h} + \frac{0.195w_1}{h} - 0.357 + 0.0283y\right)L_{w2}\ \text{nH/m,} \tag{4.25}$$

where

$$x = \frac{0.12w_1}{h} - 0.47,$$

$$y = \sin\left(\frac{\pi w_1}{h} - 0.75\pi\right),$$

L_{w1} = inductance per unit length of Z_{o1} microstrip line,
L_{w2} = inductance per unit length of Z_{o2} microstrip line.

The discontinuity reactances given by Eqs. (4.23) through (4.25) can be converted to equivalent line lengths and the results used to compensate the T-junction, but this is a somewhat involved procedure. Dydyk [8] proposed a relatively simple compensation arrangement for microstrip T-junctions using short matching sections of transmission line to achieve a minimum mismatch at the junction, leading to the general shape of junction shown in Figure 4.12a. However, the Dydyk technique leads to additional step discontinuities in the compensated T-junction, which can detract from the performance of the circuit. Chadha and Gupta [9] proposed a simple compensation technique, shown in Figure 4.12b, which involved removing a triangular section of microstrip track at the junction. No generalized design equations were given for this technique, but published data showed it gave good results up to frequencies around 8 GHz. More recently, Rastogi and colleagues [10] used Sonnet® software to optimize the performance of microstrip T-junctions compensated with the triangular cut-out and showed excellent results up to 10 GHz.

4.3 Microstrip Enclosures

The open nature of a microstrip line makes it very easy to surface mount active and passive components. However, some caution is needed when enclosing a microstrip circuit within a metallic package to avoid the fringing field from the microstrip coupling to the metal walls of the package. Figure 4.13 shows the cross-section of a microstrip line surrounded by a metallic case. The metallic case provides shielding for the microstrip line.

(a)   (b)

Figure 4.12 Compensated microstrip T-junctions: (a) Dydyk technique and (b) Notch compensation.

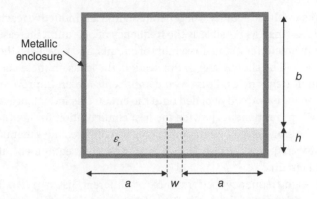

Figure 4.13   Cross-section of microstrip line surrounded by a metallic enclosure.

It is normally assumed that the enclosure will have little effect on the electromagnetic field surrounding the microstrip line if $a > 5w$ and $b > 5h$. If these conditions are not met, then the presence of the enclosure will affect both the characteristic impedance (Z_o) and the effective dielectric constant $(\varepsilon_{r,eff}^{MSTRIP})$ of the microstrip line. March [11] has provided design equations to enable the modified values of Z_o and $\varepsilon_{r,eff}$ to be calculated for an enclosure of given dimensions.

In addition to modifying the transmission properties of a microstrip line, the presence of a metallic enclosure can lead to undesirable cavity modes. These modes refer to the electromagnetic field patterns within the enclosure, which have similar properties to those in a resonant cavity. The occurrence of these modes is due to spurious radiation from a microstrip circuit, primarily from circuit discontinuities. These modes can be suppressed by including some absorbing material within the metal package. Williams and Paananen [12] showed that this suppression can be achieved very effectively by including a dielectric substrate coated with a resistive film on the underside of the lid of the package. In general, including any lossy material within the enclosure will suppress unwanted modes, but care is needed to avoid the material causing excess loss in the microstrip circuit itself.

4.4   Packaged Lumped-Element Passive Components

4.4.1   Typical Packages for RF Passive Components

Component packaging is an important issue for components used at RF and microwave frequencies, as parasitic resistances and reactances associated with the package can have a significant effect on the electrical performance. Figure 4.14 shows the three most common types of package for RF passive components.

One of the main differences between the various packages is the configuration of the connecting leads. Connecting leads, even straight sections of wire, will have self-inductance. This inductance is small, and negligible at low frequencies, but at

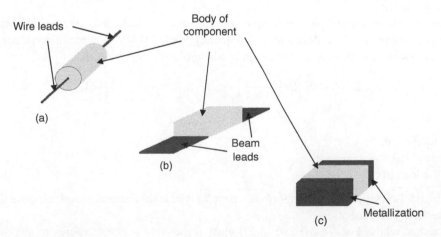

Figure 4.14   Typical packages for high-frequency passive components: (a) wire ended, (b) beam lead, and (c) Chip.

RF and microwave frequencies this small inductance will generate significant inductive reactance. So, as a general principle, component leads should be made as small as possible as the frequency of operation increases.

Wire-ended components are convenient for manual assembly of circuits, and have been the traditional type of component used up to the UHF region. But, as will be shown later in this section, the leads on these components contribute unacceptable amounts of inductance at high frequencies. Beam-lead components are an improvement, as the flat leads have less self-inductance, and if the component is correctly bonded onto the circuit this inductance can effectively be absorbed into the feed line to the component. Chip components provide the best configuration for high-frequency usage, because there is minimal lead inductance. But the small size of chip components usually precludes manual assembly, and accurate positioning and bonding of chip passives requires the use of precise surface-mount equipment, although this type of equipment is ideal for automatic assembly of circuits.

It is useful to compare the high-frequency performance of different package configurations by comparing the self-inductance of circular wires with that of flat leads. Wadell [13] quotes the following expression for calculating the self-inductance of a conducting circular wire lead

$$L = 0.002l \left[\ln \left(\frac{4l}{d} \right) - 1 + \frac{d}{2l} + \frac{\mu_r T(x)}{4} \right] \, \mu H, \tag{4.26}$$

where

l is length of the wire in cm
d is the wire diameter in cm
μ_r is the relative permeability of the wire material (usually 1.0),

and

$$T(x) = \frac{0.873 + 0.00186x}{1 - 0.278x + 0.128x^2}, \tag{4.27}$$

where

$$x = 2\pi r \sqrt{\frac{2\mu f}{\sigma}}, \tag{4.28}$$

and where

r is the radius of the wire in cm
μ is the permeability of the wire material (usually μ_o)
f is the frequency in Hz
σ is the conductivity of the wire material.

It should be noted that the self-inductance of a circular wire is influenced by the skin depth in the material, and consequently the self-inductance is frequency dependent as can be seen from Eq. (4.26). However, for wire diameters used for discrete components up to frequencies around 5 GHz, the effect of the frequency-dependent term $T(x)$ in Eq. (4.26) is relatively small. Above frequencies around 5 GHz discrete components with circular wire leads are not normally used, so as to avoid unwanted parasitics.

At frequencies above 5 GHz, active and passive components are normally packaged with very thin beam leads, where the lead self-inductance is considered to be independent of frequency. Wadell [13] also quotes a corresponding expression for the self-inductance of a flat lead, such as would be found in a beam-lead component, as

$$L = 0.002l \left[\ln \left(\frac{2l}{w+t} \right) + 0.5 + 0.2235 \left(\frac{w+t}{l} \right) \right] \, \mu H, \tag{4.29}$$

where

l is the length of the lead in cm
w is the width of the lead in cm
t is the thickness of the lead in cm.

In Example 4.7, which follows, the lead inductances of a circular wire and a flat lead are calculated, using lead dimensions typically found in practical components.

It is worth noting that whilst Eqs. (4.26) and (4.29) yield reasonably accurate values of self-inductance for leads with regular shapes, the leads bonded in a practical circuit have non-uniform shapes that may significantly affect their inductance. Ndip et al. [14] addressed the issue of non-uniform bonding structures of various shapes, and developed analytical models that showed good agreement between theory and practical measurement.

Example 4.7 Compare the self-inductances at 1 GHz of the leads, assumed to be made of copper, on each of the following components:

(i) A wire-ended component, of the form shown in Figure 4.14a, which has 5 mm long circular leads, each of diameter 0.6 mm.
(ii) A beam-lead component, of the form shown in Figure 4.14b, in which each lead has a length of 210 μm, a width of 115 μm, and a thickness of 10 μm.

Comment upon the results.
Take the resistivity of copper as $1.56 \times 10^{-8}\ \Omega\,m$

Solution

(i) Using Eq. (4.28): $x = 2\pi \times 0.03 \times \sqrt{\dfrac{2 \times 4\pi \times 10^{-7} \times 10^{9}}{(1.56 \times 10^{-8})^{-1}}} = 1.18 \times 10^{-3}$.

Substituting in Eq. (4.27):

$$T(x) = \frac{0.873 + (0.00186 \times 1.18 \times 10^{-3})}{1 - (0.278 \times 1.18 \times 10^{-3}) + (0.128 \times [1.18 \times 10^{-3}]^{2})} \simeq 0.873.$$

Substituting in Eq. (4.26):

$$L\,(\text{cir.wire}) = 0.002 \times 0.5 \times \left[\ln\left(\frac{4 \times 0.5}{0.06}\right) - 1 + \frac{0.06}{2 \times 0.5} + \frac{1 \times 0.873}{4}\right]\ \mu H = 2.78\ \text{nH}.$$

(ii) Using Eq. (4.29):

$L\,(\text{beam lead})$

$$= 0.002 \times 0.021 \times \left[\ln\left(\frac{2 \times 0.021}{0.0115 + 0.0010}\right) + 0.5 + 0.2235\left(\frac{0.0115 + 0.0010}{0.021}\right)\right]\ \mu H$$

$$= 0.078\ \text{nH}.$$

Comment: A flat beam-lead has significantly less self-inductance than a circular wire. Also, Eq. (4.26) indicates that the self-inductance of a circular wire will increase with frequency, whereas Eq. (4.29) shows that the inductance of a flat beam-lead will not.

4.4.2 Lumped-Element Resistors

There are a number of methods available for implementing resistors in high-frequency circuits. Resistors using carbon compositions as the resistive element can be formed into any of the package configurations shown in Figure 4.14. However, the granular nature of carbon composition resistors tends to degrade the performance at high frequencies. This degradation in performance has been attributed to parasitic capacitance effects between the granules of the composition. Wire-wound resistors are not suitable for high frequency work, because the winding tends to generate excess inductance. Metal-film resistors are the best type of resistor for high-frequency work, although in situations where thick-film technology is used to fabricate a circuit printed thick-film resistors are another good option. Thick-film resistors can also be laser-trimmed after printing, where it is necessary to have precise values of resistance.

In general, all lumped-element RF resistors (apart from printed resistors) will suffer from packaging parasitics which become more significant as the frequency increases. Figure 4.15 shows the equivalent circuit which is normally used to represent a high-frequency resistor with a nominal value of **R**.

The circuit shown in Figure 4.15 includes two series inductors, each of value L_S, to represent the lead inductances, and a capacitor, C_{pk}, to represent the packaging capacitance between the ends of the structure. The presence of both inductance and capacitance in the equivalent circuit means that the structure will exhibit a resonant frequency. Although the

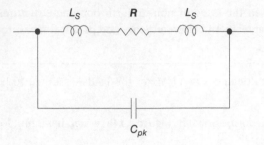

Figure 4.15 Equivalent circuit representing a packaged high-frequency resistor.

resonant frequency is normally quite high, it can become a significant issue for circuits working in the microwave frequency range. For these frequencies, careful consideration must be given to the type of resistor used so as to minimize the parasitics.

Example 4.8 A packaged 5 kΩ resistor has the equivalent circuit shown in Figure 4.15. The self-inductance of each lead is 2.8 nH and the packaging capacitance is 0.14 pF. Determine the impedance of the resistor at the following frequencies: 5, 50, and 500 MHz. Comment upon the result.

Solution
Referring to Figure 4.15, the total impedance, Z_T, is given by

$$\frac{1}{Z_T} = \frac{1}{R + j2\omega L_S} + j\omega C_{pk},$$

which can be rearranged to give

$$Z_T = \frac{R + j2\omega L_S}{1 - 2\omega^2 C_{pk} L_S + j\omega C_{pk} R}.$$

At 5 MHz:

$$\omega L_S = 2\pi \times 5 \times 10^6 \times 2.8 \times 10^{-9} = 0.088\ \Omega,$$

$$\omega^2 C_{pk} L_S = (2\pi \times 5 \times 10^6)^2 \times 0.14 \times 10^{-12} \times 2.8 \times 10^{-9} \simeq 0,$$

$$\omega C_{pk} R = 0.022\ \Omega,$$

$$Z_T = \frac{5000 + j2 \times 0.088}{1 + j0.022}\ \Omega = (4997.51 - j109.77)\ \Omega.$$

At 50 MHz:

$$\omega L_S = 0.088 \times 10\ \Omega = 0.88\ \Omega,$$

$$\omega^2 C_{pk} L_S \simeq 0,$$

$$\omega C_{pk} R = 0.022 \times 10\ \Omega = 0.22\ \Omega,$$

$$Z_T = \frac{5000 + j2 \times 0.88}{1 + j0.22}\ \Omega = (4771.36 - j1047.94)\ \Omega.$$

At 500 MHz:

$$\omega L_S = 0.088 \times 100\ \Omega = 8.8\ \Omega,$$

$$\omega^2 C_{pk} L_S = 0.004,$$

$$\omega C_{pk} R = 0.022 \times 100\ \Omega = 2.2\Omega,$$

$$Z_T = \frac{5000 + j2 \times 8.8}{1 + j2.2}\ \Omega = (858.30 - j1885.74)\ \Omega.$$

Summary:

Frequency (MHz)	Z_T (Ω)
5	4997.51$-j$109.77
50	4771.36$-j$1047.94
500	858.30$-j$1885.74

Comment: At high frequencies within the RF range, relatively small capacitive and inductive parasitics begin to have a very significant effect on the impedance of the resistor. At 500 MHz the reactive component of the impedance is larger than the resistive component.

4.4.3 Lumped-Element Capacitors

RF lumped-element capacitors are straightforward developments of the parallel-plate capacitor, in which the capacitance is obtained from two oppositely charged parallel conducting plates separated by a dielectric. The capacitance of the structure is then given simply by

$$C = \frac{\varepsilon A}{d},$$ (4.30)

where ε is the permittivity of the dielectric, A is the area of the plates, and d is the distance between the plates. RF and microwave capacitors come in several configurations, although some traditional methods of construction such as those using wound strips of foil and dielectric are not suitable because they have excess inductance. Generally, capacitors for high-frequency usage have a planar or chip format, and are selected for the RF and microwave properties of the dielectric. Suitable dielectrics will have low-loss, a reasonably high dielectric constant, and good thermal properties, which means a low temperature coefficient. The temperature coefficient represents the change in capacitance value for a given change in temperature, and is specified in ppm/°C. Two of the most popular dielectric materials are ceramic and mica. Ceramic is a good choice because the material is normally low loss, has good thermal characteristics, and the component can be made small in size due to the high dielectric constants available with ceramic materials. Moreover, it is possible to obtain ceramic materials that have either positive or negative temperature coefficients. For example, magnesium titanate is a ceramic material with a low dielectric constant and positive temperature coefficient, whereas calcium titanate has a negative temperature coefficient. By mixing these two materials a ceramic dielectric is obtained that has high temperature stability. Mixed dielectric materials are often referred to as NPO (negative–positive–zero) dielectrics, and used in circuits such as resonators and filters, where stability is essential. Mica is also a popular choice for high-frequency capacitors, because it is low loss and has a very low temperature coefficient. It is usually employed in film capacitors in which a conductor such as silver is deposited onto a mica film. Like NPO dielectric capacitors, silvered mica film capacitors are used in situations where temperature stability is critical. However, one slight drawback to the use of mica capacitors is that mica has a relatively low dielectric constant, around 6, and so the capacitor tends to be physically quite large.

The equivalent circuit of a capacitor is shown in Figure 4.16a, where **C** is the nominal capacitance value, R_s is the series resistance of each lead, R_d is the resistance of the dielectric, L_S is the inductance of each lead, or the equivalent inductance

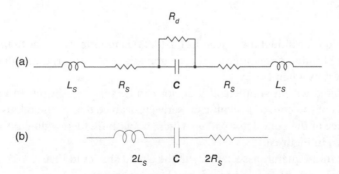

Figure 4.16 Equivalent circuits representing capacitors.

of the metallization if the capacitor is in a chip format. Since the resistance of the dielectric is normally extremely high, the equivalent circuit can be redrawn as a simple series resonant circuit shown in Figure 4.16b.

The presence of lead inductance together with significant capacitance means that the capacitor will have a resonant frequency, and care must be taken to ensure this is not within the working RF range of the component.

Example 4.9 What is the resonant frequency of a 10 pF capacitor, in which each conducting lead has an effective inductance of 0.7 nH? Comment upon the result.

Solution
The capacitor forms a series resonant circuit in which the capacitance is 10 pF and the inductance is (2×0.7) nH. The resonant frequency is then:

$$f_o = \frac{1}{2\pi\sqrt{LC}}$$

$$= \frac{1}{2\pi\sqrt{1.4 \times 10^{-9} \times 10 \times 10^{-2}}} \text{Hz}$$

$$= 1.34 \text{ GHz}.$$

Comment: Realistic values were given in the question for C and L_S and these led to a relatively low resonant frequency, indicating that some caution is needed in the selection and application of lumped-element capacitors in high-frequency circuit designs.

Another useful parameter to specify the performance of an RF capacitor is the quality factor, represented by Q. Using the normal definition of the quality factor of a series resonant circuit containing a capacitor we have

$$Q = \frac{1/\omega C}{R} = \frac{1}{\omega C R}, \tag{4.31}$$

where C is capacitance and R is the effective series resistance (ESR). The expression for Q in Eq. (4.31) is often written in the form

$$Q = \frac{1/\omega C}{ESR}. \tag{4.32}$$

The value of ESR will depend on the resistance of the dielectric and the resistance of the capacitor leads.

4.4.4 Lumped-Element Inductors

The traditional method of constructing an inductor is to wind a coil of wire on a cylindrical former. The inductance is then given by the well-known formula

$$L = \frac{0.394 r^2 N^2}{9r + 10l} \text{ μH}, \tag{4.33}$$

where

$r =$ radius of coil (cm)
$N =$ number of turns on the coil
$l =$ length of coil (cm).

The accuracy of the expression for inductance given in Eq. (4.33) is reportedly better than 1% when $l > 0.7r$. Eq. (4.33) indicates that there is a range of inductor geometries (i.e. combinations of r and l) that will give a particular inductance, but an optimum value of Q[1] results when $l = 2r$.

At RF, the parasitics associated with a simple coil inductor can be very significant. In particular, there is significant capacitance between the turns of the coil, and significant series resistance due to the relatively long length of conductor forming the coil. The resistance of the conductor will also increase with frequency due to the skin effect. The equivalent circuit for an inductor is shown in Figure 4.17.

Clearly there will be a resonant frequency associated with the circuit shown in Figure 4.17. Below resonance the component will be primarily inductive, and above resonance it will look capacitive.

---

1 Since an inductor will have both reactive and resistive components in series we can refer to the Q of an inductor simply as $Q = \omega L/R$.

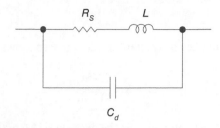

Figure 4.17 Equivalent circuit representing an inductor.

The use of a ferrite core within the inductor is one method of improving the performance. The presence of the ferrite means that fewer turns are required to achieve a particular inductance, which means less ohmic loss, and hence higher Q. Also, the use of a ferrite core will reduce the size of the component. However, some caution is needed in the use of ferrite cores, as the core itself may introduce additional loss.

A popular method of constructing a ferrite inductor for RF applications is to use a toroid, which is a small diameter ring of ferrite on which the turns of the coil are wound. This can form a very compact, high Q component. A useful in-depth discussion of the practical design issues relating to toroidal inductors is given by Bowick [15].

4.5 Miniature Planar Components

The unwanted parasitic reactances associated with the packaging and encapsulation of RF and microwave lumped passive components were discussed in Section 4.4. A different approach to the implementation of passive components in hybrid RF and microwave circuits is to use distributed planar components, of the type normally found in MMICs (monolithic microwave integrated circuits). In this section we will introduce some of the main types of distributed passive components, and discuss their principal high-frequency features.

4.5.1 Spiral Inductors

The typical layout of a microstrip rectangular spiral inductor is shown in Figure 4.18. The conducting track is formed into a spiral, which may be square, circular, or sometimes hexagonal. A dielectric bridge is used to connect with the inner end of the spiral. Using thick-film technology, the bridge is easily formed by printing a strip of dielectric over the turns of the track, as shown in Figure 4.18, and then over-printing this with a strip of conductor to connect with the inner end of the spiral.

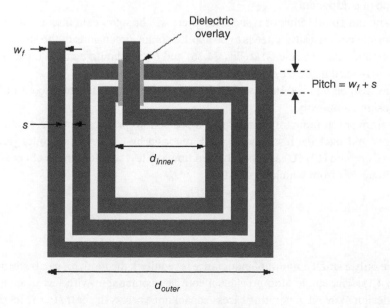

Figure 4.18 Spiral inductor.

Points to note:

(i) An accurate expression for the inductance of a square spiral in free space was developed by Mohan et al. [16]

$$L = K_1 \mu_o \frac{N^2 d_{avg}}{1 + K_2 \rho} \text{ nH,} \qquad (4.34)$$

where N is the number of turns on the spiral, d_{avg} is the average side length of the spiral in μm, ρ is the fill factor defined by

$$\rho = \frac{d_{outer} - d_{inner}}{d_{outer} + d_{inner}} \qquad (4.35)$$

and K_1 and K_2 are factors determined by the geometry of the spiral. Mohan and colleagues [16] gave these values as $K_1 = 2.34$ and $K_2 = 2.75$ for a square spiral, and $K_1 = 2.25$ and $K_2 = 3.55$ for an octagonal spiral.

(ii) In a microstrip configuration the presence of the ground plane beneath the spiral will tend to reduce the inductance, usually by around 10%. Gupta and colleagues [2] quote an expression for a correction factor, Kg, to represent the effect of the ground plane as

$$K_g = 0.57 - 0.145 \ln\left(\frac{w_f}{h}\right), \qquad (4.36)$$

where w_f is the width of the spiral track and h is the thickness of the substrate. The effective inductance of the microstrip spiral is then given by

$$L_{eff} = K_g L, \qquad (4.37)$$

where L is the free-space inductance given by Eq. (4.34).

(iii) A spiral inductor will exhibit some self-capacitance, made up of the capacitance between adjacent turns and the capacitance to the ground plane.

(iv) It follows from the last point that the spiral inductor will have a self-resonant frequency – the inductor should be operated well below this value.

(v) The inductor will also have a significant series resistance, whose value will be determined by the number of turns and the narrowness of the track. In order to make a compact component the width of the track will normally be small, leading to high ohmic loss. This high loss is a particular feature of many spiral inductors, and leads to relatively low Q values for this type of component.

(vi) The shape of a rectangular spiral inductor is such that there will be significant discontinuities, particular at the track corners, and appropriate compensation needs to be applied to the dimensions of the spiral.

(vii) For designs at frequencies above a few GHz, Eqs. (4.36) and (4.37) should be used only to give a prototype design, which should then be refined using electromagnetic simulation.

(viii) Spiral inductors may also be realized in a circular format. This will reduce the discontinuities, but at the cost of a slight increase in design complexity.

(ix) Spurious radiation from spiral inductors can also be a problem in practical circuits, leading to unwanted coupling between components and package resonances. Work reported by Caratelli and colleagues [17] using a full-wave finite-difference time-domain (FDTD) analysis showed how the level of radiation is affected both by the geometry of the spiral, and the materials from which it is made.

4.5.2 Loop Inductors

In circuit situations where only a small amount of inductance is required, the inductor can be formed from a simple, single loop as shown in Figure 4.19. The single loop provides a compact component with few discontinuities, although some caution is needed in the design to avoid generating excess capacitance across the entry to the loop.

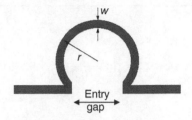

Figure 4.19 Loop inductor.

Points to note:

(i) For a single loop inductor the free-space inductance is given approximately by:

$$L = 12.57r \left(\ln \left[\frac{8\pi r}{w} \right] - 2 \right) \text{ nH},$$ (4.38)

where

L = inductance

r = mean radius of loop in cm

w = width of track forming loop in cm.

Whilst this expression for the inductance of a loop is frequently quoted in the literature, it takes no account of the size of the gap at the entry to the loop. Thus, it should be regarded as very approximate, and used to establish an initial value in a practical design, which should then be refined using electromagnetic simulation.

(ii) In order to avoid unwanted capacitance across the entry to the loop, the size of the gap is normally made of the order of $5w$, where w is the width of the track.

(iii) The self-capacitance of the loop inductor tends to be less than for a spiral inductor. This capacitance is essentially due to the capacitance per unit length of the conductor. Thus, the self-resonant frequency of the loop is higher than for a spiral inductor.

Example 4.10 Determine the inductance provided by each of the following structures, and comment upon the results. Each of the structures is formed from a flat (ribbon) conductor of width 0.8 mm of negligible thickness.

(i) A three-turn rectangular planar spiral inductor with a pitch of 1.4 mm and an inner dimension of 4 mm.

(ii) A single-loop inductor of radius 5 mm. (Neglect the effect of the entry gap.)

(iii) A straight 12 mm length of conductor.

Solution

(i)
$$\text{Pitch} = w + s = 0.8 + s = 1.4 \quad \Rightarrow \quad s = 0.6 \text{ mm}.$$

Referring to Figure 4.18:

$$d_{outer} = d_{inner} + 6w + 4s = (4 + 6 \times 0.8 + 4 \times 0.6) \text{ mm} = 11.2 \text{ mm},$$

$$d_{avg} = 0.5 \times (11.2 + 4) \text{ mm} = 7.6 \text{ mm}.$$

The fill factor for the spiral is found using Eq. (4.35):

$$\rho = \frac{11.2 - 4}{11.2 + 4} = 0.474.$$

Substituting in Eq. (4.34), and using the K-factors for a square spiral:

$$L(\text{spiral}) = 2.34(4\pi \times 10^{-7}) \times \frac{3^2 \times 7.6}{1 + (1.75 \times 0.474)} \text{H} = 87.34 \text{ nH}.$$

(ii) Using Eq. (4.38):

$$L(\text{loop}) = 12.57 \times 0.5 \times \left(\ln \left(\frac{8\pi \times 0.5}{0.08} \right) - 2 \right) \text{ nH}$$

$$= 19.21 \text{ nH}.$$

(Continued)

(Continued)

(iii) Using Eq. (4.26):

$$L(\text{straight ribbon}) = 0.002 \times 1.2 \times \left[\ln\left(\frac{2 \times 1.2}{0.08}\right) + 0.5 + 0.2235 \times \left(\frac{0.08}{1.2}\right)\right] \mu H$$

$$= 9.40\,\text{nH}.$$

Summary:

Circuit	Inductance (nH)
Spiral	87.34
Loop	19.21
Straight ribbon	9.40

Comment: The spiral and loop inductors specified in this example would occupy approximately the same area of substrate, but the spiral configuration gives substantially more inductance. The theory used makes no allowance for the entry gap of the loop; if a practical gap were considered then the inductance of the loop would be rather less than the 19.21 nH that was calculated. As would be expected, the self-inductance of the straight ribbon is less than the inductance of the other configurations, but is still around 50% of that of the loop. This suggests that using straight lengths of line to provide small values of inductance may be preferable to using a loop, which would in practice have more self-capacitance and hence a more troublesome self-resonant frequency.

4.5.3 Interdigitated Capacitors

The conventional configuration of conductors forming an interdigitated capacitor is shown in Figure 4.20. The structure employs a number of fingers to couple two metallic conductors. The use of fingers effectively creates a long coupling gap between the two conductors. Normally, the widths and spacing of the fingers is the same, i.e. $w_f = s$. This type of capacitor was originally developed for use in MMICs, but it can also be conveniently implemented in hybrid circuits by forming the coupled fingers in a microstrip conductor.

A comprehensive analysis of interdigitated capacitors was reported by Alley [18], who developed the following approximate expression for the total capacitance of the structure

$$C = (\varepsilon_r - 1)l[(N - 3)A_1 + A_2]\,\text{pF}, \tag{4.39}$$

where ε_r is the dielectric constant of the substrate, l is the overlapping finger length in μm, N is the number of fingers, and A_1 and A_2 are correction factors which are functions of h/w_f where h is the thickness of the substrate and w_f is the width of each finger. Alley [18] provided curves from which A_1 and A_2 could be obtained for a given ratio of h/w_f. Subsequently,

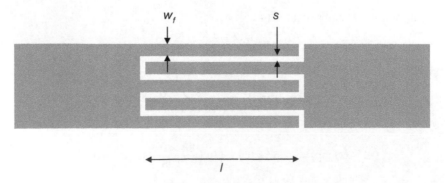

Figure 4.20 Interdigitated capacitor.

Bahl [19] used a curve-fitting technique to obtain the following closed-form equations from which values of A_1 and A_2 can be calculated:

$$A_1 = 4.409 \tanh\left[0.55\left(\frac{h}{w_f}\right)^{0.45}\right] \times 10^{-6} \text{ pF/m}. \tag{4.40}$$

$$A_2 = 9.92 \tanh\left[0.52\left(\frac{h}{w_f}\right)^{0.5}\right] \times 10^{-6} \text{ pF/m}. \tag{4.41}$$

Points to note about interdigitated capacitors:

(i) They are normally used to provide values of capacitance up to around 1 pF. For values above 1 pF the dimensions of the components can become rather large.

(ii) The overall dimensions are normally made less than one quarter wavelength, so they can be regarded electrically as lumped components.

(iii) They suffer from a number of discontinuities and so the expression given in Eq. (4.39) for the total capacitance should be regarded as an approximation.

(iv) In a microstrip line the interdigitated capacitor can be designed for a good series match.

(v) They are precise components and are normally used where precise values of capacitance are important, such as in filters and matching networks.

(vi) The Q-factor of an interdigitated capacitor can be increased by increasing the ratio of finger width (w_f) to gap size (s).

(vii) Various CAD models for interdigitated capacitors that include the effects of losses are available in the literature, for example, Zhu and Wu [20]. These models permit precise design of low-loss interdigitated capacitors in a microstrip format.

4.5.4 Metal–Insulator–Metal Capacitor

In situations where the capacitance needed is greater than can be provided by an interdigitated structure, then a metal–insulator–metal (MIM) capacitor can be used. This type of capacitor, sometimes referred to as an overlay capacitor, can provide relatively high capacitance, but with less precision than an interdigitated capacitor. The MIM capacitor is suitable for inclusion in either monolithic or hybrid RF integrated circuits. The structure of a MIM capacitor in thick-film microstrip is shown in Figure 4.21; for clarity, the plan view of the component is shown in Figure 4.21a, and the side view in Figure 4.21b. In this structure a dielectric is printed over the open-end of the microstrip line forming the lower electrode of the capacitor. Another microstrip line is then printed over the dielectric to form the upper electrode. Normally three layers of dielectric are printed successively to avoid pin holes forming in the dielectric. If pin holes exist in the dielectric,

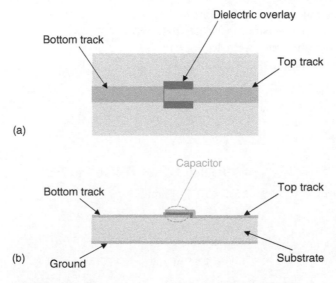

Figure 4.21 MIM capacitor: (a) plan view and (b) side view.

then these will fill with the conductor metal when the upper electrode is printed, thereby shorting the two electrodes together. It should be noted that the width of the upper electrode is chosen to maintain the correct impedance match to the microstrip line.

The approximate capacitance of a MIM capacitor can be found using simple parallel-plate capacitor theory as

$$C = \frac{\varepsilon_m \varepsilon_o A}{d}, \tag{4.42}$$

where ε_m is the dielectric constant of the printed dielectric between the electrodes, ε_o is the permittivity of a vacuum, A is the overlapping area of the electrodes, and d is the thickness of the printed dielectric.

Substituting $\varepsilon_o = 8.854 \times 10^{-12}$ F/m in Eq. (4.42) gives the capacitance of the MIM capacitor as

$$C = \frac{8.854 \varepsilon_m A}{d \times 10^6} \text{ pF}, \tag{4.43}$$

where d is in μm and A is in (μm)^2.

Points to note about MIM capacitors:

(i) MIM capacitors are useful as de-coupling capacitors, where relatively high capacitance values are needed, but without high precision.

(ii) Equation (4.42) is an approximation as it takes no account of the fringing effects at the edges of the electrodes.

(iii) Whilst MIM capacitors can conveniently be made using thick-film technology, there can be reliability problems associated with the need to print the upper electrode over the edge of the dielectric. This is known as the 'print-over edge effect', and relates to the thinning of the conductor as it is printed over the edge of the dielectric.

(iv) Since the resistance of the dielectric layer between the electrodes is extremely high, the only source of ohmic loss in a MIM capacitor is the resistance of the electrodes. Consequently the Q of this type of capacitor can be relatively high.

(v) If the MIM capacitor is made using thick-film technology, some trimming of the structure is possible to control the capacitance value. But if precise capacitance values are required, a better solution is to use an interdigitated structure.

Example 4.11   Calculate the capacitance of each of the following structures, and comment upon the results:

(i) A microstrip interdigitated capacitor with the following specification:
Number of fingers = 4
Finger width = 50 μm
Finger spacing = 50 μm
Overlapping finger length = 2 mm
Substrate thickness = 250 μm
Substrate relative permittivity = 9.8

(ii) A MIM capacitor with the following specification:
Relative permittivity of dielectric separating the electrodes = 7.8
Thickness of dielectric separating the electrodes = 24 μm
Area of electrodes = 350 × 2000 (μm)^2

Solution

(i)
$$\frac{h}{w_f} = \frac{250}{50} = 5.$$

Using Eq. (4.40):

$$A_1 = 4.409 \tanh[0.55 \times (5)^{0.45}] \times 10^{-6} \text{ pF/m}$$
$$= 3.583 \times 10^{-6} \text{ pF/m}.$$

Using Eq. (4.41):

$$A_2 = 9.92 \tanh[0.52 \times (5)^{0.5}] \times 10^{-6} \text{ pF/m}$$
$$= 8.15 \times 10^{-6} \text{ pF/m}.$$

Substituting in Eq. (4.39) to find the total capacitance of the interdigitated capacitor:

$$C = (9.8 - 1) \times 2000 \times [(4 - 3) \times 3.586 + 8.15] \times 10^{-6} \text{ pF}$$

$$= 0.206 \text{ pF}.$$

(ii) Substituting in Eq. (4.43) to find the total capacitance of the MIM capacitor:

$$C = \frac{8.854 \times 7.8 \times (350 \times 2000)}{24 \times 10^6} \text{pF}$$

$$= 2.018 \text{ pF}.$$

Comment: Each capacitor structure occupies approximately the same substrate area, but the MIM capacitor gives around 20 times the capacitance of the interdigitated capacitor. However, the MIM capacitor cannot normally be fabricated with the same accuracy as the interdigitated capacitor.

Appendix 4.A Insertion Loss Due to a Microstrip Gap

We know from Section 4.2.5 that a microstrip gap can be modelled as a π-network consisting of three capacitors. The insertion loss of the gap can then be determined from simple circuit theory using conventional network parameters, which are defined in Chapter 5.

Figure 4.22 shows a π-network of three admittances.

Pozar [21] gives the following *ABCD* matrix for this network as

$$\begin{bmatrix} A & B \\ C & D \end{bmatrix} = \begin{bmatrix} 1 + \dfrac{Y_2}{Y_3} & \dfrac{1}{Y_3} \\ Y_1 + Y_2 + \dfrac{Y_1 Y_2}{Y_3} & 1 + \dfrac{Y_1}{Y_3} \end{bmatrix}. \tag{4.44}$$

Using the conversions given in Appendix 5.A we can determine the transmission loss through the π-network in terms of the *S*-parameter, S_{21}, i.e.

$$S_{21} = \frac{2}{A + BY_o + CZ_o + D}$$

$$= \frac{2}{\left(1 + \dfrac{Y_2}{Y_3}\right) + \left(\dfrac{1}{Y_3}\right)Y_o + \left(Y_1 + Y_2 + \dfrac{Y_1 Y_2}{Y_3}\right)Z_o + \left(1 + \dfrac{Y_1}{Y_3}\right)}$$

$$= \frac{2Y_3 Y_o}{Y_o^2(1 + Y_3) + (Y_1 + Y_2 + Y_3)Y_o + Y_1 Y_2 + Y_2 Y_3 + Y_3 Y_1}, \tag{4.45}$$

where $Y_o = 1/z_o$ and Z_o is the impedance terminating the input and output of the network.

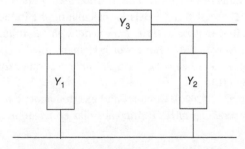

Figure 4.22 π-Network of admittances.

For our π-network of capacitors we have $Y_1 = Y_2 = j\omega C_p$ and $Y_3 = j\omega C_s$. Substituting for Y_1, Y_2, and Y_3 in Eq. (4.45), and simplifying, gives

$$S_{21} = \frac{2}{2 + \dfrac{2C_p}{C_s} + j\left(2\omega C_p Z_0 + \dfrac{\omega C_p^2 Z_0}{C_s} - \dfrac{1}{\omega C_s Z_0}\right)}. \tag{4.46}$$

The insertion loss of the gap in dB is then given by

$$\begin{aligned}(IL)_{dB} &= 20\log|S_{21}|^{-1}\\ &= 20\log|0.5(a + jb)|, \end{aligned} \tag{4.47}$$

where

$$a = 2 + \frac{2C_p}{C_s} \tag{4.48}$$

and

$$b = 2\omega C_p Z_0 + \frac{\omega C_p^2 Z_0}{C_s} - \frac{1}{\omega C_s Z_0}. \tag{4.49}$$

References

1 Edwards, T.C. and Steer, M.B. (2016). *Foundations for Microstrip Circuit Design*, 4e. Chichester, UK: Wiley.

2 Gupta, K.C., Garg, R., Bahl, I., and Bhartia, P. (1996). *Microstrip Lines and Slotlines*. Norwood, MA: Artech House.

3 Hammerstad, E.O. and Bekkadal, F. (1975). *A Microstrip Handbook*. Norway: University of Trondheim.

4 Fooks, E.H. and Zakarevicius, R.A. (1990). *Microwave Engineering Using Microstrip*. Australia: Prentice Hall.

5 Easter, B., Gopinath, A., and Stephenson, I.M. (1978). Theoretical and experimental methods for evaluating discontinuities in microstrip. *Radio and Electronic Engineer* 48 (1/2): 73–84.

6 Garg, R. and Bahl, I.J. (1978). Microstrip discontinuities. *International Journal of Electronics* 45.

7 Alexopoulos, N. and Wu, S.-C. (1994). Frequency-independent equivalent circuit model for microstrip open-end and gap discontinuities. *IEEE Transactions on Microwave Theory and Techniques* 42 (7).

8 Dydyk, K.M. (1972). Master the T-junction and sharpen your MIC designs. *IEEE Microwave Magazine* (May 1972).

9 Chadha, R. and Gupta, K.C. (1982). Compensation of discontinuities in planar transmission lines. *IEEE Transactions on Microwave Theory and Techniques* 30 (12).

10 Rastogi, A.K., Bano, M., and Sharma, S. (2014). Design and simulation model for compensated and optimized T-junctions in microstrip line. *International Journal of Advanced Research in Computer Engineering and Technology* 31 (12).

11 March, S.L. (1981). Empirical formulas for the impedance and effective dielectric constant of covered microstrip for use in the computer-aided design of microwave integrated circuits. *Proceedings of 11th European Microwave Conference*, Amsterdam, Netherlands (7–11 September 1981), pp. 671–676

12 Williams, D.F. and Paananen, D.W. (1989). Suppression of resonant modes in microwave packages. *MTT-S Microwave Symposium Digest*, Long Beach, CA (13–15 June 1989), pp. 1263–1265

13 Wadell, B. (1998). Modelling circuit parasitics – part 2. *IEEE Instrumentation and Measurement Magazine* 1 (2).

14 Ndip, I., Oz, A., Reichl, H. et al. (2015). Analytical models for calculating the inductances of bond wires in dependence on their shapes, bonding parameters, and materials. *IEEE Transactions on Electromagnetic Compatibility* 57 (2): 241–249.

15 Bowick, C. (1982). *RF Circuit Design*. Newton, MA: Butterworth-Heinemann.

16 Mohan, S.S., Hershenson, M., Boyd, S.P., and Lee, T.H. (1999). Simple accurate expressions for planar spiral inductances. *IEEE Journal of Solid-State Circuits* 34 (10).

17 Caratelli, D., Cicchetti, R., and Faraone, A. (2006). Circuital and electromagnetic performances of planar microstrip inductors for wireless applications. *Proceedings of IEEE Antennas and Propagation Symposium*, Albuquerque (July 2006).

18 Alley, G.D. (1970). Interdigital capacitors and their application to lumped-element microwave integrated circuits. *IEEE Transactions on Microwave Theory and Techniques* MTT-18 (12).

19 Bahl, I. (2003). *Lumped Elements for RF and Microwave Circuits*. Artech House.

20 Zhu, L. and Wu, K. (1998). A general-purpose circuit model of interdigital capacitor for accurate design of low-loss microstrip circuits. *Proceedings of IEEE MTT-Symposium*, Baltimore, pp. 1755–1758.

21 Pozar, D.M. (2001). *Microwave and RF Design of Wireless Systems*. New York: Wiley.

5

S-Parameters

5.1   Introduction

Scattering parameters, usually abbreviated to *S*-parameters, are network parameters used to specify the performance of linear networks and devices at microwave frequencies.

At low frequencies (that is below microwave), the properties of multi-port linear networks are normally expressed in terms of *Z*-parameters, *Y*-parameters, or *ABCD* parameters. These parameters, which are defined in the appendices of this chapter, require standard terminating impedances such as short circuits and open circuits, which are readily provided in practice at low frequencies. At higher (microwave) frequencies, the provision of short or open circuits is not so easy. At high frequencies, short circuits tend to have residual inductance, and open circuits have residual capacitance, and so to use them as reference terminations requires that these stray reactances are determined from subsidiary experiments.

Also, at low frequencies (below microwave), measurement of current and voltage associated with the characterization of linear networks is relatively straightforward, and consequently the measurement of *Z*, *Y*, and *ABCD* parameters is relatively easy. But at microwave frequencies, current and voltage are more difficult to measure, and so the accurate determination of *Z*, *Y*, and *ABCD* parameters is again correspondingly more difficult. For these reasons, a different approach is adopted at microwave frequencies and the network is specified in terms of the microwave power which is reflected from a given network port, and also passed to other network ports, when microwave power is incident upon the given port. Thus, the power applied to the given port may be regarded as being scattered, partly as reflected power, and partly as power transmitted to the other ports, hence the name scattering parameters.

As in the case of *Z*, *Y*, and *ABCD* parameters, a reference impedance is required. Specific resistive impedances are readily available at microwave frequencies. Normally, a 50 Ω reference impedance is used.

5.2   *S*-Parameter Definitions

The *S*-parameters of an *n*-port, linear network are defined in terms of an incident wave, V_r^+, on the *r*th port. This is related to a normalized parameter, a_r, through the expression.

$$a_r = \frac{V_r^+}{\sqrt{Z_o}},$$

(5.1)

where Z_o is the impedance terminating the port. (Note that the wave has been normalized such that the amplitude of the a_r parameter represents the square root of the wave power.) The resulting normalized reflected wave at the *r*th port is denoted by

$$b_r = \frac{V_r^-}{\sqrt{Z_o}}.$$

(5.2)

RF and Microwave Circuit Design: Theory and Applications, First Edition. Charles E. Free and Colin S. Aitchison.
© 2022 John Wiley & Sons Ltd. Published 2022 by John Wiley & Sons Ltd.
Companion website: www.wiley.com/go/free/rfandmicrowave

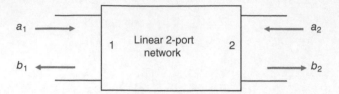

Figure 5.1 Power wave voltage nomenclature for a two-port network.

The incident wave on the rth port will also generate waves leaving the remaining ports, and in general we may write the normalized wave leaving an arbitrary ith port as

$$b_i = \frac{V_i^-}{\sqrt{Z_0}},$$

(5.3)

where Z_0 is the impedance terminating the ith port. The S-parameters of a linear network are defined as the ratio of these normalized parameters; we can demonstrate this by considering a simple two-port network as shown in Figure 5.1, and where we have assumed that both ports are terminated in a matched load, Z_0.

The electrical performance of the two-port network is completely represented by the four S-parameters shown in Eq. (5.4).

$$S_{11} = \frac{b_1}{a_1}\bigg|_{a_2=0},$$

$$S_{12} = \frac{b_1}{a_2}\bigg|_{a_1=0},$$

$$S_{21} = \frac{b_2}{a_1}\bigg|_{a_2=0},$$

$$S_{22} = \frac{b_2}{a_2}\bigg|_{a_1=0}.$$

(5.4)

Through inspection of Eq. (5.4), we can see that S_{11} and S_{22} are the reflection coefficients at ports 1 and 2, respectively, when the ports are terminated in matched loads. Also, S_{21} and S_{12} are the forward and reverse voltage gains, respectively, when the ports are terminated in matched loads. For clarity, the significance of the four S-parameter terms has been summarized in Eq. (5.4), for a linear two-port network with the ports terminated in matched loads.

$$S_{11} = \text{reflection coefficient at port 1,}$$

$$S_{12} = \text{reverse voltage gain,}$$

$$S_{21} = \text{forward voltage gain,}$$

$$S_{22} = \text{reflection coefficient at port 2.}$$

(5.5)

The S-parameters may conveniently be expressed in matrix form as

$$[S] = \begin{bmatrix} S_{11} & S_{12} \\ S_{21} & S_{22} \end{bmatrix}.$$

(5.6)

It is important to note that the convention for writing the order of the suffices of individual S-parameters identifies the input and output ports. Thus, for an S-parameter written in the form S_{mn}, m refers to the output port and n to the input port.

Points to note:

(i) S-parameters represent both magnitude, and phase, and are therefore complex quantities.
(ii) S-parameters are specified with the ports of the network or device terminated in a specific impedance, usually 50 Ω.
(iii) S-parameters vary with frequency.
(iv) When the ports of a network are terminated by impedances other than 50 Ω, we must use modified S-parameters to find the reflection coefficient at the ports of the network (modified S-parameters are explained in Section 5.5).

Example 5.1 Write the S-matrix representing an ideal amplifier that has matched input and output ports, and a gain of 15 dB.

Solution
Since the input and output ports are matched, the reflection coefficient at these ports will be zero, i.e. $S_{11} = 0$ and $S_{22} = 0$. Also, if the amplifier is ideal, there will be zero reverse gain, and hence $S_{12} = 0$.

$$15 = 20 \ \log|S_{21}| \quad \Rightarrow \quad |S_{21}| = 10^{15/20} \quad \Rightarrow \quad |S_{21}| = 5.62.$$

And so the S-matrix will be

$$[S] = \begin{bmatrix} 0 & 0 \\ 5.62\angle 0° & 0 \end{bmatrix}.$$

Notes:

(1) *We have assumed that in the ideal case the transmission phase is zero (or some multiple of 360°).*
(2) *When converting from dB to an S-parameter value we use 20 log|S_{21}|, since S-parameters are voltage ratios, as can be seen from the definition in Eq. (5.1).*

Example 5.2 Determine the S-matrix representing a 20 cm length of matched coaxial cable at 800 MHz. The cable has a lossless dielectric with a relative permittivity of 2.3.

Solution

$$S_{11} = S_{22} = 0 \quad \text{since the cable is matched,}$$

$$|S_{12}| = |S_{21}| = 1 \quad \text{since the cable is lossless.}$$

At 800 MHz, the cable wavelength is

$$\lambda_C = \frac{c}{f\sqrt{\varepsilon_r}} = \frac{3 \times 10^8}{800 \times 10^6 \times \sqrt{2.3}} \ \text{m} = 0.247 \ \text{m}.$$

Transmission phase change, ϕ, is given by

$$\phi = \beta l = \frac{2\pi}{\lambda_C}l = \frac{2\pi}{0.247} \times 0.2 \ \text{rad} = 5.088 \ \text{rad}$$

$$\equiv 291.50°.$$

Therefore, the S-matrix of the specified length of cable is

$$[S] = \begin{bmatrix} 0 & 1\angle -291.50° \\ 1\angle -291.50° & 0 \end{bmatrix}.$$

Note that the transmission phase is negative due to the time delay incurred by the signal passing through the cable.

The concept of S-parameters can be applied to networks or devices with more than two ports. For a four-port network, the S-matrix is

$$[S] = \begin{bmatrix} S_{11} & S_{12} & S_{13} & S_{14} \\ S_{21} & S_{22} & S_{23} & S_{24} \\ S_{31} & S_{32} & S_{33} & S_{34} \\ S_{41} & S_{42} & S_{43} & S_{44} \end{bmatrix}. \tag{5.7}$$

Example 5.3 Write the *S*-matrix for an ideal 3 dB branch line coupler, with the port configuration shown in Figure 2.4.

Solution
Notes:

(i) An ideal coupler will be matched at the four ports and so the diagonal elements of the *S*-matrix, S_{11} through S_{44}, which represent the reflection coefficients at the four ports, will be zero.
(ii) A 3 dB power split means the input voltage to each arm will be reduced by $1/\sqrt{2}$.

Thus, we have

$$[S] = \begin{bmatrix} 0 & 0 & \frac{1}{\sqrt{2}}\angle-180° & \frac{1}{\sqrt{2}}\angle-90° \\ 0 & 0 & \frac{1}{\sqrt{2}}\angle-90° & \frac{1}{\sqrt{2}}\angle-180° \\ \frac{1}{\sqrt{2}}\angle-180° & \frac{1}{\sqrt{2}}\angle-90° & 0 & 0 \\ \frac{1}{\sqrt{2}}\angle-90° & \frac{1}{\sqrt{2}}\angle-180° & 0 & 0 \end{bmatrix},$$

i.e.

$$[S] = -\frac{1}{\sqrt{2}}\begin{bmatrix} 0 & 0 & 1 & j \\ 0 & 0 & j & 1 \\ 1 & j & 0 & 0 \\ j & 1 & 0 & 0 \end{bmatrix}.$$

Note that in the second matrix we have replaced $\angle-90°$ by $-j$, using simple complex number theory: $1\angle-90° = \cos(-90°)+j\sin(-90°) = 0-j = -j$.

Example 5.4 Write the *S*-matrix for an ideal microstrip hybrid ring, with the port configuration shown in Figure 2.23.

Solution

$$[S] = \begin{bmatrix} 0 & \frac{1}{\sqrt{2}}\angle-90° & 0 & \frac{1}{\sqrt{2}}\angle-270° \\ \frac{1}{\sqrt{2}}\angle-90° & 0 & \frac{1}{\sqrt{2}}\angle-90° & 0 \\ 0 & \frac{1}{\sqrt{2}}\angle-90° & 0 & \frac{1}{\sqrt{2}}\angle-90° \\ \frac{1}{\sqrt{2}}\angle-270° & 0 & \frac{1}{\sqrt{2}}\angle-90° & 0 \end{bmatrix},$$

i.e.

$$[S] = \frac{1}{\sqrt{2}}\begin{bmatrix} 0 & -j & 0 & j \\ -j & 0 & -j & 0 \\ 0 & -j & 0 & -j \\ j & 0 & -j & 0 \end{bmatrix}.$$

Example 5.5 Write the *S*-matrix for an ideal 15 dB microstrip directional coupler having the port configuration shown in Figure 2.16.

Solution
The first step is to find the coupling coefficient, k

$$10 \log \left(\frac{1}{k}\right)^2 = 15 \quad \Rightarrow \quad k = 0.178,$$

$$\sqrt{1 - k^2} = 0.984.$$

Using Eq. (2.16) the *S*-matrix can now be written directly as

$$[S] = \begin{bmatrix} 0 & -j0.984 & 0 & 0.178 \\ -j0.984 & 0 & 0.178 & 0 \\ 0 & 0.178 & 0 & -j0.984 \\ 0.178 & 0 & -j0.984 & 0 \end{bmatrix}.$$

5.3 Signal Flow Graphs

One method of analyzing circuits described in terms of *S*-parameters is to make use of a signal flow graph. In general, this type of graph consists of a number of nodes, interconnected by directed branches, and provides a graphical representation of a mathematical expression. In our case, the nodes represent the variables in an *S*-parameter expression, and the branches the values of the *S*-parameters. The generation of a signal flow graph can be explained by deriving the flow graph to represent the simple two-port network shown in Figure 5.1. Here we have four variables, namely a_1, b_1, a_2, and b_2, so our flow graph will have four nodes, representing these four variables. We know the *S*-parameter relationships between these variables from Section 5.2, and we use these relationships to form the connecting branches between the nodes, as shown in Figure 5.2. The directions of the arrows on the branches is obtained by logical deduction from the *S*-parameter definitions; for example, we know that S_{21} is the forward gain, and therefore the arrow for this branch must be directed from the input node, a_1, to the output node, b_2.

Using the arrows we can sum the signals entering a given node and so generate the following *S*-parameter equations.

$$b_1 = S_{11}a_1 + S_{12}a_2,$$

$$b_2 = S_{22}a_2 + S_{21}a_1. \tag{5.8}$$

Once a network has been represented by the appropriate signal flow graph it may be analyzed using Mason's non-touching loop rule [1].

5.4 Mason's Non-touching Loop Rule

Mason [1] developed a simple rule, based on closed-path loops within a signal flow graph that enables the overall gain of a network to be determined.

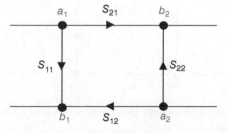

Figure 5.2 Signal flow graph for a linear two-port network.

A flow graph can be regarded as made up of a number of closed-path loops which are defined as follows:

(a) A first-order loop is a closed, sequential path from a given node returning to that given node without passing through that node more than once. The gain of the loop is equal to the product of the S-parameters around the loop.

(b) A second-order loop refers to the combination of two first-order loops that do not touch at any point. The gain of a second-order loop is equal to the product of the gains of the two first-order loops.

(c) A third-order loop refers to the combination of three first-order loops that do not touch at any point. The gain of a third-order loop is equal to the product of the gains of the three first-order loops.

Mason's rule provides a convenient method of analyzing flow graphs; it gives the ratio, T, of the amplitude of a wave leaving a particular port to the amplitude of a wave entering that, or any other port, and can be expressed as [1].

$$T = \frac{\sum_{n=1}^{n} P_n \Delta_n}{\Delta},$$

(5.9)

where

n = number of forward paths linking the two ports being considered, i.e. the input and output ports

P_n = value of the product of the S-parameters and reflection coefficients around the nth path

Δ = 1 − (sum of the values of all first-order loops) + (sum of the values of all second-order loops) − (sum of the values of all third-order loops) + ...

Δ_n = the value of Δ for that portion of the flow graph not touching the nth path

5.5 Reflection Coefficient of a Two-Port Network

Mason's rule can be used to calculate S_{11} for a linear two-port network terminated in an impedance, Z_L. The network is shown in Figure 5.3, with the corresponding signal flow graph in Figure 5.4. In order to maintain consistent units in the flow graph, the load impedance is represented by its reflection coefficient, Γ_L.

Through inspection of Figure 5.4, it can be seen that there is only one first-order loop, represented by $S_{22}\Gamma_L$ and no-higher order loops, and so we have

$$\Delta = 1 - S_{22}\Gamma_L.$$

(5.10)

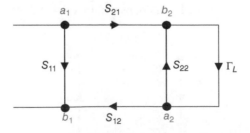

Figure 5.3 A linear two-port network.

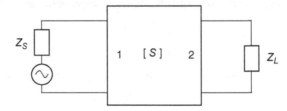

Figure 5.4 Flow graph for a two-port network terminated with a load having a reflection coefficient, Γ_L.

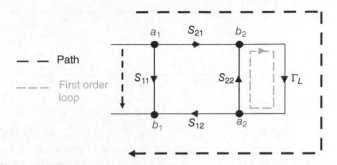

Figure 5.5 Identification of paths and first-order loop for the flow graph of Figure 5.4.

Also, it can be seen that there are two paths from node a_1 to node b_1, namely S_{11} and $S_{21}\Gamma_L S_{12}$. For clarity, the first-order loop, and the two paths are shown in Figure 5.5, as broken lines.

Applying Mason's non-touching loop rule to find the input reflection coefficient (S'_{11}) gives

$$S'_{11} = \frac{b_1}{a_1} = \frac{S_{11}(1 - S_{22}\Gamma_L) + S_{21}\Gamma_L S_{12}}{1 - S_{22}\Gamma_L}, \tag{5.11}$$

i.e.

$$S'_{11} = S_{11} + \frac{S_{21}\Gamma_L S_{12}}{1 - S_{22}\Gamma_L}. \tag{5.12}$$

S'_{11} is often referred to as a modified S-parameter, and shows how the input reflection coefficient of the two-port network is influenced by the load if S_{12} is non-zero. (Note that we have followed the normal convention of denoting a modified S-parameter by a primed quantity.)

Following a similar procedure the modified output reflection coefficient can be found as

$$S'_{22} = S_{22} + \frac{S_{12}\Gamma_S S_{21}}{1 - S_{11}\Gamma_S}, \tag{5.13}$$

where Γ_S is the reflection coefficient of the source impedance.

Example 5.6 The following S-matrix represents the performance of a two-port non-reciprocal network at a particular frequency, i.e. a network that has different transmission properties in the two directions.

$$[S] = \begin{bmatrix} 0.12\angle - 24° & 0.14\angle 98° \\ 0.93\angle - 171° & 0.19\angle 31° \end{bmatrix}.$$

If the device is terminated with an impedance $(160-j22)\ \Omega$, determine the reflection coefficient of the load, and the reflection coefficient at the input to the device.

Comment on the answers.

Solution

From Chapter 1 we know that the reflection coefficient, Γ, of an impedance, Z, is given by

$$\Gamma = \frac{Z - Z_0}{Z + Z_0},$$

and so for the load impedance we have

$$Z_L = (160 - j22)\ \Omega \quad \Rightarrow \quad \Gamma_L = \frac{160 - j22 - 50}{160 - j22 + 50}$$

$$\equiv 0.52\angle - 5.3°.$$

(Continued)

(Continued)

Using Eq. (5.12):

$$S'_{11} = (0.12\angle - 24°) + \frac{(0.93\angle - 171°) \times (0.52\angle - 5.3°) \times (0.14\angle 98°)}{1 - (0.19\angle 31°) \times (0.52\angle - 5.3°)}$$

$$= 0.18\angle - 43.4°.$$

Comment: The magnitude of the reflection coefficient at the input to the device is significantly less than the magnitude of the load reflection coefficient. This is because the device has the properties of an isolator, whose non-reciprocal behaviour reduces reflections from the load. The properties of isolators are covered in more detail in Chapter 6.

Example 5.7   The performance of an RF FET (Field Effect Transistor) is represented by the following *S*-matrix:

$$[S] = \begin{bmatrix} 0.30\angle 160° & 0.03\angle 62° \\ 6.10\angle 65° & 0.40\angle - 38° \end{bmatrix}.$$

Suppose that a two-stage amplifier is made by connecting two of the FETs in cascade, and terminating the output with a load having a reflection coefficient of $0.35\angle + 24°$. Draw the flow diagram for the circuit and hence find the reflection coefficient at the input terminals of the amplifier terminated in the specified load.

Solution

The cascade of two FETs, with the second terminated with a load having a reflection coefficient Γ_L is shown in Figure 5.6.

Figure 5.6   Two FETs in cascade terminated in load Z_L, with reflection coefficient, Γ_L.

Flow diagram:

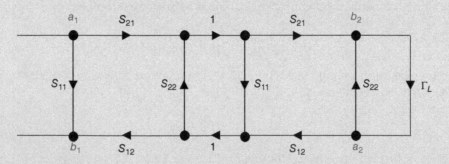

Figure 5.7   Flow diagram for two FETs in cascade, terminated in Γ_L.

From inspection of Figure 5.7, we see that there are three first-order loops and three paths from a_1 to b_1, as shown in Figure 5.8. Since two of the first-order loops are non-touching there will be one second-order loop.

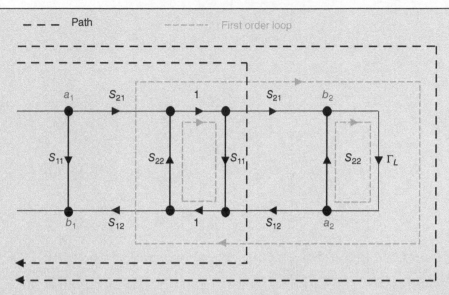

Figure 5.8 Identification of paths and first-order loops for the flow diagram shown in Figure 5.7.

Note that on the signal flow graph in Figure 5.8 where there is a direct path connection between two nodes we have shown the value of the path as 1.

Value of path 1: S_{11},

Value of path 2: $S_{21}S_{11}S_{12}$,

Value of path 3: $S_{21}S_{21}\Gamma_L S_{12}S_{12}$,

Values of first order loops: $S_{22}S_{11}$ $S_{22}\Gamma_L$ $S_{21}\Gamma_L S_{12}S_{22}$,

Value of second order loop: $S_{22}S_{11} \times S_{22}\Gamma_L$.

Applying Mason's rule:

$$\Gamma_{in} = \frac{b_1}{a_1} = \frac{S_{11}\Delta + S_{21}S_{11}S_{12}(1 - S_{22}\Gamma_L) + S_{21}S_{21}\Gamma_L S_{12}S_{12}}{\Delta},$$

where

$$\Delta = 1 - (S_{22}S_{11} + S_{22}\Gamma_L + S_{21}\Gamma_L S_{12}S_{22}) + (S_{22}S_{11}S_{22}\Gamma_L).$$

After substituting the data from the question we obtain

$$\Delta = 0.94\angle - 4.7°$$

and

$$\Gamma_{in} = 0.26\angle 170.3°.$$

Note that the flow diagram could have been simplified by representing the second FET by its modified input reflection coefficient as shown in Figure 5.9, where from Eq. (5.12) the modified input reflection coefficient, S'_{11} of the second FET is

$$S'_{11} = S_{11} + \frac{S_{21}\Gamma_L S_{12}}{1 - S_{22}\Gamma_L}$$

(Continued)

(Continued)

and hence

$$\Gamma_{in} = S_{11} + \frac{S_{21} S'_{11} S_{12}}{1 - S_{22} S'_{11}}.$$

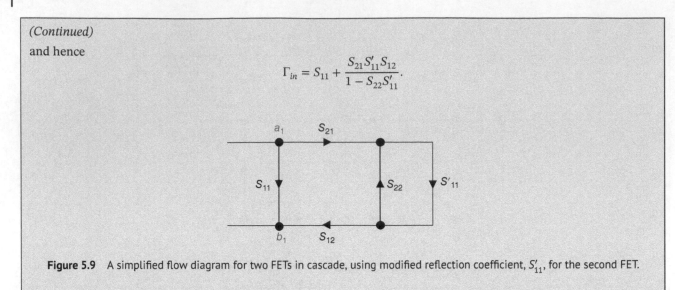

Figure 5.9 A simplified flow diagram for two FETs in cascade, using modified reflection coefficient, S'_{11}, for the second FET.

5.6 Power Gains of Two-Port Networks

A two-port network connected between a source having an impedance, Z_S, and a load having an impedance, Z_L, is shown in Figure 5.10.

There are three definitions of power gain that can be applied to the two-port network shown in Figure 5.10:

Operating Power Gain (G_p)
  This is the ratio of the power dissipated in the load (Z_L) to the actual input power to the network.
Available Power Gain (G_A)
  This is the ratio of the available power at the output of the network to the power available from the source. This gain assumes conjugate impedance matching of the source to the input of the network, and the output of the network to the load, i.e. it is the maximum power gain that is possible under fully matched conditions.
Transducer Power Gain (G_T)
  This is the ratio of the power delivered to the load to the power that is available from the source.

The three gains can be expressed in terms of the *S*-parameters of the network, and these are given in Eqs. (5.14) through (5.16). The derivations of these equations are given in a number of textbooks; Pozar [2] derives the expressions by considering voltage waves at the input and output of the network, and Glover et al. [3] obtain the same expressions through application of Mason's rule, which was discussed in Section 5.4.

$$G_p = \frac{P_L}{P_{in}} = \frac{|S_{21}|^2 (1 - |\Gamma_L|^2)}{(1 - |\Gamma_{in}|^2) \times |1 - S_{22}\Gamma_L|^2}, \tag{5.14}$$

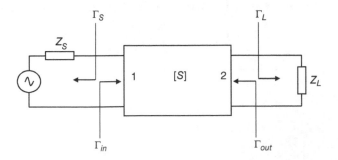

Figure 5.10 A linear active two-port device connected between source, Z_S, and load, Z_L.

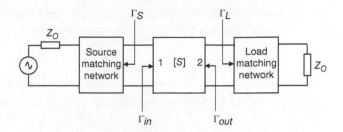

Figure 5.11 Unilateral device with matching networks.

$$G_A = \frac{P_{avail,o/p}}{P_{avail,source}} = \frac{|S_{21}|^2(1-|\Gamma_S|^2)}{(1-|\Gamma_{out}|^2)\times|1-S_{11}\Gamma_S|^2},\tag{5.15}$$

$$G_T = \frac{P_L}{P_{avail,source}} = \frac{|S_{21}|^2(1-|\Gamma_S|^2)(1-|\Gamma_L|^2)}{|1-\Gamma_S\Gamma_{in}|^2\times|1-S_{22}\Gamma_L|^2}.\tag{5.16}$$

If the reverse gain of the two-port network is small enough to be neglected, i.e. $S_{12}\simeq 0$, it follows from Eq. (5.12) that $\Gamma_{in}=S_{11}$, and the transducer power gain can be written as

$$G_{TU} = \frac{|S_{21}|^2(1-|\Gamma_S|^2)(1-|\Gamma_L|^2)}{|1-\Gamma_S S_{11}|^2\times|1-S_{22}\Gamma_L|^2},\tag{5.17}$$

where G_{TU} is referred to as the unilateral transducer power gain.

A very common high-frequency, two-port network is the amplifier, in which the amplifying active device such as an FET is operating between source and load impedances each having a value of Z_o, where the value of Z_o is usually 50 Ω. In this case matching networks are normally inserted at the input and output of the device, as shown in Figure 5.11, to provide the designer of the amplifier with control of the values of Γ_S and Γ_L.

In order to obtain maximum unilateral transducer power gain we require conjugate matching conditions at the input and output of the device, i.e. we require

$$\Gamma_S = S_{11}^* \quad \text{and} \quad \Gamma_L = S_{22}^*.\tag{5.18}$$

With the matching conditions stated in Eq. (5.18) we have[1]

$$|\Gamma_S| = |S_{11}| \quad \text{and} \quad |\Gamma_L| = |S_{22}|,\tag{5.19}$$

$$S_{11}\Gamma_S = |S_{11}|^2 \quad \text{and} \quad S_{22}\Gamma_L = |S_{22}|^2.\tag{5.20}$$

Substituting the matching conditions into Eq. (5.17) gives the maximum unilateral transducer power gain as

$$G_{TU,max} = \frac{|S_{21}|^2(1-|S_{22}|^2)(1-|S_{11}|^2)}{|(1-|S_{11}|^2)(1-|S_{22}|^2)|^2},\tag{5.21}$$

which can be rearranged as

$$G_{TU,max} = \frac{1}{1-|S_{11}|^2}|S_{21}|^2\frac{1}{1-|S_{22}|^2}.\tag{5.22}$$

It is convenient to write Eq. (5.22) in the form

$$G_{TU,max} = G_S \times G_o \times G_L,\tag{5.23}$$

where

G_o is the inherent gain of the device; *the inherent gain of the device is the power gain of the device with the ports directly terminated in 50 Ω, which are the conditions under which the S-parameters are measured, and so $G_o = |S_{21}|^2$.*

G_S is the effective gain of the source matching network.

G_L is the effective gain of the load matching network.

---

1 Note that we have made use of one of the basic properties of a complex number, namely that the product of a given complex number with its conjugate value is equal to the magnitude of the complex number squared, i.e. if $Z = a + jb$ then $Z \times Z^* = (a+jb)\times(a-jb) = a^2 + b^2 = |Z|^2$.

It is important to specify the *effective* gains of the source and load matching networks since these networks will normally be constructed from passive components, and they only have gain in the sense that they allow more signal power to flow into and out of the device.

Example 5.8 The performance of a two-port amplifying device is represented by the following *S*-matrix

$$[S] = \begin{bmatrix} 0.69\angle -34° & 0.01\angle -27.8° \\ 3.12\angle 132° & 0.71\angle 57° \end{bmatrix}.$$

Assuming the device to exhibit unilateral behaviour, determine:

(i) The inherent power gain of the device in dB.
(ii) The effective power gains of source and load matching networks, assuming the source and load impedances are 50 Ω.
(iii) The overall gain of the device, with matching network, when the source and load impedances are 50 Ω.

Comment on the results for parts (ii) and (iii).

Solution

(i) $G_T = 10 \ \log (3.12)^2 \ \text{dB} = 9.88 \ \text{dB},$

(ii) $G_S = 10 \ \log \left(\frac{1}{1-(0.69)^2} \right) \ \text{dB} = 2.81 \ \text{dB},$

$G_L = 10 \ \log \left(\frac{1}{1-(0.71)^2} \right) \ \text{dB} = 3.05 \ \text{dB},$

(iii) $G_{UP,max} = (9.88 + 2.81 + 3.05) \ \text{dB} = 15.74 \ \text{dB}.$

Comment: A significant portion of the overall gain comes from the matching networks, showing the importance of including matching networks in a practical design.

5.7 Stability

It follows from the earlier discussion on modified reflection coefficients, Eqs. (5.12) and (5.13), that an active network will become unstable if $S'_{11} = 1$ or $S'_{22} = 1$, since the reflected voltage will be greater than the incident voltage. The potential instability of a transistor that is to be used for an amplifier circuit can be determined from the *S*-parameters of the transistor using the Rollet Stability Factor (*K*). This factor is calculated from [4]

$$K = \frac{1 + |D|^2 - |S_{11}|^2 - |S_{22}|^2}{2 \ |S_{12}S_{21}|}, \tag{5.24}$$

where $D = S_{11}S_{22} - S_{12}S_{21}$.

If $K > 1$, and $|D| < 1$, then the transistor is unconditionally stable for any combination of source and load impedance. The second requirement, namely $|D| < 1$, is almost always true for high-frequency transistors and is often omitted from stability specifications.

If $K < 1$, the transistor is potentially unstable, and will oscillate with certain combinations of source and load impedance. *(Normally a different transistor would be chosen for the particular application if K was less than 1; if this is not possible, then care is needed to check for possible instability over the whole of the expected working frequency range.)*

The easiest way of dealing with situations where *K* is less than 1 is to make use of stability circles. These are circles drawn on the Smith chart that represent the boundary between those values of source or load impedance that cause instability and those which result in stable operation. Stability circles will be considered in more detail in Chapter 9, where the stability of small-signal amplifiers is discussed.

5.8 Supplementary Problems

Q5.1 Write the S-matrix representing an ideal amplifier that has a gain of 22 dB.

Q5.2 Determine the S-matrix representing a 25 mm length of 50 Ω microstrip line at 1 GHz. The line has been fabricated on a substrate that has a relative permittivity of 9.8. The line may be assumed to be lossless.

Q5.3 The following S-matrix represents an RF amplifier:

$$[S] = \begin{bmatrix} 0.12\angle 24° & 0.08\angle -42° \\ 8.17\angle 130° & 0.21\angle 53° \end{bmatrix}.$$

Determine the input reflection coefficient of the amplifier if the output is terminated:
(i) With 50 Ω.
(ii) With a load impedance $(74 - j14)\,\Omega$.

Q5.4 An amplifying device represented by the following S-matrix is connected between a 50 Ω source and a 50 Ω load.

$$[S] = \begin{bmatrix} 0.71\angle -35° & 0.01\angle 77° \\ 4.52\angle -130° & 0.83\angle 91° \end{bmatrix}.$$

Making any reasonable assumptions determine how much additional gain would be obtained by using source and load matching networks.

Q5.5 A particular RF device has two states represented by the following S-matrices.

$$\text{State 1: } [S] = \begin{bmatrix} 0.08\angle 17° & 0.95\angle -62° & 0.01\angle 154° \\ 0.94\angle -63° & 0.03\angle -53° & 0.01\angle 75° \\ 0.01\angle 152° & 0.01\angle 76° & 0.05\angle 91° \end{bmatrix}.$$

$$\text{State 2: } [S] = \begin{bmatrix} 0.07\angle 19° & 0.01\angle 72° & 0.97\angle -64° \\ 0.01\angle 73° & 0.04\angle -51° & 0.01\angle 75° \\ 0.96\angle -65° & 0.01\angle 76° & 0.06\angle 87° \end{bmatrix}.$$

Deduce the function of the device by inspection of the S-parameter matrices.

Q5.6 The performance of a particular FET is represented by the following S-matrix:

$$[S] = \begin{bmatrix} 0.71\angle -35° & 0.01\angle 77° \\ 4.52\angle -130° & 0.83\angle 91° \end{bmatrix}.$$

Determine the overall S_{21} of two of the specified FETs connected in cascade.

Q5.7 Three two-port networks with S-matrices $[S]^A$, $[S]^B$, and $[S]^C$, are connected in cascade. Show that the overall S_{21} of the cascade is given by

$$S_{21} = \frac{S_{21}^A \times S_{21}^B \times S_{21}^C}{1 - (S_{22}^A \times S_{11}^B) - (S_{22}^B \times S_{11}^C) - (S_{21}^B \times S_{11}^C \times S_{12}^B \times S_{22}^A) + (S_{22}^A \times S_{11}^B \times S_{22}^B \times S_{11}^C)}.$$

Q5.8 The following S-matrix applies to a general purpose microwave FET that is connected between a source with impedance 30 Ω and a load having an impedance of 75 Ω.

$$[S] = \begin{bmatrix} 0.77\angle 47° & 0.18\angle -52° \\ 1.41\angle -66° & 0.46\angle 38° \end{bmatrix}.$$

Determine:
  (i) Operating power gain in dB
  (ii) Available power gain in dB
  (iii) Transducer power gain in dB
  (iv) Unilateral transducer power gain in dB.

Q5.9 Suppose that the FET specified in Q5.8 was connected between a 50 Ω source and a 50 Ω load, and that source and load matching networks were used to provide conjugate impedance matching at the input and output of the FET. Calculate the maximum unilateral transducer power gain.

Appendix 5.A Relationships Between Network Parameters

S-parameters are by far the most convenient and useful network parameters for RF and microwave work. However, it is also useful to know how to convert from S-parameters to other common network parameters, and vice versa. In this appendix relationships between S-parameters and three other parameter sets, namely $ABCD$-parameters, Y-parameters, and Z-parameters, are stated. In all cases it is assumed that the input and output impedances, Z_o, are equal and real. The derivations of the various relationships are straightforward, but somewhat lengthy, and can be found in a number of textbooks, for example [5, 6], and have not been repeated here.

A more extensive table of parameter relationships is given by Frickey [7], who also considers the cases of unequal complex input and output impedances. However, Frickley's paper should be read in conjunction with the comments of Marks and Williams [8], who provide useful clarification on Frickley's system of wave definitions.

5.A.1 Transmission Parameters (*ABCD* Parameters)

The conventional voltage and current directions for a two-port network are shown in Figure 5.12.
Referring to Figure 5.12 the transmission parameters are defined by

$$\begin{bmatrix} V_1 \\ I_1 \end{bmatrix} = \begin{bmatrix} A & B \\ C & D \end{bmatrix} \begin{bmatrix} V_2 \\ -I_2 \end{bmatrix},$$

where

$$A = \left. \frac{V_1}{V_2} \right|_{I_2=0} \qquad B = \left. \frac{V_1}{-I_2} \right|_{V_2=0},$$

$$C = \left. \frac{I_1}{V_2} \right|_{I_2=0} \qquad D = \left. \frac{I_1}{-I_2} \right|_{V_2=0},$$

and the relationships with S-parameters are

$$S_{11} = \frac{A + BY_0 - CZ_0 - D}{A + BY_0 + CZ_0 + D} \qquad S_{12} = \frac{2(AD - BC)}{A + BY_0 + CZ_0 + D};$$

$$S_{21} = \frac{2}{A + BY_0 + CZ_0 + D} \qquad S_{22} = \frac{-A + BY_0 - CZ_0 + D}{A + BY_0 + CZ_0 + D};$$

$$A = \frac{(1 + S_{11})(1 - S_{22}) + S_{12}S_{21}}{2S_{21}} \qquad B = \frac{(1 + S_{11})(1 + S_{22})Z_0 - S_{12}S_{21}Z_0}{2S_{21}};$$

$$C = \frac{(1 - S_{11})(1 - S_{22})Y_0 - S_{12}S_{21}Y_0}{2S_{21}} \qquad D = \frac{(1 - S_{11})(1 + S_{22}) + S_{12}S_{21}}{2S_{21}}.$$

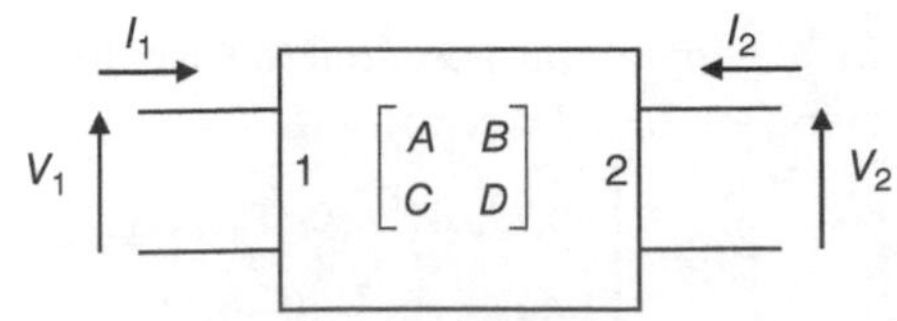

Figure 5.12 Voltage and current directions for a two-port network.

5.A.2 Admittance Parameters (*Y*-Parameters)

Referring to the voltage and current directions specified in Figure 5.12 the admittance parameters are defined by

$$\begin{bmatrix} I_1 \\ I_2 \end{bmatrix} = \begin{bmatrix} Y_{11} & Y_{12} \\ Y_{21} & Y_{22} \end{bmatrix} \begin{bmatrix} V_1 \\ V_2 \end{bmatrix},$$

where

$$Y_{11} = \left(\frac{I_1}{V_1} \right)_{V_2=0} \quad Y_{12} = \left(\frac{I_1}{V_2} \right)_{V_1=0},$$

$$Y_{21} = \left(\frac{I_2}{V_1} \right)_{V_2=0} \quad Y_{22} = \left(\frac{I_2}{V_2} \right)_{V_1=0},$$

and the relationships with *S*-parameters are

$$S_{11} = \frac{(1 - Y_{11}Z_0)(1 + Y_{22}Z_0) + Y_{12}Y_{21}Z_0^2}{(1 + Y_{11}Z_0)(1 + Y_{22}Z_0) - Y_{12}Y_{21}Z_0^2} \quad S_{12} = \frac{-2Y_{12}Z_0}{(1 + Y_{11}Z_0)(1 + Y_{22}Z_0) - Y_{12}Y_{21}Z_0^2},$$

$$S_{21} = \frac{-2Y_{21}Z_0}{(1 + Y_{11}Z_0)(1 + Y_{22}Z_0) - Y_{12}Y_{21}Z_0^2} \quad S_{22} = \frac{(1 + Y_{11}Z_0)(1 - Y_{22}Z_0) + Y_{12}Y_{21}Z_0^2}{(1 + Y_{11}Z_0)(1 + Y_{22}Z_0) - Y_{12}Y_{21}Z_0^2},$$

$$Y_{11} = \frac{(1 - S_{11})(1 + S_{22}) + S_{12}S_{21}}{(1 + S_{11})(1 + S_{22}) - S_{12}S_{21}} \times \frac{1}{Z_0} \quad Y_{12} = \frac{-2S_{12}}{(1 + S_{11})(1 + S_{22}) - S_{12}S_{21}} \times \frac{1}{Z_0},$$

$$Y_{21} = \frac{-2S_{21}}{(1 + S_{11})(1 + S_{22}) - S_{12}S_{21}} \times \frac{1}{Z_0} \quad Y_{22} = \frac{(1 + S_{11})(1 - S_{22}) + S_{12}S_{21}}{(1 + S_{11})(1 + S_{22}) - S_{12}S_{21}} \times \frac{1}{Z_0}.$$

5.A.3 Impedance Parameters (*Z*-Parameters)

Referring to the voltage and current directions specified in Figure 5.12, the impedance parameters are defined by

$$\begin{bmatrix} V_1 \\ V_2 \end{bmatrix} = \begin{bmatrix} Z_{11} & Z_{12} \\ Z_{21} & Z_{22} \end{bmatrix} \begin{bmatrix} I_1 \\ I_2 \end{bmatrix},$$

where

$$Z_{11} = \left(\frac{V_1}{I_1} \right)_{I_2=0} \quad Z_{12} = \left(\frac{V_1}{I_2} \right)_{I_1=0},$$

$$Z_{21} = \left(\frac{V_2}{I_1} \right)_{I_2=0} \quad Z_{22} = \left(\frac{V_2}{I_2} \right)_{I_1=0},$$

and the relationships with *S*-parameters are

$$S_{11} = \frac{(Z_{11} - Z_0)(Z_{22} + Z_0) - Z_{12}Z_{21}}{(Z_{11} + Z_0)(Z_{22} + Z_0) - Z_{12}Z_{21}} \quad S_{12} = \frac{2Z_{12}Z_0}{(Z_{11} + Z_0)(Z_{22} + Z_0) - Z_{12}Z_{21}},$$

$$S_{21} = \frac{2Z_{21}Z_0}{(Z_{11} + Z_0)(Z_{22} + Z_0) - Z_{12}Z_{21}} \quad S_{22} = \frac{(Z_{11} + Z_0)(Z_{22} - Z_0) - Z_{12}Z_{21}}{(Z_{11} + Z_0)(Z_{22} + Z_0) - Z_{12}Z_{21}},$$

$$Z_{11} = \frac{(1 + S_{11})(1 - S_{22})Z_0 + S_{12}S_{21}Z_0}{(1 - S_{11})(1 - S_{22}) - S_{12}S_{21}} \quad Z_{12} = \frac{2S_{12}Z_0}{(1 - S_{11})(1 - S_{22}) - S_{12}S_{21}},$$

$$Z_{21} = \frac{2S_{21}Z_0}{(1 - S_{11})(1 - S_{22}) - S_{12}S_{21}} \quad Z_{22} = \frac{(1 - S_{11})(1 + S_{22})Z_0 + S_{12}S_{21}Z_0}{(1 - S_{11})(1 - S_{22}) - S_{12}S_{21}}.$$

References

1 Mason, S.I. (1953). Feedback theory – some properties of signal flow graphs. *Proceedings of the Institute of Radio Engineers* 41: 1144–1156.

2 Pozar, D.M. (2012). *Microwave Engineering*, 4e. New York: Wiley.

3 Glover, I.A., Pennock, S.R., and Shepherd, P.R. (2005). *Microwave Devices, Circuits and Subsystems*. Chichester, UK: Wiley.

4 Rollett, J. (1962). Stability and power-gain invariants of linear two ports. *IRE Transactions on Circuit Theory* 9 (1): 29–32.

5 Matthai, G., Young, L., and Jones, E.M.T. (1980). *Microwave Filters, Impedance-Matching Networks, and Coupling Structures*. Norwood, MA: Artech House.

6 Collin, R.E. (1992). *Foundations for Microwave Engineering*, 2e. New York: McGraw-Hill.

7 Frickey, D.A. (1994). Conversions between *S,Z, Y, h, ABCD*, and *T* parameters which are valid for complex source and load impedances. *IEEE Transactions on Microwave Theory and Techniques* 42 (2): 205–2011.

8 Marks, R.B. and Williams, D.F. (1995). Comments on "conversions between *S, Z, Y, h, ABCD*, and *T* parameters which are valid for complex source and load impedances". *IEEE Transactions on Microwave Theory and Techniques* 43 (4): 914–915.

6

Microwave Ferrites

6.1 Introduction

Ferrite materials are important constituents in a number of microwave circuits, primarily because of their potential to provide passive non-reciprocal transmission when magnetized with an external DC magnetic field. Non-reciprocal transmission is fundamental to the performance of two very important passive microwave components, namely isolators and circulators. The traditional method of fabricating ferrite-based microwave devices was to mount the ferrite material in a metallic waveguide, and to use a permanent magnet to provide the required DC magnetic field. Ferrite pucks have also been used in planar microstrip and stripline circuits, again with a permanent magnet to provide the magnetizing field. However, the use of a permanent magnet makes planar circuits bulky and costly. More recently, the need for permanent magnets has been overcome by incorporating magnetizing windings within multilayer packages. This makes the integration of ferrite devices within planar circuits more practical, and is an important aspect of the development of system-in-package (SiP) devices. Also, the development of ferrite materials that can be screen printed or deposited as thin films has made the inclusion of non-reciprocal components within hybrid microwave integrated circuits more feasible. In related work, ferrite LTCC tapes have been developed that permit non-reciprocal components to be included within low-cost, highly integrated multilayer packages. Research in ferrite materials for RF and microwave devices is still a very active area, and among recent advances has been the development of self-biasing ferrites that obviate the need for an external magnetic field.

This chapter commences with a review of the principal properties of ferrites, sufficient to understand their use in important RF and microwave devices. A detailed description of the chemical composition of ferrites and the associated manufacturing techniques is outside the scope of the present textbook, but readers interested in these aspects are recommended to consult the extended review paper by Harris [1]. This paper provides an excellent, in-depth description of modern microwave ferrites, and is supported by a copious list of references.

The chapter concludes with an introduction to some specific examples of ferrite devices, both in waveguide and planar circuits.

6.2 Basic Properties of Ferrite Materials

6.2.1 Ferrite Materials

Ferrites refer to a class of insulating materials that exhibit strong magnetic effects. The behaviour of these materials at microwave frequencies can be understood through consideration of the properties of spinning electrons.

The electrons within the atoms of all materials are spinning about their own axes. The rotation of electric charge associated with this spinning motion creates a magnetic moment such that the spinning electron is equivalent to a small magnetic dipole, with the direction of the dipole coincident with the axis of spin, as shown in Figure 6.1.

In most materials the spin axes of the electrons are aligned and occur in anti-parallel pairs, so there is no net magnetic effect. In magnetic materials, however, there are some unpaired electron spins within the atoms, and so the materials exhibit some magnetic properties.

RF and Microwave Circuit Design: Theory and Applications, First Edition. Charles E. Free and Colin S. Aitchison.
© 2022 John Wiley & Sons Ltd. Published 2022 by John Wiley & Sons Ltd.
Companion website: www.wiley.com/go/free/rfandmicrowave

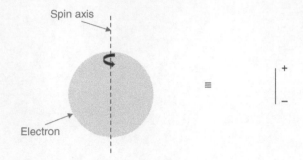

Figure 6.1 (a) Spinning electron and (b) equivalent magnetic dipole.

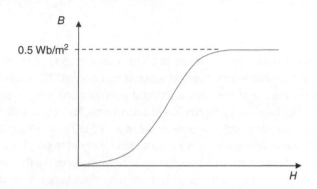

Figure 6.2 Non-linear B-H characteristic of ferrites.

Materials may be broadly classified into three types:

Ferromagnetic. In these materials there is strong coupling between the atoms, which results in all the spin axes of the unpaired electrons being aligned. The equivalent magnetic dipoles then act in the same direction, and consequently the material exhibits strong magnetic effects. Iron is an example of a ferromagnetic material.

Ferrimagnetic. The effect of coupling between atoms in these materials is to divide the electron spins into two groups having oppositely oriented spins. But the number of electron spins within each group is not equal, and so the material still exhibits some magnetic properties.

Antiferromagnetic. In these materials the number of electron spins within each group of oppositely directed spins is equal, and so there is no net magnetic field. These materials are therefore said to be non-magnetic.

Ferrites are examples of low-loss ferrimagnetic materials, and most have a polycrystalline structure, with very high resistivity. The high resistivity means that losses associated with eddy currents are very small. They have a non-linear B-H characteristic, which saturates at around $0.5 \, \text{Wb/m}^2$ as shown in Figure 6.2.

Ferrite materials are available in a number of forms. The traditional method of manufacturing ferrites is to sinter together a mixture of metallic oxide (MO) and ferric oxide (Fe_2O_3) powders, where M is a divalent metal such as magnesium or zinc. The resulting mixture is then represented by the general chemical formula $MOFe_2O_3$. This type of manufacturing technique is used to provide a bulk ferrite material for use in a range of passive microwave devices, such as metallic waveguide isolators and circulators, and for producing ferrite pucks for surface-mounting in hybrid microwave integrated circuits. However, due to advances in manufacturing techniques, ferrite is now available in a wide range of formats for planar circuit applications. These include thin-film deposition, and pastes for thick-film printing. Low-loss ferrite substrates can now be produced with very low surface roughness. More recently, ferrite tapes have been produced for use in multilayer, LTCC structures.

6.2.2 Precession in Ferrite Materials

If ferrite material is placed in a DC magnetic field, H_o, the magnetic dipoles (i.e. the spin axes) will try and align themselves with the DC field. However, the spin axes will not initially move directly into alignment with the external field because the spinning motion of the electrons exert a torque, which causes the individual spin axes to precess gyroscopically around the direction of the DC magnetic field as shown in Figure 6.3. They eventually align with the DC magnetic field.

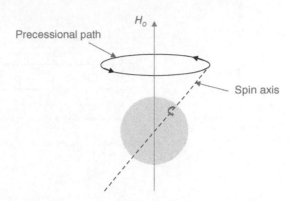

Figure 6.3 Precessional motion of an electron.

The free angular precessional frequency, i.e. the angular precessional frequency when the material is not subject to any external AC magnetic fields, is known as the Larmor angular frequency, ω_o, and given by

$$\omega_o = 2\pi\gamma H_o, \tag{6.1}$$

where γ is the gyromagnetic ratio, and H_o is the DC magnetic field strength. The gyromagnetic ratio is the ratio of the magnetic moment, m, of an electron, to its angular momentum, P, i.e.

$$\gamma = \frac{m}{P}. \tag{6.2}$$

For most magnetic materials, $\gamma = 35.18$ kHz.m/A. (*Note that the value of γ is often expressed in Oersted [Oe] units, where 1 Oe = 79.58 A/m. Then in Oersted units, $\gamma = 2.8$ MHz/Oe.*)

Since there are frictional losses and other damping forces within the ferrite material, the precessional motion will not continue in a circle, as shown in Figure 6.3, but rather will spiral in towards the centre as the kinetic energy of the electrons is converted into heat, and eventually the spin axis of each electron will align itself with the external DC magnetic field.

An interesting (and useful) situation arises if a small AC magnetic field, H_{AC}, is applied to the ferrite in a direction perpendicular to that of H_o, with an angular frequency equal to ω_o. If the AC magnetic field is in the form of a square wave, then the resultant magnetic field, H_R, will periodically switch between two directions X and Y, as shown in Figure 6.4.

If we assume that at time $t = 0$ the precessional path around direction X has reached point 1, and that the square wave signal is switched so that the resultant lies in direction Y, the spin axis will start to precess around direction Y, with an increased displacement. After half a period the spin axis will have precessed to point 2, whereupon the direction of the AC field is switched again and the spin axis will start to precess around direction X, reaching point 3 after a further half period. Thus, the amplitude of the precessional path will build up until an equilibrium situation is reached where the friction and damping energy losses within the material equal the energy received from the precessing electrons.

For the purpose of illustration, Figure 6.4 shows the precessional path in the form of circles around the two directions X and Y. However, as the equilibrium position is approached the precession path takes the form of a spiral because the displacement of the path is gradually reduced until there is a balance between the energy dissipated and that supplied from the AC magnetic field. When the frequency of the disturbing AC field is equal to the natural precession frequency, then

$$T = \frac{2\pi}{\omega_o}, \tag{6.3}$$

and the AC field effectively induces resonance within the ferrite, and energy is taken from the perturbing AC field and dissipated as heat within the ferrite. This is known as resonance absorption, and is the theoretical basis of the waveguide resonance isolator, which will be described in a later section of this chapter.

6.2.3 Permeability Tensor

The very high resistivities of ferrite materials means that they are essentially insulators, and so electromagnetic waves can penetrate the material. There will be an interaction between the electron spins and the applied high-frequency waves. In

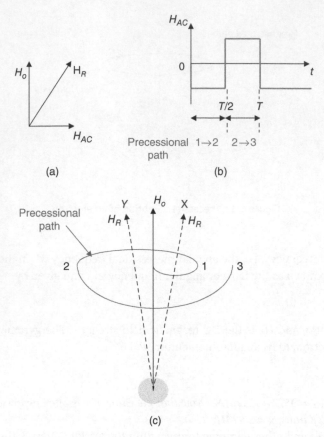

Figure 6.4 Forced precession in ferrites: (a) orthogonal AC magnetic field added to H_o, (b) AC magnetic field in form of a square wave, and (c) precessional path due to disturbing AC field.

the case of a perfect, homogeneous, non-magnetic insulator, the relationship between the magnetic flux density, **B**, and the applied magnetic field, **H**, is simply

$$B = \mu_0 \mu_r H, \tag{6.4}$$

where μ_o is the permeability of free space, and μ_r is the relative permeability of the insulator material. It should be noted that **B** and **H** are shown in bold to indicate that they are vector quantities, which means in this case that they have both magnitude and direction.

It can be shown [2] that if a high-frequency plane wave is propagated through a ferrite material in the same direction as the DC magnetic field, the interaction between the spinning electrons and the wave causes RF field components in one transverse direction to generate additional orthogonal components with a 90° phase difference. For example, if a microwave plane wave is propagating in the z-direction (the same direction as H_o) then

$$B_x = \mu H_x - jk H_y, \tag{6.5}$$

$$B_y = jk H_x + \mu H_y, \tag{6.6}$$

$$B_z = \mu_0 H_z, \tag{6.7}$$

where

$$\mu = \mu_0 \left(1 + \frac{\omega_0 \omega_m}{\omega_0^2 - \omega^2} \right), \tag{6.8}$$

$$k = \mu_0 \frac{\omega_m \omega}{\omega_0^2 - \omega^2}, \tag{6.9}$$

$$\omega_m = \frac{2\pi \gamma M_0}{\mu_0}, \tag{6.10}$$

and ω_o is the Larmor frequency, ω is the angular frequency of the microwave signal, and M_o is the saturation magnetization within the ferrite. Magnetization refers to the density of the magnetic field and is expressed in the units of Weber per square metre (Wb/m^2). (*Sometimes the magnetization of a ferrite is expressed in the units of Tesla (T): 1 T $\equiv$ 1 Wb/m^2.*)

Equations (6.5) through (6.7) can be combined as

$$B = \mu H, \tag{6.11}$$

where μ is known as the permeability tensor, and represented by the matrix

$$\mu = \begin{bmatrix} \mu & -jk & 0 \\ jk & \mu & 0 \\ 0 & 0 & \mu_o \end{bmatrix}. \tag{6.12}$$

Example 6.1 The following data apply to a ferrite in a DC magnetizing field of magnitude 80 kA/m: $\gamma = 35.18$ kHz.m/A; $\omega_m = 2\pi \times 2$ GHz. Determine: (i) the Larmor angular frequency; (ii) the values of the elements of the permeability tensor when a 1.5 GHz microwave signal is applied to the ferrite.

Solution

(i) $\omega_o = 2\pi \times 35.18 \times 10^3 \times 80 \times 10^3$ rad/s

$\qquad = 17.69 \times 10^9$ rad/s.

(ii) $\omega_o^2 - \omega^2$

$\qquad = [(17.69 \times 10^9)^2 - (2\pi \times 1.5 \times 10^9)^2]$ rad^2/s^2

$\qquad = 2.24 \times 10^{20}$ rad^2/s^2.

$\mu = \mu_o \left(1 + \dfrac{17.69 \times 10^9 \times 2\pi \times 2 \times 10^9}{2.24 \times 10^{20}} \right)$

$\qquad = \mu_o(1 + 0.99) = 1.99\mu_o.$

$k = \mu_o \left(\dfrac{2\pi \times 2 \times 10^9 \times 2\pi \times 1.5 \times 10^9}{2.24 \times 10^{20}} \right)$

$\qquad = \mu_o \left(\dfrac{1.18 \times 10^{20}}{2.24 \times 10^{20}} \right) = 0.53\mu_o.$

$\mu = \begin{bmatrix} 1.99\mu_o & -j0.53\mu_o & 0 \\ j0.53\mu_o & 1.99\mu_o & 0 \\ 0 & 0 & \mu_o \end{bmatrix}$

$\quad = \mu_o \begin{bmatrix} 1.99 & -j0.53 & 0 \\ j0.53 & 1.99 & 0 \\ 0 & 0 & 1 \end{bmatrix}.$

6.2.4 Faraday Rotation

When a linearly polarized (LP) wave propagates through ferrite in the same direction as an applied DC magnetic field, the plane of polarization of the wave will be rotated. This phenomenon is known as Faraday rotation, and can be explained through consideration of the behaviour of circularly polarized (CP) waves within the material.

Any LP wave can be decomposed into the sum of two CP waves with opposite hands of rotation, as depicted in Figure 6.5.

Consider an LP wave applied to a cylinder of ferrite material together with an axial DC magnetic field, of strength H_o, as shown in Figure 6.6.

We know that the spinning electrons within the ferrite will be precessing around the direction of the DC magnetic field. With the direction of H_o shown in Figure 6.6, the electrons will be precessing in a clockwise direction. Also, we know that the LP wave can be decomposed into two CP waves rotating in opposite directions. Thus, in the ferrite the CP wave with clockwise rotation will be closely coupled to the precessional motion of the electrons, whilst the CP wave with counter-clockwise

Figure 6.5 Linearly polarized wave decomposed into the sum of two circularly polarized waves.

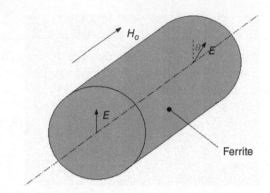

Figure 6.6 Linearly polarized wave applied to ferrite.

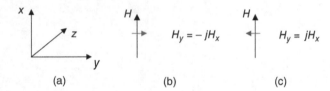

Figure 6.7 Magnetic field vector rotation in x–y plane: (a) coordinate system, (b) clockwise rotation, and (c) counter-clockwise rotation.

rotation will be essentially unaffected by the precessional motion of the electrons. This means that each of the CP waves will have a different velocity of propagation within the ferrite, and hence a different transmission phase. Thus, when the two CP waves recombine at the output of the ferrite, the resulting LP wave will appear to have been rotated through some angle θ_1, as shown in Figure 6.6. The value of θ will depend on the properties of the ferrite material, the strength of the axial magnetic field, and the length of the ferrite cylinder. (This rotation is called Faraday rotation, after Michael Faraday, who first observed this phenomenon through experiments in optics.)

We can obtain an expression for θ in terms of the elements of the permeability tensor, by first comparing Eq. (6.5) with the magnetic field relationships for a CP wave. Figure 6.7 shows a magnetic field vector, H, propagating in the z-direction and rotating in the x–y plane.

For clockwise, or positive rotation, the y component of magnetic field will always lead the x component by 90°, and so $H_y = jH_x$. Substituting this relationship into Eq. (6.5) gives, for a loss-less ferrite material

$$B_x = \mu H_x - jk H_y$$
$$= \mu H_x - k H_x$$
$$= (\mu - k)H_x \tag{6.13}$$

and so the effective permeability of a clockwise rotating CP wave is

$$(\mu_{eff})_+ = \mu - k. \tag{6.14}$$

Similarly, substituting $H_y = -jH_x$ into Eq. (6.5) to represent a counter-clockwise rotating CP wave gives

$$B_x = (\mu + k)H_x \tag{6.15}$$

and

$$(\mu_{eff})_- = (\mu + k), \tag{6.16}$$

where in Eqs. (6.13) and (6.15) we have used the accepted convention of subscripts + and − to represent clockwise and counter-clockwise CP waves, respectively.

The phase propagation constants for the clockwise and counter-clockwise CP waves are then

$$\beta_+ = \frac{2\pi}{\lambda_+} = \frac{2\pi f}{(v_p)_+} = \frac{\omega}{c}\sqrt{\varepsilon_r \frac{(\mu_{eff})_+}{\mu_0}} = \frac{\omega}{c}\sqrt{\varepsilon_r \frac{\mu - k}{\mu_0}},$$

i.e.

$$\beta_+ = \omega\sqrt{\varepsilon}\sqrt{\mu - k} \tag{6.17}$$

and similarly

$$\beta_- = \omega\sqrt{\varepsilon}\sqrt{\mu + k} \tag{6.18}$$

where ε and ε_r are the permittivity and relative permittivity, respectively, of the ferrite material, and the other symbols have their usual meanings. Thus, through a length, l, of ferrite the phases of the clockwise and counter-clockwise CP waves will have changed by $\beta_+ l$ and $\beta_- l$, respectively.

When the two CP waves recombine after transmission through the ferrite the plane of polarization of the LP wave will appear to have been rotated by

$$\theta = \frac{\beta_+ l - \beta_- l}{2} = \frac{\omega\sqrt{\varepsilon}}{2}\left(\sqrt{\mu - k} - \sqrt{\mu + k}\right)l. \tag{6.19}$$

The sign of θ, i.e. whether the plane of polarization is rotated to the right or left, will depend on the value of k, which can be positive or negative depending on whether the frequency of operation is below or above the resonant frequency of the ferrite material, i.e. the Lamor frequency, which was defined in Section 6.2.2.

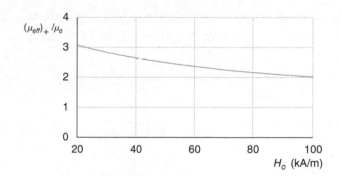

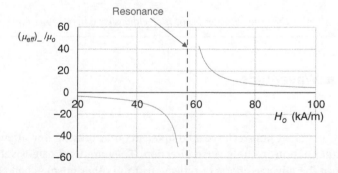

Figure 6.8 Variation of effective permeabilities of CP waves in ferrite as a function of magnetizing field at 2 GHz. Ferrite is magnetized in the direction of propagation ($\gamma = 35$ kHz.m/A $\quad \omega_m = 2\pi \times 5.6 \times 10^9$ rad/s).

The transmission phase change through the cylinder of ferrite is given by

$$\phi = \frac{\beta_+ l + \beta_- l}{2} = \frac{\omega\sqrt{\varepsilon}}{2}\left(\sqrt{\mu - k} + \sqrt{\mu + k}\right)l, \tag{6.20}$$

i.e.

$$\phi = \frac{\omega\sqrt{\varepsilon}}{2}[(\mu_{eff})_+ + (\mu_{eff})_-]l. \tag{6.21}$$

The values of the effective permeabilities $(\mu_{eff})_+$ and $(\mu_{eff})_-$ are dependent on the frequency of operation, the value of the DC magnetic field, and the parameters of the ferrite material. As an illustration, Figure 6.8 shows the variation of these permeabilities as a function of magnetizing field strength, above saturation, in the region of resonance. It can be seen that the effective permeability of CP waves with clockwise rotation remains substantially constant, but those waves with counter-clockwise rotation experience significant changes in permeability on either side of the resonance position. In particular, it can be seen that there is a change of sign of the effective permeability of counter-clockwise rotating waves on either side of resonance.

Example 6.2 The permeability tensor at 6 GHz of a cylinder of ferrite magnetized in the axial direction (i.e. z-direction) is given by

$$\mu = \mu_o \begin{bmatrix} 2.43 & -j1.02 & 0 \\ j1.02 & 2.43 & 0 \\ 0 & 0 & 1 \end{bmatrix}.$$

Determine the Faraday rotation in a 1 cm length of the cylinder at 6 GHz, given that the ferrite has a dielectric constant of 12.

Solution
Comparing the given permeability tensor with Eq. (6.12) gives

$$\mu = 2.43\mu_o = 2.43 \times 4\pi \times 10^{-7} \text{ H/m} = 30.54 \times 10^{-7} \text{ H/m},$$
$$k = 1.02\mu_o = 1.02 \times 4\pi \times 10^{-7} \text{ H/m} = 12.82 \times 10^{-7} \text{ H/m},$$
$$\mu - k = (30.54 - 12.82) \times 10^{-7} \text{ H/m} = 17.72 \times 10^{-7} \text{ H/m},$$
$$\mu + k = (30.54 + 12.82) \times 10^{-7} \text{ H/m} = 43.36 \times 10^{-7} \text{ H/m}.$$

Using Eq. (6.19):

$$\theta = \frac{\omega\sqrt{\varepsilon_o \varepsilon}}{2}(\sqrt{\mu - k} - \sqrt{\mu + k})l$$

$$= \frac{2\pi \times 6 \times 10^9 \times \sqrt{8.84 \times 10^{-12} \times 12}}{2} \times \left(\sqrt{17.72 \times 10^{-7}} - \sqrt{43.36 \times 10^{-7}}\right) \times 0.01 \text{ rad}$$

$$= -1.46 \text{ rad}$$

$$\equiv -83.64°.$$

Comment: The negative sign indicates that the rotation is counter-clockwise when viewed in the direction of magnetization.

6.3 Ferrites in Metallic Waveguide

6.3.1 Resonance Isolator

Isolators are passive, two-port, matched non-reciprocal components that provide low attenuation in the forward direction and very high matched attenuation in the reverse direction. They are used extensively in microwave circuits and sub-systems, primarily to reduce the adverse effects of reflections from other mismatched components. For example, an isolator can be inserted between a frequency source and a reactive load to prevent the reactance of the load from pulling the frequency of the source off-tune. Another good example of their use is where an isolator is inserted between the output

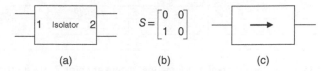

Figure 6.9 Two-port isolator: (a) two-port isolator, (b) ideal *S*-matrix, and (c) circuit symbol.

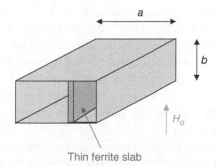

Figure 6.10 Waveguide isolator.

of a high power amplifier and a mismatched load to prevent high power reflections from the load causing damage to the output stage of the amplifier.

An ideal isolator can be represented by the *S*-parameter matrix given in Figure 6.9, which shows zero transmission loss in the forward direction from port 1 to port 2, and infinite loss in the reverse direction from port 2 to port 1, with both ports of the isolator being matched.

Very effective isolators can be fabricated in rectangular metallic waveguide by using an off-centred thin slab of ferrite placed between the broad walls of the waveguide, as shown in Figure 6.10. Also shown in Figure 6.10 is the direction of a DC magnetic field, with a magnitude H_o, which is applied across the ferrite slab. In practical devices the magnetic field is provided by a permanent magnet, assembled around the waveguide.

The operation of the waveguide isolator can be understood by considering the changing magnetic field due to the dominant TE_{10} mode propagating within the waveguide. As was described in Chapter 1, the magnetic field of the dominant mode takes the form of closed loops in the transverse plane. These loops of magnetic field are shown schematically in Figure 6.11. Also shown in this figure is the DC magnetic field, H_o, which is perpendicular to the plane of the RF magnetic field loops, and directed out of the plane of the paper. Thus the free precessional motion of the electrons in the ferrite will be counter-clockwise.

Figure 6.11 shows a point *P* which is off-centre in the waveguide, and in the position occupied by the ferrite slab. If we consider the rotation of the RF magnetic field at *P*, then for a wave propagating in the forward direction from port 1 to port 2 the loops of magnetic field are moving left to right and the magnetic field at *P* is effectively CP, with rotation in the clockwise direction. Similarly, for a wave propagating from port 2 to port 1, the magnetic field at *P* appears to be

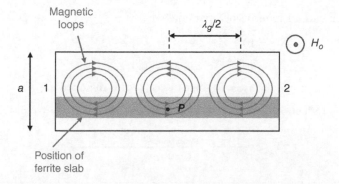

Figure 6.11 Magnetic field of TE_{10} mode in a rectangular waveguide.

	$t = 0$	$t = T/4$	$t = T/2$	$t = 3T/4$
$1 \rightarrow 2$	$\rightarrow$	$\downarrow$	$\leftarrow$	$\uparrow$
$2 \rightarrow 1$	$\rightarrow$	$\uparrow$	$\leftarrow$	$\downarrow$

Figure 6.12 Circular polarization of an RF magnetic field at point *P* as a function of time.

rotating in the counter-clockwise direction. For clarity, the details of the rotations are shown in Figure 6.12, where time $t = 0$, corresponds to the magnetic field distribution shown in Figure 6.11.

Thus, for waves travelling from port 1 to port 2, the clockwise rotation of the RF field opposes the precessional rotation of the electrons in the ferrite, and there is little interaction between the RF field and the precessing electrons, and consequently little transmission loss. But for waves travelling from port 2 to port 1, the counter-clockwise rotation of the RF field reinforces the precessional motion of the electrons, and if the frequency of the RF wave is equal to the precessional frequency, resonance can be induced as described in Section 6.2.2. At resonance energy will be absorbed from the RF wave and dissipated as heat within the ferrite, thus giving high transmission loss from port 2 to port 1. This type of isolator is often referred to as a waveguide resonance isolator.

Waveguide resonance isolators are extremely effective, non-reciprocal components, and typically have less than 0.4 dB attenuation in the forward direction, and greater than 25 dB attenuation in the reverse direction, together with a good match, and are easily manufactured in different waveguide sizes covering the microwave and mm-wave frequency range. In practical components the ends of the ferrite slab are normally tapered to minimize reflections at the input and output ports, and typically the VSWR of waveguide isolators is less than 1.3.

Example 6.3 A signal source that has an output impedance of 50 Ω is connected to an antenna that has an input impedance of $(75 - j47)\,\Omega$.

(i) Determine the VSWR at the output of the signal source.
(ii) Determine the VSWR at the output of the signal source if a matched isolator with the following specification is connected between the signal source and the antenna, with the forward path being that from the source to the antenna:

   Forward path: Loss = 0.27 dB Transmission phase = $-34°$
   Reverse path: Loss = 26.12 dB Transmission phase = $-57°$.

Comment upon the result.

Solution

(i)
$$\Gamma_{antenna} = \frac{Z_{antenna} - Z_o}{Z_{antenna} + Z_o} = \frac{75 - j47 - 50}{75 - j47 + 50}$$

$$= \frac{25 - j47}{125 - j47} = \frac{53.24\angle - 62°}{133.54\angle - 20.61°} = 0.40\angle - 41.39°,$$

$$VSWR = \frac{1 + |\rho_{antenna}|}{1 - |\rho_{antenna}|} = \frac{1 + 0.4}{1 - 0.4} = 2.33.$$

(ii) Writing the isolator specifications in terms of *S*-parameters:

$$S_{21} = 10^{-0.27/20}\angle - 34° = 0.97\angle - 34°,$$

$$S_{12} = 10^{-26.12/20}\angle - 57° = 0.05\angle - 57°,$$

$$S_{11} = S_{22} = 0 \text{ (since the isolator is matched)}.$$

Using Eq. (5.9) to find the reflection coefficient at the input to the isolator:

$$\Gamma_{in} = \frac{S_{11}(1 - S_{22}\rho_{antenna}) + S_{21}\rho_{antenna}S_{12}}{1 - S_{22}\rho_{antenna}}.$$

Substituting data:

$$\Gamma_{in} = 0.97\angle -34° \times 0.40\angle -41.39° \times 0.05\angle -57°$$

$$= 0.0194\angle -132.39°,$$

$$VSWR = \frac{1 + 0.0194}{1 - 0.0194} = 1.04.$$

Comment: The presence of the isolator has significantly reduced the mismatch at the output of the signal source.

6.3.2 Field Displacement Isolator

The field displacement isolator is similar in construction to the resonance isolator discussed in Section 6.2.2, in that it employs a length of ferrite, located off-centre in a rectangular waveguide. The difference is that in the field displacement device the DC magnetic field biases the ferrite to some point below resonance. The function of the ferrite is then simply to modify the field distribution within the waveguide, such that there is a low concentration of field in the vicinity of the ferrite for forward travelling waves, and high field concentration around the ferrite for reverse travelling waves. A resistance card, usually made from mylar with a resistive coating, is positioned next to the ferrite to absorb energy from the reverse travelling waves, thus providing non-reciprocal operation. The cross-section of a field displacement isolator is shown in Figure 6.13.

Since the field displacement isolator does not rely on resonance within the ferrite material it has a much wider bandwidth than the resonance isolator.

6.3.3 Waveguide Circulator

The basic microwave circulator is a matched three-port device in which signals applied at one port circulate around and emerge from only one of the two adjacent ports. In the circulator shown in Figure 6.14a the direction of circulation is clockwise, and so signals applied at port 1 emerge at port 2, signals applied at port 2 emerge at port 3, and signals applied at port 3 emerge at port 1. This behaviour leads to the ideal S-matrix for a three-port circulator shown in Figure 6.14b. There are many uses for circulators in microwave communication systems, and one familiar arrangement is to use the circulator to enable a transmitter and a receiver to share a common antenna, as shown in Figure 6.14c.

In the arrangement shown in Figure 6.14c, the output power from the transmitter is supplied to the antenna, and the non-reciprocal properties of the circulator mean that the output stage of the transmitter is protected from reflections due to antenna mismatch. Also, the sensitive input of the receiver is protected from the relatively high power output of the transmitter, because there is no transmission from port 1 to port 3. Signals received by the antenna are rotated around the circulator and appear at the input of the receiver.

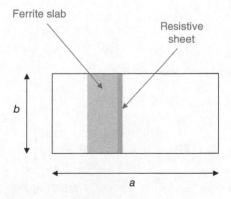

Figure 6.13 View of cross-section of a field displacement isolator in a rectangular waveguide.

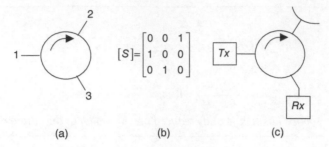

Figure 6.14 Three-port circulator: (a) three-port circulator, (b) ideal *S*-matrix, and (c) typical use.

Example 6.4 A three-port circulator has the following symmetrical performance data:

Insertion loss = 0.34 dB
Isolation = 24.6 dB
VSWR = 1.23

Determine the *S*-matrix for the circulator.

Solution
We will consider the circulator to have the port nomenclature and direction of circulation shown in Figure 6.14a.

Insertion loss:

$$-0.34 = 20\log|S_{21}|,$$
$$\log|S_{21}| = -0.017,$$
$$|S_{21}| = 10^{-0.017} = 0.962,$$

and since the circular has symmetrical performance:

$$|S_{21}| = |S_{32}| = |S_{13}| = 0.962.$$

Isolation:

$$-24.6 = 20\log|S_{12}|,$$
$$\log|S_{12}| = -1.23,$$
$$|S_{12}| = 10^{-1.23} = 0.059,$$

and since the circular has symmetrical performance:

$$|S_{12}| = |S_{23}| = |S_{31}| = 0.059.$$

Matching:

$$VSWR = 1.23 = \frac{1 + |S_{11}|}{1 - |S_{11}|},$$
$$|S_{11}| = 0.103,$$

and since the circular has symmetrical performance:

$$|S_{11}| = |S_{22}| = |S_{33}| = 0.103.$$

Final *S*-matrix:

$$[S] = \begin{bmatrix} 0.103 & 0.059 & 0.962 \\ 0.962 & 0.103 & 0.059 \\ 0.0.059 & 0.962 & 0.103 \end{bmatrix}.$$

Comment: For convenience we have written the elements of the S-matrix of the circulator just in terms of their magnitudes, but it must be appreciated that the S-parameters of a real isolator will be complex, with the angles representing the transmission phases between the various ports.

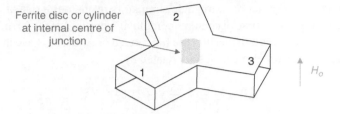

Figure 6.15 *Y*-junction waveguide circulator.

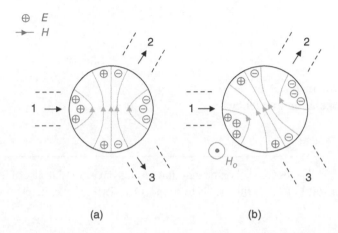

Figure 6.16 Standing wave patterns in a ferrite disc: (a) unmagnetized and (b) magnetized with DC field.

A common configuration for waveguide circulators is the *Y*-junction circulator shown in Figure 6.15. It is formed from the *H*-plane junctions of three rectangular waveguides, which have an angular spacing of 120°. The ferrite is located at the internal centre of the junction, and usually takes the form of a cylinder which occupies the full height of the waveguide. The ferrite is magnetized off-resonance with a DC magnetic field, H_o, which is applied parallel to the axis of the ferrite cylinder. Sometimes the ferrite is in the form of a disc, which is cemented to one of the broad wall faces at the centre of the junction, but does not occupy the full height of the waveguide.

When the dominant TE_{10} mode enters the junction from one of the waveguide ports, two counter-rotating modes are generated within the ferrite cylinder. These two modes interfere to cause a standing wave pattern, with maximum and minimum field intensities at various angular positions around the ferrite disc, which can be considered to act as a dielectric resonator. Figure 6.16a shows the standing wave pattern in the transverse plane of the cylinder when the disc is un-magnetized, and with signal applied at port 1. The field pattern shown in this figure corresponds to the lowest resonant frequency, and is therefore the dominant resonant mode. In this situation, the signal entering from port 1 generates equal amplitude output signals at ports 2 and 3, since the magnitudes of the resonant field in the ferrite are the same adjacent to these ports. When the DC magnetic field is applied the spin axes of the electrons precess around the direction of the DC magnetic field, as described in Section 6.2.2. One of the rotating modes is then closely coupled to the precessional motion, whilst the other is not. This means that the two rotating modes within the ferrite travel with different angular velocities, which causes an angular displacement of the standing wave pattern. By adjusting the strength of the magnetic field, the standing wave pattern can be rotated until a minimum of the pattern is opposite one of the ports, which is then isolated. This is shown in Figure 6.16b where the resultant field in the ferrite is zero adjacent to port 3. Then the signal at port 1 will be coupled to port 2, with port 3 isolated. Since the device is symmetrical about the axis of the ferrite disc, the same behaviour will be observed when a signal is applied to any of the three ports, i.e. the signal will only be coupled only to the adjacent port.

The operation of the *Y*-junction waveguide circulator is such that the electromagnetic field within the ferrite is strongly coupled to the input and output fields within the rectangular waveguides. This means that the device has a low-*Q* and therefore broad bandwidth. The *Y*-junction circulator can also function as a very effective two-port isolator, with the third port terminated with a matched load.

The principal design parameters for Y-junction waveguide circulators relate to the side of the ferrite disc. Helszajn and Tan [3] investigated the performance of this type of circulator employing partial-height ferrite resonators. They concluded that the best performance was obtained when the ferrite disc functioned as a quarter-wave dielectric resonator supporting the dominant TM_{11} mode, where the ferrite dimensions are related by

$$\beta_o R = \frac{1}{\sqrt{\varepsilon_r}} \left[\left(\frac{\pi R}{2t} \right)^2 + (1.84)^2 \right]^{0.5}, \tag{6.22}$$

where

$$\beta_o = \frac{2\pi}{\lambda_o} \tag{6.23}$$

and

R = radius of ferrite disc
t = thickness of ferrite disc
ε_r = relative permittivity of ferrite material
λ_o = free-space wavelength at operating frequency.

Example 6.5   A 10 GHz ferrite disc resonator is to be manufactured from a circular disc of ferrite which has a thickness of 2.5 mm and a dielectric constant of 13. If the disc is to support the TM_{11} mode, determine the required diameter of the disc.

Solution
10 GHz $\Rightarrow \lambda_o = 30$ mm.
Rearranging Eq. (6.22):

$$(\beta_o R)^2 = \frac{1}{\varepsilon_r} \left[\left(\frac{\pi R}{2t} \right)^2 + (1.84)^2 \right],$$

$$R^2 \left(\beta_o^2 - \frac{\pi^2}{4\varepsilon_r t^2} \right) = \frac{(1.84)^2}{\varepsilon_r},$$

$$R = \frac{1.84}{\sqrt{\varepsilon_r}} \left(\beta_o^2 - \frac{\pi^2}{4\varepsilon_r t^2} \right)^{-0.5}.$$

Substituting data:

$$R = \frac{1.84}{\sqrt{13}} \left(\left(\frac{2\pi}{30} \right)^2 - \frac{\pi^2}{4 \times 13 \times (2.5)^2} \right)^{-0.5} \text{ mm} = 4.40 \text{ mm},$$

Diameter = 2×4.40 mm = 8.80 mm.

The above discussion of the Y-junction circulator has been based upon the use of a circular ferrite cylinder. However, the use of a ferrite cylinder leads to relatively high VSWR values at the three ports of the circulator, due to reflections from the face of the cylinder as waves enter the junction. In many practical devices the ferrite cylinder is replaced by triangular-shaped ferrite, as shown in Figure 6.17. With this arrangement ferrite presents a tapered discontinuity to the waveguide ports, which reduces the reflections and hence reduces the VSWR at the three ports.

The triangular-shaped ferrite behaves in a similar way to the cylindrical ferrite, in that counter-rotating modes lead to a resonant standing-wave pattern. A more extensive discussion of the use of triangular-shaped ferrites is given by Linkhart [4], who also provides practical design information on ferrite dimensions. The use of a triangular shape has the additional advantage of preventing coupling to higher-order modes in broadband operation.

Two additional points to note concerning waveguide circulators:

(i) Section 6.3.3 has focussed on three-port waveguide circulators, since these are the most common types of circulator. But four-port circulators can also be designed and fabricated using various techniques; Collin [2] describes two types

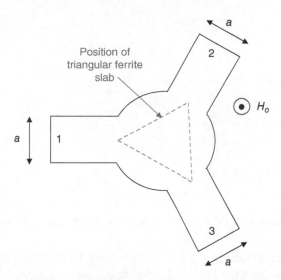

Figure 6.17 Triangular ferrite in a Y-junction circulator.

of four-port circulator, one using a combination of rectangular waveguide junctions and a gyrator,[1] and the other using a circular waveguide.

(ii) A three-port circulator with one port terminated in a matched load behaves as an isolator.

6.4 Ferrites in Planar Circuits

6.4.1 Planar Circulators

The principles of the three-port waveguide circulator using a magnetized ferrite disc can be applied directly to microstrip circuits. The common method of construction of a microstrip Y-junction circulator is shown in Figure 6.18.

In this arrangement, a ferrite disc is embedded in a circular cavity which has been laser-cut in the surface of a low-loss ceramic substrate. The top surface of the disc is metallized and is flushed with the top surface of the substrate. The metallization is bonded to microstrip tracks to form the continuous microstrip layer depicted in Figure 6.18, with the three ports having a mutual angular spacing of 120°. A DC magnetic field, H_o, is applied in a direction perpendicular to the plane of the substrate to magnetize the ferrite disc. This field is normally obtained from a small permanent magnet attached to the substrate.

A recent development in planar circuits is the integration of ferrite Y-junction circulators within an LTCC structure as part of a SiP transceiver. Van Dijk and colleagues [5] embedded a pre-sintered ferrite disc into a cavity within a multilayer

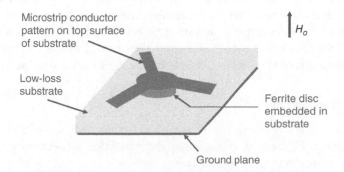

Figure 6.18 Y-junction circulator in a microstrip.

---

1 A gyrator is a lossless non-reciprocal two-port component with a 180° difference between the transmission phases in the forward and reverse directions.

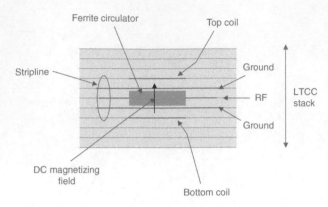

Figure 6.19 LTCC ferrite circulator with integrated winding.

LTCC structure. In their work the shrinkage of the LTCC and the dimensions of the ferrite disc were matched so that a tight fit was achieved between the ferrite and the LTCC cavity after firing, without introducing stress in the finished package. After manufacturing the LTCC stack with the integrated ferrite, magnets were mounted on the surface of the structure to provide the required DC magnetizing field. Measured *S*-parameter data were reported in [5] for circulator devices at C-band and Ku-band. The data showed the typical insertion loss and isolation at C-band to be around 0.5 and 22 dB, respectively. Similar data at Ku-band gave the mid-band insertion loss and isolation as 1 and 24 dB, respectively. These data demonstrate the viability of integrating ferrite components within an LTCC package and significantly enhance the capability of LTCC SiP structures.

One of the drawbacks to integrating ferrites within planar packages has been the need for magnets to provide the DC magnetizing field. These magnets are relatively heavy and bulky, and detract from the lightweight, compact structures that are possible using LTCC and multilayer thick-film technology. Yang and colleagues [6] developed an innovative structure that incorporated planar windings within an LTCC package to provide the DC magnetic field, thus overcoming the need for magnets. The basic concept of the new structure is illustrated in Figure 6.19.

In the LTCC stack shown in Figure 6.19, the ferrite component is formed in stripline, with the ferrite material between the two ground planes. Two planar coils are positioned on LTCC layers immediately above and below the ground planes, and these provide a vertical DC magnetic field through the ferrite. Positioning the coils outside the stripline ensures that there is no interaction between the conductors forming the coils and the RF signal propagating through the ferrite component. Data presented in [6] show that a circulator fabricated with the new structure had an insertion loss of 3 dB, an isolation of 8 dB, and a return loss better than 20 dB at 14 GHz. Although this performance data is not as good as that achievable with more conventional planar structures employing magnets, it does show a significant potential advance in ferrite planar devices.

There is, however, one problem associated with using embedded windings to provide the DC magnetization, and that is the heat generated by the coils. The heat reduces the saturation magnetization of the ferrite and this will affect the performance of the RF component. Arabi and Shamim [7] investigated the effect of this heat on the performance of a tuneable ferrite LTCC filter with embedded windings. They used a combination of simulation and measurement to show that the heating effect of the windings could cause temperature variations between 25 and 190 °C, which resulted in a downward shift of the centre frequency of the filter by around 10%. However, their work did show that the effect is predictable through simulation, and hence the consequences of internal heat generation can be included in the LTCC package design.

6.4.2 Edge-Guided-Mode Propagation

When a wide microstrip line ($w/h \gg 1$) is positioned on a magnetized ferrite substrate the RF energy tends to be concentrated on one side of the line, as shown in Figure 6.20. This is due to the interaction between the magnetic field of the dominant microstrip mode and the precessing electrons in the ferrite.

Since most of the propagating energy is concentrated along one edge of the microstrip, it is often referred to as an edge-guided-mode. With propagation in the z-direction, and with the direction of DC magnetizing field shown in Figure 6.20, the energy is concentrated along the left-hand edge of the microstrip. If the direction of propagation is reversed, the field will be concentrated along the right-hand edge of the microstrip.

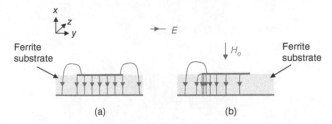

Figure 6.20 E-field distribution on a wide microstrip line on a ferrite substrate: (a) unmagnetized and (b) magnetized.

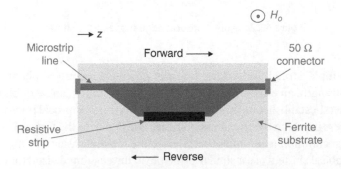

Figure 6.21 Edge-guided-mode microstrip isolator.

Hines [8] analyzed the propagation along wide microstrip lines on ferrite and showed that with weak magnetizing fields the phase propagation constant (β_z) for the edge-guided-mode is given by

$$\beta_z = 2\pi f \sqrt{\mu_0 \varepsilon_0 \varepsilon_r}, \tag{6.24}$$

where f is the frequency of operation, and other symbols have their usual meanings. The significance of this expression is that it indicates that edge-mode propagation is dispersion-free, since the velocity of propagation is constant, i.e.

$$v_p = \frac{\omega}{\beta_z} = \frac{2\pi f}{2\pi f \sqrt{\mu_0 \varepsilon_0 \varepsilon_r}} = \frac{1}{\sqrt{\mu_0 \varepsilon_0 \varepsilon_r}}. \tag{6.25}$$

6.4.3 Edge-Guided-Mode Isolator

The edge-guided-mode can be exploited to make a planar field displacement isolator by depositing a resistive (lossy) film along one edge of the microstrip line, as shown in Figure 6.21.

In Figure 6.21, the width of the microstrip line is shown increased in the vicinity of the resistive layer in order to emphasize the field displacement effect. With this structure the field beneath the resistive strip is weak for propagation in the forward direction, and so the forward attenuation is relatively low. But for propagation in the reverse direction the high concentration of field beneath the lossy strip leads to high attenuation. The edge-guided-mode isolator has very broad bandwidth performance, although the insertion loss is relatively high, typically of the order of 1 dB, because not all the RF field can be eliminated from beneath the resistive layer when propagation is in the forward direction. In an interesting paper Elshafiey and colleagues [9] performed full-wave analyses of edge-mode microstrip isolators on ferrite substrates, and considered in particular situations where an additional dielectric layer was added either above or below the ferrite so as to form a multilayer structure. The data presented in this paper showed that with appropriate choice of dielectric thickness and permittivity this type of multilayer structure could theoretically increase the non-reciprocity and isolation of edge-mode isolators to around 20 dB at X-band.

6.4.4 Phase Shifters

Continuous or discrete phase shifters can be implemented using a microstrip line fabricated on a ferrite substrate, which is subjected to a DC magnetic field. Varying the current in the electromagnet providing the magnetic field will change the effective permeability of the substrate, as discussed in the previous sections, and this in turn will vary the transmission phase

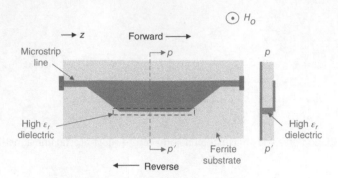

Figure 6.22 Non-reciprocal edge-mode phase shifter.

along the microstrip line. A simple, straight microstrip line will provide a reciprocal phase shifter, and with appropriate control of the magnetic field strength, an effective low-loss, digital phase shifter can be implemented.

There are also a number of well-established techniques for producing non-reciprocal microstrip phase shifters on ferrite. These are two-port phase shifters in which the transmission phase is different in the forward and reverse directions through the phase shifter. The two most common techniques use: (i) edge-mode propagation; (ii) meander lines.

Non-reciprocal edge-mode phase shifters use a similar principle to the edge-mode isolator discussed in Section 6.4.3, but with the resistive strip replaced by a low-loss, high permittivity dielectric is inserted beneath one edge of the wide microstrip line, as shown in Figure 6.22.

As described in Section 6.2.2, the RF fields are confined to the edges of the wide microstrip line. In the forward direction, these fields will travel primarily in the ferrite substrate with some fringing into the air. In the reverse direction, the fields will travel through the opposite edge of the wide section, and through the dielectric insert which has a high dielectric constant. Thus, the transmission phases in the two directions will be different, giving the phase shifter its non-reciprocal property. Since both the ferrite and dielectric can be screen printed it is relatively straightforward to fabricate this type of isolator, and in a multilayer structure the magnetizing field can be generated from windings buried within the layers.

The use of microstrip meandered lines on a ferrite substrate can also provide an effective non-reciprocal phase shifter. Figure 6.23 shows a typical arrangement, with closely spaced microstrip lines in the meandered section, and the DC magnetizing field, H_o, applied longitudinally, and parallel to the long side of the meander.

The lines forming the meander are $\lambda_s/4$ long, where λ_s is the microstrip wavelength on the ferrite substrate. The centre-to-centre spacing, s, of the lines in the meander is made sufficiently small that there is reasonably strong coupling between adjacent lines. This type of structure was originally investigated by Roome and Hair [10] who deduced the transmission phase characteristics of the structure from the interaction between the CP field produced by the coupled lines and the DC magnetizing field. Figure 6.23 includes an enlarged view of the magnetic field lines between two adjacent lines within the meander. The view shows the situation at the centre of the meander line section, i.e. in the plane pp'. Since the meandered sections are $\lambda_s/4$ long, there will be a delay of 90° between currents at the centres of adjacent lines, and so the resulting microwave magnetic field will be CP. The enlarged view in Figure 6.21 shows the magnetic fields, H_m and H_n, due to the adjacent lines in the meander. It should be noted that circular polarization is only produced at the centre of the structure, and at positions off-centre the phase difference is either greater or less than 90° (i.e. the path difference

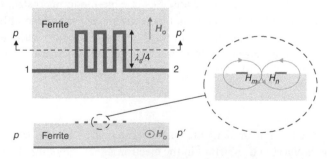

Figure 6.23 Non-reciprocal phase shifter using meander lines on ferrite.

is either greater or less than $\lambda_s/4$), and the polarization will be elliptical. If waves are travelling from port 1 to port 2, the direction of rotation of the polarized wave will be clockwise. This will be in the same direction as the precessional motion of the electrons with the direction of DC magnetizing field shown in Figure 6.23. If the waves travel from port 2 to port 1, the rotation of the polarized RF wave will be counter-clockwise, thus opposing the precessional motion of the electrons. Consequently, the device will exhibit a non-reciprocal transmission phase change since the effective permeability of the ferrite will be different for RF waves travelling in opposite directions.

The implementation of microwave phase shifters using newer fabrication technologies such as LTCC and SIW has been the subject of extensive recent research. Bray and Roy [11] first demonstrated that a ferrite-filled phase shifter could be successfully embedded within an LTCC multilayer structure. They reported a switchable non-reciprocal phase shift of 52.8° at 36 GHz, with a total insertion loss of 3.6 dB, from a prototype multilayer structure whose dimensions were 35 mm (length) × 5 mm (width) × 1 mm (height). This corresponds to a figure of merit (FoM) of ~15°/dB. (*Note that one of the figures of merit used to represent the performance of phase shifters is the ratio of the phase change in degrees to the insertion loss in dB.*) These results demonstrated the feasibility of the packaging technique and showed the potential for close integration of ferrite devices with other components in an LTCC SiP structure. These initial work advances in LTCC materials technology and SIW design techniques have enabled ferrite-based phase shifters to be implemented in still smaller packages, and with improved performance. An example of more recent work is that of Kagita and colleagues [12] who have reported a ferrite-based SIW phase shifter integrated in an LTCC package that occupied a substrate area of approximately 11 mm × 5 mm. A 10-layer LTCC stack was used in their work, and included embedded windings to provide the necessary magnetization for the ferrite. They reported a measured FoM of 100°/dB for the SIW section of a non-reciprocal phase shifter at X-band. This represents a very good figure-of-merit for the device, although it excludes the effects of losses in the launcher sections, which will tend to reduce the FoM.

6.5 Self-Biased Ferrites

Self-biased ferrite materials are those which exhibit a preferential magnetic polarization without the need for an external DC magnetic field. Eliminating the need for an external magnetic field leads to a significant reduction in weight and cost, and is particularly advantageous for integrated circuit devices. The development of ferrite materials with self-biased properties has been the subject of extensive research, and typical materials used for high-frequency applications are barium hexaferrite (BaM) and strontium hexaferrite ($SrFe_{12}O_{19}$). A detailed discussion of the chemical processes involved in the preparation of these materials, and their high-frequency behaviour, is outside the scope of the present textbook, but an informative overview can be found in Reference [13]. It suffices to say that these materials have essentially the same properties as traditional, externally magnetized ferrites, and can be used to fabricate a range of microwave devices, including isolators and circulators. Of particular interest for RF and microwave SiP designs is the opportunity to screen print self-biased ferrite materials, which can lead to low-weight, compact components with high functionality. Thick-film ferrite BaM coatings are prepared using a basic screen printing technique, but with the inclusion of a DC magnetic field applied between the printing and sintering stages to impart internal magnetization to the coating.

Numerous examples of self-biased ferrite devices are now reported in the literature, with performances approaching that of more traditional structures using external magnetization. Good examples are Wang and colleagues [14] who described a self-biased ferrite circulator with 21 dB isolation and 1.52 dB insertion loss at 13.6 GHz, and O'Neil and Young [15] who reported a microstrip circulator working in the range 21–23 GHz with an isolation better than 15 dB and an insertion loss ~1.5 dB.

6.6 Supplementary Problems

Q6.1 The gyromagnetic ratio of a particular ferrite sample is 35 kHz.m/A. What is the precessional frequency if a DC magnetic bias of 87.6 kA/m is applied to the material?

Q6.2 The value of γ for a particular ferrite material is 2.8 MHz/Oe. What value of DC magnetic field is needed to produce a precessional frequency of 2.6 GHz?

Q6.3 The following data apply to a ferrite material: $\gamma = 35\,\text{kHz.m/A}$; $\omega_m = 4\pi \times 10^9\,\text{rad/s}$. A LP RF plane wave is applied to the material in the same direction as a DC magnetic field, $H_o = 80\,\text{kA/m}$. Plot a graph of the effective permeabilities, $(\mu_{eff})_+$ and $(\mu_{eff})_-$, of the material for RF frequencies between 0 and 5 GHz.

Q6.4 A 2 GHz LP plane wave is applied to a ferrite material with the following properties: $\gamma = 35\,\text{kHz.m/A}$; $\omega_m = 11.2\pi \times 10^9\,\text{rad/s}$; $\varepsilon_r = 12$. If the material is magnetized with a DC field of 80 kA/m applied in the same direction as the propagating wave, determine:

  (i) The wavelengths of the two hands of circular polarization within the material.
  (ii) The rotation per unit length of the plane of polarization of the 2 GHz wave.
  (iii) The angle and direction of rotation of the plane of polarization of the 2 GHz wave in a 10 mm length of the ferrite.

Q6.5 A signal generator with an output impedance of 50 Ω is connected to a load that has an impedance of $(142 + j76)\,\Omega$. Determine:

  (i) The VSWR at the output of the generator.
  (ii) The VSWR at the output of the generator if an isolator represented by the following S-matrix is connected between the generator and the load.

$$[S] = \begin{bmatrix} 0.09\angle 51° & 0.07\angle - 124° \\ 0.93\angle - 77° & 0.11\angle - 72° \end{bmatrix}.$$

Q6.6 Figure 6.24 shows a circulator connecting the output of a microwave power source, an antenna, and a matched load, Z_o.

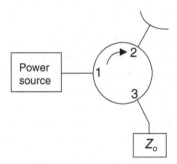

Figure 6.24 Circuit for supplementary problem Q6.6.

The power source has an output impedance of 50 Ω, and the antenna has an impedance of $(77 - j54)\,\Omega$. The performance of the circulator is represented by the following S-matrix:

$$[S] = \begin{bmatrix} 0.07\angle 27° & 0.04\angle - 61° & 0.91\angle - 55° \\ 0.93\angle - 44° & 0.05\angle 33° & 0.05\angle - 78° \\ 0.06\angle - 54° & 0.92\angle - 58° & 0.06\angle 84° \end{bmatrix}.$$

Determine the VSWR at the output of the power source.

References

1 Harris, V.G. (2012). Modern microwave ferrites. *IEEE Transactions on Magnetics* 48 (3): 1075–1104.
2 Collin, R.E. (1992). *Foundations for Microwave Engineering*, 2e. McGraw Hill: New York.

3 Helszajn, J. and Tan, F.C. (1975). Design data for radial-waveguide circulators using partial-height ferrite resonators. *IEEE Transactions on Microwave Theory and Techniques* 23 (3): 288–298.

4 Linkhart, D.K. (1989). *Microwave Circulator Design*. Norwood, MA: Artech House.

5 Van Dijk, R., van der Bent, G., Ashari, M., and McKay, M. (2014). Circulator integrated in low temperature co-fired ceramics technology. *Proceedings of 44th European Microwave Conference*, Rome, Italy (6–9 October 2014), pp. 1544–1547.

6 Yang, S., Vincent, D., Bray, J., and Roy, L. (2015). Study of a ferrite LTCC multifunctional circulator with integrated winding. *IEEE Transactions on Components, Packaging and Manufacturing Technology* 5 (7): 879–886.

7 Arabi, E. and Shamim, A. (2015). The effect of self-heating on the performance of a tuneable filter with embedded windings in a ferrite LTCC package. *IEEE Transactions on Components, Packaging and Manufacturing Technology* 5 (3): 365–371.

8 Hines, M.E. (1971). Reciprocal and nonreciprocal modes of propagation in ferrite stripline and microstrip devices. *IEEE Transactions on Microwave Theory and Techniques* 19 (5): 442–451.

9 Elshafiey, T.M.F., Aberle, T.L., and El-Sharawy, E. (1996). Full wave analysis of edge-guided mode microstrip isolator. *IEEE Transactions on Microwave Theory and Techniques* 44 (12): 2661–2668.

10 Roome, G.T. and Hair, H.A. (1968). Thin ferrite devices for microwave integrated circuits. *IEEE Transactions on Electron Devices* ED-15 (7): 473–482.

11 Bray, J.R. and Roy, L. (2004). Development of a millimeter-wave ferrite-filled antisymmetrically biased rectangular waveguide phase shifter embedded in low-temperature cofired ceramic. *IEEE Transactions on Microwave Theory and Techniques* 52 (7): 1732–1739.

12 Kagita, S., Basu, A., and Koul, S.K. (2017). Characterization of LTCC-based ferrite tape in X-band and its application to electrically tuneable phase shifter and notch filter. *IEEE Transactions on Magnetics* 53 (1).

13 Harris, V.G. (2010). The role of magnetic materials in RF, microwave, and mm-wave devices: the quest for self-biased materials. *Proceedings of IEEE 2010 National Aerospace and Electronics Conference*, Fairborn, OH (14–16 July 2010).

14 Wang, J., Yang, A., Chen, Y. et al. (2011). Self biased Y-junction circulator at Ku band. *IEEE Microwave and Wireless Components Letters* 21 (6): 292–294.

15 O'Neil, B.K. and Young, J.L. (2009). Experimental investigation of a self-biased microstrip circulator. *IEEE Transactions on Microwave Theory and Techniques* 57 (7): 1669–1674.

7

Measurements

7.1 Introduction

As the operating frequency of electronic circuits increases into the RF and Microwave region measurement techniques become more complex, and additional care is needed both in using the correct measurement procedure and in the care of measurement instrumentation. Whilst there is an extensive range of specialist high-frequency test instrumentation available for measuring specific parameters such as frequency and power, the core measurement system in any RF or microwave laboratory is the network analyzer. The primary function of a network analyzer is to measure and display the S-parameters of a passive or active circuit. The majority of network analyzers are two-port instruments, which can measure the properties of a two-port device-under-test, although multi-port analyzers are available for special applications. In general, network analyzers may be separated into two types, namely scalar and vector instruments. As the names imply, a scalar network analyzer will only measure the magnitudes of the S-parameters of the device being tested, whereas a vector instrument can measure both magnitude and phase. In design and development laboratories it is usually necessary to have a vector network analyzer (VNA), since both magnitude and phase information is necessary for most design tasks. In order to make precise high-frequency measurements using a VNA, it is necessary to use a calibration procedure to eliminate systematic errors, and the chapter will discuss the sources of these errors and the various calibration techniques that are available.

The chapter commences with a discussion of RF and microwave connectors. All high-frequency measurements require some form of connector, primarily to connect the device-under-test to the measuring instrument, yet the influence of connectors in measurement systems is often given insufficient attention. The choice of appropriate connectors, and the maintenance of connectors, can have significant influence on measurement performance. A review of some common connectors is given, together with a discussion of the best-practice procedures that should be followed to ensure good connector performance.

7.2 RF and Microwave Connectors

Connectors used in high-frequency applications normally have a coaxial structure, with most connectors having the conventional pin (male) and socket (female) formats. The two key parameters used to specify the performance of a high-frequency connector are the characteristic impedance, and the maximum usable frequency. The values of these parameters are determined by the dimensions of the connector, and the permittivity of the dielectric filling, and can be evaluated using the equations given in Appendix 1.A of Chapter 1.

A selection of high-frequency connectors is shown in Figure 7.1, with associated data in Table 7.1.

Type-N connectors are used on a wide range of RF and microwave instrumentation, and as terminations on low-loss interconnect cables. They are available in general or precision versions with the precision version giving good performance up to 18 GHz. The connector has an outer diameter of 7 mm, with a centre pin for the male connector, and a hollow cylinder in the female connector to receive the centre pin. The hollow cylinder of the female connector normally has four or six spring-loaded contacts to take up the tolerances of the male centre pin. With correct maintenance this type of connector is capable of giving good repeatability for multiple connections, although the absence of a flat on the body of the connector means that the tightness of the connector depends only on the user, rather than being determined more precisely through the use of a torque-spanner.

RF and Microwave Circuit Design: Theory and Applications, First Edition. Charles E. Free and Colin S. Aitchison.
© 2022 John Wiley & Sons Ltd. Published 2022 by John Wiley & Sons Ltd.
Companion website: www.wiley.com/go/free/rfandmicrowave

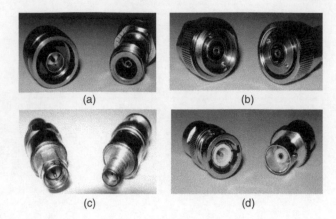

Figure 7.1 Common RF/microwave connectors: (a) type-N, (b) APC-7, (c) SMA, and (d) BNC.

Table 7.1 Data on common RF/microwave connectors.

Description	Outer diameter (mm)	Maximum frequency (GHz)
APC-7	7	20
Type-*N* (general)	7	12
Type-*N* (precision)	7	18
SMA	3.5	20
2.4 mm coaxial (K)	2.4	50
1 mm coaxial	1	110
BNC		~1.5

APC-7 connectors are 7 mm precision, gender neutral, connectors. (The acronym APC-7 stands for **A**mphenol **P**recision **C**onnector – **7** mm.) Rather than a pin-and-socket construction, the centre and outer contacts simply butt together. To improve connectivity the mating faces of the centre conductor have a number of spring-loaded contacts. Two mating connectors are tightened by an arrangement of threaded sleeves. The benefit of this arrangement is that there is no tendency to bend the centre pin when two connectors are joined. This method of construction contributes to the APC-7 connector having the lowest reflection coefficient, and most repeatable performance of all 20 GHz connectors. The return loss of two properly joined connectors is normally of the order of 30 dB. This type of connector was common on older microwave instrumentation, but because of its relatively high cost, and rather awkward tightening arrangement, it is no longer widely used.

SMA connectors have an outer diameter of 3.5 mm, with the male and female versions having a conventional pin and socket construction. (The acronym SMA stands for **S**ub-**M**iniature **A**.) This is a low-cost connector, primarily designed for one-off connections, and consequently the repeatability of connection is poor. It is frequently used on semi-rigid coaxial cable, particularly for interconnecting microwave sub-assemblies. In addition, it has wide application in panel mounting formats for connecting to planar circuits such as a microstrip.

K connectors are high-performance 2.92 mm air-interface coaxial connectors that give mode-free operation to 40 GHz. They are similar in appearance to SMA connectors from which they can be distinguished by the absence of a dielectric interface.

BNC connectors are low-cost, general purpose coaxial connectors widely used for low RF applications. (The acronym BNC stands for **B**ayonet **N**avy **C**onnector.) The repeatability of this type of connector is poor, and the usual recommendation is that it should not be used above 1 GHz.

7.2.1 Maintenance of Connectors

The importance of correct maintenance of high-frequency connectors is often overlooked. Poorly maintained connectors can lead to high VSWR at the connector interface, together with excess loss, and these factors can dominate a high-frequency measurement. Recommended best practice in the measurement laboratory requires:

(i) The conducting parts of the connector, particularly the centre pin, should be clean, undamaged, and free of corrosion.

(ii) Connectors should always be joined with the correct torque (tightness) using a torque spanner with the correct torque setting.

(iii) Connectors should always be cleaned before each measurement. It is important that the cleaning procedure and materials (chemicals) do not damage the connector. Dirt, notably metallic dust, can build up on the connector, particularly on the face of the dielectric insert. Initially, compressed air should be used to clean both the threads and dielectric faces of the connectors. If necessary, particles on the face of the dielectric should be removed with a cotton bud or lint-free cloth dipped in a suitable solvent such as isopropyl alcohol (IPA). This cleaning procedure should be followed prior to each measurement.

7.2.2 Connecting to Planar Circuits

Measurements on two-port microstrip or coplanar circuits can be made using a universal test fixture of the type shown in Figure 7.2.

In Figure 7.2, the circuit-under-test is shown clamped between two jaws of the test fixture. One jaw is fixed, and the other is movable in both the lateral and transverse directions. Each jaw has a K-connector, which is used to connect the test fixture through a high-frequency coaxial cable to the measurement instrumentation. The K-connectors are terminated in small metallic tabs which make contact with the ends of the circuit being tested. Using K-connectors this type of fixture provides repeatable measurements up to 40 GHz.

For measurements on planar circuits in the millimetre-wave region coplanar probes mounted in a probe station are normally used. A typical coplanar probe is shown in Figure 7.3.

The probe shown in Figure 7.3 has a conventional description of 100 GSG: the 100 refers to the centre-to-centre spacing of adjacent tips in the coplanar probe (in micrometres), and GSG stands for ground-signal-ground, i.e. the centre contact carries the signal with the outer contacts being grounded. In the example shown, the coplanar probe is fed from a 1 mm coaxial connector, which enables the probe to be used up to 110 GHz. The probes are normally mounted on a probe station, an example of which is shown in Figure 7.4.

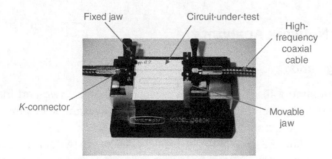

Figure 7.2 Two-port universal test fixture.

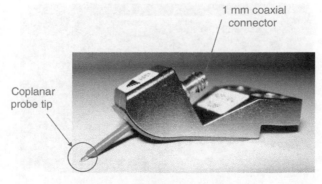

Figure 7.3 100 GSG coplanar probe.

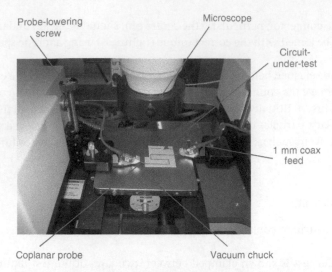

Figure 7.4 Typical 110 GHz probe station.

The probe station shown in Figure 7.4 employs a vacuum chuck to hold the circuit-under-test in a fixed position. The two probes are mounted on micro-manipulated stages that provide the user with precise control of the probe positions in the $X-Y$ (horizontal) plane. The probes are manually lowered into position using vertical screw drives. A microscope can be moved over the point of probe contact to ensure correct positioning with respect to the coplanar lines on the circuit. For convenience, the microscope is normally connected to a video screen to allow the user to more easily position the probes. It is important to ensure that all three probes tips are in contact with the circuit and this is achieved by watching for all three probes to slip on the conductors on the circuit. Some probe stages employ an additional, orthogonally positioned microscope to allow the user to visually inspect the contacts between the probes and the circuit. The probe station shown in Figure 7.4 is intended for use up to 110 GHz, and so 1 mm coaxial cables are used to feed the probes. These coaxial lines are connected to the two ports of a VNA.

7.3 Microwave Vector Network Analyzers

7.3.1 Description and Configuration

The conventional VNA found in almost all RF and microwave laboratories is a two-port instrument capable of measuring and displaying *S*-parameter data over a broad range of frequencies. An example of a modern high-performance VNA, manufactured by Keysight Technologies, is shown in Figure 7.5.

Two coaxial test ports are provided on the front of the instrument. There are two stages in making a measurement using a VNA: firstly, known impedance standards are connected to the coaxial ports to calibrate the instrument (calibration is covered in more detail in Section 7.3.3); secondly, the two-port device-under-test (DUT) is connected between the coaxial

Figure 7.5 Modern vector network analyzers © Keysight Technologies, Inc. *Source*: Courtesy of Keysight Technologies, Inc.

ports, and the complex *S*-parameters of the DUT are displayed as a function of frequency. The frequency range, and the format of the display, can be set by push-buttons on the front-panel. The display format includes a Smith chart representation, and for convenience all four *S*-parameters can normally be displayed simultaneously. In normal use the analyzer is also connected to a computer which can record and further process the measured *S*-parameter data.

Whilst modern VNAs are highly complex instruments, their basic operation can be understood through reference to the simple functional diagram shown in Figure 7.6.

The six basic internal parts of a VNA shown in Figure 7.6 are:

1. *Synthesized signal source.* This permits the precise selection of low-noise, individual frequencies, or a range of frequencies when operating in a swept mode. (The structure and performance of synthesized signal sources are discussed in detail in Chapter 11.) The signal source may also incorporate an isolator in the output to prevent variations in load impedance pulling the frequency of the source off-tune.
2. *Test set.* This is the key high-frequency unit within a VNA, and is the interface between the VNA and the device-under-test. It contains switches which determine the direction of signal propagation through the DUT, and directional couplers to measure reflections from the two ports of the DUT. The internal configuration of a typical test set is outlined in Figure 7.7.

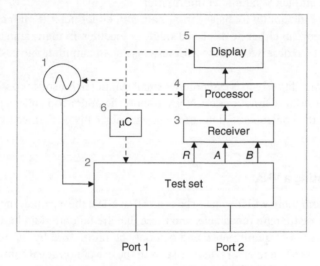

Figure 7.6 Functional diagram of a typical VNA.

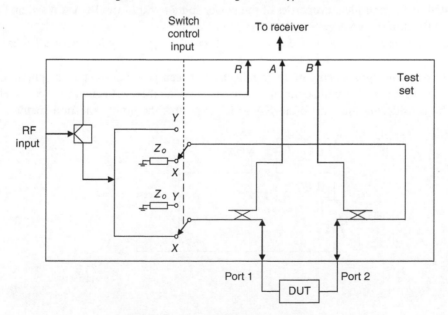

Figure 7.7 Structure of a typical test set.

In the arrangement shown, the RF signal entering the test set from the synthesized source is divided by means of a power splitter, with the majority of the signal being directed to port 1 of the DUT, and the remainder forming a reference signal, R, for the receiver. The two switches determine the direction of the signal within the test set, and through the DUT. The switches are normally ganged together, and their position selected by the user, either from the front panel of the VNA or through the computer control. With the two switches in the X positions shown in Figure 7.7, the RF signal travels through the DUT from port 1 to port 2, with the output port being terminated in an internal matched load, Z_0. Under these conditions the ratio of the voltage signals at ports B and R (B/R) is proportional to S_{21} and the ratio of the signals at ports A and R (A/R) is proportional to S_{11}. Similarly, when the switches are in the Y positions, the RF signal travels from port 2 to port 1 through the DUT, and A/R is proportional to S_{12} and B/R is proportional to S_{22}. Thus, by making appropriate compensations for the known losses in the test set, the S-parameter matrix representing the device-under-test can be determined.

3. *Receiver*. The receiver down-converts the RF or microwave signals at ports R, A, and B to a lower frequency for ease of processing, whilst maintaining the correct amplitude and phase information. Demodulators are included within the receiver to recover the amplitude and phase data in a baseband format.

4. *Processor*. The processor stores data from the calibration and uses this in conjunction with the measured data from the DUT to determine required complex S-parameter information.

5. *Display*. VNAs normally have the ability to display information in either polar (Smith chart) or rectangular Cartesian formats. The four S-parameters can either be displayed singly or together. In many situations, such as amplifier design, it is very convenient to be able to observe how the gain response of an amplifier varies simultaneously with the match at the two ports of the device.

6. *Micro-controller*. A micro-controller, which is essentially a small computer, is used to control the functions of the VNA. It is convenient for illustrating the basic functions of a VNA to show the micro-controller as a separate unit, but in modern instruments the functions of the controller and the processor are normally performed by what is effectively an internal PC and processor.

7.3.2 Error Models Representing a VNA

In a practical measurement system there will inevitably be errors that affect the accuracy of the measurement. These errors can broadly be classified into those that are repeatable, and those that are random. Random errors, such as those resulting from poor or worn connectors, are not quantifiable, and can only be minimized by careful measurement management. However, there will be some errors (system errors) that arise from the use of non-ideal components and connectors within a VNA system that are repeatable, and quantifiable. To illustrate the origin and effects of systematic errors we will consider two cases, both related to the non-ideal properties of the power splitter within the VNA test set, and use the concept of multiple reflections [1] within the VNA system to quantify the errors.

CASE 1: *The effect of a mismatch at port 2 (see Figure 7.8) of the power splitter with all other components being considered ideal.*

This mismatch will result in signals being multiple-reflected between port 2 of the power splitter and the DUT. Thus, the measured reflection from the DUT will be the cumulative result of all these reflections. We can see how this will affect the measured data by considering that part of a VNA which measures the input reflection coefficient of the DUT; the

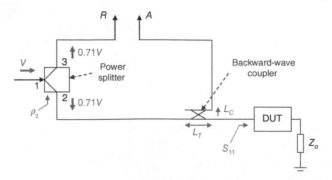

Figure 7.8 Basic reflectometer circuit of a VNA.

essential features of this part of a VNA are shown in Figure 7.8, and the arrangement is often referred to as a reflectometer. This figure shows the RF input applied to a 3 dB power splitter, which splits the power by a factor of $1/2$, but splits the voltage by a factor of $1/\sqrt{2}$, since power is proportional to voltage squared. Thus, if the amplitude of the input voltage to the power splitter is V, the amplitudes of the output voltages will be 0.71 V. Reflections from the DUT are coupled-off through a backward-wave coupler and applied to port A. In the diagram shown in Figure 7.8, L_T is the voltage coefficient representing the through loss of the direction coupler, and L_C is the voltage coefficient representing the coupling loss (see Example 7.1). After compensating the voltage at port R for the known coupling loss through the backward-wave coupler, and the known through loss of the coupler, the reflection coefficient of the DUT is given by the ratio of the voltages at A and R, i.e. V_A/V_R.

For clarity, Figure 7.9 shows the multiple reflections that occur between the DUT and the mismatched port of the power splitter [1]. The sum (V_r) of the voltage reflections at the input of the DUT will be

$$V_r = 0.71V \ L_T e^{-j\beta d} S_{11} + 0.71V \ L_T^3 e^{-j3\beta d} \rho_2 S_{11}^2 + 0.71V \ L_T^5 e^{-j5\beta d} \rho_2^2 S_{11}^3 + \dots$$

$$= 0.71V \ L_T e^{-j\beta d} S_{11}(1 + L_T^2 e^{-j2\beta d} \rho_2 S_{11} + L_T^4 e^{-j4\beta d} \rho_2^2 S_{11}^2 + \dots),$$

i.e.

$$V_r = 0.71V \ L_T e^{-j\beta d} S_{11} \sum_{n=0}^{n=\infty} (L_T^2 e^{-j2\beta d} \rho_2 S_{11})^n, \tag{7.1}$$

where β is the phase propagation coefficient of the line connecting the power splitter to the DUT, and ρ_2 is the reflection coefficient of port 2 of the power splitter.

Since both $|\rho_2|$ and $|S_{11}|$ are <1, Eq. (7.1) represents a convergent power series, where

$$\sum_{n=0}^{n=\infty} x^n = \frac{1}{1-x} \tag{7.2}$$

and so we may rewrite Eq. (7.1) as

$$V_r = \frac{0.71V \ L_T e^{-j2\beta d} S_{11}}{1 - L_T^2 e^{-j2\beta d} \rho_2 S_{11}}. \tag{7.3}$$

Referring to Figure 7.8, the voltage at port A will be

$$V_A = V_r L_C = \frac{0.71V \ L_T e^{-j2\beta d} S_{11}}{1 - L_T^2 e^{-j2\beta d} \rho_2 S_{11}} \times L_C, \tag{7.4}$$

where L_C is the coupling loss factor through the backward wave coupler, and L_T is the through loss of the coupler.

We must also compensate the reference signal at port R, for the known losses, L_C and L_T, in the path of the reflected signal, and also the phase change due to the distance, d, between the power splitter and the DUT, and so the compensated voltage at port R will be

$$V_R = 0.71V \times L_C \times L_T \times e^{-j\beta d}. \tag{7.5}$$

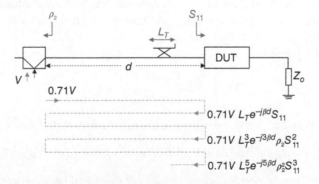

Figure 7.9 Multiple reflections in a VNA test set.

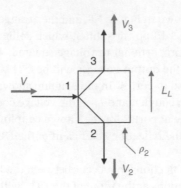

Figure 7.10 Additional power splitter imperfections.

The measured reflection coefficient will then be given by

$$(S_{11})_{measured} = \frac{V_A}{V_R} = \frac{\dfrac{0.71V \; L_T e^{-j\beta d} S_{11}}{1 - L_T^2 e^{-j2\beta d} \rho_2 S_{11}} \times L_C}{0.71V \times L_C \times L_T \times e^{-j\beta d}}$$

$$= \frac{1}{1 - L_T^2 e^{-j2\beta d} \rho_2 S_{11}} \times S_{11}$$

$$= \varepsilon \times S_{11}, \tag{7.6}$$

where ε is the error ratio, and given by

$$\varepsilon = \frac{1}{1 - L_T^2 e^{-j2\beta d} \rho_2 S_{11}}. \tag{7.7}$$

The error ratio is a measure of the uncertainty in the measurement due to imperfections in the system. For the ideal case, when $\rho_2 = 0$, we have $\varepsilon = 1$ and

$$(S_{11})_{measured} = S_{11}. \tag{7.8}$$

CASE 2: *The additional effects of leakage and tracking errors in the power splitter.*

Three potential sources of error within a power splitter are identified in Figure 7.10. These are: (i) the non-zero reflection coefficient, ρ_2, at port 2; (ii) the unequal voltages at ports 2 and 3; and (iii) the leakage, L_L, from port 2 to port 3.

The effects of mismatch at port 2 have already been discussed. The tracking error refers to the unequal split of power between the output ports of the power splitter, and the tracking ratio error, k, at any given frequency will be

$$k = \frac{V_2}{V_3}, \tag{7.9}$$

where V_2 and V_3 are the voltages appearing at ports 2 and 3, respectively, due to a voltage V incident on port 1. If there is any leakage from port 2 to port 3 of the splitter, i.e. if L_L is non-zero, each of the multiple reflections between the DUT and port 2 of the splitter will generate an additional voltage component travelling from port 3 of the splitter to the reference port, R.

Rewriting Eq. (7.4) in terms of V_2 we have

$$V_A = \frac{V_2 \; L_T e^{-j2\beta d} S_{11}}{1 - L_T^2 e^{-j2\beta d} \rho_2 S_{11}} \times L_C. \tag{7.10}$$

Considering the leakage from port 2 to port 3 the voltage at the reference port will now be

$$V_R = V_3 + V_2 L_T^2 e^{-j2\beta d} S_{11} L_L + V_2 L_T^4 e^{-j4\beta d} \rho_2 S_{11}^2 L_L + V_2 L_T^6 e^{-j6\beta d} \rho_2^2 S_{11}^3 L_L + \cdots \tag{7.11}$$

It should be noted that if the reference port is correctly matched, there will be no reflections from this port and so any mismatch at port 3 of the power splitter will have no effect.

Rearranging Eq. (7.11) we have

$$V_R = V_3 + V_2 L_T^2 e^{-j2\beta d} S_{11} L_L (1 + L_T^2 e^{-j2\beta d} \rho_2 S_{11} + L_T^4 e^{-j4\beta d} \rho_2^2 S_{11}^3 + \ldots)$$

$$= V_3 + V_2 L_T^2 e^{-j2\beta d} S_{11} L_L \sum_{n=0}^{n=\infty} (L_T^2 e^{-j2\beta d} \rho_2 S_{11})^n$$

$$= V_3 + \frac{V_2 L_T^2 e^{-j2\beta d} S_{11} L_L}{1 - L_T^2 e^{-j2\beta d} \rho_2 S_{11}}. \tag{7.12}$$

The measured reflection coefficient will be

$$(S_{11})_{measured} = \frac{V_A}{V_R L_T L_C e^{-j\beta d}}, \tag{7.13}$$

where we have compensated the reference voltage for the known losses, L_T and L_C, in the measurement arm, and also for the distance between the power splitter and the DUT. Combining Eqs. (7.9), (7.11), and (7.13), and rearranging gives

$$(S_{11})_{measured} = \frac{k e^{-j\beta d} S_{11}}{1 + L_T^2 S_{11} e^{-j2\beta d}(k L_L - \rho_2)} = \varepsilon S_{11}. \tag{7.14}$$

In this case the error ratio is

$$\varepsilon = \frac{k e^{-j\beta d}}{1 + L_T^2 S_{11} e^{-j2\beta d}(k L_L - \rho_2)}. \tag{7.15}$$

Example 7.1 The performance of a non-ideal power splitter is represented by the following *S*-matrix:

$$[S] = \begin{bmatrix} 0.05\angle 10° & 0.70\angle 52° & 0.78\angle 50° \\ 0.70\angle 52° & 0.08\angle -20° & 0.10\angle -130° \\ 0.78\angle 50° & 0.10\angle -130° & 0.09\angle -28° \end{bmatrix}.$$

This power splitter is used in the reflectometer circuit shown in Figure 7.8 to measure the input reflection coefficient of a DUT whose S_{11} value is $0.73 \angle 140°$. Determine the error ratio if a 20 dB microstrip directional coupler is used in the reflectometer. At the measurement frequency the electrical distance of the lossless line between port 2 of the power splitter and the input of the DUT is 0.10λ. All components in the reflectometer circuit, other than the power splitter, may be considered to be ideal.

Solution
To find the error ratio we need to evaluate ε using Eq. (7.15).

(i) *To find L_T for directional coupler shown schematically in Figure* 7.11:

$$P_1 = P_2 + P_3,$$

$$1 = \frac{P_2}{P_1} + \frac{P_3}{P_1} \quad \Rightarrow \quad \frac{P_2}{P_1} = 1 - \frac{P_3}{P_1},$$

$$20 \text{ dB} \quad \Rightarrow \quad \frac{P_3}{P_1} = \frac{1}{100} = 0.01.$$

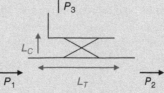

Figure 7.11 Directional coupler losses.

(Continued)

(Continued)

Thus,

$$\frac{P_2}{P_1} = 1 - 0.01 = 0.99,$$

$$|L_T| = \sqrt{\frac{P_2}{P_1}} = \sqrt{0.99} = 0.99.$$

From Chapter 2, we know that the voltage at the coupled port of a directional coupler lags that at the input port by 90°, and so we have

$$L_T = 0.99\angle - 90°.$$

(ii) *To find the transmission phase change between power splitter and DUT:*

$$d = 0.10\lambda \quad \Rightarrow \quad \beta d = \frac{2\pi}{\lambda} \times 0.10\lambda \ \text{rad} = 0.20\pi \ \text{rad} \equiv 36°.$$

(iii) *Deducing L_L, ρ_2 and k from the given S-matrix:*

$$L_L = S_{32} = 0.10 \angle - 130°,$$

$$\rho_2 = S_{22} = 0.08 \angle - 20°.$$

From Eq. (7.9):

$$k = \frac{V_2}{V_3} = \frac{S_{21}}{S_{31}} = \frac{0.70\angle 52°}{0.78\angle 50°} = 0.90\angle 2°.$$

Substituting the data into Eq. (7.15):

$$kL_L - \rho_2 = (0.90\angle 2° \times 0.10\angle - 130°) - (0.08\angle - 20°) = 0.14\angle - 161.30°,$$

$$\varepsilon = \frac{0.90\angle 2° \times 1\angle - 36°}{1 + [(0.99\angle - 90°)^2 \times (0.73\angle 140°) \times (1\angle - 72°) \times (0.14\angle - 161.30°)]}$$

$$= \frac{0.90\angle - 34°}{1 + [0.10\angle - 27.33°]} = \frac{0.90\angle - 34°}{1.09\angle - 2.63°}$$

$$= 0.83\angle - 31.37°.$$

Comment: The result shows that the effect of systematic errors (multiple reflections) within a measurement system can have a significant effect on the result. This emphasizes the need for calibration to remove these systematic errors.

Whilst the summing of multiple reflections is very useful in illustrating how errors occur in a VNA system, it is impractical to analyze a real system in this way since there are many sources of error, and consequently too many multiple reflection paths to be realistically summed.

The normal approach to dealing with VNA errors is to recognize that the overall effect of the many errors within the instrument is to create a finite number of error paths in the system. These paths are normally represented by the well-known 12-term error model [2]. This error model, shown in Figure 7.12, is essentially an enhanced signal flow graph that shows all the additional signal paths arising from the system errors.

The error model is divided into two flow graphs: one representing forward transmission through the DUT and the other the reverse transmission. Each graph shows the three measured parameters, and the six error paths resulting from systematic errors. There are 12 error paths in total, giving rise to the name 12-term error model. The 12 systematic errors associated with each error path are, for forward transmission:

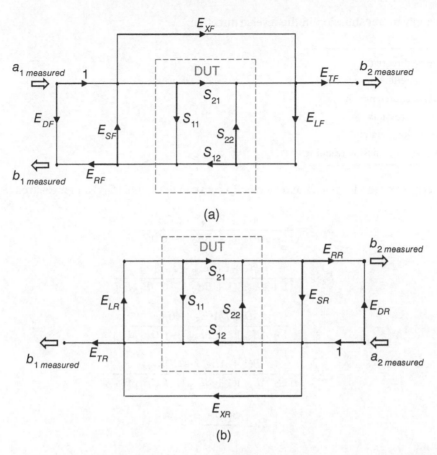

Figure 7.12 Twelve-term error model for a VNA: (a) forward direction and (b) reverse direction.

E_{DF}	Forward Directivity. *This error is due to the finite directivity of the directional coupler connected to port 1 of the DUT (see Figure 7.8), and results in some of the power from the source being directly coupled to the measurement port A. This means that the signal measured at B is not only dependent on the reflection from port 1 of the DUT.*
E_{RF}	Forward Reflection Tracking. *This error is related to the frequency responses of the receivers measuring the R and A signals. Since the VNA is measuring the ratio A/R, any variation with frequency in the performance of the two receivers measuring the individual R and A signals will cause an error. The relative performance of two receivers as a function of frequency is known as tracking.*
E_{SF}	Forward Source Match. *This error is due to the interaction between the source impedance and the input impedance of the DUT, which results in multiple reflections between the source and DUT as shown in Figure 7.9.*
E_{XF}	Forward Crosstalk. *This is due to unwanted coupling and leakage between ports 1 and 2 of the DUT. It is normally due to proximity coupling between the cables and circuits within the VNA.*
E_{LF}	Forward Load Match. *This error is due to the interaction between the load impedance (Z_o) within the VNA and the output impedance of the DUT, which results in multiple reflections.*
E_{TF}	Forward Transmission Tracking. *The reason for this error is similar to that for Forward Reflection Tracking, and relates to the relative frequency responses of the receivers measuring the B/R ratio.*

Similar definitions apply to transmission in the reverse direction:

E_{DR}	Reverse directivity
E_{RR}	Reverse reflection tracking
E_{SR}	Reverse source match
E_{XR}	Reverse crosstalk
E_{LR}	Reverse load match
E_{TR}	Reverse transmission tracking

Analysis of the 12-term error model gives the relationship between the actual (S_{ijA}) and measured (S_{ijM}) S-parameters of the DUT as [3]

$$S_{11A} = \frac{A(1 + BE_{SR}) - CDE_{LF}}{(1 + AE_{SF})(1 + BE_{SR}) - CDE_{LF}E_{LR}}, \tag{7.16}$$

$$S_{21A} = \frac{C + CB(E_{SR} - E_{LF})}{(1 + AE_{SF})(1 + BE_{SR}) - CDE_{LF}E_{LR}}, \tag{7.17}$$

$$S_{12A} = \frac{D + DA(E_{SF} - E_{LR})}{(1 + AE_{SF})(1 + BE_{SR}) - CDE_{LF}E_{LR}}, \tag{7.18}$$

$$S_{22A} = \frac{B(1 + AE_{SF}) - CDE_{LR}}{(1 + AE_{SF})(1 + BE_{SR}) - CDE_{LF}E_{LR}}, \tag{7.19}$$

where

$$A = \frac{S_{11M} - E_{DF}}{E_{RF}}, \tag{7.20}$$

$$B = \frac{S_{22M} - E_{DR}}{E_{RR}}, \tag{7.21}$$

$$C = \frac{S_{21M} - E_{XF}}{E_{TF}}, \tag{7.22}$$

$$D = \frac{S_{12M} - E_{XR}}{E_{TR}} \tag{7.23}$$

and

$$S_{11M} = \frac{b_{1measured}}{a_{1measured}}, \tag{7.24}$$

$$S_{21M} = \frac{b_{2measured}}{a_{1measured}}, \tag{7.25}$$

$$S_{12M} = \frac{b_{1measured}}{a_{2measured}}, \tag{7.26}$$

$$S_{22M} = \frac{b_{2measured}}{a_{2measured}}. \tag{7.27}$$

By replacing the DUT with known impedance standards, such as short-circuits and open-circuits, the values of the error paths can be calculated, and the appropriate corrections applied to measurements of the DUT. This process is known as calibration, and is covered in more detail in the next section.

7.3.3 Calibration of a VNA

Calibration is an essential part of any measurement using a VNA. It involves terminating the analyzer with known impedances of very high quality. Since the analyzer knows the expected values of these impedances, it can calculate the values of the error paths in the stored error models and subsequently apply corrections to measurements on a DUT.

Many of the system errors are frequency dependent and it is essential that the process of calibration is carried out at all frequencies within the range of a proposed measurement.

VNA manufacturers supply calibration kits (known simply as Cal Kits) which consist of sets of precisely defined standards, in the form of short-circuits, open-circuits, matched loads, and through lines. These impedance standards are available in various connection formats, the most common being coaxial, waveguide, and on-wafer. When the calibration mode is invoked using touch-sensitive buttons on the front panel of the VNA, algorithms within the instrument will prompt the user to connect the standards to the ports of the analyzer in a particular order. Various calibration algorithms are available, and the most common are:

(i) SOLT (Short-Open-Load-Through)

The four standards are connected to the analyzer in a pre-determined sequence, and the calibration data stored for subsequent use with measurements on the DUT. Normally, there will be high-quality coaxial cables attached to the ports of the VNA and the impedance standards are connected to the ends of these cables. Thus, the cables are effectively part of the VNA and the reference planes (measurement planes) are established at the ends of the cables. This calibration technique is normally available on all commercial VNAs. The technique requires very high quality standards, and also requires the VNA to have accurate models of the various impedance standards; for example, it is essential for the VNA to have precise information on the fringing capacitance of the open-circuit standards, and information on the inductance of the short-circuits. Accurate modelling of the standards becomes particularly important for microwave and millimetre-wave measurements.

(ii) TRL (Though-Reflect-Load)

This technique requires the use of a standard through line of constant characteristic impedance, and non-zero length. Also needed is a reflection standard, which can be either a short-circuit, or open-circuit, and has the advantage that it does not need to be an ideal component. The drawback to the TRL technique is that the through line must be long enough to contain a number of cycles of transmission phase, which means it can be inconveniently long for measurement frequencies at the lower end of the RF spectrum. Long reference lines may also give problems due to variations in characteristic impedance with distance.

The need for careful, precise calibration cannot be over-emphasized, particularly for measurements at microwave and millimetre-wave frequencies. System errors associated with a VNA can vary with time and physical conditions, and whilst calibration data can be stored within the instrument it is good measurement practice to calibrate the VNA at the beginning of each measurement sequence. Calibration techniques are constantly evolving, and a useful overview of the development and application of the various calibration procedures is provided by Rumiantsev and Ridler [4].

7.4 On-Wafer Measurements

One important aspect of high-frequency measurements is that of characterizing devices which are embedded within planar integrated circuits. Typical circuits may either be in hybrid or monolithic formats, with the device under investigation mounted in a coplanar line. Measurements in these circumstances are usually referred to as on-wafer measurements.

Calibration is an essential part of on-wafer measurements, and the measurement standards are provided in planar formats. If space permits, the standards should be included on the actual wafer or planar circuit being measured. Figure 7.13 shows the normal footprints of on-wafer calibration standards. Probe manufacturers such as FormFactor™ (formerly Cascade Microtech) also provide calibration tiles, which are small (~15 × 15 mm) squares of very high quality alumina substrate that contain a large array of calibration standards; detailed information on the sizes of the standards, together with their electrical characteristics are available on the manufacturer's website.[1]

The short-circuit is provided by a conducting pad which shorts the three probe tips together, whereas an open-circuit is obtained by suspending the three probes in the air. As with the coaxial open-circuit standards, appropriate compensation must be applied to account for the capacitive fringing at the ends of the probes. A nominal load can be obtained by positioning the probes on a resistive pad that has the appropriate resistance per unit square. Finally, a through ('thru') line is provided by a short, matched (normally 50 Ω) coplanar line. Usually, through lines of different lengths are provided in a typical on-wafer set of calibration standards.

---

1 www.formfactor.com.

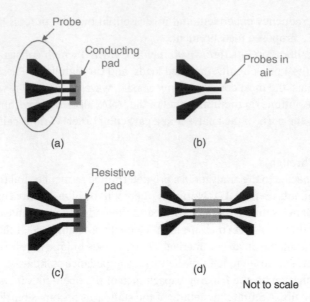

Figure 7.13 Footprints of on-wafer calibration standards: (a) short circuit, (b) open circuit, (c) load, and (d) thru line.

A number of precautions need to be observed when making on-wafer measurements, particularly at millimetre-wave frequencies:

(1) The probes must be carefully positioned (centred) with respect to the coplanar line, both for the actual measurement and for contacting the calibration standards.

(2) All three probe fingers must touch the circuit – the normal technique is to observe (through a microscope) all three fingers skating (slipping) on the conductors when the probes are brought into contact with the conducting pads. Occasionally, wafer probing stations are equipped with an additional microscope that provides a 'sideways' view, such that actual contact can be observed.

(3) As mentioned previously, the calibration standards should be included on the actual wafer/circuit being measured, to obtain more precise calibration.

(4) At millimetre-wave frequencies the TRL calibration technique is most appropriate, partly because of the difficulty of making precise open and short standards required by the SOLT method, and also because the high frequency means that only relatively short through lines are required for wideband calibration.

(5) Often the device-under-test will have been wire or ribbon bonded into place, and appropriate compensation must be applied to compensate for the parasitics associated with the bonding.

(6) On-wafer circuits should always be calibrated over the full frequency range of interest, since losses and dispersion vary very significantly with frequency in the millimetre-wave region.

(7) It is important to be aware of evanescent modes. These are localized fields that surround discontinuities, such as probe tips. These fields are not a problem unless the evanescent modes couple, due to close spacing of the discontinuities, and so cause resonances. This indicates that care should be taken in placing probes near discontinuities or components, and it also indicates that the circuit designer should be careful about the layout of a 'closely-spaced' design.

(8) It is also important to be aware of the possibility of multiple reflections if there is a mismatch at the point where the probes contact the wafer/circuit. In these circumstances, reflections from the device-under-test will cause multiple reflections between the probe and the DUT, so that the measured reflection coefficient will be the sum of all these reflections, i.e. the measured reflection coefficient will be a function of the probe mismatch.

(9) Accurate positioning of the probe tips is also crucial for millimetre-wave measurements. For example, at 100 GHz the free-space wavelength is only 3 mm, and if the substrate wavelength is 1.2 mm an error of 50 μm in the position of a probe will lead to a phase error of 15°.

Figure 7.14 shows an idealized view of a device embedded within a coplanar line, with two coplanar probes attached to the ends of the line.

Suppose that the arrangement shown in Figure 7.14 is to be used to measure the input reflection coefficient, S_{11}, of the embedded device, by connecting the probes to a VNA via an appropriate wafer-probing station. We can show the effects of a

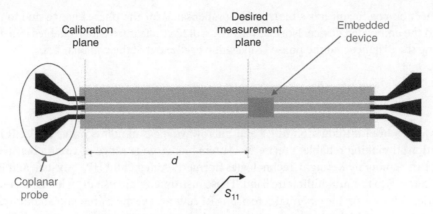

Figure 7.14 Device embedded within a coplanar line.

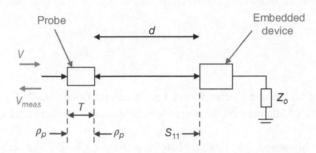

Figure 7.15 Embedded device with a mismatched input probe.

mismatch between the probes and the coplanar line using the concept of multiple reflections which was discussed in Section 7.3.2. For clarity, a simplified diagram showing the essential features relevant to a reflection coefficient measurement is given in Figure 7.15.

In the arrangement shown in Figure 7.15, we have assumed that the embedded device is correctly terminated by a matched load, Z_0. The reflection coefficient, ρ_p, represents the mismatch due to poor contact between the probe and the coplanar line, and T represents the transmission loss through the probe contact. The input voltage from the VNA is V, and the measured reflected voltage from the circuit is $V_{measured}$. Using the concept of multiple reflections between the mismatched probe and the embedded device, and assuming the coplanar line to be lossless, the reflected voltage can be written as

$$
\begin{aligned}
V_{measured} &= \rho_p V + S_{11} e^{-j2\beta d} TV + S_{11}^2 \rho_p e^{-j4\beta d} TV + S_{11}^3 \rho_p^2 e^{-j6\beta d} TV + \dots \\
&= \rho_p V + S_{11} e^{-j2\beta d} TV (1 + S_{11} \rho_p e^{-j2\beta d} + S_{11}^2 \rho_p^2 e^{-j4\beta d} + \dots) \\
&= \rho_p V + S_{11} e^{-j2\beta d} TV \sum_{n=0}^{n=\infty} (S_{11} \rho_p e^{-j2\beta d})^n \\
&= \rho_p V + \frac{S_{11} e^{-j2\beta d} TV}{1 - S_{11} \rho_p e^{-j2\beta d}}.
\end{aligned}
\tag{7.28}
$$

The measured reflection coefficient is then

$$
(S_{11})_{measured} = \frac{V_{measured}}{V} = \rho_p + \frac{S_{11} e^{-j2\beta d} T}{1 - S_{11} \rho_p e^{-j2\beta d}}.
\tag{7.29}
$$

It can be seen from Eq. (7.29) that if the probe makes ideal contact with the coplanar line, i.e. if $\rho_p = 0$ and $T = 1$, then

$$
(S_{11})_{measured} = S_{11} e^{-j2\beta d},
\tag{7.30}
$$

and the measured reflection coefficient has simply to be compensated for the phase change due to the known distance between the probe and the embedded device. Normally a VNA will have a feature which enables this to be done automatically by simply entering the distance and the phase propagation coefficient of the coplanar line.

7.5 Summary

This chapter has provided a brief introduction to RF and microwave measurements using VNAs. It has emphasized the importance of calibration in making reliable, precise measurements. A wide range of comprehensive VNAs is available from various manufacturers, notably Keysight Technologies (formerly Agilent and HP), Anritsu, and Rohde and Schwarz, covering frequencies from RF to the sub-millimetre band. These instruments are usually wideband, and contain a comprehensive range of facilities. VNAs form the essential core of any RF/microwave measurement laboratory, but their versatility and complexity is reflected in very high cost. A relatively recent innovation at the lower frequencies has been the introduction of low cost analyzers, from companies such as LA Techniques, which contain all of the high-frequency instrumentation, primarily the signal source and test set, but which can be connected to a standard PC to perform all the data processing and display functions.

References

1 Bryant, G.F. (1988). *Principles of Microwave Measurements*. London: Peter Peregrinus Ltd.
2 Dunsmore, J.P. (2012). *Handbook of Microwave Component Measurements: With Advanced VNA Techniques*. Sussex, UK: Wiley.
3 Glover, I.A., Pennock, S.R., and Shepherd, P.R. (2005). *Microwave Devices, Circuits and Subsystems*. Sussex, UK: Wiley.
4 Rumiantsev, A. and Ridler, N. (2008). VNA calibration. *IEEE Microwave Magazine* 9 (3): 86–99.

8

RF Filters

8.1 Introduction

Filters are essential circuits in all RF and microwave systems. Their primary function is to control which frequencies can propagate through a particular part of a system. Signals should be transmitted within the desired pass-band of the filter with minimum loss, and with minimum change to the phase relationships between the frequency components passing through the filter. Whilst filters are relatively straightforward circuits, it is important that they are designed correctly to avoid unwanted transmission impairments.

This chapter will introduce the concepts of RF filter design using lumped components. It will subsequently be shown how the lumped designs can be extended into distributed formats, and how modern fabrication processes permit the design of compact, high-quality planar filters.

8.2 Review of Filter Responses

There are four common types of filters, indicated by the idealized amplitude-frequency responses in Figure 8.1, which show the insertion loss (L) in dB as a function of frequency (f). The four types are known as low-pass, high-pass, band-pass, and band-stop filters.

In practice, there are six key aspects of filter responses which must be addressed when selecting and designing filters:

(1) The shape of the amplitude-frequency response within the pass-band.
(2) The low insertion loss of low-pass and band-pass filters within their pass-bands.
(3) The high insertion loss of band-stop and high-pass filters within their stop bands.
(4) The phase-frequency response within the pass-band.
(5) The rate at which the filter develops attenuation at the edges of the pass-band.
(6) The return loss (or VSWR) of the filter within the pass-band.

8.3 Filter Parameters

In Figure 8.2, a filter is shown as a two-port network, terminated with source and load impedances Z_s and Z_L, respectively. Also shown in the figure are the available power from the generator, P_{in}, the reflected power, P_r, and the available power from the output of the filter, P_{out}.

There are six main parameters that determine the performance of a filter at a particular frequency.

Insertion loss (IL): *This is the sum of the dissipative and reflective losses introduced when a filter is inserted into a system.*

$$IL = 10 \log \left(\frac{P_{out}}{P_{in}} \right) \text{ dB.} \tag{8.1}$$

Return loss (RL): *This is the loss due to reflections at a specified port of a filter.*

$$RL = 10 \log \left(\frac{P_{in}}{P_r} \right). \tag{8.2}$$

RF and Microwave Circuit Design: Theory and Applications, First Edition. Charles E. Free and Colin S. Aitchison.
© 2022 John Wiley & Sons Ltd. Published 2022 by John Wiley & Sons Ltd.
Companion website: www.wiley.com/go/free/rfandmicrowave

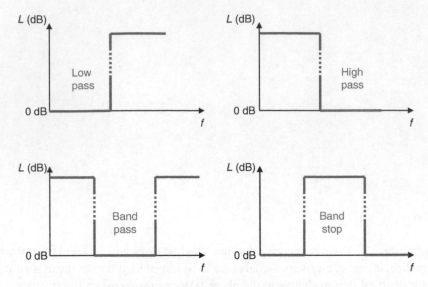

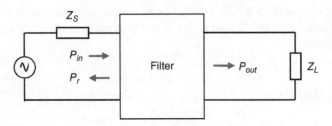

Figure 8.2 Filter as a two-port network.

Bandwidth (B): *This is the bandwidth between frequencies at which the filter loss is at a specified level (usually 3 dB) higher than the nominal level in the pass-band.*

$$B = \Delta f_{3\,dB}, \tag{8.3}$$

where Δf is the frequency span between the 3 dB positions.

Phase response (ϕ): *This is the transmission phase through the filter and can be specified in terms of input and output voltages as*

$$\phi = \mathrm{Arg}\left(\frac{V_{out}}{V_{in}}\right). \tag{8.4}$$

Group delay (τ): *The significance of group delay was explained in Section 1.2.2. It is measured in seconds and represented by*

$$\tau = \frac{d\phi}{d\omega}. \tag{8.5}$$

Roll-off: *This is a measure of how quickly a filter develops attenuation between the pass-band and the stop-band. It is specified in dB/Hz, and often as dB/octave.*

The insertion loss and return loss can also be written in terms of S-parameters as

$$IL = 10\log\left|\frac{1}{S_{21}}\right|^2 = 10\log\left|\frac{1}{S_{12}}\right|^2, \tag{8.6}$$

$$RL(\text{port }1) = 10\log\left|\frac{1}{S_{11}}\right|^2, \tag{8.7}$$

$$RL(\text{port }2) = 10\log\left|\frac{1}{S_{22}}\right|^2. \tag{8.8}$$

8.4 Design Strategy for RF and Microwave Filters

There are three basic steps in the design of an RF filter:

(1) Design a low-pass prototype having the required response.
(2) Transform the low-pass prototype to the required type of filter, with the appropriate centre and cut-off frequencies.
(3) Implement the filter in a lumped or distributed format.

8.5 Multi-Element Low-Pass Filter

The basic lumped-element configuration for a low-pass filter consists of a series inductor and a shunt capacitor, as shown in Figure 8.3.

As the frequency increases the reactance of the inductor will increase and that of the capacitor will decrease, and since the two reactances form a simple potential divider across the input the output voltage will decrease giving a low-pass response. The output voltage is given by Eq. (8.9):

$$V_{out} = \frac{1/j\omega C}{j\omega L + 1/j\omega C} \, V_{in} = \frac{1}{1 - \omega^2 LC} \, V_{in}. \tag{8.9}$$

It is often more convenient to convert the circuit shown in Figure 8.3 to a symmetrical π-network through the addition of an extra identical shunt capacitor, as shown in Figure 8.4. Since the additional capacitor is in parallel with the original circuit, the essential low-pass characteristic still pertains. The π-network has the advantage that the input and output impedances are identical. (It should be noted that the circuit shown in Figure 8.3 could be converted to a symmetrical T-network by adding a series inductor, and this would also create identical input and output impedances without changing the essential low-pass characteristic.)

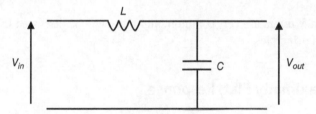

Figure 8.3 Basic low-pass circuit configuration.

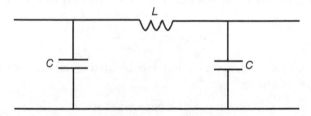

Figure 8.4 Basic low-pass circuit configuration using a π-network.

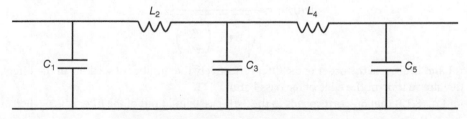

Figure 8.5 Lumped-element low-pass filter.

Greater control of the shape of the filter response can be obtained by adding additional low-pass sections, as shown in Figure 8.5. The relative values of the inductors and capacitors in a multi-element configuration determine the precise frequency response of the filter, as will be shown later.

8.6 Practical Filter Responses

The ideal frequency responses shown in Figure 8.1, with zero loss in the pass-band, and infinitely fast roll-off, cannot be realized with practical circuit components. There are four types of filter response that are commonly encountered in practice.

Butterworth: *Sometimes referred to as maximally flat, this response provides maximum amplitude flatness within the pass-band, but suffers from a relatively poor roll-off.*
Chebyshev: *This response has a small amplitude ripple within the pass-band, but a much better (faster) roll-off compared with a Butterworth filter.*
Elliptical: *This provides a very sharp transition between the pass-band and the stop-band, but exhibits a small amplitude ripple within both the pass-band and the stop-band.*
Bessel: *This type of response has a maximally flat phase response in the pass-band, but suffers from a rather poor amplitude response, with the worst (slowest) roll-off of the four filter responses mentioned here.*

Which type of filter response is chosen in a particular situation depends on the amplitude and phase characteristics of the frequency components passing through the filter, and on the associated system requirements. For example, the Butterworth response is good for signals which have been analogue modulated, and where amplitude variations within the filter would cause distortion. Digital signals are more tolerant of amplitude variations during transmission, and so in these cases Chebyshev or Elliptical responses are better choices because they offer a much better roll-off. Having good (fast) roll-off is particularly important for filters used for channel separation in communication systems, where there is a need to minimize crosstalk between adjacent channels.

The Butterworth and Chebyshev filters are very common, and these will be discussed in more detail in the subsequent sections to illustrate filter design principles.

8.7 Butterworth (or Maximally Flat) Response

8.7.1 Butterworth Low-Pass Filter

The behaviour of RF filters is usually described in terms of their amplitude-frequency transfer function, designated by $|H(j\omega)|$ where

$$|H(j\omega)| = \left| \frac{V_{out}(j\omega)}{V_{in}(j\omega)} \right|,$$ (8.10)

and $V_{in}(j\omega)$ and $V_{out}(j\omega)$ are the complex voltages at the input and output of the filter, at an angular frequency ω. In logarithmic notation the transfer function will be

$$|H(j\omega)|_{dB} = 20\log|H(j\omega)|.$$ (8.11)

Mathematically, the amplitude-frequency transfer function of a Butterworth filter is given by Eq. (8.12) [1].

$$|H(j\omega)| = \frac{1}{\sqrt{1 + \varepsilon^2 \left(\frac{\omega}{\omega_p} \right)^{2n}}},$$ (8.12)

where ω_p is the pass-band limit, n is the order of the filter (effectively the number of sections in the filter), and ε is a factor which determines the attenuation at the edge of the pass-band.

It can be seen from Eq. (8.12) that ω_p corresponds to the 3 dB pass-band limit of the filter, since when $\omega = \omega_p$

$$|H(j\omega)| = \frac{1}{\sqrt{2}} \quad \text{or} \quad |H(j\omega)|^2 = \frac{1}{2}.$$

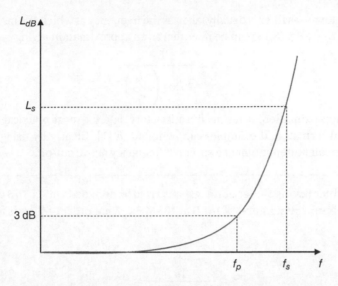

Figure 8.6 Sketch of a Butterworth low-pass frequency response.

From Eq. (8.12) the insertion loss, L_{dB}, of the filter in dB can be written as

$$L_{\mathrm{dB}} = 10\log\left[1 + \varepsilon^2\left(\frac{\omega}{\omega_p}\right)^{2n}\right]\ \mathrm{dB} = 10\log\left[1 + \varepsilon^2\left(\frac{f}{f_p}\right)^{2n}\right]\ \mathrm{dB}.$$

The insertion loss of a Butterworth low-pass filter, as a function of frequency, is sketched in Figure 8.6. The maximum loss in the pass-band occurs when $f = f_p$, giving

$$L_{max} = 10\log(1 + \varepsilon^2),$$

i.e.

$$\frac{L_{max}}{10} = \log(1 + \varepsilon^2) \quad \Rightarrow \quad \varepsilon = \sqrt{10^{L_{max}/10} - 1}.$$

If we specify a particular loss, L_S, at a frequency, f_S, in the stop-band, then

$$L_S = 10\log\left[1 + \varepsilon^2\left(\frac{\omega_S}{\omega_p}\right)^2\right],$$

and substituting for ε gives

$$L_S = 10\log\left[1 + \left(10^{L_{max}/10} - 1\right)\left(\frac{\omega_S}{\omega_p}\right)^{2n}\right],$$

and rearranging gives

$$n = \frac{\log\left(\dfrac{10^{L_S/10} - 1}{10^{L_{max}/10} - 1}\right)}{2\log\left(\dfrac{\omega_S}{\omega_p}\right)}. \tag{8.13}$$

Eq. (8.13) gives a relationship between the loss at the edge of the pass-band and the loss at a particular frequency in the stop-band, and can be used to determine the required order of filter needed to provide a particular roll-off between the pass-band and the stop-band.

The working pass-band of a low-pass filter is usually taken as the frequency at which the insertion loss is 3 dB, i.e. $L_{max} = 3$ when $f = f_c$. Then with $L_S \gg L_{max}$, Eq. (8.13) can be rewritten as an approximation giving

$$n \approx \frac{L_S}{20 \log \left(\dfrac{\omega_S}{\omega_p} \right)}. \tag{8.14}$$

Although Eq. (8.14) is an approximation, it results in satisfactory data for most practical situations. Equation (8.14) is often presented in a graphical format, and examples can be found in [1]. Clearly by using a higher-order filter we can achieve a faster roll-off with greater attenuation at a specified frequency above cut-off.

Example 8.1   A low-pass filter having Butterworth response is to be designed with a 3 dB cut-off frequency of 80 MHz. The filter is to exhibit a minimum attenuation of 16 dB at 105 MHz. What order of filter is required?

Solution
Using Eq. (8.14):

$$n = \frac{16}{20 \log \left({}^{105}\!/_{80} \right)} = 6.774.$$

Since the order of the filter must be an integer number (i.e. there must be an integer number of sections in the filter) we require a seventh-order filter.

Example 8.2   Referring to the filter defined in Example 8.1, find how much the attenuation at 105 MHz would increase if an eleventh-order filter were used.

Solution
Using Eq. (8.14):

$$11 = \frac{L_{dB}}{20 \log \left({}^{105}\!/_{80} \right)} \quad \Rightarrow \quad L_{dB} = 25.98 \text{ dB},$$

$$\text{Increase} = (25.98 - 16) \text{ dB} = 9.98 \text{ dB}.$$

It is the values of the reactances in the filter that determine the precise shape of the frequency response. These values for a low-pass filter are often presented in terms of a normalized filter specification, with terminating impedances of 1 Ω and a cut-off angular frequency of 1 rad/s.

The values of the normalized parameters can be obtained by writing the transfer function, $H(j\omega)$, in terms of the L and C parameters and equating to the value of the transfer function specified in Eq. (8.12). The normalized parameter values (g_k) to give a Butterworth response can be calculated from

$$g_k = 2 \sin \left[\frac{(2k - 1)\pi}{2n} \right] \qquad k = 1, 2, \dots n,$$

where k denotes the position of the filter element and n is the total number of elements. For convenience, the values of g_k to give a Butterworth response are shown in Table 8.1.

Table 8.1 Butterworth normalized parameters.

	$n = 2$	$n = 3$	$n = 4$	$n = 5$	$n = 6$	$n = 7$
$k = 1$	1.4142	1.0000	0.7654	0.6180	0.5176	0.4550
$k = 2$	1.4142	2.0000	1.8478	1.6180	1.4142	1.2470
$k = 3$		1.0000	1.8478	2.0000	1.9319	1.8019
$k = 4$			0.7654	1.6180	1.9319	2.0000
$k = 5$				0.6180	1.4142	1.8019
$k = 6$					0.5176	1.2470
$k = 7$						0.4450

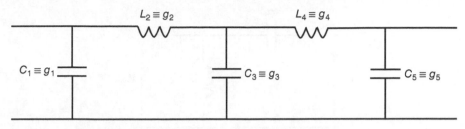

Figure 8.7 Normalized low-pass filter ($C_1 \equiv g_1, L_2 \equiv g_2, ...$).

A fifth-order filter is shown in Figure 8.7.

For a normalized filter specification with $\omega_p = 1$ rad/s and $Z_o = 1\ \Omega$, the values of the reactances are (using data from Table 8.1):

$$C_1 = g_1 = 0.6180 \text{ F},$$
$$L_2 = g_2 = 1.6180 \text{ H},$$
$$C_3 = g_3 = 2.0000 \text{ F},$$
$$L_4 = g_4 = 1.6180 \text{ H},$$
$$C_5 = g_5 = 0.6180 \text{ F}.$$

Obviously the values obtained are unrealistic because we are using a normalized (unrealistic) filter specification; to obtain values for a real fifth-order filter we need to scale the data according to the required filter specification.

To find the component values for a filter with terminating impedances, Z_o, and a cut-off frequency, f_p, the following scaling is applied:

$$C_k \quad \Rightarrow \quad C_k \times \frac{1}{2\pi f_p} \times \frac{1}{Z_o},$$
$$L_k \quad \Rightarrow \quad L_k \times \frac{1}{2\pi f_p} \times Z_o. \tag{8.15}$$

Note that in Eq. (8.15) we have divided the prototype (normalized) values by ω_p to scale the frequency, and effectively multiplied the prototype reactance by Z_o to scale the impedance.

Example 8.3 Design a five-section lumped-element low-pass filter with a 3 dB point at 950 MHz. The filter is to have a Butterworth response, and is to work between terminating impedances of 50 Ω.

Solution

Considering the fifth-order filter configuration shown in Figure 8.7, we have:

$$C_1 = C_5 = 0.6180 \times \frac{1}{50 \times 2\pi \times 950 \times 10^6} \text{ F} = 2.07 \text{ pF},$$

$$L_2 = L_4 = 1.618 \times \frac{50}{2\pi \times 950 \times 10^6} \text{ H} = 13.55 \text{ nH},$$

$$C_3 = 2.000 \times \frac{1}{50 \times 2\pi \times 950 \times 10^6} \text{ F} = 6.70 \text{ pF}.$$

The final circuit is shown in Figure 8.8.

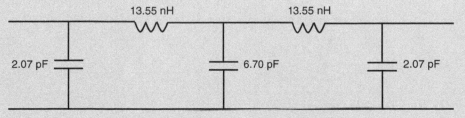

Figure 8.8 Summary of solution to Example 8.3.

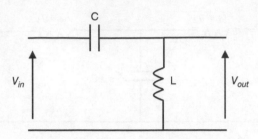

Figure 8.9 Basic high-pass circuit configuration.

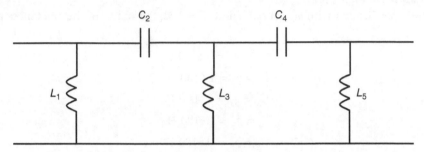

Figure 8.10 Configuration of a five-element high-pass filter.

8.7.2 Butterworth High-Pass Filter

The basic lumped-element configuration for a high-pass filter consists of a series capacitor and a shunt inductor, as shown in Figure 8.9.

The voltage at the output of the high-pass configuration shown in Figure 8.9 is given by Eq. (8.16).

$$V_{out} = \frac{j\omega L}{j\omega L + 1/j\omega C} V_{in} = \frac{\omega^2 LC}{\omega^2 LC - 1} V_{in}. \tag{8.16}$$

It is clear from inspection of Eq. (8.16) that a high-pass characteristic is obtained, since V_{out} approaches V_{in} as the frequency increases, and V_{out} approaches zero as the frequency tends to zero. A multi-element high-pass filter is thus obtained by interchanging the capacitive and inductive elements in a low-pass filter to form the circuit shown in Figure 8.10. The element values are then found using the same general procedure as for the low-pass filter, but using the reciprocal of the values from the appropriate table of normalized coefficients.

Example 8.4   Design a fifth-order high-pass filter with a cut-off frequency of 450 MHz. The filter is to have a Butterworth response and work between source and load impedances of 50 Ω.

Solution
Referring to the circuit shown in Figure 8.10:

$$L_1 = \frac{1}{0.618} \times \frac{50}{2\pi \times 450 \times 10^6} \text{ H} = 28.61 \text{ nH},$$

$$C_2 = \frac{1}{1.618} \times \frac{1}{50 \times 2\pi \times 450 \times 10^6} \text{ F} = 4.37 \text{ pF},$$

$$L_3 = \frac{1}{2.000} \times \frac{50}{2\pi \times 450 \times 10^6} \text{ H} = 8.84 \text{ nH},$$

$$C_3 = C_2 = 4.37 \text{ pF},$$

$$L_5 = L_1 = 28.61 \text{ nH}.$$

The final circuit is shown in Figure 8.11.

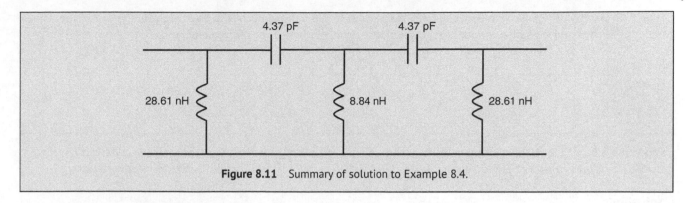

Figure 8.11 Summary of solution to Example 8.4.

8.7.3 Butterworth Band-Pass Filter

The response of a band-pass filter with a Butterworth response is sketched in Figure 8.12, showing the centre frequency, f_O, and upper and lower cut-off frequencies f_2 and f_1, respectively. The two cut-off frequencies define the 3 dB bandwidth of the filter, i.e. the pass-band frequency range over which the insertion loss is less than 3 dB; this is also shown as Δf in Figure 8.12.

The band-pass response is obtained through a frequency transformation from the equivalent low-pass filter response. It can be shown [2] that this is achieved by adding series and shunt components to the low-pass filter. A series inductor is added to the capacitor to form a series resonant circuit, with a resonant frequency f_O. A parallel capacitor is added to the inductor to form a parallel resonant circuit, also with a resonant frequency, f_O. The resulting band-pass filter is shown in Figure 8.13.

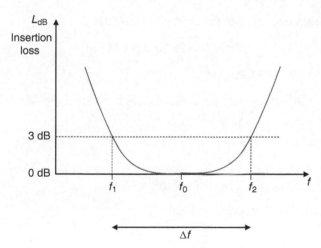

Figure 8.12 Sketch of Butterworth response of a band-pass filter.

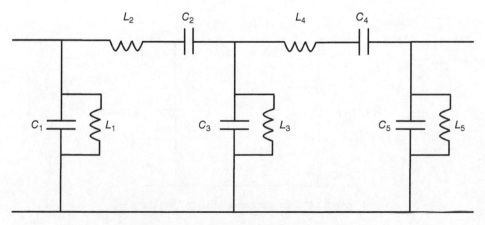

Figure 8.13 Lumped-element, band-pass filter.

The 3 dB bandwidth of the band-pass filter is then equal to the bandwidth of the low-pass prototype, and the centre frequency of the band-pass filter is equal to the geometric mean of the two cut-off frequencies, i.e.

$$B = f_p = f_2 - f_1, \tag{8.17}$$

$$f_O = \sqrt{f_1 f_2}. \tag{8.18}$$

Example 8.5 Design a third-order band-pass filter with upper and lower cut-off frequencies of 95 and 85 MHz, respectively. The filter is to work between source and load impedances of 50 Ω, and is to have a Butterworth response.

Solution
(i) The first step is to design the low-pass prototype:

$$f_p = (95 - 85) \text{ MHz} = 10 \text{ MHz}.$$

Referring to the low-pass configuration shown in Figure 8.4:

$$C_1 = 1.000 \times \frac{1}{2\pi \times 10 \times 10^6 \times 50} \text{ F} = 318.27 \text{ pF},$$

$$L_2 = 2.000 \times \frac{50}{2\pi \times 10 \times 10^6} \text{ H} = 1.59 \text{ μH},$$

$$C_3 = C_1 = 318.27 \text{ pF}.$$

(ii) Calculate the geometric mean frequency of the band-pass filter (BPF):

$$f_O = \sqrt{85 \times 95} \text{ MHz} = 89.86 \text{ MHz}.$$

(iii) Calculate L_1 in first parallel resonant circuit of BPF:

$$89.86 \times 10^6 = \frac{1}{2\pi \sqrt{L_1 \times 318.27 \times 10^{-12}}} \quad \Rightarrow \quad L_1 = 9.85 \text{ nH}.$$

Calculate C_2 in series resonant circuit of BPF:

$$89.86 \times 10^6 = \frac{1}{2\pi \sqrt{1.59 \times 10^{-6} \times C_2}} \quad \Rightarrow \quad C_2 = 1.97 \text{ pF}.$$

Calculate L_3 in second parallel resonant circuit of BPF:

$$L_3 = L_1 = 9.85 \text{ nH}.$$

The completed design is shown in Figure 8.14.

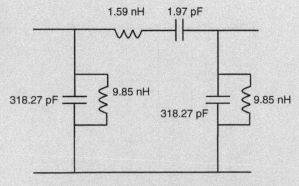

Figure 8.14 Summary of solution to Example 8.5.

8.8   Chebyshev (Equal Ripple) Response

The amplitude-frequency response of a Chebyshev filter is given by

$$H(j\omega) = \frac{1}{\sqrt{1 + \varepsilon^2 C_n^2(x)}},$$ (8.19)

where

$$\varepsilon^2 = 10^{L_r/10} - 1,$$

and L_r is the pass-band ripple in dB, and $C_n(x)$ is the Chebyshev polynomial of order n. The argument, x, is given by

$$x = \frac{f}{f_c},$$

where f is the frequency of interest and f_c is the cut-off frequency. Note that for the Chebyshev filter the cut-off frequency is defined as the band edge where the attenuation corresponds to the value of the ripple (see Figure 8.15). It should also be noted that the higher the order of the filter the more ripples exist in the pass-band, which means that the slope of the response at the edge of the pass-band increases, so the filter develops attenuation more quickly at the beginning of the stop-band.

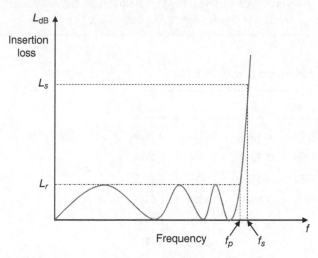

Figure 8.15   Sketch of Chebyshev low-pass filter response.

Using a similar procedure to that adopted for the Butterworth filter, it can be shown that the order of a Chebyshev filter can be expressed as

$$n = \frac{\cosh^{-1}\left(\frac{10^{L_s/10} - 1}{10^{L_r/10} - 1} \right)^{\frac{1}{2}}}{\cosh^{-1}\left(\frac{f_S}{f_p} \right)},$$ (8.20)

where L_S, L_r, f_S, and f_p are specified in Figure 8.15.

Example 8.6   An RF low-pass filter is required to meet the following specification:

Pass-band loss	<0.8 dB below 80 MHz
Stop-band loss	>20 dB above 95 MHz

(i) What order of Chebyshev filter is required to meet the specification?
(ii) What order of maximally flat filter is required to meet the same specification?
     Comment upon the result.

(Continued)

(Continued)

Solution

(i) Using Eq. (8.20):

$$n = \frac{\cosh^{-1}\left(\dfrac{10^{20/10}-1}{10^{0.8/10}-1}\right)^{\frac{1}{2}}}{\cosh^{-1}\left(\dfrac{95}{80}\right)} = \frac{\cosh^{-1}(22.13)}{\cosh^{-1}(1.1875)} = 6.29,$$

i.e. a seventh-order Chebyshev filter is required.

(ii) Using Eq. (8.13):

$$n = \frac{\log\left(\dfrac{10^{20/10}-1}{10^{0.8/10}-1}\right)}{2\log\left(\dfrac{95}{80}\right)} = \frac{2.689}{0.149} = 18.05,$$

i.e. a nineteenth-order maximally flat filter is required.

Comment: A significantly higher-order maximally flat filter is required to achieve the same degree of roll-off as a Chebyshev filter. Indeed, in this example the number of sections required in the maximally flat filter is unrealistically high, and indicates that a maximally flat filter is not a practical choice to meet the specification.

Table 8.2 Chebyshev normalized parameters (Ripple 0.01 dB).

	$n = 2$	$n = 3$	$n = 4$	$n = 5$	$n = 6$	$n = 7$
$k = 1$	0.4489	0.6292	0.7129	0.7563	0.7814	0.7969
$k = 2$	0.4078	0.9703	1.2004	1.3049	1.3600	1.3924
$k = 3$		0.6292	1.3213	1.5773	1.6897	1.7481
$k = 4$			0.6476	1.3049	1.5350	1.6331
$k = 5$				0.7563	1.4970	1.7481
$k = 6$					0.7098	1.3924
$k = 7$						0.7969

Normalized filter parameters to give a Chebyshev response are given in Table 8.2, for a response having an in-band ripple of 0.01 dB. The filter design procedure using the data in Table 8.2 is the same as that already described for Butterworth filters in Section 8.7.1, and is demonstrated in Example 8.7.

Example 8.7   Design a fifth-order band-pass filter with upper and lower cut-off frequencies of 75 and 69 MHz, respectively. The filter is to work between source and load impedances of 75 Ω, and is to have a Chebyshev response with a ripple of 0.01 dB.

Solution

(i) Referring to Figure 8.7 and calculating values for the low-pass prototype filter:

$$f_p = (75 - 69)\ \text{MHz} = 6\ \text{MHz},$$

$$C_1 = 0.7563 \times \frac{1}{2\pi \times 6 \times 10^6 \times 75}\ \text{F} = 267.45\ \text{pF},$$

$$L_2 = 1.3049 \times \frac{75}{2\pi \times 6 \times 10^6}\, H = 2.60\ \mu H,$$

$$C_3 = 1.5773 \times \frac{1}{2\pi \times 6 \times 10^6 \times 75}\, F = 557.78\ pF,$$

$$L_4 = L_2 = 2.60\ \mu H,$$

$$C_5 = C_1 = 267.45\ pF.$$

(ii) Referring to Figure 8.13 and converting the low-pass prototype to band-pass:

$$f_O = \sqrt{69 \times 75}\ \text{MHz} = 71.94\ \text{MHz}.$$

Determining the reactances needed to create parallel and series resonant circuits:

$$71.94 \times 10^6 = \frac{1}{2\pi \sqrt{L_1 \times 267.45 \times 10^{-12}}} \quad \Rightarrow \quad L_1 = 18.30\ \text{nH},$$

$$71.94 \times 10^6 = \frac{1}{2\pi \sqrt{2.60 \times 10^{-6} \times C_2}} \quad \Rightarrow \quad C_2 = 1.88\ \text{pF},$$

$$71.94 \times 10^6 = \frac{1}{2\pi \sqrt{L_3 \times 557.78 \times 10^{-12}}} \quad \Rightarrow \quad L_3 = 8.77\ \text{nH},$$

$$C_4 = C_2 = 1.88\ \text{pF},$$

$$L_5 = L_1 = 18.30\ \text{nH}.$$

The final circuit is shown in Figure 8.16.

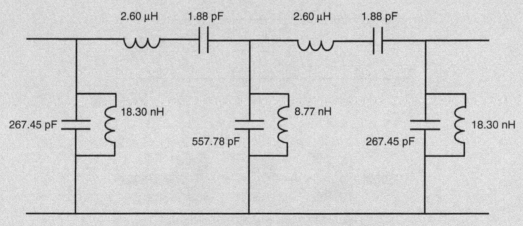

Figure 8.16 Summary of solution to Example 8.7.

8.9 Microstrip Low-Pass Filter, Using Stepped Impedances

Low-pass filters can easily, and effectively, be implemented by cascading short lengths of microstrip line as shown in Figure 8.17. Intuitively, one can reason that the shunt capacitors of the lumped-element design can be represented by rectangular 'pads' of microstrip, which form effective parallel plate capacitors with the ground plane. Similarly, reducing the width of the microstrip track to form narrow interconnecting sections would be expected to introduce series inductance, as occurs when the cross-section of any ac-carrying conductor is reduced.

We can relate the dimensions of the microstrip filter to the required values of L and C by returning to the equivalent circuits of transmission lines discussed in Appendix 1.E of Chapter 1.

Each of the narrow line sections of microstrip will have a high characteristic impedance and can be modelled as a π-network as discussed in Appendix 1.E. The π-network representing these sections is shown in Figure 8.18.

The values of L_N and C_N are derived in Appendix 1.E as

$$L_N = \frac{Z_{oN}}{\omega} \sin \beta_N l_N,$$

$$C_N = \frac{1}{\omega Z_{oN}} \tan \left(\frac{\beta_N l_N}{2} \right), \tag{8.21}$$

where we have used the suffix N to indicate the narrow microstrip line. The other parameters are defined in Appendix 1.E.

For small values of l_N the values of L_N and C_N in Eq. (8.21) can be approximated as

$$L_N = \frac{Z_{oN} \beta_N l_N}{\omega},$$

$$C_N = \frac{\beta_N l_N}{2 \omega Z_{oN}}. \tag{8.22}$$

It follows from Eqs. (8.21) and (8.22) that for short line lengths and large values of Z_{oN} the equivalent circuit will be predominantly inductive, and so the narrow section of microstrip line will represent the required series inductance, with

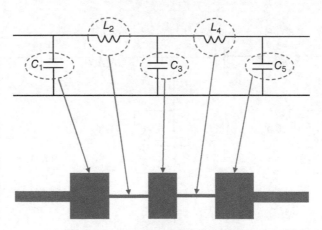

Figure 8.17   Low-pass filter in microstrip.

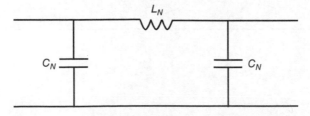

Figure 8.18 π-Network representing a narrow transmission line.

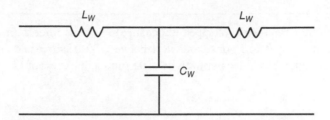

Figure 8.19 T-network representing a wide transmission line.

the capacitive components of the network being neglected. We can term these the fringing capacitances. From Eq. (8.21) the required length of the narrow section of microstrip will be

$$l_N = \frac{1}{\beta_N} \sin^{-1}\left(\frac{\omega L_N}{Z_{oN}}\right),$$ (8.23)

where L_N is the required inductance.

Using a similar procedure, each of the wide sections of low impedance microstrip can be modelled as a T-network, as shown in Figure 8.19.

It is shown in Appendix 8.A that the network will be predominantly capacitive with the capacitance, C_W, and inductance, L_W, being given by

$$C_W = \frac{1}{\omega Z_{oW}} \sin(\beta_W l_W)$$ (8.24)

and

$$L_W = \frac{Z_{oW}}{\omega} \tan\left(\frac{\beta_W l_W}{2}\right),$$ (8.25)

where the subscript W has been used to denote the wide section of line. The remaining parameters are defined in Appendix 8.A. In a similar manner to the design of the narrow section, we can, to a first approximation, neglect the effect of the inductive components, i.e. those giving the fringing inductances. It follows from Eq. (8.24) that the length, l_w, of the wide section of microstrip necessary to provide a given value of capacitance, C_W will be

$$l_W = \frac{1}{\beta_W} \sin^{-1}(\omega C_W Z_{oW}).$$ (8.26)

The design procedure for a microstrip low-pass filter using stepped-impedances is essentially an iterative procedure [2] in which the lengths l_N and l_W of the narrow and wide sections of microstrip are first calculated using chosen values of line impedance, and then refined if necessary to compensate for the values of C_N and L_W which have been neglected. The practical steps in the design procedure are listed below, and subsequently illustrated through a practical example, Example 8.8.

Design procedure for low-pass microstrip filter:

(i) Design the low-pass prototype.
(ii) Choose the impedance of the narrow section of the filter; it is normal practice to choose the highest impedance that the fabrication process will allow, i.e. an impedance that does not require excessively narrow lines that may be difficult to fabricate without breaks. Hence, calculate the width of the high impedance section(s) using microstrip design graphs or CAD.
(iii) Choose the impedance of the wide section of the filter; it is normal practice to choose the lowest viable impedance that the fabrication process will allow, i.e. an impedance that is not so low that it requires an excessively wide microstrip line, which may lead to transverse modes (see Chapter 2). Hence, calculate the width of the low impedance section(s) using microstrip design graphs or CAD.
(iv) Calculate the lengths of the narrow and wide sections of the filter using Eqs. (8.23) and (8.26).
(v) Use Eqs. (8.21) and (8.25) to calculate the values of the fringing capacitance and inductance.
(vi) If the values of the fringing reactances are significant compared to those in the prototype, then the prototype values must be compensated and step (iv) repeated.

Example 8.8 Design a fifth-order microstrip (stepped-impedance) low-pass filter with a 3 dB cut-off frequency of 1 GHz. The filter is to have a maximally flat response, and work between 50 Ω source and load impedances. The filter is to be fabricated on a substrate that has a relative permittivity of 9.8, and a thickness of 1.0 mm.

Solution

(i) Prototype design:

$$C_1 = 0.618 \times \frac{1}{2\pi \times 10^9 \times 50} \text{ F} = 1.97 \text{ pF},$$

$$L_2 = 1.618 \times \frac{50}{2\pi \times 10^9} \text{ H} = 12.87 \text{ nH},$$

$$C_3 = 2.000 \times \frac{1}{2\pi \times 10^9 \times 50} \text{ F} = 6.37 \text{ pF},$$

$$L_4 = L_2 = 12.87 \text{ nH},$$

$$C_5 = C_1 = 1.97 \text{ pF}.$$

(ii) Choose $Z_{oN} = 24 \, \Omega$
Using microstrip design graphs:

$$24 \, \Omega \quad \Rightarrow \quad \frac{w}{h} = 3.5 \quad \Rightarrow \quad w = 3.5 \times 1.0 \text{ mm} = 3.5 \text{ mm},$$

$$\frac{w}{h} = 3.5 \quad \Rightarrow \quad \varepsilon_{r,eff} = 7.52,$$

$$\lambda_W = \frac{\lambda_O}{\sqrt{7.52}} = \frac{3 \times 10^8 / 10^9}{\sqrt{7.52}} \text{ m} = 109.40 \text{ mm}.$$

(iii) Choose $Z_{oW} = 100 \, \Omega$
Using microstrip design graphs:

$$100 \, \Omega \quad \Rightarrow \quad \frac{w}{h} = 0.10 \quad \Rightarrow \quad w = 0.10 \times 1.0 \text{ mm} = 100 \, \mu\text{m}.$$

$$\frac{w}{h} = 0.10 \quad \Rightarrow \quad \varepsilon_{r,eff} = 5.98,$$

$$\lambda_N = \frac{\lambda_O}{\sqrt{5.98}} = \frac{0.300}{\sqrt{5.98}} \text{ m} = 122.68 \text{ mm}.$$

(iv) Using Eq. (8.23):
$$l_1 = \frac{109.40}{2\pi} \sin^{-1}(2\pi \times 10^9 \times 1.97 \times 10^{-12} \times 24) \text{ mm} = 5.25 \text{ mm}.$$

Using Eq. (8.21):
$$l_2 = \frac{122.68}{2\pi} \sin^{-1}\left(\frac{2\pi \times 10^9 \times 12.87 \times 10^{-9}}{100} \right) \text{ mm} = 18.39 \text{ mm}.$$

Using Eq. (8.23):
$$l_3 = \frac{109.40}{2\pi} \sin^{-1}(2\pi \times 10^9 \times 6.37 \times 10^{-12} \times 24) \text{ mm} = 22.45 \text{ mm}.$$

(v) Using Eqs. (8.21) and (8.25) to evaluate the values of the fringing capacitance and inductance:

$$L_{1,FRINGING} = \frac{24}{2\pi \times 10^9} \tan\left(\frac{2\pi \times 5.25}{2 \times 109.40} \right) \text{ H} = 0.58 \text{ nH},$$

$$C_{2,FRINGING} = \frac{1}{2\pi \times 10^9 \times 100} \tan\left(\frac{2\pi \times 18.39}{2 \times 122.68} \right) \text{ F} = 0.81 \text{ pF},$$

$$L_{3,FRINGING} = \frac{24}{2\pi \times 10^9} \tan\left(\frac{2\pi \times 22.45}{2 \times 109.40} \right) \text{ H} = 2.87 \text{ nH}.$$

(vi) Since the values of the 'fringing' reactances are significant compared to the prototype values, compensation should be applied. The compensated prototype values are:

$$C_{1,COMPENSATED} = C_1 - C_{2,FRINGING} = (1.97 - 0.81)\ \text{pF} = 1.16\ \text{pF},$$

$$L_{2,COMPENSATED} = L_2 - L_{1,FRINGING} - L_{3,FRINGING} = (12.87 - 0.58 - 2.87)\ \text{nH} = 9.42\ \text{nH},$$

$$C_{3,COMPENSATED} = C_3 - 2C_{2,FRINGING} = (6.37 - 2 \times 0.81)\ \text{pF} = 4.75\ \text{pF}.$$

Using Eq. (8.26) to find the length of the first section after compensation:

$$l_{1,COMPENSATED} = \frac{109.40}{2\pi} \sin^{-1}(2\pi \times 10^9 \times 1.16 \times 10^{-12} \times 24)\ \text{mm} = 3.06\ \text{mm}.$$

Using Eq. (8.23) to find the length of the second section after compensation:

$$l_{2,COMPENSATED} = \frac{122.68}{2\pi} \sin^{-1}\left(\frac{2\pi \times 10^9 \times 9.42 \times 10^{-9}}{100}\right)\ \text{mm} = 12.37\ \text{mm}.$$

Using Eq. (8.26) to find the length of the third section after compensation:

$$l_{3,COMPENSATED} = \frac{109.40}{2\pi} \sin^{-1}(2\pi \times 10^9 \times 4.75 \times 10^{-12} \times 24)\ \text{mm} = 13.90\ \text{mm}.$$

From considerations of symmetry:

$$l_{4,COMPENSATED} = l_{2,COMPENSATED} = 12.37\ \text{mm},$$

$$l_{5,COMPENSATED} = l_{1,COMPENSATED} = 3.06\ \text{mm}.$$

The completed design is shown in Figure 8.20.

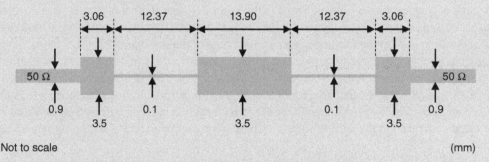

Figure 8.20 Summary of solution to Example 8.8.

Notes: (1) *All dimensions shown are in mm.*

(2) *For completeness, 50 Ω lines have been added at the input and output* $(50\ \Omega \Rightarrow \frac{w}{h} = 0.9 \Rightarrow w = 0.9\ \text{mm})$.

Whilst the microstrip stepped-impedance low-pass filter can provide a very good microwave response, it does suffer from significant discontinuities in the form of step changes in width. In practice, appropriate compensations for the step changes in width must be added to the design, using the theory given in Chapter 4.

8.10 Microstrip Low-Pass Filter, Using Stubs

Whilst the use of stepped impedances to create a microstrip low-pass filter provides a straightforward design, the filter can sometimes occupy a significant area of substrate particularly at lower frequencies. A more compact design can be obtained by using stubs to create the inductive and capacitive reactances required by the prototype circuit. There is a well-established design procedure for filters using microstrip stubs, based on the application of Richards' transformation [3] and Kuroda identities [4].

It was shown in Chapter 1 that stubs can be used to replace inductive or capacitive reactances. The process of converting reactances into stubs was formalized by Richards; the results are known as Richards' transformations, and are illustrated in Figure 8.21.

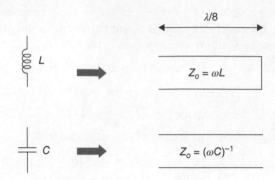

Figure 8.21 Richards' transformations.

The transformations in Figure 8.21 show that an inductor (L) can be replaced by a short-circuited stub whose length is $\lambda/8$ and whose characteristic impedance is chosen to be equal to the reactance of the inductor. Similarly, a capacitor can be replaced by an open-circuited stub of length $\lambda/8$, whose characteristic impedance is equal to the reactance of the capacitor. The application of Richards' transformations to a T-network low-pass filter is shown in Figure 8.22.

It can be seen from Figure 8.22 that the series inductor has transformed to a series-connected short-circuited stub, and the shunt capacitor to a shunt-connected open-circuited stub. The lengths of the stubs are $\lambda_c/8$, where λ_c is the wavelength at the cut-off frequency of the filter. It should be noted that the stub representation of the filter is only correct at one frequency, and whilst the stub circuit gives a good representation of the filter in the pass-band, there can be significant differences in the stop-band, particularly at frequencies significantly above the cut-off frequency. For example, at a higher frequency which makes the stub lengths $\lambda/2$, the open-circuited stub will present a shunt open-circuit to the main transmission line, and the short-circuited stubs will present series short-circuits to the main line, and there will be perfect transmission through the structure. So with stub-type filters it is necessary to check the performance well into the stop-band; this can be conveniently done with CAD software to sweep the frequency over a wide range of the stop-band.

There is an obvious problem in directly implementing in a microstrip the stub filter shown in Figure 8.22, in that it is not physically possible to make a series stub connection in a microstrip. This is where the designer needs to make use of one of Kuroda's identities. Kuroda developed a series of equivalent transmission line structures, which enable series- and shunt-connected components to be interchanged. One of these identities enables a series-connected stub to be transformed to a shunt-connected stub through the addition of a $\lambda/8$ length of transmission line of an appropriate impedance, as shown in Figure 8.23.

Notes on Figure 8.23:

(i) The required transmission line impedances are:

$$Z_2 = Z_o + \frac{Z_o^2}{Z_1},$$

$$Z_3 = Z_o + Z_1.$$

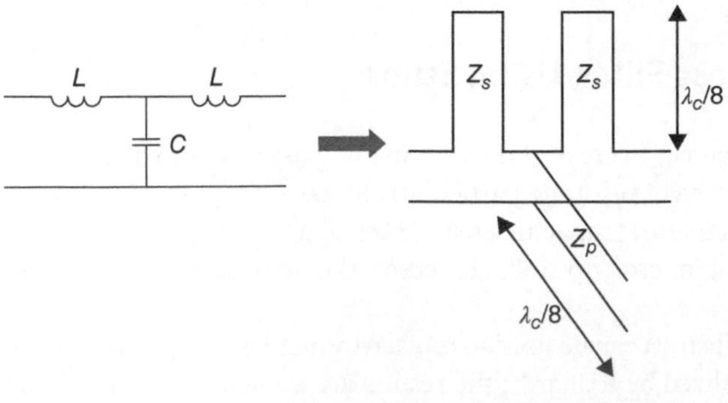

Figure 8.22 T-network low-pass filter and its stub equivalent.

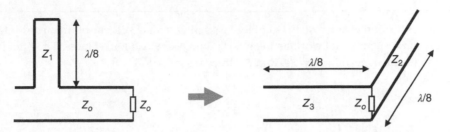

Figure 8.23 Kuroda identity for transforming a series-connected short-circuited stub into a shunt-connected open-circuited stub.

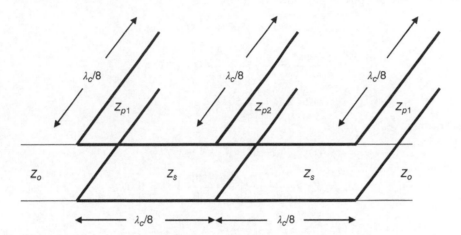

Figure 8.24 Low-pass filter using shunt-connected, open-circuited stubs.

(ii) The series-connected stub is attached to a length of matched transmission line, having a characteristic impedance of 50 Ω, but since it is matched its length is immaterial.

Thus, using a combination of Richard's transformation and Kuroda's identity the low-pass filter can be represented by the arrangement of open-circuited, $\lambda/8$ stubs shown in Figure 8.24.

Summary of design data for the $\lambda_c/8$ transmission line elements circuits shown in Figure 8.24:

$$Z_s = Z_o + \omega_c L, \tag{8.27}$$

$$Z_{p1} = Z_o + \frac{Z_o^2}{\omega_c L}, \tag{8.28}$$

$$Z_{p2} = (\omega_c C)^{-1}, \tag{8.29}$$

where

Z_o = characteristic impedance of transmission lines connected to the filter

Z_s = characteristic impedance of transmission line between the stubs

Z_{p1}, and Z_{p2} are characteristic impedances of the stubs

L and C are values from prototype, lumped-element design

$\omega_c = 2\pi f_c$

f_c = cut-off frequency of the filter.

To complete the design, the stub information must be converted into microstrip dimensions using the appropriate microstrip design data. Example 8.9 shows all of the steps in the design of a microstrip stub filter in numerical detail.

Example 8.9 Design a third-order microstrip (stub) low-pass filter with a 3 dB cut-off frequency of 1 GHz. The filter is to have a maximally flat response, and work between 50 Ω source and load impedances. The filter is to be fabricated on a substrate that has a relative permittivity of 9.8, and a thickness of 50 mil.
Comment: 50 mil ≡ 1.27 mm.

Solution
Normalized parameters from Table 8.1:

$$g_1 = 1.000 \quad g_2 = 2.000 \quad g_3 = 1.000.$$

Reactance values for T-network prototype, shown in Figure 8.22:

$$\omega_c L_1 = \omega_c \times 1.000 \times \frac{50}{\omega} \, \Omega = 50 \, \Omega,$$

$$\frac{1}{\omega_c C_2} = \frac{1}{\omega_c} \times \frac{\omega_c \times 50}{2.000} \, \Omega = 25 \, \Omega,$$

$$\omega_c L_2 \equiv \omega_c L_1 = 50 \, \Omega.$$

Using Eqs. (8.27) through (8.29):

$$Z_s = Z_0 + \omega_c L_1 = (50 + 50) \, \Omega = 100 \, \Omega,$$

$$Z_{p1} = Z_0 + \frac{Z_0^2}{\omega_c L_1} = \left(50 + \frac{(50)^2}{50}\right) \Omega = 100 \, \Omega,$$

$$Z_{p2} = (\omega_c C_2)^{-1} = 25 \, \Omega.$$

Converting impedance data into microstrip dimensions:

$$25 \, \Omega \quad \Rightarrow \quad \frac{w}{h} = 3.3 \quad \Rightarrow \quad w = 3.3 \times 1.27 \text{ mm} = 4.19 \text{ mm},$$

$$\frac{w}{h} = 3.3 \quad \Rightarrow \quad \varepsilon_{r,eff} = 7.48 \quad \Rightarrow \quad \lambda_S = \frac{\lambda_O}{\sqrt{7.48}} = \frac{300}{\sqrt{7.48}} \text{ mm} = 109.69 \text{ mm},$$

$$\frac{\lambda_S}{8} = \frac{109.69}{8} \text{ mm} = 13.71 \text{ mm},$$

$$100 \, \Omega \quad \Rightarrow \quad \frac{w}{h} = 0.10 \quad \Rightarrow \quad w = 0.10 \times 1.27 \text{ mm} = 127.0 \, \mu\text{m},$$

$$\frac{w}{h} = 0.10 \quad \Rightarrow \quad \varepsilon_{r,eff} = 5.98 \quad \Rightarrow \quad \lambda_S = \frac{\lambda_O}{\sqrt{5.98}} = \frac{300}{\sqrt{5.98}} \text{ mm} = 122.68 \text{ mm},$$

$$\frac{\lambda_S}{8} = \frac{122.68}{8} \text{ mm} = 15.34 \text{ mm},$$

$$50 \, \Omega \quad \Rightarrow \quad \frac{w}{h} = 0.9 \quad \Rightarrow \quad w = 0.9 \times 1.27 \text{ mm} = 1.14 \text{ mm}.$$

The completed microstrip design is shown in Figure 8.25.

Notes:

(i) *Where a microstrip junction is involved in a practical design, the length dimensions should always be specified to the geometric centre of the junction.*

(ii) *In Figure 8.25, the 50 Ω microstrip lines are shown chamfered at the entries to the filter section; in practice, this would be done to minimize reflections at the junction between two microstrip lines of significantly different widths.*

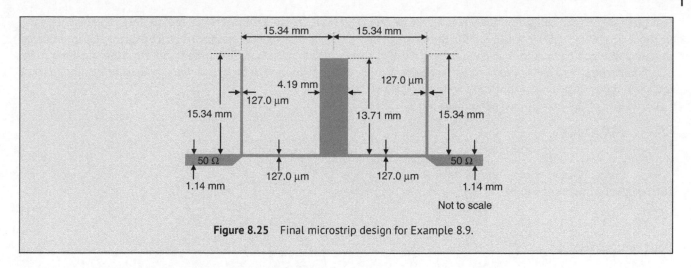

Figure 8.25 Final microstrip design for Example 8.9.

8.11 Microstrip Edge-Coupled Band-Pass Filters

Microstrip band-pass filters can be modelled as a series of coupled resonant circuits, where the resonant circuits take the form of $\lambda_S/2$ sections of open-ended microstrip lines. Figure 8.26 depicts a microstrip band-pass filter where each resonant section is edge-coupled to the adjacent sections.

The design of microstrip band-pass filters is well-established in the literature, with a very comprehensive account being given by Matthai et al. [1], and a very informative summary presented by Edwards and Steer [5]. Consequently, only the main points of the design will be stated here, to show how a practical microstrip band-pass filter can be realized from a given specification.

Each of the half-wavelength sections in the filter shown in Figure 8.26 can be regarded as a parallel resonant circuit. But we know that a band-pass filter can be modelled as a series of alternate parallel and series resonant circuits. So the half-wavelength resonators must be interconnected with immittance[1] inverters which convert a parallel resonant circuit to a series resonant circuit, and vice versa. The simplest type of immittance inverter is a quarter-wavelength of transmission line, which converts impedance to admittance (and vice versa), as was shown in Chapter 1. In the microstrip filter shown in Figure 8.26, it can be seen that adjacent resonators are edge-coupled over a $\lambda_S/4$ length, and it is these coupled sections

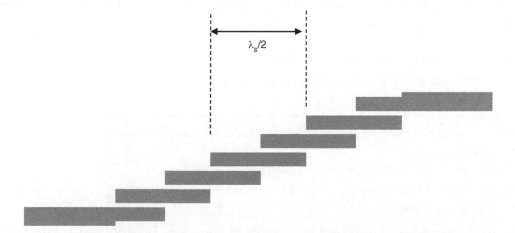

Figure 8.26 General structure of a microstrip edge-coupled band-pass filter.

---

1 Immittance is a term used to represent a complex quantity that may be either an impedance or an admittance.

that possess the required immittance inverter properties. The values of these inverters depend on the g_k coefficients, such as those given in Tables 8.1 and 8.2 for Butterworth and Chebyshev responses, respectively. Thus, it is the immittance inverters that determine the shape of the response of microstrip band-pass filters. The immittance inverters are denoted by J_{nm}, where the subscripts n and m identify a particular coupling section in the filter, and whose values determine the odd and even mode impedances that apply to the coupled sections.

The values of the immittance parameters are [1]:

Frist coupled section:

$$J_{01} = Y_O \sqrt{\frac{\pi\delta}{2g_0 g_1}}. \tag{8.30}$$

Intermediate coupled sections:

$$J_{k,k+1} = \frac{Y_O \pi\delta}{2\sqrt{g_k g_{k+1}}}. \tag{8.31}$$

Final coupled section:

$$J_{n,n+1} = Y_O \sqrt{\frac{\pi\delta}{2g_n g_{n+1}}}, \tag{8.32}$$

where Y_O is the admittance at each port of the filter, δ is the fractional bandwidth, and n is the number of resonator sections.

The geometry of the structure depends on the required values of the odd and even mode impedances, which can be calculated from:

$$(Z_{oo})_{k,k+1} = Z_0(1 - J_{k,k+1}Z_o + (J_{k,k+1}Z_o)^2), \tag{8.33}$$

$$(Z_{oe})_{k,k+1} = Z_0(1 + J_{k,k+1}Z_o + (J_{k,k+1}Z_o)^2), \tag{8.34}$$

where Z_o is the impedance at each port of the filter.

Thus, the steps in the design of a microstrip band-pass filter are:

(i) Decide on the order of the filter, i.e. the number of resonators required, based on the specification.
(ii) Calculate the immittance parameters using Eqs. (8.30) through (8.32).
(iii) Calculate the odd and even mode impedances for each coupled section, using Eqs. (8.33) and (8.34).
(iv) Determine the widths and spacings of each coupled section, based on the odd and even mode impedances. *Comment: It will be necessary to use appropriate CAD software at this stage.*
(v) Determine the length of the resonator sections. The length of each resonator will be the sum of the lengths of the adjacent quarter-wavelength coupled sections. As explained in Chapter 2, when considering coupled microstrip lines we must specify the length of the coupled region in terms of the average wavelength, where the average wavelength refers to the average of the odd and even mode wavelengths.
(vi) There will be fringing of the fields at the two ends of each resonator and so a correction must be applied to the length of each resonator to compensate for the open-end effects.

Example 8.10 illustrates steps in the practical design of a microstrip band-pass filter.

Example 8.10 Design a microstrip band-pass filter to satisfy the following specification:

Centre frequency = 10 GHz
Bandwidth = 1.2 GHz
Chebyshev response with 0.01 dB ripple
Attenuation at 11.2 GHz > 20 dB

The filter is to work between 50 Ω source and load impedances, and is to be fabricated on a 0.635 mm thick substrate that has a relative permittivity of 9.8.

Solution

(i) Low-pass prototype:
$$L_r = 0.01 \text{ dB at } 0.6 \text{ GHz}$$
$$L_S > 20 \text{ dB at } 1.2 \text{ GHz.}$$

Using Eq. (8.20) to find the required order of the filter:

$$n = \frac{\cosh^{-1}\left(\frac{10^{20/10}-1}{10^{0.01/10}-1}\right)^{\frac{1}{2}}}{\cosh^{-1}\left(\frac{1.2}{0.6}\right)} = \frac{\cosh^{-1}(207.23)}{\cosh^{-1}(2)} = \frac{6.03}{1.316} = 4.58,$$

i.e. we require a fifth-order filter.

(ii) Reading normalized parameters from Table 8.2:

$$g_1 = 0.7563 \quad g_2 = 1.3049 \quad g_3 = 1.5773 \quad g_4 = 1.3049 \quad g_5 = 0.7563.$$

Comment: $g_0 = g_6 = 1$
Fractional bandwidth: $\delta = \frac{1.2}{10} = 0.12$
Terminating impedances: $Z_o = 50 \ \Omega$
Using Eqs. (8.30) through (8.32):

$$J_{01} = \frac{1}{50}\sqrt{\frac{\pi \times 0.12}{2 \times 1 \times 0.7563}} \ S = 9.99 \text{ mS,}$$

$$J_{12} = \frac{1}{50} \times \frac{\pi \times 0.12}{2\sqrt{0.7563 \times 1.3049}} \ S = 3.80 \text{ mS,}$$

$$J_{23} = \frac{1}{50} \times \frac{\pi \times 0.12}{2\sqrt{1.3049 \times 1.5773}} \ S = 2.63 \text{ mS,}$$

$$J_{34} = \frac{1}{50} \times \frac{\pi \times 0.12}{2\sqrt{1.5773 \times 1.3049}} \ S = 2.63 \text{ mS,}$$

$$J_{45} = \frac{1}{50} \times \frac{\pi \times 0.12}{2\sqrt{1.3049 \times 0.7563}} \ S = 3.80 \text{ mS,}$$

$$J_{56} = \frac{1}{50}\sqrt{\frac{\pi \times 0.12}{2 \times 0.7563 \times 1}} \ S = 9.99 \text{ mS.}$$

(iii) Using Eqs. (8.33) and (8.34):
First coupled section:

$$Z_{oo} = 50(1 - (9.99 \times 10^{-3} \times 50) + (9.99 \times 10^{-3} \times 50)^2) \ \Omega = 37.50 \ \Omega,$$
$$Z_{oe} = 50(1 + (9.99 \times 0.001 \times 50) + (9.99 \times 0.001 \times 50)^2) \ \Omega = 87.45 \ \Omega.$$

Second coupled section:

$$Z_{oo} = 50(1 - (3.80 \times 10^{-3} \times 50) + (3.80 \times 10^{-3} \times 50)^2) \ \Omega = 42.31 \ \Omega,$$
$$Z_{oe} = 50(1 + (3.80 \times 0.001 \times 50) + (3.80 \times 0.001 \times 50)^2) \ \Omega = 61.31 \ \Omega.$$

Third coupled section:

$$Z_{oo} = 50(1 - (2.63 \times 10^{-3} \times 50) + (2.63 \times 10^{-3} \times 50)^2) \ \Omega = 44.29 \ \Omega,$$
$$Z_{oe} = 50(1 + (2.63 \times 0.001 \times 50) + (2.63 \times 0.001 \times 50)^2) \ \Omega = 57.44 \ \Omega.$$

Fourth coupled section ≡ third coupled section
Fifth coupled section ≡ second coupled section
Sixth coupled section ≡ first coupled section

(Continued)

(Continued)

(iv) *Readers wishing to check the data in this section will need to access a suitable CAD package.*

Coupled section	w (μm)	s (μm)	λ_{even} (mm)	λ_{odd} (mm)	λ_{av} (mm)	$\lambda_{av}/4$ (mm)
First	358	90	12.32	14.24	13.28	3.32
Second	580	290	11.42	14.34	12.88	3.22
Third	615	385	11.26	14.44	12.85	3.22
Fourth	615	385	11.26	14.44	12.85	3.22
Fifth	580	290	11.42	14.34	12.88	3.22
Sixth	358	90	12.32	14.24	13.28	3.32

(v) Resonator lengths:

$$l_1 = (3.32 + 3.22)\ \text{mm} = 6.54\ \text{mm},$$
$$l_2 = (3.22 + 3.22)\ \text{mm} = 6.44\ \text{mm},$$
$$l_3 = (3.22 + 3.22)\ \text{mm} = 6.44\ \text{mm},$$
$$l_4 = (3.22 + 3.22)\ \text{mm} = 6.44\ \text{mm},$$
$$l_5 = (3.32 + 3.22)\ \text{mm} = 6.54\ \text{mm}.$$

(vi) Each resonator length should be shortened to account for the open-end effects, as explained in Chapter 3, using

$$\Delta l = 0.412h \times \frac{\varepsilon_{r,eff} + 0.3}{\varepsilon_{r,eff} - 0.258} \times \frac{w/h + 0.262}{w/h + 0.813}.$$

We can find the line extensions at the ends of each resonator;

w (μm)	Δl (μm)
358	171
580	194
615	197

Compensated resonator lengths:

$$l_1 = (6.54 - 0.171 - 0.194)\ \text{mm} = 6.175\ \text{mm},$$
$$l_2 = (6.44 - 0.194 - 0.197)\ \text{mm} = 6.049\ \text{mm},$$
$$l_3 = (6.44 - 0.197 - 0.197)\ \text{mm} = 6.046\ \text{mm},$$
$$l_4 = l_2 = 6.049\ \text{mm},$$
$$l_5 = l_1 = 6.175\ \text{mm}.$$

For the purpose of manufacture we need to specify the length of each coupled section, as shown in Figure 8.27:

Coupled section	Length (mm)
First	3.22–0.171 = 3.049 ≈ 3.05
Second	3.22–0.194 = 3.026 ≈ 3.03
Third	3.22–0.197 = 3.023 ≈ 3.03
Fourth	3.023 ≈ 3.03
Fifth	3.026 ≈ 3.03
Sixth	3.049 ≈ 3.05

The completed final design is shown in Figure 8.27.

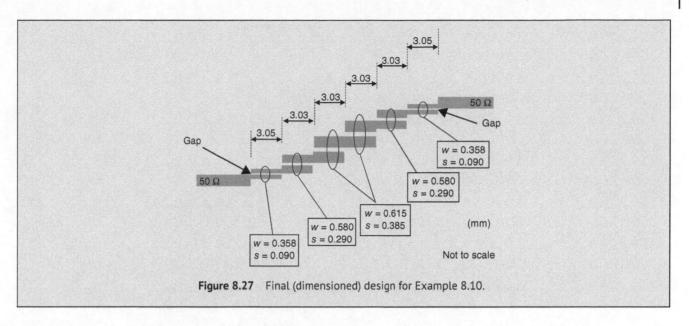

Figure 8.27 Final (dimensioned) design for Example 8.10.

8.12 Microstrip End-Coupled Band-Pass Filters

An alternative to the edge-coupled band-pass filter is the end-coupled arrangement illustrated in Figure 8.28.

The filter shown in Figure 8.28 consists of a number of in-line $\lambda/2$ resonators, with capacitive coupling between adjacent resonators. The coupling is across the end gaps between the resonators and these gaps function as immittance inverters, and consequently this type of filter can be modelled in a similar way to the edge-coupled filter described in the last section. End-coupled filters tend to be less popular than those using edge-coupling for two reasons:

(i) End-coupled filters occupy more substrate area because the resonators are coupled end-to-end.
(ii) End-coupled filters require very small gaps between the resonators in order to give adequate coupling. This is particularly true of the end sections of the filter where failure to achieve tight coupling limits the bandwidth of the filter (the required end gaps in typical microstrip filters fabricated on alumina are often of the order of 10 μm, or less).

The second limitation mentioned above, namely that of requiring very small coupling gaps, can be largely overcome by using multilayer technology to fabricate the filter. Figure 8.29 shows the end gap between two resonators in single and multilayer formats.

The practicalities of producing a multilayer coupling gap are straightforward: one conductor is deposited directly onto the substrate, then the end is over-printed with one or more layers of dielectric, and finally the second conductor is printed on top of the dielectric with the desired amount of overlap. The original physical gap has now been replaced by an overlap, giving a small amount of broad-wall coupling. Figure 8.30 shows how the coupling with this type of multilayer structure varies with the degree of overlap, for a 50 Ω microstrip line at 10 GHz.

It is clear from Figure 8.30 that a relatively small overlap results in very strong coupling, i.e. very low transmission loss. It can also be seen that it is possible to choose an amount of overlap that gives zero transmission phase; if the transmission phase is not zero then this must be taken into account in the design. It should also be noted that the presence of the dielectric will modify the characteristic impedance of the conductors, and some small change in line width will be required to maintain the correct characteristic impedance of the conducting strips in the vicinity of the dielectric.

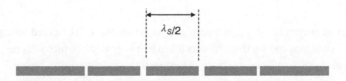

Figure 8.28 Layout of an end-coupled microstrip band-pass filter.

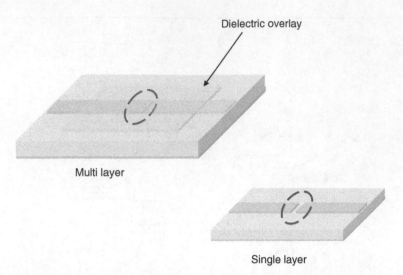

Figure 8.29 End-coupling microstrip gap in single and multilayer formats.

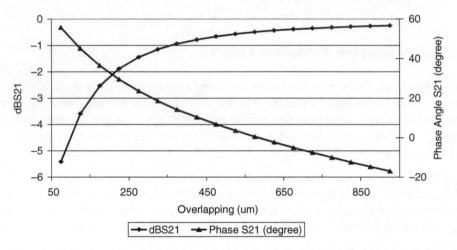

Figure 8.30 End-coupling between two overlapping conductors (applies to 50 Ω lines on 635 μm thick alumina, at 10 GHz).

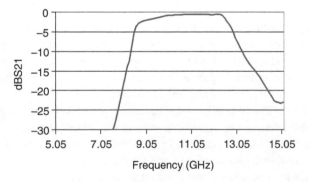

Figure 8.31 Measured response of a multilayer end-coupled band-pass filter.

Figure 8.31 shows a measured response of a microstrip end-coupled filter, fabricated using multilayer technology. The tight coupling produced by the overlapping sections has resulted in a very wide bandwidth response, with a 3 dB bandwidth approaching 45% at 1.5 GHz, and good roll-characteristics between the pass and stop bands.

Whilst the use of multilayer technology is a powerful and proven approach, it should be appreciated that the problem of fabricating small gaps in a single layer component has been exchanged for that of achieving accurate registration between the layers in a multilayer structure. However, the latter problem is really one of equipment accuracy, and modern mask aligners can easily provide the degree of position resolution required by RF and microwave components.

8.13   Practical Points Associated with Filter Design

(i) When making filters with lumped components at RF, it is important to take account of component parasitics. For example, lead inductance and packaging capacitance must be included in the circuit model at the design stage, as must additional resistance associated with skin effect.

(ii) It is useful to consider using planar components such as spiral inductors and distributed capacitors (overlay or inter-digitated) in an RF design, effectively as lumped components.

(iii) Transmission line resonators should always be considered for RF filter design, as in general they have fewer parasitics and can be designed with a greater degree of accuracy.

(iv) When using planar components, particularly at the higher RF frequencies, it is essential to consider discontinuities within the design. We saw an example of this in the stepped microstrip low-pass filter, which will naturally always include significant step discontinuities.

(v) Many modern fabrication techniques, such as those using ceramics and polymers, permit easy and cheap multilayer components to be made, and these can be used to advantage.

8.14   Summary

The basic concepts of RF filter design have been introduced, starting with lumped-component designs, and extending through to planar designs using microstrips. The role of multilayer technology in producing high-performance filters at microwave frequencies has been discussed. Whilst there is a wealth of published information, including design equations, for lumped-element and single-layer microstrip filters, the use of multilayer technology requires a three-dimensional approach to filter design, and the use of field theory based CAD to model and refine multilayer structures is essential.

8.15   Supplementary Problems

Q8.1   A lumped-element, low-pass filter having a maximally flat response is to be designed to have a bandwidth of 500 MHz, and is to provide 20 dB attenuation at 650 MHz. What is the required order of the filter?

Q8.2   Design a five-element low-pass filter with a 3 dB point at 600 MHz. The filter is to have a Butterworth response, and is to work between terminating impedances of 50 Ω.

Q8.3   A low-pass passive filter is required to meet the following specification:

Pass-band loss	<0.08 dB below 13 MHz
Stop-band loss	>40 dB above 30 MHz

(a)   Show that this response can be met using a fifth-order Chebyshev filter.
(b)   Determine the order of a Butterworth filter which will meet the specification.

Q8.4   Design a five-element high-pass filter with a 3 dB point at 600 MHz. The filter is to have a Butterworth response, and is to work between terminating impedances of 75 Ω.

Q8.5   A lumped-element, low-pass filter having a maximally flat response is to be designed to have a 3 dB bandwidth of 450 MHz, and is to provide 20 dB attenuation at 650 MHz. What is the required order of the filter?

Q8.6 A third-order, band-pass lumped-element filter is to be designed to work in a 50 Ω system. The filter is to have a Chebyshev frequency response (0.01 dB ripple) with upper and lower cut-off frequencies of 75 and 69 MHz, respectively. Determine the values of the components of the filter.

Q8.7 A seventh-order maximally flat band-pass filter is to be designed to work in a 50 Ω system. The upper and lower 3 dB cut-off frequencies of the filter are 725 and 670 MHz, respectively. Determine the values of the components of the filter.

Q8.8 What is the attenuation of the filter designed in Q8.7 at a frequency of 790 MHz?

Q8.9 A fifth-order microstrip stepped-impedance low-pass filter, having a maximally flat response and a 3 dB cut-off frequency of 2 GHz, is to be fabricated on a substrate that has a thickness of 0.8 mm and a relative permittivity of 9.8. The filter is to have port impedances of 50 Ω. Choose 24 Ω for the characteristic impedance of the wide section of the filter, and 100 Ω for that of the narrow sections, and hence determine the physical dimensions of the microstrip filter.

Q8.10 Design a third-order lumped-element low-pass filter, having a Chebyshev response with a 0.01 dB ripple, and a bandwidth of 4 GHz. The filter is to work between source and load impedances of 50 Ω.

Q8.11 Convert the filter designed in Q8.10 to a distributed design using the stepped-impedance microstrip technique in which the substrate has a relative permittivity of 9.8 and a thickness of 2 mm. Choose 22 Ω for the characteristic impedance of the wide sections of the filter, and 96 Ω for that of the narrow section.

Q8.12 Show that Eqs. 8.27 through 8.29, which are the design equations for a third-order microstrip low-pass filter using shunt-connected $\lambda/8$ stubs, can be written as

$$Z_s = Z_0(1 + g_2),$$

$$Z_{p1} = Z_0 \left(1 + \frac{1}{g_2}\right),$$

$$Z_{p2} = \frac{Z_0}{g_1},$$

where g_1, g_2, and g_3 are the normalized coefficients of the required filter response.

Hence, design a third-order microstrip filter using shunt-connected $\lambda/8$ stubs to satisfy the following specification:

    Port impedances = 40 Ω

    Cut-off frequency = 6 GHz

    Chebyshev response with 0.01 dB ripple

The filter is to be fabricated on a 1 mm thick substrate that has a relative permittivity of 9.8.

Appendix 8.A Equivalent Lumped T-Network Representation of a Transmission Line

Figure 8.32 shows a T-network consisting of three impedances Z_1, Z_2, and Z_3.

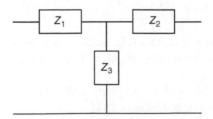

Figure 8.32 T-network of impedances.

The *ABCD* matrix representing the network shown in Figure 8.32 is [6]

$$\begin{bmatrix} A & B \\ C & D \end{bmatrix} = \begin{bmatrix} 1 + \dfrac{Z_1}{Z_3} & Z_1 + Z_2 + \dfrac{Z_1 Z_2}{Z_3} \\ \dfrac{1}{Z_3} & 1 + \dfrac{Z_2}{Z_3} \end{bmatrix}. \tag{8.35}$$

For a symmetrical network we may write $Z_1 = Z_2 = Z_s$ and for convenience $Z_3 = Z_p$. Then Eq. (8.35) becomes

$$\begin{bmatrix} A & B \\ C & D \end{bmatrix} = \begin{bmatrix} 1 + \dfrac{Z_s}{Z_p} & 2Z_s + \dfrac{Z_s^2}{Z_p} \\ \dfrac{1}{Z_p} & 1 + \dfrac{Z_s}{Z_p} \end{bmatrix}. \tag{8.36}$$

The *ABCD* matrix representing a lossless transmission line of length, l, and characteristic impedance, Z_o, is [6]

$$\begin{bmatrix} A & B \\ C & D \end{bmatrix} = \begin{bmatrix} \cos \beta l & jZ_o \sin \beta l \\ jY_o \sin \beta l & \cos \beta l \end{bmatrix}, \tag{8.37}$$

If the T-network is to represent the lossless transmission line the *ABCD* matrices in Eqs. (8.36) and (8.37) must be equal, giving

$$Z_p = (jY_o \sin \beta l)^{-1} \tag{8.38}$$

and

$$Z_s = jZ_o \tan \left(\frac{\beta l}{2} \right). \tag{8.39}$$

For the wide section of line we have

$$Z_s = j\omega L_W \quad Z_p = \frac{1}{j\omega C_W} \quad Z_o = Z_{oW} \quad \beta = \beta_W \quad l = l_W.$$

Substituting these parameters into Eqs. (8.38) and (8.39) gives

$$C_W = \frac{1}{\omega Z_{oW}} \sin(\beta_W l_W) \tag{8.40}$$

and

$$L_W = \frac{Z_{oW}}{\omega} \tan \left(\frac{\beta_W l_W}{2} \right). \tag{8.41}$$

References

1 Matthai, G.L., Young, L., and Jones, E.M.T. (1980). *Microwave Filters, Impedance Matching Networks, and Coupling Structures*. Dedham, MA: Artech House.
2 Fooks, E.H. and Zakarevicius, R.A. (1990). *Microwave Engineering Using Microstrip Circuits*. Australia: Prentice Hall.
3 Richards, P.I. (1948). Resistor-transmission-line circuits. *Proceedings of the IRE* 36: 217–220.
4 Ozaki, N. and Ishii, J. (1958). Synthesis of a class of strip-line filters. *IRE Transactions on Circuit Theory* CT-5: 104.
5 Edwards, T.C. and Steer, B. (2000). *Foundations of Interconnect and Microstrip Design*, 3e. Chichester: Wiley.
6 Pozar, D.M. (2012). *Microwave Engineering*, 4e. Wiley.

9

Microwave Small-Signal Amplifiers

9.1 Introduction

Transistors used for small-signal amplification at microwave frequencies are usually supplied by the manufacturer in beam-lead packages, with the transistor performance data in an S-parameter format. The transistors are normally GaAs MESFETs (gallium arsenide **me**tal **s**emiconductor **f**ield **e**ffect **t**ransistors) because these offer better noise performance at microwave frequencies, compared to silicon bipolar devices.

The design procedure for microwave small-signal amplifiers is largely that of designing suitable input and output matching networks to enable the transistor to give the required performance when connected between specified source and load impedances. If the source and load impedances are conjugately matched to the input and output, respectively, of the amplifier, then it follows from Eq. (5.22) of Chapter 5 that the maximum unilateral transducer power gain of the amplifier will be

$$G_{TU,max} = \frac{1}{1 - |S_{11}|^2} |S_{21}|^2 \frac{1}{1 - |S_{22}|^2}, \tag{9.1}$$

where the S-parameters are those of the transistor.

In addition to gain requirements, the required performance will normally include noise specifications and these also impinge on the design of the matching networks. The main consideration of this chapter will be on the design of suitable matching networks, and since the frequencies are in the microwave region the primary focus will be on distributed networks.

One of the major problems that can arise when designing amplifiers at microwave frequencies is that of unwanted oscillations, and the conditions for stable amplification will also be established.

9.2 Conditions for Matching

Matching of a small-signal microwave amplifier is achieved through the addition of input and output matching networks on both sides of the amplifying device, in this case a MESFET, as shown in Figure 9.1.

In Chapter 5, the maximum unilateral transducer power gain was shown to occur when

$$\Gamma_S = \Gamma_{IN}^* \quad \text{and} \quad \Gamma_L = \Gamma_{OUT}^*,$$

i.e. when

$$\Gamma_S = S_{11}^* \quad \text{and} \quad \Gamma_L = S_{22}^*. \tag{9.2}$$

In addition to providing S-parameter data, the manufacturer of a microwave MESFET will provide information on the conditions needed to achieve minimum noise performance, i.e. minimum noise figure.[1] This normally necessitates a certain degree of mismatch at the device input, and the manufacturer will specify Γ_{opt} which is the value of Γ_S required for minimum noise performance. So the matching conditions for a low-noise amplifier are

$$\Gamma_S = \Gamma_{opt} \quad \text{and} \quad \Gamma_L = S_{22}^*. \tag{9.3}$$

It is important to stress that Γ_{opt} is always specified looking away from the input of the MESFET. Also, the values of Γ_{opt} and S_{11}^* are not normally the same, and therefore the input matching network must be designed either to give maximum

---

1 Noise Figure is defined in Chapter 14.

RF and Microwave Circuit Design: Theory and Applications, First Edition. Charles E. Free and Colin S. Aitchison.
© 2022 John Wiley & Sons Ltd. Published 2022 by John Wiley & Sons Ltd.
Companion website: www.wiley.com/go/free/rfandmicrowave

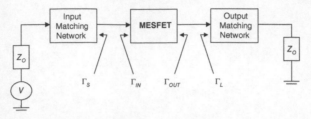

Figure 9.1 Small-signal matching of amplifiers.

gain or best noise performance, at maximum bandwidth. In general, designing the source matching network for best noise performance will involve some sacrifice in gain and bandwidth, and normal practice is to use cascaded amplifiers to obtain the combination of low noise and high gain.

Example 9.1 The following data apply to a microwave FET which is to be used as an amplifier between a 50 Ω source and a 50 Ω load:

$$[S] = \begin{bmatrix} 0.77\angle -11° & 0.01\angle 26° \\ 3.67\angle -108° & 0.83\angle 47° \end{bmatrix}.$$

$$\Gamma_{opt} = 0.53\angle 57°.$$

Determine:

(i) The insertion gain (in dB) if the FET is connected directly between source and load.
(ii) The maximum transducer power gain if source and load matching networks are introduced.
(iii) The reduction in maximum transducer power gain if the amplifier is designed for low-noise performance. Make appropriate simplifying assumptions in the calculations.

Solution

(i) From the definition of S-parameters the insertion gain is given directly as:

$$G = 10 \log (3.67)^2 \text{ dB} = 11.29 \text{ dB}.$$

(ii) Since $|S_{12}|$ is small we can assume that the performance of the FET is unilateral and use Eq. (5.22) for maximum unilateral transducer power gain:

$$G_{TU,max} = \frac{1}{1-(0.77)^2} \times (3.67)^2 \times \frac{1}{1-(0.83)^2} = 106.35 \equiv 20.27 \text{ dB}.$$

(iii) Using the general expression for unilateral transducer power gain, Eq. (5.17), with $\Gamma_S = \Gamma_{opt}$:

$$G_{TU} = \frac{(3.67)^2 \times (1-(0.83)^2) \times (1-(0.53)^2)}{|(1-(0.77\angle -11)(0.53\angle 57))(1-(0.83\angle 47)(0.83\angle -47))|^2}$$

$$= 51.95 \equiv 17.16 \text{ dB}.$$

Reduction in transducer power gain = (20.27–17.16) dB = 3.11 dB.

9.3 Distributed (Microstrip) Matching Networks

The most compact microstrip matching network to use consists of an open-circuited stub and a quarter-wave transformer, as shown in Figure 9.2, to represent the load matching network.

The function of the load matching network is to create the required value of Γ_L at the output of the FET. The real part of Γ_L is created by the $\lambda/4$ transformer and the imaginary part by the stub. For the purpose of design it is convenient to work in

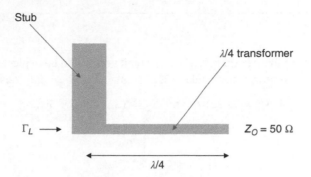

Figure 9.2 Load matching network.

the admittance plane, since we are dealing with microstrip, and microstrip stubs are necessarily in parallel with the main line. So, if Γ_L is specified as an admittance Y_L, where $Y_L = G_L \pm jB_L$, the two requirements of the matching network are:

(i) The stub must have an input admittance of $\pm jB_L$.
(ii) The characteristic impedance of the $\lambda/4$ transformer must be $\sqrt{Z_o(G_L)^{-1}}$, as discussed in Chapter 1.

Procedure on the Smith chart:

(i) Plot the required value of Γ_L at the amplifier frequency.
(ii) Convert Γ_L to y_L by rotating 180° (remembering from Chapter 1 that the reflection coefficient point also corresponds to the impedance at that point, and that converting from impedance to admittance requires a 180° rotation on the Smith chart).
(iii) Record the value of y_L ($=g_L \pm jb_L$).
(iv) Plot $0 \pm jb_L$ and move counter-clockwise around the zero conductance circle to $y_{O/C}$. The distance moved is equal to the electrical length of the stub. (Note that it has been assumed that the stub is open-circuited, which is normally the case for microstrip designs.)
(v) Calculate the impedance of the $\lambda/4$ transformer, from

$$Z_{oT} = \sqrt{Z_o \times \frac{50}{g_L}} = \sqrt{50 \times \frac{50}{g_L}} = 50\sqrt{\frac{1}{g_L}}.$$

(vi) Use microstrip design graphs (or a CAD package) to convert the data into physical dimensions.

The design technique for the source matching network follows the same general procedure as that for the load, but with Γ_L replaced with Γ_S.

Example 9.2 The following data apply to an 8 GHz microwave FET that is to be used to make an amplifier to work between source and load impedances which are both 50 Ω.

$$[S] = \begin{bmatrix} 0.69\angle - 82° & 0.01\angle - 174° \\ 4.22\angle - 220° & 0.36\angle 73° \end{bmatrix}.$$

$$\Gamma_{opt} = 0.56\angle - 51°.$$

(i) Design source and load matching networks to enable the amplifier to provide maximum gain.
(ii) Design source and load matching networks to enable the amplifier to give best gain commensurate with minimum noise performance.

The matching networks are to be fabricated in microstrip on a 0.5 mm thick substrate that has a relative permittivity of 9.8.

(Continued)

(Continued)

Solution

(i) It can be seen from the given S-matrix that $|S_{12}|$ is very small and so can be neglected. Thus, we can regard the amplifier as having unilateral performance, and use Eq. (5.18) to find the conditions for maximum gain, i.e.

$$\Gamma_S = S_{11}^* = 0.69\angle 82° \quad \text{and} \quad \Gamma_L = S_{22}^* = 0.36\angle -73°.$$

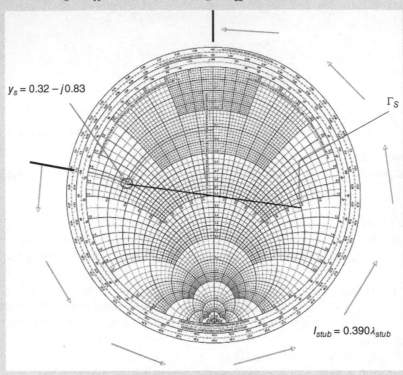

Figure 9.3 Smith chart for source matching (maximum gain).

Source matching (see Figure 9.3):

$$\text{From the Smith chart: } \Gamma_S = 0.69\angle 82° \quad \Rightarrow \quad y_S = 0.32 - j0.83,$$

$$g = 0.32,$$

$$l_{stub} = 0.390\lambda_{stub}.$$

$$\lambda/4 \text{ transformer: } Z_{oT} = \sqrt{50 \times \frac{50}{0.32}} \ \Omega = 88.39 \ \Omega.$$

Using microstrip design data:

$$88.39 \ \Omega \quad \Rightarrow \quad \frac{w}{h} = 0.16 \quad \Rightarrow \quad w = 0.08 \text{ mm} \ \varepsilon_{r,eff} = 6.05,$$

$$\lambda_{s,transformer} = \frac{\lambda_O}{\sqrt{6.05}} = \frac{3\times10^8 / 8\times10^9}{\sqrt{6.05}} \text{ m} = 15.25 \text{ mm},$$

$$l_{transformer} = \frac{15.25}{4} \text{ mm} = 3.81 \text{ mm}.$$

Stub: *The choice of characteristic impedance for the stub is arbitrary; normally a value is chosen such that the width of the stub will not cause a significant discontinuity at the junction between the stub and the main line. For convenience, we will choose the characteristic impedance to be 50 Ω.*

$$Z_{o,stub} = 50 \ \Omega \quad \Rightarrow \quad \frac{w}{h} = 0.9 \quad \Rightarrow \quad w = 0.45 \text{ mm} \ \varepsilon_{r,eff} = 6.6,$$

$$\lambda_{s,stub} = \frac{\lambda_O}{\sqrt{6.6}} = \frac{3\times10^8 / 8\times10^9}{\sqrt{6.6}} \text{ m} = 14.60 \text{ mm},$$

$$l_{stub} = 0.390 \times 14.60 \text{ mm} = 5.69 \text{ mm}.$$

The final design of source matching network is shown in Figure 9.4.

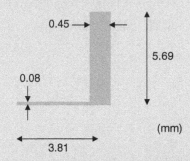

Figure 9.4 Source matching network for maximum gain (not to scale).

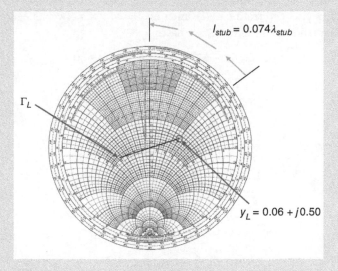

Figure 9.5 Smith chart for load matching (maximum gain).

Load matching (see Figure 9.5):

From the Smith chart: $\Gamma_L = 0.36\angle - 73° \quad \Rightarrow \quad y_S = 0.65 + j0.50,$

$$g = 0.65,$$

$$l_{stub} = 0.074\lambda_{stub}.$$

$\lambda/4$ transformer: $Z_{oT} = \sqrt{50 \times \dfrac{50}{0.65}} \; \Omega = 62.02 \; \Omega.$

Using microstrip design data:

$$62.02 \; \Omega \quad \Rightarrow \quad \frac{w}{h} = 0.52 \quad \Rightarrow \quad w = 0.26 \text{ mm} \quad \varepsilon_{r,eff} = 6.37,$$

$$\lambda_{s,transformer} = \frac{\lambda_O}{\sqrt{6.37}} = \frac{3\times10^8 / 8\times10^9}{\sqrt{6.37}} \text{ m} = 14.86 \text{ mm},$$

(Continued)

(Continued)

$$l_{transformer} = \frac{14.86}{4} \text{ mm} = 3.72 \text{ mm.}$$

Stub: $Z_{o,stub} = 50 \ \Omega \quad \Rightarrow \quad w = 0.45 \text{ mm} \quad \Rightarrow \quad \lambda_{s,stub} = 14.60 \text{ mm,}$

$$l_{stub} = 0.074 \times 14.60 \text{ mm} = 1.08 \text{ mm.}$$

The final design of load matching network is shown in Figure 9.6.

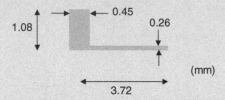

Figure 9.6 Load matching network for maximum gain (not to scale).

(ii) To achieve minimum noise we require, from Eq. (9.3), $\Gamma_S = \Gamma_{opt} = 0.56 \angle -51°$ and $\Gamma_L = S_{22}^*$. So the design of the load matching network is unchanged from part (i), and only the source network needs to be re-designed. Referring to the Smith chart shown in Figure 9.7 we have:

$$\Gamma_S = 0.56 \angle -51° \quad \Rightarrow \quad y_S = 0.34 + j0.42,$$

$$g = 0.34,$$

$$l_{stub} = 0.063 \lambda_{stub}.$$

$$\lambda/4 \text{ transformer: } Z_{oT} = \sqrt{50 \times \frac{50}{0.34}} \ \Omega = 85.75 \ \Omega.$$

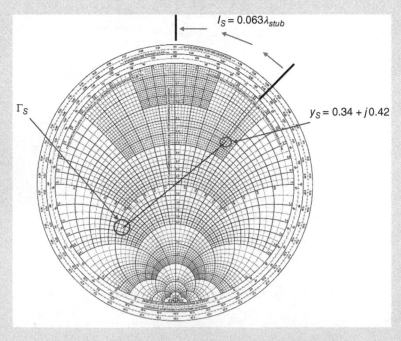

Figure 9.7 Smith chart for source matching (minimum noise).

Using microstrip design data:

$$85.75\ \Omega \quad \Rightarrow \quad \frac{w}{h} = 0.18 \quad \Rightarrow \quad w = 0.09 \text{ mm} \quad \varepsilon_{r,eff} = 6.08,$$

$$\lambda_{s,transformer} = \frac{\lambda_O}{\sqrt{6.08}} = \frac{3\times10^8 / 8\times10^9}{\sqrt{6.08}} \text{ m} = 15.21 \text{ mm},$$

$$l_{transformer} = \frac{15.21}{4} \text{ mm} = 3.80 \text{ mm},$$

$$\text{Stub: } Z_{0,stub} = 50\ \Omega \quad \Rightarrow \quad w = 0.45 \text{ mm} \quad \Rightarrow \quad \lambda_{s,stub} = 14.60 \text{ mm},$$

$$l_{stub} = 0.063 \times 14.60 \text{ mm} = 0.92 \text{ mm}.$$

The final design is shown in Figure 9.8.

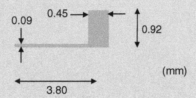

Figure 9.8 Source matching network for minimum noise (not to scale).

Example 9.3 Suppose that the amplifier designed in Example 9.2 for maximum unilateral transducer power gain was used at 9 GHz, with the same matching networks. Calculate the new unilateral transducer power gain of the amplifier at 9 GHz, given that the following S-matrix applies to the FET at 9 GHz.

$$[S] = \begin{bmatrix} 0.67\angle -84° & 0.01\angle -168° \\ 4.10\angle -230° & 0.38\angle 82° \end{bmatrix}.$$

Solution

The first steps are to find the values of Γ_S and Γ_L at 9 GHz, for the matching circuits designed in Example 9.2. Figure 9.9 shows the source matching circuit (not to scale) that was designed at 8 GHz. In this figure, y_1 and y_2 are the normalized admittances at 9 GHz of the transformer and stub, respectively.

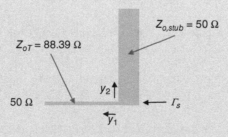

Figure 9.9 Source matching network for maximum gain (not to scale).

The new electrical length of the transformer $= \dfrac{9}{8} \times \dfrac{\lambda_{s,transformer}}{4} = 0.281\lambda_{s,transformer}.$

(Continued)

(Continued)

To find the value of y_1 we simply treat the transformer as a length of transmission line, of characteristic impedance equal to $88.39\,\Omega$, of length $0.281\lambda_{s,transformer}$, terminated with $50\,\Omega$. So we normalize $50\,\Omega$ with respect to $88.39\,\Omega$, convert to the admittance plane, and move a distance $0.281\lambda_{s,transformer}$ clockwise around the VSWR circle, as shown on the Smith chart in Figure 9.10.

Normalizing $50\,\Omega$ w.r.t $88.39\,\Omega$ $\Rightarrow$ $z_{50} = \dfrac{50}{88.39} = 0.57$.

Converting to admittance plane $\Rightarrow$ $y_{50} = 1.75$.

$$y_1 \text{ (normalized w.r.t 88.39 }\Omega) = 0.58 + j0.13,$$

$$y_1 \text{ (normalized w.r.t 50 }\Omega) = (0.58 + j0.13) \times \frac{50}{88.39} = 0.33 + j0.07.$$

At 9 GHz the new electrical length of the stub $= \dfrac{9}{8} \times 0.390\lambda_{s,stub} = 0.439\lambda_{s,stub}$.

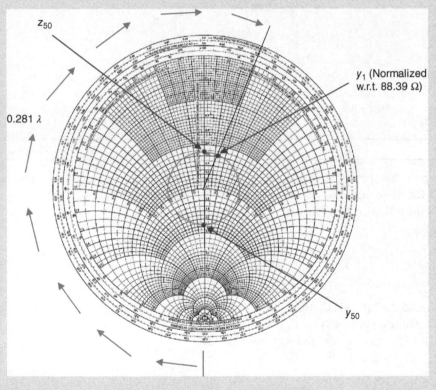

z₅₀

y_1 (Normalized w.r.t. 88.39 Ω)

0.281 λ

y_{50}

Figure 9.10 Source network data at 9 GHz.

We can then find the new normalized input admittance of the stub, y_2, by moving clockwise a distance of $0.439\lambda_{s,stub}$ from the open-circuit position around the outer circle of the Smith chart, shown in Figure 9.11.

From Figure 9.11: $y_2 = 0 - j0.40$.

The normalized input admittance, y_{in}, of the source network at 9 GHz is then

$$y_{in} = y_1 + y_2 = 0.33 + j0.07 - j0.40 = 0.33 - j0.33.$$

Using Eq. (1.43) we have

$$\Gamma_S = \frac{z_{in} - 1}{z_{in} + 1} = \frac{1/y_{in} - 1}{1/y_{in} + 1} = \frac{1 - y_{in}}{1 + y_{in}}$$

$$= \frac{1 - (0.33 - j0.33)}{1 + (0.33 - j0.33)} = \frac{0.67 + j0.33}{1.33 - j0.33}$$

$$= \frac{0.75\angle 26.22°}{1.37\angle -13.93°}$$

$$= 0.55\angle 40.15°.$$

Following the same procedure for the load network gives

$$\Gamma_L = 0.40\angle -83.03°.$$

The unilateral transducer power gain is given by Eq. (5.21), i.e.

$$G_{TU} = \frac{|S_{21}|^2 (1 - |\Gamma_L|^2)(1 - |\Gamma_S|^2)}{|(1 - S_{11}\Gamma_S)(1 - S_{22}\Gamma_L)|^2}$$

$$= \frac{(4.10)^2 (1 - (0.40)^2)(1 - (0.55)^2)}{|(1 - [0.67\angle -84°] \times [0.55\angle 40.15°])(1 - [0.38\angle 82°] \times [0.40\angle -83.03°])|^2}$$

$$= 22.65 \equiv 13.55 \text{ dB}.$$

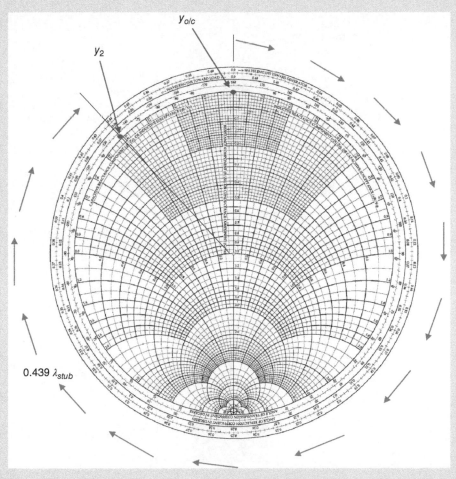

Figure 9.11 Load network data at 9 GHz.

9.4 DC Biasing Circuits

In a practical microwave FET amplifier it is necessary to provide DC connections to bias the gate and drain terminals of the FET. These connections must allow the DC current to flow to the FET, without significantly affecting the RF signals. One common form of bias network is shown in Figure 9.12.

The justification for the circuit shown in Figure 9.12 is that the open-circuit at the top of the wide pad will reflect an open-circuit to the junction with the main line since it consists of a $\lambda/2$ line, thus preventing the RF signal from propagating through the bias connection. The widths of the lines forming the bias network are not crucial; the width of the wide section should merely be sufficient to provide a convenient pad for bonding, and the width of the narrow section should be as small as possible to minimize the physical discontinuity at the junction with the main line. In a typical microstrip circuit on alumina, the wide section width would be of the order of 3–5 mm, and the width of the narrow section around 50 μm. In choosing the width of the narrow section it must always be remembered that this section must be of sufficient cross-section to carry to required level of DC current. The circuit shown in Figure 9.12 necessarily contains significant microstrip discontinuities, due to fringing from the wide pad and a large change in width at the junction of the wide and narrow sections, and these must be compensated for in practical designs. Also, the presence of the bond wire on the pad will change its behaviour slightly and often two bias networks are effectively connected in cascade, as shown in Figure 9.13, to further isolate the bond connection.

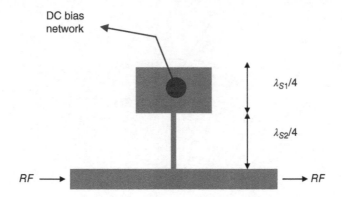

Figure 9.12   DC bias connection.

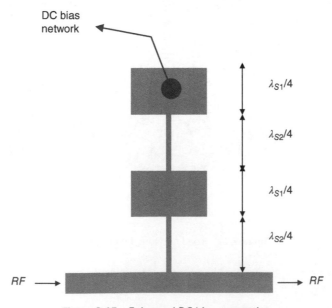

Figure 9.13   Enhanced DC bias connection.

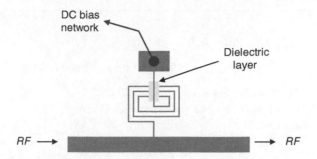

Figure 9.14 DC bias connection using inductive choke.

Whilst bias networks of the type shown in Figures 9.12 and 9.13 are simple and effective, they do suffer from two disadvantages:

(i) They tend to be narrow band, since the required behaviour depends on the sections being a quarter-wavelength long.
(ii) They occupy quite a significant area of substrate, particularly the network of Figure 9.13.

These disadvantages can be overcome by using a large series inductance to connect the DC bias to the RF circuit. The large inductance will provide a large series reactance which will block the microwave signal. The inductance could be in the form of a surface-mount component, or formed from a planar spiral. At microwave frequencies, lumped inductors in surface-mount packages suffer from significant unwanted parasitics, particularly package capacitance, and can be a source of unwanted resonances. Planar spiral inductors, of the type used in MMICs, provide a simple, compact bias solution. A typical arrangement is shown in Figure 9.14.

In Figure 9.14, a dielectric strip is shown placed over part of the spiral inductor to enable an over-pass connection to be made to the centre of the spiral.

Example 9.4 Design a microstrip 10 GHz DC bias connection with the geometry shown in Figure 9.13. The bias connection is to be made to a 50 Ω microstrip line fabricated on a substrate that has a relative permittivity of 9.8 and a thickness of 635 μm.

Solution
As discussed in Section 9.4, we may arbitrarily choose the widths of the wide and narrow sections of the bias connection; choose 2.5 mm for the widths of the pads, and 75 μm for the widths of the narrow sections.

$$10 \text{ GHz} \quad \Rightarrow \quad \lambda_O \text{ (free space wavelength)} = 30 \text{ mm}.$$

Using the microstrip design graphs:

$$2.5 \text{ mm} \quad \Rightarrow \quad \frac{w}{h} = \frac{2500}{635} = 3.94 \quad \Rightarrow \quad \varepsilon_{r,eff} = 7.62,$$

$$\lambda_s(\text{pads}) = \frac{30}{\sqrt{7.62}} \text{ mm} = 10.87 \text{ mm} \quad \Rightarrow \quad \frac{\lambda_s(\text{pads})}{4} = 2.72 \text{ mm},$$

$$75 \text{ μm} \quad \Rightarrow \quad \frac{w}{h} = \frac{75}{635} = 0.12 \quad \Rightarrow \quad \varepsilon_{r,eff} = 6.0,$$

$$\lambda_s(\text{narrow}) = \frac{30}{\sqrt{6.0}} \text{ mm} = 12.25 \text{ mm} \quad \Rightarrow \quad \frac{\lambda_s(\text{narrow})}{4} = 3.06 \text{ mm}.$$

Compensation for open-end effect (see Chapter 4):

$$l_{eo} = 0.412h \left(\frac{\varepsilon_{r,eff} + 0.3}{\varepsilon_{r,eff} - 0.258} \right) \left(\frac{w/h + 0.262}{w/h + 0.813} \right) = 249 \text{ μm}.$$

(Continued)

(Continued)

Compensation for symmetrical step (see Chapter 4):

$$l_{es} \approx l_{eo}\left(1 - {}^{w_1}/_{w_2}\right) = 249 \times (1 - {}^{75}/_{2500}) \ \mu m = 242 \ \mu m.$$

Compensated length of 'end' pad $=(2720 - 249 - 242) \ \mu m = 2229 \ \mu m.$
Compensated length of 'middle' pad $=(2720 - 242 - 242) \ \mu m = 2236 \ \mu m.$
The final design is shown in Figure 9.15.

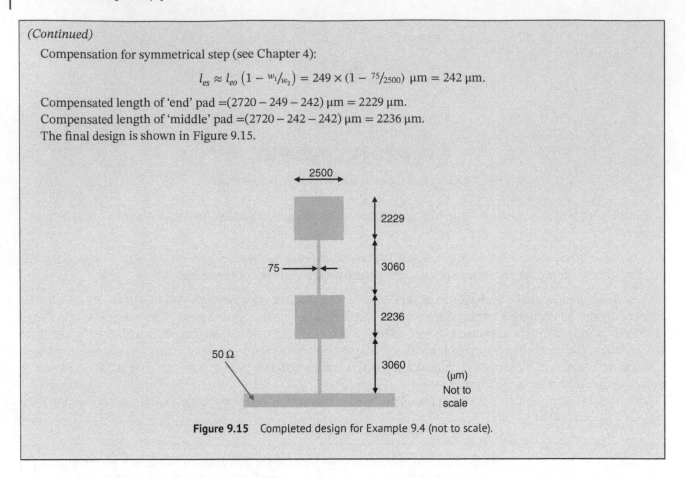

Figure 9.15 Completed design for Example 9.4 (not to scale).

9.5 Microwave Transistor Packages

The conventional package for microwave FETs used in hybrid microstrip amplifiers is shown in Figure 9.16. The package has four beam-lead connections, one each for the gate and drain terminals, and two for the source. The leads are designed to be surface bonded onto the microstrip circuit. Two leads are provided for the source in order to keep the field within the package symmetrical and improve high-frequency performance.

It is important to note that an FET manufacturer will supply S-parameter data for the transistor, specified at the point where the beam-leads join the main package. So the microstrip connections should be brought as close to the main body of the package as possible; in practice, this may involve some shaping of the ends of the microstrip to avoid unwanted coupling. For convenience, it is useful to establish a reference plane $\lambda_s/2$ away from the FET, as shown in Figure 9.16. Since the parameters on a transmission line repeat every half-wavelength, the manufacturer's S-parameter data will apply in this plane. Whilst it may be impractical to attach matching circuitry, such as stubs, very close to the FET there are no such problems with regards to the reference plane.

Errors associated with the positioning of matching circuitry with respect to the S-parameter datum line can be significant at microwave frequencies where short distances can represent significant phase changes.

Normally microwave amplifiers are operated in the common source configuration, which requires the source leads to be grounded. In a microstrip circuit the grounding can be achieved by making the microstrip tracks connected to the source terminals $\lambda_s/2$ long, and then using a VIA to connect the end of the source tracks to the ground plane. However, the FET could be far more efficiently mounted (and the source leads grounded) by using a coplanar waveguide, where the ground plane is on the same side of the substrate as the signal tracks, thus eliminating the need for through-substrate VIAs.

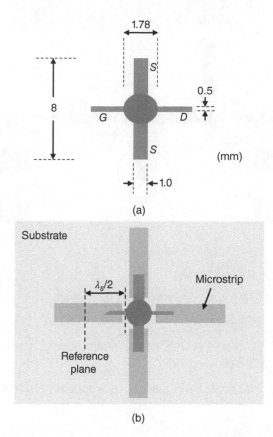

Figure 9.16 Mounting of a microwave FET in a microstrip circuit. (a) Packaged microwave FET, showing typical dimensions for a 10 GHz device; not to scale. (b) Packaged FET surface-mounted on microstrip.

9.6 Typical Hybrid Amplifier

Figure 9.17 shows the typical layout of a hybrid microwave amplifier on microstrip, with some of the key features identified.

The components included in Figure 9.17, and not so far discussed, are the DC breaks. These are necessary to prevent the DC bias to the gate and drain of the FET from affecting other devices connected to the input and output of the amplifier. They would also be needed between stages in a multi-stage amplifier. The DC break function could be provided by a surface mount capacitor connected across a suitable gap in a microstrip line. But at microwave frequencies, surface mount capacitors may introduce unwanted effects due to reactive and resistive parasitics, and the finger break gives a more reliable wideband performance.

9.7 DC Finger Breaks

The essential requirements for a DC break are that it should have low insertion loss, a predictable transmission phase, and wideband performance. The DC finger break in a microstrip was originally proposed by Lacombe and Cohen [1] and is shown in more detail in Figure 9.18. The shape of the structure is dictated by the need to provide a wideband match and this was shown to be in excess of one octave.

As has already been pointed out, it is important to be able to predict the phase change through a DC break in order to ensure that its insertion into a circuit will not introduce unwanted phase-related effects. The calculation of the phase change through the coupled section shown in Figure 9.18 is straightforward from the theory of coupled lines and is given [1] by $\angle S_{21}$, where

$$S_{21} = \frac{2Z_{21}}{(Z_{11} + 1)(Z_{22} + 1) - Z_{12}Z_{21}}$$

(9.4)

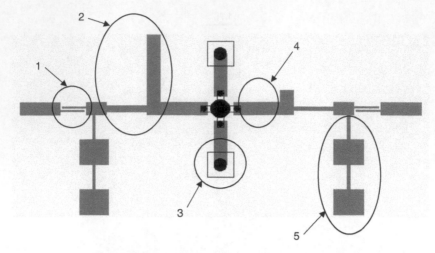

1. DC finger break

2. *I/P* matching network (stub + transformer)

3. Through-substrate conducting VIA

4. $\lambda_S/2$ section to shift reference plane

5. DC bias network

Figure 9.17 Typical layout of a hybrid microstrip amplifier.

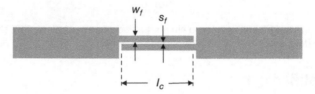

Figure 9.18 Microstrip DC finger break.

and

$$Z_{11} = Z_{22} = -j\frac{Z_{oe} + Z_{oo}}{2}\frac{\cot\theta}{Z_o}, \tag{9.5}$$

$$Z_{12} = Z_{21} = -j\frac{Z_{oe} - Z_{oo}}{2}\frac{\csc\theta}{Z_o}, \tag{9.6}$$

where the subscripts 1 and 2 refer to the input and output of the coupled line section, Z_{oe} and Z_{oo} are the even and odd mode impedances, respectively, of the coupled lines, Z_o is the impedance of the main line, and θ is the phase change through the coupled section. Also, Free and Aitchison [2] showed that there was an excess phase associated with the type of finger break depicted in Figure 9.18. The excess phase was defined as the difference between the phase change calculated using Eq. (9.4) and the phase change through the same physical length of 50 Ω microstrip line. This excess phase was shown to be significant through practical and theoretical investigations at X-band, and a theoretical model was developed [2] to account for the additional phase.

Practical measurements showed that for a 10 GHz microstrip DC break fabricated on alumina, the excess phase exhibited a quasi-linear variation from 0 to 25° over a 3 GHz bandwidth.

9.8 Constant Gain Circles

Constant gain circles are contours on the Smith chart that show how the gain of an amplifier varies with the source and load impedances. Two sets of circles can be drawn, one showing how the gain varies with the input impedance, and the other how the gain varies with the output impedance. The circles have two important practical applications: (i) they show how selective mismatching can be used to control the gain, which is important for broad-band design; (ii) when plotted in conjunction with constant noise circles (discussed in the next section), they show how a trade-off can be made between gain and noise performance. We shall consider the special case of an amplifier with unilateral gain (which is the common situation), and in this case the circles are sometimes referred to as circles of constant unilateral gain.

Equation (5.17) of Chapter 5 gives the unilateral transducer power gain of an amplifying device as

$$G_{TU} = \frac{|S_{21}|^2(1 - |\Gamma_L|^2)(1 - |\Gamma_S|^2)}{|(1 - S_{11}\Gamma_S)(1 - S_{22}\Gamma_L)|^2}. \tag{9.7}$$

Equation (9.7) can be rewritten as

$$G_{TU} = \frac{(1 - |\Gamma_S|^2)}{|(1 - S_{11}\Gamma_S)|^2} \times |S_{21}|^2 \times \frac{(1 - |\Gamma_L|^2)}{|(1 - S_{22}\Gamma_L)|^2}, \tag{9.8}$$

i.e.

$$G_{TU} = G_S \times G_D \times G_L, \tag{9.9}$$

where G_S and G_L are the effective power gain factors provided by the source and load matching networks, respectively. Two points should be noted in respect of G_S and G_L:

(i) These are described as *effective* power gain factors because the matching networks are passive, and therefore cannot individually contribute a power gain. However, since they permit more power to flow into and out of the active device, they can effectively increase the value of G_{TU}.

(ii) In general, the values of G_S and G_L will be greater than 1, since they reduce the mismatch losses at the input and output of the active device.

From Eqs. (9.8) and (9.9) we have

$$G_S = \frac{(1 - |\Gamma_S|^2)}{|(1 - S_{11}\Gamma_S)|^2} \tag{9.10}$$

and

$$G_L = \frac{(1 - |\Gamma_L|^2)}{|(1 - S_{22}\Gamma_L)|^2}. \tag{9.11}$$

We know that the effective gains G_S and G_L will be a maximum when $\Gamma_S = S_{11}^*$ and $\Gamma_L = S_{22}^*$, respectively, giving

$$G_{S,max} = \frac{1}{1 - |S_{11}|^2} \tag{9.12}$$

and

$$G_{L,max} = \frac{1}{1 - |S_{22}|^2}. \tag{9.13}$$

Considering the source matching network, the normalized gain factor, g_S, can be defined as

$$g_S = \frac{G_S}{G_{S,max}}. \tag{9.14}$$

Substituting for G_S and $G_{S,max}$ gives

$$g_S = \frac{(1 - |\Gamma_S|^2)(1 - |S_{11}|^2)}{|(1 - S_{11}\Gamma_S)|^2}. \tag{9.15}$$

Equation (9.15) can be rearranged [3] in the form of the equation of a circle[2] whose centre, c_{gS}, lies along a vector through S_{11}^* and is given by

$$c_{gS} = \frac{g_S|S_{11}|}{1 - |S_{11}|^2(1 - g_S)}, \tag{9.16}$$

---

2 The general equation of a circle in the x–y coordinate plane is $(x - a)^2 + (y - b)^2 = c^2$, where a and b are the coordinates of the centre of the circle and c is the radius of the circle.

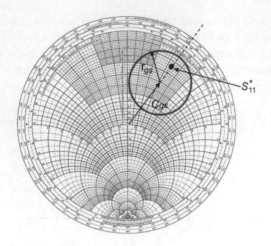

Figure 9.19 Plotting a constant gain circle in the input plane.

and whose radius, r_{gS}, is given by

$$r_{gS} = \frac{(1 - |S_{11}|^2)\sqrt{1 - g_S}}{1 - |S_{11}|^2(1 - g_S)}.$$ (9.17)

Since Eqs. (9.16) and (9.17) were derived in the reflection coefficient plane the values of the centre and radius of the constant gain circles can be taken directly from the refection coefficient scale on the Smith chart. An example of plotting a constant gain circle in the input plane is shown in Figure 9.19.

Following a similar procedure, the centre and radius of the constant gain circles for the output side of the amplifier are

$$c_{gL} = \frac{g_L|S_{22}|}{1 - |S_{22}|^2(1 - g_L)},$$ (9.18)

$$r_{gL} = \frac{(1 - |S_{22}|^2)\sqrt{1 - g_L}}{1 - |S_{22}|^2(1 - g_L)},$$ (9.19)

where the centres of the circles lie on a vector through S_{22}^*.

Example 9.5   The S_{11} value of a packaged FET at a particular frequency is $0.72\angle112°$. Assuming that the FET exhibits unilateral performance, draw the 2 dB, 1 dB, and 0 dB unilateral constant gain circles for the source.

Solution

$$G_{S,max} = \frac{1}{1 - (0.72)^2} = 2.08 \equiv 3.18 \text{ dB.}$$

G_S	g_S	c_{gS}	r_{gS}
2 dB ≡ 1.58	$\dfrac{1.58}{2.08} = 0.76$	0.62	0.27
1 dB ≡ 1.26	$\dfrac{1.26}{2.08} = 0.61$	0.55	0.38
0 dB ≡ 1.00	$\dfrac{1.00}{2.08} = 0.48$	0.47	0.48

The three required constant gain circles are plotted in Figure 9.20.

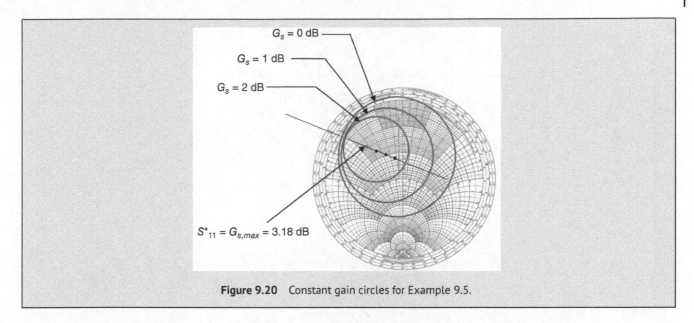

Figure 9.20 Constant gain circles for Example 9.5.

9.9 Stability Circles

In the previous section, we saw how the gain of a transistor amplifier could be controlled through selective mismatching of the source and load matching networks. But it is crucial in any practical design to ensure that the intended amplifier will not oscillate with particular values of source and load impedance. Therefore, the stability of the amplifier should always be examined once the matching networks have been designed.

The Rollett stability factor (K) was introduced in Chapter 5 as a method of using S-parameter data to determine whether a device such as a transistor is stable for arbitrary values of source and load impedance. If the value of K is less than 1, the transistor is only conditionally stable and may oscillate for certain combinations of source and load impedance.

The easiest way of dealing with situations where K is less than 1 is to make use of stability circles. These are circles drawn on the Smith chart that represent the boundary between those values of source or load impedance that cause instability and those which represent stable operation.

Two sets of stability circles can be drawn, one in the Γ_L plane, and the other in the Γ_S plane. The output stability circle is the loci of Γ_L which makes the magnitude of the modified input reflection coefficient unity; if $|S'_{11}| \geq 1$ then more energy is reflected from the input than is incident, and therefore the circuit is oscillating. Using Eq. (5.12), we have

$$S'_{11} = 1 = S_{11} + \frac{S_{21}\Gamma_L S_{12}}{1 - S_{22}\Gamma_L}. \tag{9.20}$$

Equation (9.20) can be rearranged to give

$$\left| \Gamma_L - \frac{(S_{22} - [S_{11}S_{22} - S_{12}S_{21}]S^*_{11})^*}{|S_{22}|^2 - |S_{11}S_{22} - S_{12}S_{21}|^2} \right| = \frac{|S_{12}S_{21}|}{|S_{22}|^2 - |S_{11}S_{22} - S_{12}S_{21}|^2}. \tag{9.21}$$

For convenience, Eq. (9.21) can be written as

$$\left| \Gamma_L - \frac{(S_{22} - DS^*_{11})^*}{|S_{22}|^2 - |D|^2} \right| = \frac{|S_{12}S_{21}|}{|S_{22}|^2 - |D|^2}, \tag{9.22}$$

where

$$D = S_{11}S_{22} - S_{12}S_{21}. \tag{9.23}$$

Equation (9.22) is in the form of the equation of a circle with centre, c_{sL}, and radius, r_{sL}, where

$$c_{sL} = \frac{(S_{22} - DS^*_{11})^*}{|S_{22}|^2 - |D|^2} \tag{9.24}$$

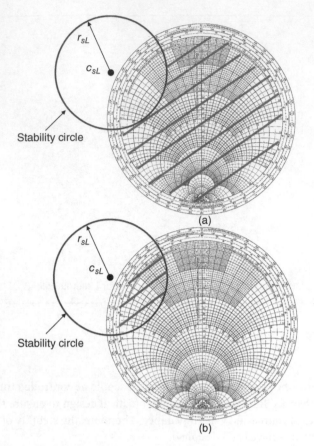

Figure 9.21 Examples of output stability circles, showing stable regions (shaded). (a) $|S_{11}|<1$. (b) $|S_{11}|>1$.

and

$$r_{sL} = \frac{|S_{12}S_{21}|}{|S_{22}|^2 - |D|^2}. \tag{9.25}$$

Similarly, the input stability circle is the loci of Γ_S which makes the modified output reflection coefficient unity, which gives a circle with centre, c_{sS}, and radius, r_{sS}, where

$$c_{sS} = \frac{(S_{11} - DS_{22}^*)^*}{|S_{11}|^2 - |D|^2} \tag{9.26}$$

and

$$r_{sS} = \frac{|S_{12}S_{21}|}{|S_{11}|^2 - |D|^2}. \tag{9.27}$$

Having drawn a stability circle, it is necessary to decide which side of the circle represents the stable operation, and which the unstable region. If we consider Eq. (5.12): if $\Gamma_L = 0$ then $|S'_{11}| = |S_{11}|$ and if $|S_{11}| < 1$ the input reflection coefficient must also be less than 1 and the amplifier is stable. Since $\Gamma_L = 0$ is the centre of the Smith chart, the region of the chart that includes this point must be the stable region. If $|S_{11}| > 1$ when $\Gamma_L = 0$ then the input reflection coefficient is greater than 1, and the centre of the chart represents an unstable point, and the region which includes this point is the unstable region. Examples of the stable and unstable regions are shown in Figure 9.21, with the stable regions shaded. Note that the centres of the stability circles can be outside the working area[3] of the Smith chart, i.e. $|c_{sL}|$ may be greater than 1.

---

3 The working area of the Smith chart refers to the area within the $r = 0$ circle, since values of resistance less than zero have no physical meaning.

9.10 Noise Circles

The noise performance of an FET in the circuit shown in Figure 9.1 is given by the expression [3].

$$F = F_{min} + \frac{4r_n|\Gamma_S - \Gamma_{opt}|^2}{(1 - |\Gamma_S|^2)(|1 + \Gamma_{opt}|^2)}, \tag{9.28}$$

where

F = noise figure at a particular value of Γ_S
r_n = equivalent noise resistance of the FET, normalized to Z_o
Z_o = reference characteristic impedance, normally 50 Ω
F_{min} = minimum noise figure (i.e. when $\Gamma_S = \Gamma_{opt}$)

and the other parameters are as defined previously. (The significance of noise figure and equivalent noise resistance are explained in Chapter 14.)

Equation (9.28) can be re-arranged to give

$$\frac{(F - F_{min})(|1 + \Gamma_{opt}|^2)}{4r_n} = \frac{|\Gamma_S - \Gamma_{opt}|^2}{(1 - |\Gamma_S|^2)}. \tag{9.29}$$

The left-hand side of Eq. (9.29) involves parameters that specifically relate to the FET, rather than the external matching networks, and this is called the noise figure parameter, N_i. Therefore, we have

$$N_i = \frac{(F - F_{min})(|1 + \Gamma_{opt}|^2)}{4r_n} = \frac{|\Gamma_S - \Gamma_{opt}|^2}{(1 - |\Gamma_S|^2)}. \tag{9.30}$$

Equation (9.30) can be re-arranged [3] to give

$$\left|\Gamma_S - \frac{\Gamma_{opt}}{1 + N_i}\right|^2 = \frac{N_i^2 + N_i(1 - |\Gamma_{opt}|^2)}{(1 + N_i)^2}. \tag{9.31}$$

It can be seen that Eq. (9.31) is in the form of an equation of a circle, in terms of Γ_S, with a centre, c_{ni}, given by

$$c_{ni} = \frac{\Gamma_{opt}}{1 + N_i}, \tag{9.32}$$

and a radius, r_{ni}, given by

$$r_{ni} = \frac{\sqrt{N_i^2 + N_i(1 - |\Gamma_{opt}|^2)}}{1 + N_i}. \tag{9.33}$$

Since Γ_S is a reflection coefficient the noise circle can be plotted on the Smith chart for any given value of N_i, i.e. for any given value of F. The circles are called constant noise circles because for any value of Γ_S that lies on a particular circle, the noise figure of the FET is constant.

The value of noise circles to the designer of high-frequency amplifiers is that they can be viewed in conjunction with constant gain circles so the designer can see how the noise figure can be traded against the amplifier gain as Γ_S is varied.

Example 9.6 The following data apply to a microwave FET at a particular frequency:

$\Gamma_{opt} = 0.32\angle 82°$

$F_{min} = 0.7\,\text{dB}$

$r_n = 0.34$

Plot the 1.5 dB and 3 dB constant noise circles on the Smith chart.

Solution

$$F_{min} = 0.7\ \text{dB} \equiv 10^{0.07} = 1.17.$$

(Continued)

(Continued)

Calculating values for the 1.5 dB noise circle:

$$F = 1.5 \text{ dB} \equiv 10^{0.15} = 1.41,$$

$$N_i = \frac{(F - F_{min})(|1 + \Gamma_{opt}|^2)}{4r_n} = \frac{(1.41 - 1.17)(|1 + 0.32\angle 82°|^2)}{4 \times 0.34} = 0.21,$$

$$c_{ni} = \frac{0.32\angle 82°}{1 + 0.21} = 0.26\angle 82°,$$

$$r_{ni} = \frac{\sqrt{(0.21)^2 + 0.21(1 - (0.32)^2)}}{1 + 0.21} = 0.40.$$

Calculating values for the 3 dB noise circle:

$$F = 3 \text{ dB} \equiv 10^{0.3} = 2.00,$$

$$N_i = \frac{(2.00 - 1.17)(|1 + 0.32\angle 82°|^2)}{4 \times 0.34} = 0.73,$$

$$c_{ni} = \frac{0.32\angle 82°}{1 + 0.73} = 0.18\angle 82°,$$

$$r_{ni} = \frac{\sqrt{(0.73)^2 + 0.73(1 - (0.32)^2)}}{1 + 0.73} = 0.63.$$

The 1.5 and 3.0 dB constant noise circles are shown in Figure 9.22.

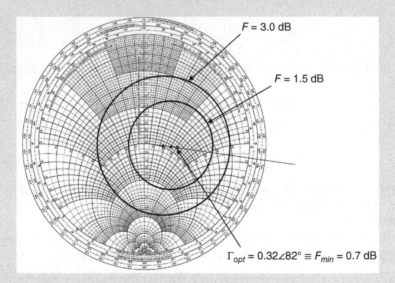

Figure 9.22 Constant noise circles for Example 9.6.

9.11 Low-Noise Amplifier Design

The primary requirement for many microwave small-signal amplifiers is for low-noise performance, particularly for receiver pre-amplifiers. As was pointed out in Section 9.2, minimum noise performance from an FET amplifier requires the source matching network to be designed such that Γ_S is equal to Γ_{opt}, rather than S_{11}^*, which is the condition for maximum gain. Normally, Γ_{opt} and S_{11}^* are different, and so the design of the source network is a compromise between providing high gain and low noise. The trade-off between these two parameters can most easily be visualized by overlaying constant gain circles and noise circles on the same Smith chart. The subsequent design procedure will be illustrated through Example 9.7.

Example 9.7   The following data apply to an 8 GHz packaged FET:

$$[S] = \begin{bmatrix} 0.798\angle-110° & 0.006\angle10° \\ 2.482\angle70° & 0.618\angle-87° \end{bmatrix},$$

$$F_{min} = 0.8 \text{ dB},$$

$$\Gamma_{opt} = 0.55\angle165°,$$

$$r_n = 0.2.$$

Using the above data find the value of Γ_S needed to design a low-noise 12 dB amplifier, and determine the corresponding noise figure of the amplifier.

Solution

Since $|S_{12}|$ is very small we may regard the amplifier as having unilateral performance. Using Eq. (5.22) we can determine the maximum unilateral gain as:

$$G_{TU,max} = \frac{1}{1-(0.798)^2} \times (2.482)^2 \times \frac{1}{1-(0.618)^2},$$

$$= 2.753 \times 6.160 \times 1.618,$$

$$\equiv 4.398 \text{ dB} + 7.895 \text{ dB} + 2.089 \text{ dB},$$

$$= 14.382 \text{ dB}.$$

The specification requires 12 dB gain for the amplifier, and this can be achieved by reducing the gain of the source network. (*Note that by reducing the gain of the **source** network the noise performance of the amplifier will be improved.*) The required effective gain factor, G_S, of the source network will therefore be

$$G_S = 4.398 \text{ dB} - (14.382 - 12) \text{ dB} = 2.016 \text{ dB} \equiv 1.591.$$

Calculating normalized gain:

$$g_S = \frac{G_S}{G_{S,max}} = \frac{1.591}{2.753} = 0.578.$$

Using Eqs. (9.16) and (9.17) to calculate centre and radius of 2.016 dB constant gain circle:

$$c_{gS} = \frac{0.578 \times 0.798}{1-(0.798)^2(1-0.578)} = 0.631,$$

$$r_{gS} = \frac{(1-(0.798)^2)\sqrt{1-0.578}}{1-(0.798)^2(1-0.578)} = 0.323.$$

The constant gain circle can now be plotted on the Smith chart; see Figure 9.23. Any value of Γ_S which lies on this circle will give the required gain, but we now need to plot the noise circles to establish the value of Γ_S which gives minimum noise.

(Continued)

(Continued)

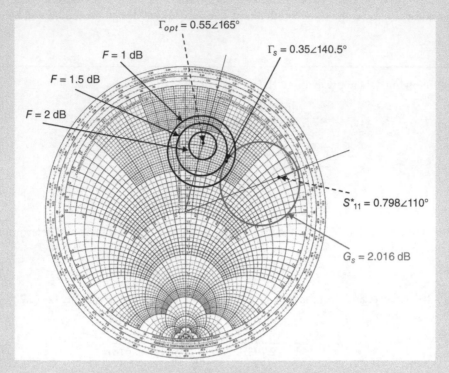

Figure 9.23 Smith chart solution for Example 9.7.

We will start by plotting the circles corresponding to noise figures of 1 dB, 1.5 dB, and 2 dB, using Eqs. (9.29) through (9.33).

1 dB noise circle: $F = 1$ dB $\equiv 1.259$ and $F_{min} = 0.8$ dB $\equiv 1.202$,

$$N_i = \frac{(1.259 - 1.202) \times (|1 + 0.55\angle165°|^2)}{4 \times 0.2} = 0.017,$$

$$c_{ni} = \frac{0.55\angle165°}{1 + 0.017} = 0.541\angle165°,$$

$$r_{ni} = \frac{\sqrt{(0.017)^2 + 0.017 \times (1 - (0.55)^2)}}{1 + 0.017} = 0.108,$$

1.5 dB noise circle: $F = 1.5$ dB $\equiv 1.413$ and $F_{min} = 0.8$ dB $\equiv 1.202$.

$$N_i = \frac{(1.413 - 1.202) \times (|1 + 0.55\angle165°|^2)}{4 \times 0.2} = 0.063,$$

$$c_{ni} = \frac{0.55\angle165°}{1 + 0.063} = 0.517\angle165°,$$

$$r_{ni} = \frac{\sqrt{(0.063)^2 + 0.063 \times (1 - (0.55)^2)}}{1 + 0.063} = 0.206,$$

2 dB noise circle: $F = 2$ dB $\equiv 1.585$ and $F_{min} = 0.8$ dB $\equiv 1.202$,

$$N_i = \frac{(1.585 - 1.202) \times (|1 + 0.55\angle165°|^2)}{4 \times 0.2} = 0.115,$$

$$c_{ni} = \frac{0.55\angle165°}{1 + 0.115} = 0.490\angle165°,$$

$$r_{ni} = \frac{\sqrt{(0.115)^2 + 0.115 \times (1 - (0.55)^2)}}{1 + 0.115} = 0.274.$$

> The three noise circles are plotted on the Smith chart shown in Figure 9.23.
>
> To find the best value for Γ_S we look for the point where the gain circle is tangential to the noise circle with the lowest noise figure. It can be seen that there is a point where the gain circle is very close[4] to the 1.5 dB noise circle, and this must give the required value of Γ_S, i.e. $\Gamma_S = 0.35\angle140.5°$.

9.12 Simultaneous Conjugate Match

In Section 9.2, we considered the special case of a unilateral device where the input and output matching networks could be designed independently. However, in the more general case, where $S_{12} \neq 0$, the design of the input network will be affected by the output, and vice versa, and so we need to make use of modified reflection coefficients as derived in Chapter 5 and shown in Eqs. (5.12) and (5.13). The requirements for conjugate impedance matching are thus

$$\Gamma_S = \left(S_{11} + \frac{S_{12}S_{21}\Gamma_L}{1 - S_{22}\Gamma_L} \right)^* \tag{9.34}$$

and

$$\Gamma_L = \left(S_{22} + \frac{S_{12}S_{21}\Gamma_S}{1 - S_{11}\Gamma_S} \right)^*. \tag{9.35}$$

Equations (9.34) and (9.35) must be solved as a pair of simultaneous equations, which leads to the result

$$\Gamma_S = \frac{B_1 \pm \sqrt{B_1^2 - 4|C_1|^2}}{2C_1} \tag{9.36}$$

and

$$\Gamma_L = \frac{B_2 \pm \sqrt{B_2^2 - 4|C_2|^2}}{2C_2}, \tag{9.37}$$

where

$$D = S_{11}S_{22} - S_{12}S_{21},$$
$$B_1 = 1 + |S_{11}|^2 - |S_{22}|^2 - |D|^2,$$
$$C_1 = S_{11} - DS_{22}^*,$$
$$B_2 = 1 + |S_{22}|^2 - |S_{11}|^2 - |D|^2,$$
$$C_2 = S_{22} - DS_{11}^*. \tag{9.38}$$

Note that using a negative sign in front of the square root in Eqs. (9.36) and (9.37) leads to a solution that is appropriate for passive matching networks where $\Gamma_S < 1$ and $\Gamma_L < 1$.

Furthermore, it can be shown that the maximum available transducer power gain (MAG) from a device with simultaneous conjugate matching, and $K > 1$, is given by

$$G_{MAG} = \frac{|S_{21}|}{|S_{12}|}(K - \sqrt{K^2 - 1}), \tag{9.39}$$

where K is the stability factor, which was defined in Chapter 5.

Another figure-of-merit that is often quoted is the maximum stable transducer power gain (MSG), which is defined as the limiting case of maximum available gain (MAG) when $K = 1$, i.e.

$$G_{MSG} = \frac{|S_{21}|}{|S_{12}|}. \tag{9.40}$$

---

4 *Comment: We have chosen the matching point where the gain circle and the 1.5 dB noise circle are very close; by plotting more noise circles we could obtain a point where the gain circle touches a noise circle, but this is a refinement best achieved using CAD software.*

Example 9.8 The following data apply to a packaged FET at 15 GHz

$$[S] = \begin{bmatrix} 0.586\angle138° & 0.076\angle-46° \\ 1.418\angle-45° & 0.664\angle169° \end{bmatrix}.$$

Calculate:

(i) The Rollett stability factor, and state if the FET is stable
(ii) The values of Γ_S and Γ_L required for a simultaneous conjugate match
(iii) The maximum available gain (MAG)
(iv) The maximum unilateral gain

Solution

(i)
$$D = (0.586\angle138° \times 0.664\angle169°) - (0.076\angle-46° \times 1.418\angle-45°)$$
$$= 0.311\angle-41°$$

$$|D| = 0.311.$$

Using Eq. (5.24):

$$K = \frac{1 - (0.586)^2 - (0.664)^2 + (0.311)^2}{2 \times 0.076 \times 1.418} = 1.45.$$

Since $K > 1$ and $|D| < 1$ the FET is unconditionally stable.

(ii) From Eq. (9.36):

$$\Gamma_S = \frac{B_1 \pm \sqrt{B_1^2 - 4|C_1|^2}}{2C_1}.$$

Evaluating the parameters in Eq. (9.38):

$$B_1 = 1 + (0.586)^2 - (0.664)^2 - (0.311)^2 = 0.806,$$

$$C_1 = 0.586\angle138° - [(0.311\angle-41°) \times (0.664\angle-169°)]$$
$$= 0.385\angle-48.37°.$$

Then,

$$\Gamma_S = \frac{0.806 - \sqrt{(0.806)^2 - 4 \times (0.385)^2}}{2 \times 0.385\angle-48°} = 0.738\angle48.37°.$$

Similarly, to find the load reflection coefficient we use Eq. (9.37) i.e.

$$\Gamma_L = \frac{B_2 \pm \sqrt{B_2^2 - 4|C_2|^2}}{2C_2},$$

where

$$B_2 = 1 + (0.664)^2 - (0.586)^2 - (0.311)^2 = 1.000,$$

$$C_2 = 0.664\angle169° - (0.311\angle-41° \times 0.586\angle-138°)$$
$$= 0.489\angle164.54°,$$

giving

$$\Gamma_L = \frac{1.000 - \sqrt{1.000 - 4 \times (0.489)^2}}{2 \times 0.489\angle164.54°} = 0.809\angle-164.54°.$$

(iii) Using Eq. (9.39):

$$G_{MAG} = \frac{1.418}{0.076} \times (1.45 - \sqrt{(1.45)^2 - 1}) = 7.46 \equiv 8.73 \text{ dB}.$$

(iv) Using Eq. (5.22):

$$G_{TU,max} = \frac{1}{1 - (0.586)^2} \times (1.418)^2 \times \frac{1}{1 - (0.664)^2}$$

$$= 5.48 \equiv 7.39 \text{ dB}.$$

9.13 Broadband Matching

One technique for designing a microwave broadband amplifier is to make use of selective mismatching to control the overall gain of the amplifier; the technique will be demonstrated through discussion of a practical example. Table 9.1 shows S-parameter data for a packaged microwave FET.

Since S_{12} is small, we may consider the FET to exhibit unilateral performance, and assuming the FET to have input and output matching networks as shown in Figure 9.1, the transducer power gains of the various stages may be computed as shown in Table 9.2.

If the requirement was for an amplifier with a nominally constant gain of 9 dB over the frequency band 6–10 GHz, this could be achieved by selectively mismatching one, or both, of the matching networks. If we assume that the output network has to be designed to give maximum gain over the required band, then the required gains of the input network, along with data for drawing the constant gain circles are given in Table 9.3.

Using the data in Table 9.3, three constant gain circles can be plotted corresponding to the gains required at the three spot frequencies. These constant gain circles are shown in Figure 9.24.

The source network must be designed such that the locus of Γ_S intersects the appropriate constant gain circle at the desired frequency. The network can be a combination of series- and shunt-lumped reactive components, or in distributed form as a combination of quarter-wave transformers and shunt stubs. A typical example of the required variation of Γ_S is included in Figure 9.24, and shows Γ_S intersecting the constant gain circles at the required spot frequencies. For wideband matching, networks consisting of a combination of only two frequency-dependent components do not normally offer sufficient degrees

Table 9.1 S-parameter data for a packaged FET at X-band.

Frequency (GHz)	S_{11}	S_{21}	S_{12}	S_{22}
6	$0.81\angle-112°$	$2.14\angle85°$	$0.01\angle37°$	$0.71\angle-53°$
8	$0.76\angle-150°$	$1.94\angle55°$	$0.01\angle37°$	$0.69\angle-70°$
10	$0.72\angle-178°$	$1.77\angle30°$	$0.01\angle46°$	$0.70\angle-84°$

Table 9.2 Network gains.

Frequency (GHz)	$G_{S,max}$ (dB)	G_D (dB)	$G_{L,max}$ (dB)	$G_{TU,max}$ (dB)
6	4.64	6.61	3.05	14.30
8	3.74	5.76	2.81	12.31
10	3.17	4.96	2.92	11.05

Table 9.3 Source gain requirements.

Frequency (GHz)	G_S (dB)	g_S	c_{gS}	r_{gS}
6	$4.64-5.30 = -0.66$	0.30	0.45	0.53
8	$3.74-3.31 = 0.43$	0.46	0.51	0.45
10	$3.17-2.05 = 1.12$	0.62	0.56	0.37

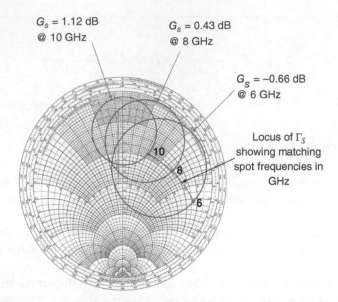

G_S = 1.12 dB
@ 10 GHz

G_S = 0.43 dB
@ 8 GHz

G_S = –0.66 dB
@ 6 GHz

Locus of Γ_S
showing matching
spot frequencies in
GHz

Figure 9.24 Source constant gain circles required for broadband matching.

of design freedom to allow the correct locus for Γ_S to be designed, and three or more components are required. In this situation, the manual design of a suitable network using a Smith chart is impractical, and the use of appropriate CAD software is essential. Using CAD techniques has the further advantage that the designer can easily check that noise and stability requirements are not being violated as the frequency response of the source network changes.

9.14 Summary

The basic concepts underlying the design of hybrid microwave amplifiers have been introduced in this chapter. The issue of stability has been emphasized, and the construction of stability circles should be a first step in any practical design. Constant gain and constant noise circles have also been introduced, and the procedures for plotting these on the Smith chart have been established. Whilst individual sets of gain and noise circles are useful, their real benefit comes when they are overlaid on the same Smith chart, so the circuit designer can see the trade-off between gain and noise performance for a particular design.

The manual use of the Smith chart in designing microwave small-signal amplifiers provides an initial solution, which can subsequently be refined using appropriate software. The role of CAD software in amplifier design cannot be over-emphasized. The ability of software to simultaneously display gain and noise circles, and to show how these vary over a specified bandwidth is an invaluable aid to efficient design. As was mentioned in Section 9.13, it is really not practical to design wideband amplifiers without the assistance of appropriate software.

9.15 Supplementary Problems

Q9.1 The following data apply to a microwave transistor at 2 GHz

$$[S] = \begin{bmatrix} 0.18\angle135° & 0.02\angle60° \\ 4.7\angle50° & 0.41\angle-47° \end{bmatrix}.$$

(i) Calculate the value of K (the Rollett stability factor). Is the transistor stable?
(ii) Calculate the maximum unilateral transducer power gain of the transistor in dB.

Q9.2 Design a 14 GHz DC bias connection having the geometry shown in Figure 9.12. The bias connection is to be fabricated in microstrip using a substrate that has a relative permittivity of 9.9 and a thickness of 0.5 mm. The bias connection is to be made to a 50 Ω microstrip line.

Q9.3 Suppose that the bias connection designed in Q9.2 was used at 15 GHz. Calculate the VSWR on the 50 Ω microstrip line, assuming that this line is terminated in matched impedances.

Q9.4 A 15 GHz FET amplifier is to be fabricated as a hybrid microwave integrated circuit using a microstrip. The value of S_{11} for the FET at 15 GHz is $0.83\angle132°$. Assuming the transistor to have unilateral performance, design a matching circuit consisting of a 50 Ω open-circuit stub and a quarter-wave transformer to connect the input of the FET to a 50 Ω source; the matching circuit has the configuration shown in Figure 9.4. (Microstrip substrate details: $\varepsilon_r = 9.8$, $h = 250\,\mu m$.)

Q9.5 Using the dimensions of the matching circuit calculated in Q9.4, determine the value of the reflection coefficient at the output of the matching circuit at 17 GHz, assuming that the source impedance is still 50 Ω.

Q9.6 The following data apply to an 18 GHz microwave transistor

$$[S] = \begin{bmatrix} 0.44\angle94° & 0.08\angle-72° \\ 1.34\angle72° & 0.67\angle140° \end{bmatrix}.$$

Determine the value of Γ_S and Γ_L required to provide a simultaneous conjugate match.

Q9.7 Draw the input and output stability circles on a Smith chart for a transistor with the following S-parameters, and indicate the regions corresponding to stable operation:

$$[S] = \begin{bmatrix} 0.865\angle-78° & 0.04\angle43° \\ 2.37\angle112° & 0.74\angle-35° \end{bmatrix}.$$

Q9.8 The following data apply to an 4 GHz packaged FET:

$$[S] = \begin{bmatrix} 0.48\angle79° & 0.01\angle35° \\ 2.85\angle107° & 0.61\angle-58° \end{bmatrix},$$

$$F_{min} = 0.6 \text{ dB},$$

$$\Gamma_{opt} = 0.45\angle-165°,$$

$$r_n = 0.4.$$

   (i) Assuming the FET to give unilateral performance, design distributed source and load matching networks to give maximum transducer power gain when the source and load impedances are both 50 Ω; the source and load matching circuits are to have the configurations shown in Figures 9.4 and 9.6, respectively. The characteristic impedance of the stub should be 50 Ω (The distributed circuit is to be fabricated in a microstrip using a substrate that has a relative permittivity of 9.8 and a thickness of 1 mm.)
   (ii) Determine the noise figure of the amplifier designed in part (i).
   (iii) Repeat the design for maximum unilateral transducer power gain, but using lumped inductors.

Q9.9 Using the Smith chart, and the concept of constant noise circles and constant gain circles, determine the maximum unilateral transducer power gain that could be obtained using the 4 GHz FET specified in Q9.8 if the maximum noise figure of the amplifier is to be 2.2 dB.

Q9.10 A 6 GHz microstrip amplifier is to be designed using the layout shown in Figure 9.17. The S-parameters of the FET at 6 GHz are given by:

$$[S] = \begin{bmatrix} 0.56\angle23° & 0.01\angle-179° \\ 7.12\angle-84° & 0.44\angle93° \end{bmatrix}.$$

Assuming the amplifier is designed to give maximum unilateral transducer power gain, determine the dimensions of all the microstrip lines in Figure 9.17 if the circuit is fabricated on a substrate with $\varepsilon_r = 9.8$ and $h = 1$ mm.

References

1 Lacombe, D. and Cohen, J. (1972). Octave-band microstrip DC blocks. *IEEE Transactions on MTT-20* 20: 555–556.

2 Free, C.E. and Aitchison, C.S. (1984). Excess phase in microstrip DC blocks. *Electronics Letters* 20 (21): 222–223.

3 Gonzalez, G. (1984). *Microwave Transistor Amplifiers, Analysis and Design*. Englewood Cliffs, NJ: Prentice Hall.

10

Switches and Phase Shifters

10.1 Introduction

A phase shifter is essentially a two-port network through which the transmission phase can be changed either continuously or in discrete steps, but without significant change in the transmission loss or in the match of the network. Phase shifters in which the transmission phase can be changed continuously are known as analogue phase shifters, and those in which the phase can be changed in discrete steps are known as digital phase shifters.

Phase shifters have many important applications in modern RF and microwave communication systems, notably for use in phase modulation circuits and phased array antennas. Digital phase shifters have wider application than their analogue counterparts and it is digital phase shifters that will be the main focus of this chapter.

Digital phase shifters require the use of high-frequency control components that can be biased to provide ON and OFF states, where the two states correspond ideally to zero and infinite impedance. In practical circuits the control components are connected to provide the functions of high-frequency switches. Figure 10.1 depicts a single pole single throw (SPST) switch, and its implementation in microstrip.

With an SPST switch, transmission from port 1 to port 2 is controlled by opening and closing the switch. In microstrip, this function is achieved by mounting a control component across a gap in the microstrip line, so that switching the control component ON corresponds to closing the switch, and with the control component OFF there is ideally no transmission between ports 1 and 2, and the switch is open. The control component may also be mounted in shunt (parallel) between the microstrip line and the ground plane. In this configuration, turning the control component ON corresponds to having the switch open, since any power travelling from port 1 to port 2 will be completely reflected, and turning the control component OFF will allow transmission and this corresponds to closing the switch. Series-mounted control components tend to give low insertion loss in the ON state, but poor isolation since with the control element OFF some power can still propagate across the gap or through the package of the particular control component. The converse situation exists with shunt-mounted control components, in that most of the power is reflected when the control component is ON, leading to high isolation. But with shunt-mounted control elements the presence of the component even when it is OFF tends to absorb or reflect power, leading to poor insertion loss. In practice, however, there is no real choice between the series- and shunt-mounted techniques, since it is physically difficult to make shunt connections with adequate high-frequency performance, particularly in hard ceramic substrates, and so most control components are mounted in series.

For many high-frequency phase shifter designs it is convenient to use single pole double throw (SPDT) switches. This type of switch allows the high-frequency power to be routed along different paths, as shown in Figure 10.2.

Changing the states of the control elements allows the power to be routed between ports 1 and 2, or between ports 1 and 3.

There are a number of high-frequency devices that can perform the required control functions, and the chapter will commence with a review of the three most common control components used in hybrid circuits, namely PIN diodes, micro-electromechanical switches (MEMS), and field effect transistors (FETS). In addition, a brief introduction to phase-change switches will be given, as these devices represent an emerging technology that is rapidly becoming popular for a variety of high-frequency switch applications.

The chapter concludes with a discussion of the various circuit configurations used in digital phase shifters.

RF and Microwave Circuit Design: Theory and Applications, First Edition. Charles E. Free and Colin S. Aitchison.
© 2022 John Wiley & Sons Ltd. Published 2022 by John Wiley & Sons Ltd.
Companion website: www.wiley.com/go/free/rfandmicrowave

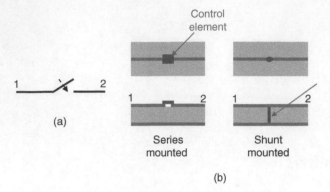

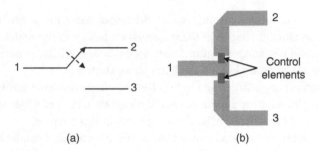

Figure 10.1 Microstrip SPST switches: (a) SPST switch and (b) its microstrip implementation.

Figure 10.2 Microstrip SPDT switches: (a) SPDT switch and (b) its microstrip implementation.

10.2 Switches

10.2.1 PIN Diodes

The acronym PIN derives from the semiconductor structure of this type of diode, which is shown schematically in Figure 10.3.

The diode has two heavily doped semiconductor regions, P$^+$ and N$^+$, separated by a very lightly doped intrinsic region of high resistivity, denoted by i in Figure 10.3 (note that the superscript $^+$ has been used to denote a heavily doped region of semiconductor). If the intrinsic region has a light P-type doping the diode is referred to as p-π-n, and if lightly N-doped as p-v-n. The idealized doping profiles of the two types are shown in Figure 10.4.

Figure 10.3 Structure of a PIN diode.

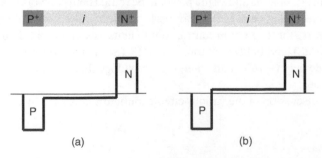

Figure 10.4 Idealized doping profiles of PIN diode types: (a) p-π-n and (b) p-v-n.

However, both types of diode exhibit the same microwave behaviour, and the general term PIN is normally used for both cases.

If a low-frequency AC voltage is applied across a PIN diode, it will behave in the same way as a conventional PN junction. During the positive half-cycle (P-region positive with respect to the N-region), electrons will be injected into the i-layer from the N$^+$ region, and holes injected from the P$^+$ region. The resistivity of the i-layer will decrease, and this decrease will continue as the value of forward bias is increased, due to the increase in the carrier concentration in the i-region. During the negative half-cycle the carriers are swept out from the i-layer to create a layer of very high resistivity. However, at frequencies greater than ~0.1 GHz, the period of the AC signal becomes short compared to the relaxation time[1] of the dielectric, and there is insufficient time for the carriers to be swept out of the i-layer during the negative half cycle. Under these circumstances, a plasma of carriers becomes trapped in the i-layer, whose resistance then depends on the carrier concentration. The resistance of the i-layer can then be controlled by applying a DC bias to the diode. The diode will then function as a variable resistor, with the value of resistance controlled by the magnitude of the DC bias. This has useful applications in some microwave circuits where a voltage-controlled attenuator is required. For switching purposes, however, the value of DC bias should be set so that in the ON state the diode has a low forward resistance, and in the OFF state a very high reverse resistance corresponding to an absence of carriers in the i-region. When the diode is reverse biased there will be a depletion layer, and the bias value should be such that this extends the width of the i-layer to minimize the capacitance; the depletion layer will also penetrate into the P$^+$ and N$^+$ regions, but the depth of penetration into these layers will be very small due to the very high carrier concentration.

One of the parameters used to specify the performance of a PIN diode used for switching purposes is the 'switching ratio (SR),' which is defined as the maximum ratio of i-layer resistivity between the forward and reverse bias states. This ratio is normally in excess of 5000 for PIN diodes. It must be emphasized that the practical significance of the SR is also affected by the circuit conditions surrounding the diode.

In the OFF state the diode can be regarded as having a capacitance C_j, which can be calculated in a similar way to that of a parallel plate capacitor, i.e.

$$C_j = \frac{\varepsilon A}{L},$$ (10.1)

where

ε = permittivity of the semiconductor material
A = cross-sectional area of the junction
L = length of the i-region

If the i-region is relatively long the junction capacitance in the OFF state will be very small. Since the diode behaviour is very different under forward and reverse bias conditions, it is necessary to have different equivalent circuits to represent each state. Under forward bias conditions the PIN diode can be represented by the equivalent circuit shown in Figure 10.5.

Since there is no junction capacitance under forward bias the semiconductor part of the diode can be represented simply by a resistor, whose value is the sum of the bulk resistance of the heavily doped regions, plus that of the low resistivity i-region. In addition to this resistance there will be some lead inductance, and also some capacitance associated with the

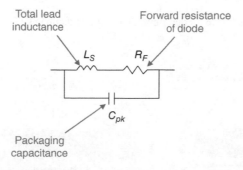

Figure 10.5 Equivalent circuit of PIN diode in ON state.

---

1 When a dielectric is subjected to an external electric field the material becomes polarized. The relaxation time is the time taken for the material to return to an equilibrium state after the field is removed.

packaging of the diode. The lead inductance acts in series with the diode resistance, and the packaging capacitance is normally specified as the parallel capacitance acting between the end of the leads. It follows from Figure 10.5 that the impedance of the PIN diode under forward bias will be

$$(Z_D)_{ON} = \frac{R_F + j\omega L_s}{(1 - \omega^2 C_{pk} L_s) + j\omega C_{pk} R_F}. \tag{10.2}$$

When the diode is reverse-biased, the equivalent circuit must include the capacitance of the *i*-region, as shown in Figure 10.6.

In Figure 10.6, R_j represents the resistance of the charge-depleted *i*-region. The value of R_j will be very large compared to the low capacitive reactance of the *i*-region, and so R_j is often omitted from the OFF state equivalent circuit. The diode impedance in the OFF state can be found from Figure 10.6 using simple circuit analysis as

$$(Z_D)_{OFF} = \frac{R_s + j(\omega L_s - [\omega C_j]^{-1})}{1 - \omega^2 C_{pk} L_s + C_{pk} C_j^{-1} + j\omega C_{pk} R_s}. \tag{10.3}$$

In practice, PIN diodes for hybrid circuits are supplied in packages that minimize the circuit parasitics, i.e. the lead inductances and packaging capacitance. Beam-lead packages, which were introduced in Chapter 4, are often used for PIN diodes intended for mounting in microstrip circuits. This type of package is shown schematically in Figure 10.7, and consists of a plastic body that encapsulates the semiconductor, and two thin flat leads. These leads are referred to as beam leads, giving the package its familiar name, and have the specific advantages of reducing lead inductance and packaging capacitance to very low levels.

Beam lead packages are designed for mounting across gaps in planar lines, as shown in Figure 10.7b for a microstrip line. Normally the gap (*g*) in the microstrip line is made slightly larger than the body (*p*) of the diode. Typical values of *p* for microwave diodes are in the range 250–500 μm. The leads of the diode are usually made of gold, and can be wedge-bonded to a microstrip gold track or attached with conducting epoxy to copper tracks. With care, this type of package can be mounted in a microstrip circuit with negligible mismatch introduced by the mounting technique.

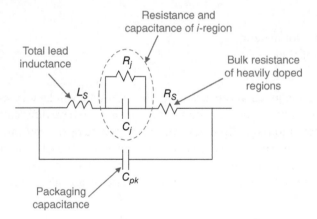

Figure 10.6 Equivalent circuit of PIN diode in OFF state.

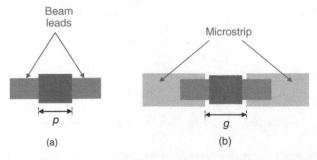

Figure 10.7 Mounting of beam-lead PIN diodes: (a) beam-lead PIN diode and (b) diode mounted across a microstrip gap.

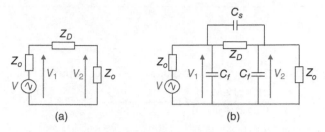

Figure 10.8 Equivalent circuits of series-mounted PIN diodes: (a) without gap compensation and (b) with gap compensation.

Three of the key parameters used to specify the performance of a PIN diode switch are the insertion loss, the isolation, and the mismatch. The insertion loss is simply the loss introduced by the switch in the ON state, and the isolation is the insertion loss in the OFF state. Expressions for the insertion loss can be found through a simple analysis of the equivalent circuits shown in Figure 10.8. The mismatch is a measure of the reflection at the input and output of the mounted PIN diode.

In Figure 10.8, the impedance of the PIN diode is represented by Z_D, and the characteristic impedance of the microstrip line by Z_0. Figure 10.8b includes the equivalent circuit of the microstrip gap, which is represented by the conventional π-network of capacitors (see Chapter 4).

Referring to Figure 10.8a we have, without the PIN diode,

$$V_1 = \frac{V}{2},$$ (10.4)

and with the PIN diode (Figure 10.8b)

$$V_2 = \frac{Z_0}{Z_D + 2Z_0} V.$$ (10.5)

The insertion loss, IL, of the PIN diode is given by

$$IL = \frac{V_1}{V_2} = \frac{Z_D + 2Z_0}{2Z_0}.$$ (10.6)

The magnitude of the insertion loss in dB is then

$$(IL)_{dB} = 20 \log \left| \frac{Z_D + 2Z_0}{2Z_0} \right|.$$ (10.7)

Therefore, in the ON state we have:

$$\text{Insertion loss} = 20 \log \left| \frac{(Z_D)_{ON} + 2Z_0}{2Z_0} \right| \text{dB}.$$

In the OFF state we have:

$$\text{Isolation} = 20 \log \left| \frac{(Z_D)_{OFF} + 2Z_0}{2Z_0} \right| \text{dB}.$$

Example 10.1 The following data apply to a 5 GHz PIN diode biased to operate as a SPST switch:

$L_s = 0.4\,\text{nH}$	$C_{pk} = 0.03\,\text{pF}$	$C_j = 0.04\,\text{pF}$
$R_j = 14\,\text{k}\Omega$	$R_s = 0.3\,\Omega$	$R_F = 1.1\,\Omega$

The PIN diode is mounted across a gap in a 50 Ω microstrip line. Neglecting the capacitive effects of the microstrip gap determine:

(i) The insertion loss in the ON state.
(ii) The isolation in the OFF state.

(Continued)

(Continued)

Solution

(i) Substituting the given data into Eq. (10.2) to find the diode impedance in the ON state gives:

$$(Z_D)_{ON} = \frac{1.1 + j12.57}{(1 - 0.012) + j0.001}\Omega = \frac{1.1 + j12.57}{0.988}\Omega = (1.11 + j12.72)\,\Omega.$$

The insertion loss is then given by Eq. (10.7):

$$(IL)_{dB} = 20\log\left|\frac{1.11 + j12.72 + 100}{100}\right|\,dB$$

$$= 0.164\,dB.$$

(ii) Substituting the given data into Eq. (10.3) to find the diode impedance in the OFF state gives:

$$(Z_D)_{OFF} = \frac{0.3 + j(12.57 - 795.67)}{1 - 0.016 + 0.75 + j0.0003}\Omega = \frac{0.3 - j783.1}{1.73 + j0.0003}\Omega$$

$$= (0.17 - j452.66)\,\Omega.$$

The isolation is then given by Eq. (10.7):

$$\text{Isolation} = 20\log\left|\frac{0.17 - j452.66 + 100}{100}\right|\,dB$$

$$= 13.32\,dB.$$

In Example 10.1, we neglected the capacitive effects of the microstrip gap, but some caution is needed in practice to ensure that transmission through the gap does not have a deleterious effect. In the diode ON state the gap has very little effect, since it is effectively shorted-out by the diode. But in the OFF state it is important that the transmission loss through the gap is large compared to the transmission loss through the diode. Reference to Figure 4.10 shows that for a typical microstrip line, gap widths less than 300 μm result in insertion losses less than ~15 dB, which is comparable with the isolation of a PIN diode in the OFF state.

Further aspects of PIN diode switches:

(i) *Switching speed.* The switching speed is defined as the time taken to move from 10 to 90% of the new transmission condition. This time interval depends on a number of different factors, and there is no simple mathematical expression for computing the switching speed. One of the main factors to consider when switching from the ON state to the OFF state is the time taken to remove the stored charge from the *i*-region. This will depend primarily on the mobility of the carriers and the width of the *i*-region. Obviously the greater the width the slower the switching speed, but a wider region means lower junction capacitance (C_j) in the OFF state, and consequently higher *i*-region impedance and better isolation. The bias circuitry is also a major factor in determining the switching speed. The circuitry must be able to provide a rapid transition from a moderately high forward bias current (I_{bias}), usually in the order of 10–20 mA, to a reverse bias voltage (V_R) typically in the range 10–20 V.

(ii) *Self-resonant frequencies.* Since the equivalent circuits of the PIN diode include both capacitive and inductive elements, the diode will exhibit resonant frequencies. It is important to ensure that these frequencies are outside the working range of the switch. In the forward bias condition the self-resonant frequency is given approximately by

$$(f_r)_{ON} = \frac{1}{2\pi\sqrt{L_s C_{pk}}} \tag{10.8}$$

and in the OFF state by

$$(f_r)_{OFF} = \frac{1}{2\pi\sqrt{L_s(C_j + C_{pk})}}. \tag{10.9}$$

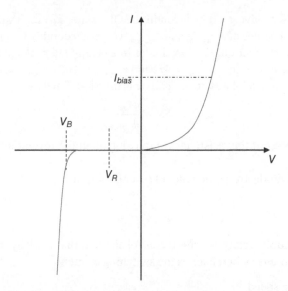

Figure 10.9 Typical I–V characteristics for a PIN diode.

(iii) *Bias conditions.* An idealized V–I characteristic for a PIN diode is shown in Figure 10.9, with the forward and reverse bias points indicated.

In the ON state the diode should be forward biased into the quasi-linear region of the characteristic so as to minimize the diode resistance, and so minimize the forward insertion loss. In the OFF state the diode should be biased into the high resistance region between 0 V and the breakdown voltage, V_B.

The value of V_R is chosen to be large so that the depletion region extends right across the i-layer, but not so large that when an alternating voltage (RF voltage) is superimposed on V_R peak excursions of the AC signal drive the net voltage into the breakdown region. This is not usually a problem for RF and microwave circuit applications where the magnitudes of the AC voltages are much less than the reverse bias value. However, it can become a serious issue when PIN diode switches are used in the 'transmit' stages of a transceiver, for example, to switch relatively high-power signals to different antenna elements.

(iv) *Power rating.* In the reverse bias condition, the maximum permissible power is determined primarily by the need to avoid the peak voltage excursions of the RF signal causing breakdown, as discussed in section (iii). This suggests that the DC bias value for reverse bias should be set at $V_B/2$, and that the peak value of the RF signal is limited to $V_B/2$. However, it is possible to allow a greater amplitude of the RF signal by using a reverse bias that is slightly less than $V_B/2$. This would allow a greater RF peak amplitude before reaching breakdown, but would mean that in the nominal OFF state the diode voltage would enter the forward bias region for part of the RF cycle. This is permissible provided the time in the forward bias region is insufficient to allow significant injection of carriers into the i-region, which can subsequently transit the i-layer. A related consideration is that if $V_R < V_B/2$ then under peak RF voltage conditions the diode is working over a non-linear portion of the V–I characteristic, which may mean the generation of unwanted harmonics.

In the forward bias condition the diode power is limited by straightforward thermal considerations. The average power dissipated in the diode under forward bias conditions will be $I_{bias}^2 \times R_F$ where R_F is the forward resistance of the diode at the selected bias point (this is given by the slope of the V–I characteristic at I_{bias}). A PIN diode will have a power dissipation rating usually specified in terms of the thermal resistance, which gives the temperature rise of the diode per Watt of applied RF power.

(v) *Figure of Merit (FoM).* The quality of PIN diode used for switching purposes is often represented by the FoM, defined as

$$FoM = \frac{1}{2\pi \times r_{i,forward} \times c_{i,reverse}}, \tag{10.10}$$

where $r_{i,forward}$ is the resistance of the i-layer under forward bias and $c_{i,reverse}$ is the capacitance of the i-layer under reverse bias. We know from the previous discussion that $r_{i,forward}$ should be small to give low insertion loss in the ON state, and $c_{i,reverse}$ should be low to give good isolation in the OFF state, so a high value of FoM indicates a good diode for switching purposes. Typical values of $r_{i,forward}$ and $c_{i,reverse}$ vary considerably for practical diodes, and depend on the bias current and the intended application. For PIN diodes in a beam-lead package that are intended for switching at microwave frequencies, typical values are $r_{i,forward} = 4\,\Omega$ and $c_{i,reverse} = 0.03\,\text{pF}$, giving a FoM of 1326 GHz. An alternative term used to specify the performance of a switching diode is the SR defined as

$$SR = \frac{\rho_{i,reverse}}{\rho_{i,forward}}, \tag{10.11}$$

where $\rho_{i,reverse}$ and $\rho_{i,forward}$ represent the resistivities of the i-region under reverse and forward bias conditions, respectively.

Typical values of SR for a PIN diode are of the order of several thousand.

10.2.2 Field Effect Transistors

Whilst the PIN diode provides a simple, relatively cheap control element that is easily mounted in microstrip or coplanar lines, FETs offer a number of advantages when used as the switching element:

- Fast (sub-nanosecond) switching speed
- Very low DC power consumption
- Bidirectional performance
- Wide bandwidth performance

An FET can be implemented in a number of different structures, the common type at RF and microwave frequencies being the MESFET (Metal Semiconductor Field Effect Transistor), so-called because it incorporates metal–semiconductor junctions. The semiconductor normally used in MESFETs intended for RF and microwave applications is gallium arsenide (GaAs) because this material has high electron mobility, and so has the potential for fast operation. The basic structure of a GaAs MESFET is shown schematically in Figure 10.10.

Essential points relating to the construction and operation of a MESFET:

(i) A layer of lightly doped N-type GaAs is grown on a semi-insulating substrate. This lightly doped layer is the main transmission channel for carriers (electrons) through the device.

(ii) The MESFET has three metal–semiconductor junctions that provide contact pads for the source, gate, and drain. Metal–semiconductor junctions can either be ohmic or rectifying, depending on the relative work functions of the metal and the semiconductor. An in-depth discussion of the physics of metal–semiconductor junctions is outside the scope of the present text, and it suffices to say if the metal is aluminium and deposited on highly doped (N^+) GaAs, an ohmic contact is formed and if deposited on lightly doped GaAs, a rectifying junction, known as a Schottky junction is

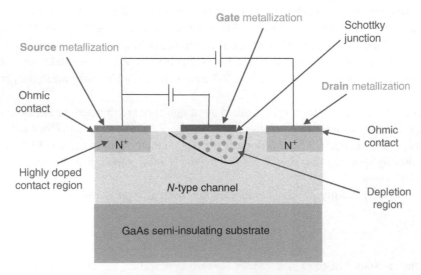

Figure 10.10 Construction of a GaAs MESFET.

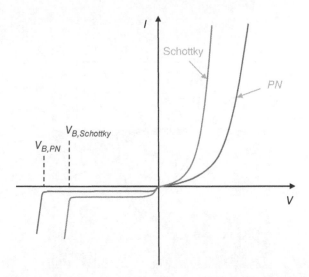

Figure 10.11   Comparison of I–V characteristics of Schottky and PN junctions.

formed. Thus, in the manufacture of a MESFET two N⁺ regions are provided beneath source and drain metallization. These provide ohmic contacts which exhibit a linear I–V relationship with low loss. The metallization for the gate is deposited directly on the lightly doped N-type GaAs and this forms a rectifying junction with similar properties to a PN junction. Like a PN junction the Schottky junction will have a voltage-controlled depletion layer when reverse biased. The main differences between a Schottky junction and a PN junction are that the Schottky junction has a higher leakage current when reverse biased, a lower built-in potential which means a lower turn-on voltage under forward bias, and a smaller avalanche breakdown voltage. A comparison of the I–V characteristics of Schottky and PN junctions is given in Figure 10.11; in Figure 10.11 the breakdown voltages of the PN and Schottky diodes are shown as $V_{B,PN}$ and $V_{B,Schottky}$, respectively.

(iii) The drain of a MESFET has a positive DC bias voltage with respect to the source, as shown in Figure 10.10. This causes electrons to flow through the N-type channel. If the gate has a negative bias with respect to the source a depletion region will exist in the channel beneath the gate. The presence of the depletion region will inhibit the flow of electrons from source to drain, and the gate voltage will act as an effective control voltage. Since current is flowing along the conducting channel from drain to source there will be a corresponding voltage drop along the channel, which causes the gate to become increasingly reverse biased moving from source to drain. This results in a widening of the depletion layer at the drain end of the gate, as shown in Figure 10.10. (Due to the asymmetry of the depletion region the inter-electrode space between the gate and the drain is often made greater than that between the gate and source.) As the gate bias voltage is made more negative, the thickness of the depletion layer will increase until it occupies the entire channel width, and blocks the flow of electrons from source to drain. The gate voltage at which this occurs is known as the pinch-off voltage, V_{po}. The depletion layer at pinch-off is depicted in Figure 10.12.

When used as a switch the gate voltage of the MESFET must be switched from a low-resistance forward bias condition to a reverse bias voltage greater than V_{po}.

(iv) Noise performance is an important issue when selecting a MESFET for RF or microwave applications. Fukui [1] developed an expression for the minimum noise factor of MESFET as

$$F_{min} = 1 + \left(\frac{5\pi C_{gs} f}{g_m} \right) \sqrt{g_m(R_g + R_s)}, \tag{10.12}$$

where
  f is the operating frequency
  R_g is the AC gate series resistance
  R_s is the source resistance
  g_m is the transconductance
  C_{gs} is the gate-source capacitance.

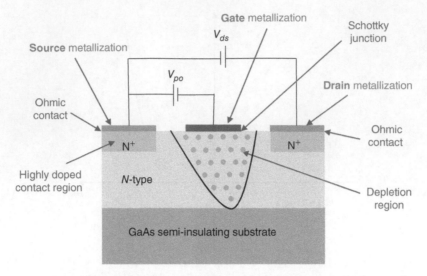

Figure 10.12 MESFET depletion layer at pinch-off.

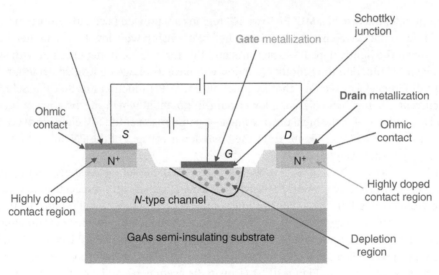

Figure 10.13 Recessed-gate MESFET.

The value of F_{min} can be reduced by reducing the source resistance, R_s. This can be achieved by using a recessed gate structure as shown in Figure 10.13. Most MESFETs used for high-frequency applications employ recessed gate technology. This type of structure is normally manufactured by first growing an N-type planar layer on the semi-insulating substrate and then etching the recess prior to depositing the metallization for the three contact pads.

Equation (10.12) shows that $(F_{min}-1)$ for the MESFET increases linearly with frequency. It is interesting to note that a corresponding expression for a bipolar junction transistor (BJT) is [1]

$$F_{min} - 1 \approx bf^2 \left(1 + \sqrt{1 + \frac{2}{bf^2}} \right),$$ (10.13)

where b is a factor that depends on the BJT parameters. This shows that $(F_{min}-1)$ for the BJT increases with the square of the frequency and so has a significantly worse noise performance compared with the MESFET at high frequencies.

When MESFETs are used for RF and microwave switching they may be connected in either a passive or active mode. Passive MESFET switches use the N-type channel simply as a transmission line between the source and drain, which is open or closed by applying a control voltage to the gate. Active MESFET switches use the transistor in a conventional amplifying configuration, with the RF input signal applied to the gate and the output RF signal taken from the drain. The active switch is open or closed by applying the appropriate DC bias voltage to the gate, and has the advantage of providing gain when switch is ON.

Passive MESFET switches are the direct equivalent of PIN diode switches, and may be series or parallel mounted as shown in Figure 10.14.

In order to find the insertion loss and isolation of MESFET switches it is necessary to establish which parameters of the MESFET equivalent circuit are most significant under forward and reverse gate bias. Figure 10.15 shows the small-signal equivalent circuit of a MESFET, which is widely used in the literature.

Significance of elements in the equivalent circuit:

R_{DS} represents the resistance of channel between drain and source
C_{DS} represents the fringing between drain and source
C_{GS} represents the gate-source capacitance
C_{GD} represents the gate-drain capacitance
R_i represents resistance of channel between gate and source
C_{DC} represents the capacitance of the dipole layer formed by the depletion region under the gate
R_G, R_D, R_S represent the resistances of the gate, drain, and source metallizations, plus bond wire/lead resistances
L_G, L_D, L_S represent the inductances of the connections to the gate, drain, and source pads

Also shown in the equivalent circuit in Figure 10.15 is a current generator where

g_m is the transconductance of the FET
V_G is the gate voltage

The equivalent circuit shown in Figure 10.15 does not include any capacitance to represent the effect of encapsulating the semiconductor structure. If necessary, the additional packaging capacitance can be added to the value of C_{DS}. For additional clarity, the components of the equivalent circuit are often shown drawn on an outline of the MESFET structure to indicate their physical origin, as in Figure 10.16.

Typical data for the equivalent circuit parameters of a low-noise GaAs MESFET are given in Table 10.1.

The data given in Table 10.1 show the relative magnitudes of the equivalent circuit parameters, and they can be used to predict the equivalent circuits of a MESFET switch in the ON and OFF conditions.

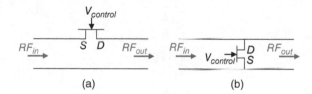

(a)   (b)

Figure 10.14   Mounting of passive MESFET switches: (a) series and (b) Parallel.

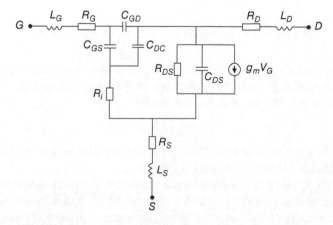

Figure 10.15   Small-signal equivalent circuit of a MESFET.

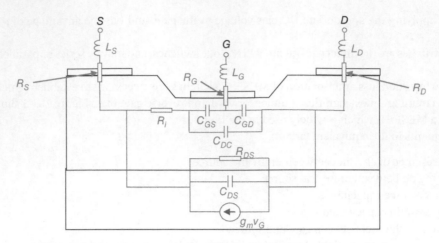

Figure 10.16 Location of components of a MESFET equivalent circuit.

Table 10.1 Typical equivalent circuit data for a GaAs MESFET.

$R_{DS} = 2.8\,\text{k}\Omega$	$R_D = 0.9\,\Omega$
$C_{DS} = 0.10\,\text{pF}$	$R_S = 0.9\,\Omega$
$C_{GS} = 0.25\,\text{pF}$	$L_G = 0.1\,\text{nH}$
$C_{GD} = 0.25\,\text{pF}$	$L_D = 0.1\,\text{nH}$
$R_i = 2.8\,\Omega$	$L_S = 0.1\,\text{nH}$
$C_{DC} = 0.02\,\text{pF}$	$g_m = 45\,\text{mS}$
$R_G = 1.4\,\Omega$	

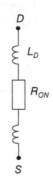

Figure 10.17 Equivalent circuit for MESFET in ON state.

In the ON state, there is a low-impedance path through the N-type channel between source and drain, and the MESFET can be modelled simply as a resistor, R_{ON}, with the appropriate parasitics due to contact/lead inductance and packaging capacitance, as shown in Figure 10.17. The value of R_{ON} is given by

$$R_{ON} = R_S + R_d + R_D, \tag{10.14}$$

where R_d is the resistance of the channel when the MESFET is forward-biased into the low-impedance state. Typically, $R_d \cong 1\,\Omega$, making $R_{ON} \cong 2.8\,\Omega$ for the MESFET specified in Table 10.1.

The situation is slightly more complicated in the OFF state where the N-type channel presents a high impedance, and we must consider the components shunting this channel, including the RF impedance of the gate. However, in practical switching circuits, external connections to the gate are normally designed to present a nominal RF open circuit, and so the effect of the gate impedance can be ignored. The resulting equivalent circuit for the OFF state is shown in Figure 10.18a.

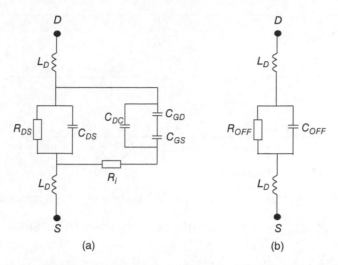

Figure 10.18 Equivalent circuits for MESFET in OFF state: (a) equivalent circuit in OFF state and (b) simplified equivalent circuit.

This circuit can be reduced to the simplified form shown in Figure 10.18b, where [2]

$$R_{OFF} = \frac{2R_{DS}}{2 + R_{DS}\omega^2 C_{GD}^2 R_G} \tag{10.15}$$

and

$$C_{OFF} = C_{DS} + \frac{C_{GS}}{2}. \tag{10.16}$$

Example 10.2 Suppose that the MESFET specified by the data in Table 10.1 was connected as a series switch in a 50 Ω microstrip line. Determine the insertion loss and isolation at 5 GHz, assuming that $R_d = 1\,\Omega$, and neglecting parasitic effects associated with the switch mounting.

Solution

Using Eq. (10.14):

$$R_{ON} = (0.9 + 1.0 + 0.9)\,\Omega = 2.8\,\Omega.$$

Calculating reactances of drain and source inductances:

$$X_L(\text{drain}) = X_L(\text{source}) = \omega L = 2 \times \pi \times 5 \times 10^9 \times 0.1 \times 10^{-9}\,\Omega = 3.14\,\Omega.$$

The impedance of the MESFET in the ON state is then

$$(Z_{FET})_{ON} = (2.8 + j6.28)\,\Omega.$$

The insertion loss is then given by Eq. (10.7)

$$|IL|_{dB} = 20 \log \left| \frac{2.8 + j6.28 + 100}{100} \right|\,dB$$

$$= 20 \log(1.03)\,dB$$

$$= 0.26\,dB.$$

Using Eqs. (10.15) and (10.16) to find impedance of MESFET in the OFF state:

$$R_{OFF} = \frac{2 \times 2.8 \times 10^3}{2 + 2.8 \times 10^3 \times (2\pi \times 5 \times 10^9)^2 \times (0.25 \times 10^{-12})^2 \times 1.4}\,\Omega$$

$$= 2.50 \times 10^3\,\Omega;$$

$$C_{OFF} = \left(0.1 + \frac{0.25}{2} \right)\,pF = 0.225\,pF.$$

(Continued)

(Continued)

Then,

$$\frac{1}{Z_{OFF}} = \frac{1}{R_{OFF}} + j\omega C_{OFF} \quad \Rightarrow \quad Z_{OFF} = \frac{R_{OFF}(1 - j\omega C_{OFF} R_{OFF})}{1 + (\omega C_{OFF} R_{OFF})^2}.$$

Substituting data,

$$Z_{OFF} = \frac{2.50 \times 10^3 \times (1 - j2\pi \times 5 \times 10^9 \times 0.225 \times 10^{-12} \times 2.50 \times 10^3)}{1 + (2\pi \times 5 \times 10^9 \times 0.225 \times 10^{-12} \times 2.50 \times 10^3)^2} \Omega$$

$$= (7.98 - j141.03)\,\Omega.$$

The total impedance of the MESFET in the OFF state (including effect of contact inductances) is then

$$(Z_{FET})_{OFF} = (7.98 - j141.03 + j6.28)\,\Omega = (7.98 - j134.75)\,\Omega.$$

Using Eq. (10.7):

$$\text{Isolation} = 20 \log \left| \frac{7.98 - j134.75 + 100}{100} \right| \text{ dB}$$

$$= 4.74\,\text{dB}.$$

The results from Example 10.2 show that the switched MESFET provides low insertion loss in the ON state, but poor isolation in the OFF state. The isolation can be improved by connecting an inductor between drain and source to provide parallel resonance with C_{OFF}. This improves the isolation by making the value of $(Z_{FET})_{OFF}$ close to R_{OFF}, which is normally of the order of 2–3 kΩ for typical MESFETS. Whilst the use of a shunt inductor can significantly improve the isolation (see Example 10.3), it tends to reduce the bandwidth, thus detracting from one of the principal advantages of MESFET switches, namely wideband performance.

Example 10.3

1. Determine the value of inductance needed between drain and source to resonate with C_{OFF} in Example 10.2.
2. Find the resulting improvement in isolation due to using the inductor.
3. Determine if the presence of the inductor has any detrimental effect on the insertion loss in the ON state.

Solution

(i). Resonance condition:

$$f = \frac{1}{2\pi\sqrt{LC_{OFF}}},$$

i.e

$$5 \times 10^9 = \frac{1}{2\pi\sqrt{L \times 0.225 \times 10^{-12}}} \quad \Rightarrow \quad L = 4.50\,\text{nH}.$$

(ii). With the inductor in place we have,

$$(Z_{FET})_{OFF} = R_{OFF} + j6.28 = (2500 + j6.28)\,\Omega.$$

Substituting in Eq. (10.7) gives

$$\text{Isolation} = 20 \log \left| \frac{2500 + j6.28 + 100}{100} \right| \text{ dB}$$

$$= 28.30\,\text{dB}.$$

So the isolation has improved by (28.30–4.74) dB = 23.56 dB.

(iii). To find the effect of the inductor on the insertion loss we must consider the impedance of the parallel combination of $L = 4.50\,\text{nH}$ and $R_d = 1\,\Omega$.

At 5 GHz, the reactance of the inductor will be

$$\omega L = (2\pi \times 5 \times 10^9 \times 4.50 \times 10^{-9})\,\Omega = 141.39\,\Omega.$$

Since $|\omega L| \gg R_d$ the impedance of the parallel combination will be approximately equal to R_d, which means that the presence of the inductor will not have any significant effect on the insertion loss.

10.2.3 Microelectromechanical Systems

MEMS switches provide an alternative to semiconductor switches such as PIN diodes and FETS. These switches are fabricated using micromachining techniques. They provide the same basic switching function as PIN diodes and FETS, but with the principal advantage of lower insertion loss, particularly at higher microwave frequencies in the range 8–100 GHz. MEMS switches are available with similar packaging to PIN diodes for use in hybrid circuits, and they can also be easily implemented in monolithic form for MMIC devices. There are two common configurations of MEMS switch, namely the cantilever switch and the capacitive switch.

A schematic view of a cantilever MEMS SPST switch is shown in Figure 10.19.

The figure shows a metallic cantilever beam anchored to a metallic pad on the left-hand end of the substrate. The other end of the beam forms a contact which is suspended in air above a metallic pad on the right-hand end of the substrate. RF connections are made to the two metallic pads on the substrate, and RF signals flow along the cantilever beam when it is closed. In the condition shown in Figure 10.19, the switch is open, and there is nominally no transmission across the gap between the two contacts. If a DC voltage is applied between the actuator pad and the beam an electrostatic field is created between the two, and the resulting electrostatic attraction deflects the cantilever beam downwards and closes the switch. When the DC voltage is removed the beam acts like a spring and returns to its original open position. In hybrid packages the contact pads at the end to the substrate are connected to external leads, and the device can be conveniently mounted across a gap in a microstrip line to act as a series ON–OFF switch.

The essential stages in the production of a MEMS cantilever switch are illustrated in Figure 10.20.

As with the PIN diode, the quality of a cantilever switch can be represented in terms of the OFF-state capacitance, C_t, (i.e. capacitance when the cantilever beam is up) and the ON-state resistance, R_c, (i.e. series resistance with the contacts closed). These two parameters can be expressed as [3].

$$C_t = C_p + \frac{\varepsilon_o A_c}{g_c} \tag{10.17}$$

and

$$R_c = \frac{\rho_c}{A_c}, \tag{10.18}$$

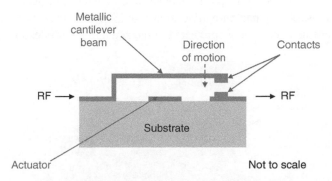

Figure 10.19 Cantilever MEMS switch.

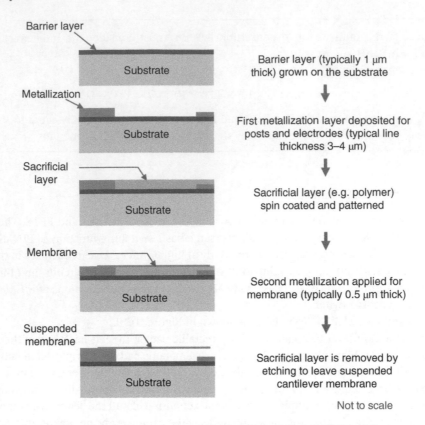

Figure 10.20 Stages in the production of a cantilever MEMS switch.

where

A_c = common contact area of the contact pair
g_c = open gap separation between the contacts
C_p = parasitic capacitance between the switch and RF line
ρ_c = surface resistivity of the metal contacts
ε_o = permittivity of free space

The capacitive MEMS switch provides an alternative to the cantilever switch, and generally gives better performance at high microwave frequencies. The principal features of a capacitive MEMS switch are shown in Figure 10.21.

In the capacitive MEMS switch, a thin metallic membrane is suspended between two metal posts, with an actuator electrode beneath the membrane. When a DC voltage is applied between the membrane and the actuator, the electrostatic force of attraction deflects the membrane until it makes contact with the actuator, and closes the switch. When the DC voltage is removed the membrane recovers to its original position, opening the switch. The geometry of this type of switch makes it very suitable for mounting across a coplanar waveguide, as shown in Figure 10.22. The metal posts are connected to the outer, earthed lines, and the central signal line is connected to the actuator electrode. So with the switch open, there is a

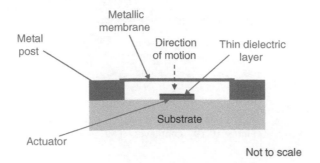

Figure 10.21 Capacitive MEMS switch.

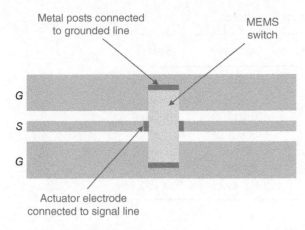

Figure 10.22 Capacitive MEMS switch mounted across a CPW line.

low-loss transmission path through the switch, but with the switch closed the deflected membrane contacts the signal line, and effectively connects it to earth. Thus the switch is operating in a reflective mode, and reflects the signal when closed. It is important to note that the upper surface of the actuator is covered with a thin layer of dielectric. This prevents the membrane making ohmic contact with the actuator and hence shorting the DC supply, which would destroy the electrostatic forces and open the switch. Since the dielectric layer is very thin there will be strong capacitive coupling between the membrane and the actuator in the closed position, thus forming an effective RF short circuit.

In both the cantilever and capacitive forms of the MEMS switch the through path in the ON state is effectively a continuous metal transmission line, and this contributes significantly to the major advantage of MEMS switches, namely very low loss in the ON state. Values of the up-state (OFF state) capacitance for MEMS switches can be as low as low as 1–3 fF, which provide very high isolation. Also, low contact resistances significantly less than 1 Ω are possible in the down-state (ON state), which means low insertion loss. The performance of MEMS switches is usually specified in terms of the FoM. However, using the same definition as for a PIN diode, Eq. (10.10), leads to an unreasonably large figure due to the very small value of capacitance in the OFF, of the order of a few fF, which is typical for MEMS structures. Hoivik and Ramadoss writing in [3] suggest that a more practical FoM for MEMS switches is

$$FoM = BW \times \frac{\text{Isolation}_{\text{OFFstate}}}{\text{Insertion loss}_{\text{ONstate}}}. \tag{10.19}$$

The physical structure of MEMS switches with a continuous metallic membrane will exhibit some series inductance. This inductance is often much less than 100 pH, and so is normally ignored in the analysis of MEMS switch performance. However, significant levels of series inductance can affect the match of the device, and simulated results in [3] show that the $(S_{11})_{dB}$ values for a 1 Ω contact resistance MEMS switch deteriorate by around 7 dB when the inductance increases from 30 to 80 pH.

Example 10.4 The following data apply to a MEMS switch: Capacitance in the OFF state = 3 fF; Series resistance in the ON state = 0.85 Ω. Determine the insertion loss and isolation at 5 GHz of the MEMS switch if it was series mounted across a suitable gap in a 50 Ω planar transmission line. (Neglect the effect of the gap reactances, and also the switch inductance.)

Solution
Using Eq. (10.7):

$$\text{Insertion loss} = 20 \log \left| \frac{0.85 + 100}{100} \right| \text{dB} = 0.074 \, \text{dB},$$

$$X_c = \frac{1}{2\pi \times 5 \times 10^9 \times 3 \times 10^{-15}} \, \Omega = 10608.95 \, \Omega,$$

$$\text{Isolation} = 20 \log \left| \frac{100 - j10608.95}{100} \right| \text{dB} = 40.51 \, \text{dB}.$$

Table 10.2 Typical performance data for RF switches.

	PIN diodes	MESFETs	MEMS
Insertion loss (dB)	0.3–1.0	0.4–2	0.05–2
Isolation (dB)	>25	>25	>40
Voltage (V)	5–10	5–30	20–80
Current (mA)	3–20	~0	0
Switching time (s)	1–100 ns	1–20 ns	1–40 μs
Frequency range	→40 GHz	→40 GHz	→100 GHz
Power handling (W)	→200	→10	→1

A comparative summary of typical performance data for PIN diode, MESFET, and MEMS switches is given in Table 10.2.

It can be seen from Table 10.2 that although MEMS switches offer significant advantages over semiconductor switches in terms of lower insertion loss and better isolation, there is a penalty in terms of lower switching speed. The choice of switch depends very much on the application. For example, where switches are used in multi-bit digital phase shifters, the transmission path through the phase shifter may involve a significant number of switches in series and so lower insertion loss is of primary importance. But for systems that require fast switching, such as the circuitry for some modern radar systems, MEMS switching speeds of the order of micro-seconds are far too slow.

The increasing popularity of MEMS switches, primarily for use in digital phase shifters, is reflected in the wealth of recent technical literature on this subject. Particularly useful discussions on RF MEMS performance can be found in references [4–10].

10.2.4 Inline Phase Change Switch Devices

Inline phase change switch (IPCS) devices exploit the changes in resistance that occur when certain phase-change materials are subjected to localized heating. Phase-change materials are those which possess two phases, namely a crystalline phase and an amorphous phase, and the material can be switched between the two phases through the application of heat. The two phases have very different electrical properties, particularly in terms of resistance. The phase-change material that has been the subject of a number of recent investigations for RF switch applications is germanium telluride (GeTe). When this material is packaged with an integrated heater line it forms a simple, small, and highly effective switch.

Particular advantages of IPCS technology are:

- small size, which means low RF parasitics
- very high FoM
- low insertion loss
- high isolation
- excellent linearity
- relatively good switching speed
- low switching voltage of MEMS switches

In a recent work, Borodulin and colleagues [10] reported measured data at 10 GHz for an IPCS SPDT switch with an insertion loss ~0.55 dB and an isolation ~53 dB. With switching speeds typically ~1 μs, this loss performance makes IPCS technology very attractive for a range of RF/microwave switch applications. One such application is reconfigurable filters; Wang and colleagues [11] provide a useful overview of this application, which includes an informative description of the fabrication process for the GeTe RF switch.

10.3 Digital Phase Shifters

10.3.1 Switched-Path Phase Shifter

This is the simplest type of digital phase shifter to visualize, and it is shown in Figure 10.23.

The circuit is made up of two transmission paths of unequal length, with four series SPST switches, D1 through D4, to route the signal. With D1 and D2 closed, and D3 and D4 open, the signal is routed through the transmission line of length, L_1, with a transmission phase given by

$$\phi_1 = \beta L_1, \tag{10.20}$$

where β is the phase propagation constant for the line. Conversely, with D1 and D2 open, and D3 and D4 closed, the signal is routed through the L_2 line with a transmission phase given by

$$\phi_2 = \beta L_2. \tag{10.21}$$

For the purpose of illustration the switches shown in Figure 10.23 are diodes, but they could equally well be MESFET or MEMS devices. The type of phase shifter shown in Figure 10.23 can be implemented easily in a number of transmission line formats, such as a coaxial cable, a microstrip, or a coplanar waveguide. As an example, a typical microstrip implementation of the basic switched line phase shifter is shown in Figure 10.24. Also included in this figure are the necessary DC bias connections, which were discussed in an earlier chapter in relation to biasing transistors in small-signal amplifiers.

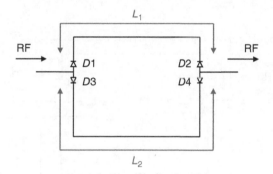

Figure 10.23 Basic switched-line phase shifter.

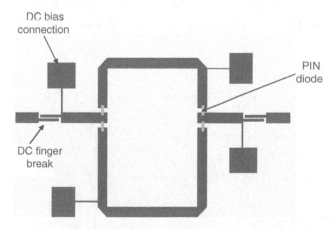

Figure 10.24 Switched-line phase shifter in microstrip.

Whilst the concept of the switched line phase shifter is very straightforward, there can be problems due to the limited isolation of the switches in the OFF state. With the switches biased to select one transmission path, it is possible for some signal to be coupled across the OFF switch into the nominally isolated path. Since the isolated path is effectively open-circuited at both ends there will be frequencies at which it will exhibit resonance. Therefore, if signals leak across the OFF state switches, the isolated path behaves as a resonant circuit that is loosely coupled to the through (ON) path. Consequently, small resonant peaks appear in the transmission responses of the ON path. This can be avoided by using SPDT switches that terminate the isolated path in matched loads, rather than leaving the ends open. This arrangement is shown schematically in Figure 10.25, which also shows the two different line lengths L_1 and L_2, giving the two different phase states.

The arrangement of four PIN diodes at each end of the phase shifter shown in Figure 10.25 enables one of the two possible phase state (paths) to be selected, whilst terminating each end of the unused path in a matched impedance. This is illustrated in Figure 10.26 where the polarity of the bias is such as to select the phase change through the L_2 path; in this figure the ON diodes have been replaced by short-circuits, and the OFF diodes by open-circuits.

The digital phase shifters shown in Figures 10.23 through 10.26 may be regarded as single-bit phase shifters since they provide two phase states. By cascading a number of these sections, a multi-bit digital phase shifter can be produced. The layout of a three-bit microstrip switched-line phase shifter is shown in Figure 10.27, with the lengths of the various sections denoted by L_n. For clarity, the DC bias connections and finger breaks have been omitted. With this type of arrangement it is necessary to have DC breaks between each section so that the switches for each section can be biased independently.

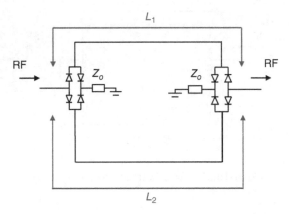

Figure 10.25 Switched-line phase shifter with terminated OFF line.

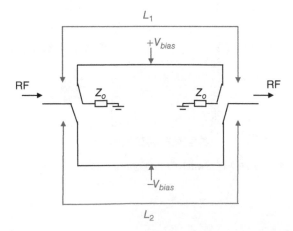

Figure 10.26 Switched-line phase shifter with terminated OFF line, and through path L_2 selected.

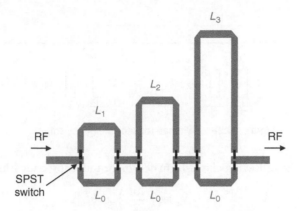

Figure 10.27   Microstrip three-bit phase shifter.

Table 10.3   Transmission phases through a three-bit phase shifter.

Path	Transmission phase (°)
$L_0 L_0 L_0$	0
$L_1 L_0 L_0$	10
$L_0 L_2 L_0$	20
$L_1 L_2 L_0$	30
$L_0 L_0 L_3$	40
$L_1 L_0 L_3$	50
$L_1 L_2 L_3$	60

A total of 12 SPST switches are used to route the signal through the circuit (alternatively, these could be replaced with 6 SPDT switches). Suppose that the electrical lengths of the various sections are: $L_0 = 360°$; $L_1 = 360° + x°$; $L_1 = 360° + y°$; $L_1 = 360° + z°$. By choosing appropriate values for x, y, and z we can produce any desired pattern of phase shifts. For example, if $x = 10$, $y = 20$, and $z = 40$, we can obtain transmission phase intervals of 10° by activating the switches to route signal through the paths shown in Table 10.3.

One of the problems with multi-bit switched line phase shifters is that a comparatively large number of switches are required, and the loss of the phase shifter is the sum the insertion losses of the individual switches. Since the RF signal passes through the same number of switches in all states of the phase shifter, this may not be a serious problem if the individual switch losses are small. A more significant problem is that of mismatch at the input to the phase shifter. Since the electrical distances between the switches change with phase selection, the reflections from individual switches, when summed at the input to the phase shifter, will be a function of the phase state (path) selected. Thus, the VSWR at the phase shifter input will change with phase selection. The problems identified above can be minimized by using switches with low insertion loss and high isolation. Consequently, MEMS switches are becoming increasingly popular for RF and microwave phase shifter applications.

10.3.2   Loaded-Line Phase Shifter

The loaded-line phase shifter employs two reactances connected in shunt with a transmission line of characteristic impedance Z_o. The transmission phase between the two reactances is θ, and they are separated by a transmission line of characteristic impedance Z_T. The basic circuit arrangement is shown in Figure 10.28, which also includes two switches to enable the circuit to be switched between two transmission phase states. The transmission phase in each state is a function of b_i, θ, and Z_T, where b_i represents the relevant normalized shunt reactance.

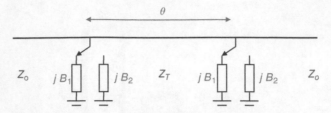

Figure 10.28 Switched loaded-line phase shifter.

Bahl and Gupta [12] developed the following closed-form design formulae for loaded-line phase shifters

$$Y_T = Y_o \sec\left(\frac{\Delta\phi}{2}\right)\sin\theta, \tag{10.22}$$

$$B_1 = Y_o\left[\sec\left(\frac{\Delta\phi}{2}\right)\cos\theta + \tan\left(\frac{\Delta\phi}{2}\right)\right], \tag{10.23}$$

$$B_2 = Y_o\left[\sec\left(\frac{\Delta\phi}{2}\right)\cos\theta - \tan\left(\frac{\Delta\phi}{2}\right)\right], \tag{10.24}$$

where $\Delta\phi$ is the difference in transmission phase between the two states, and the other symbols have their usual meanings. Full details of the circuit analysis that underpins Eqs. (10.22) through (10.24) are given by Atwater [13].

The spacing between the two reactances is normally made equal to $\lambda_T/4$, where λ_T is the wavelength on the transmission line, so that reflections from the two reactances will cancel at the input. With this separation the design Eqs. (10.22)–(10.24) become

$$Y_T = Y_o \sec\left(\frac{\Delta\phi}{2}\right), \tag{10.25}$$

$$B_1 = Y_o \tan\left(\frac{\Delta\phi}{2}\right), \tag{10.26}$$

$$B_2 = -Y_o \tan\left(\frac{\Delta\phi}{2}\right). \tag{10.27}$$

From Eqs. (10.26) and (10.27), we see that the magnitudes of the reactances in the two states are equal, but of opposite sign, i.e.

$$jB_1 = -jB_2. \tag{10.28}$$

This result is supported by the work of Garver [14], who, in one of the classic papers on phase shifters, used the results of computer simulation to show that the widest bandwidth with this type of phase shifter occurs when the magnitudes of the reactances in each state are equal.

The equations presented are based on ideal switches, and take no account of the impedances of the switch in the ON and OFF states. These impedances, obtained from the appropriate switch equivalent circuit, will modify the values of B_1 and B_2 and provide a more precise design. This has been addressed in some detail by Koul and Bhat [15], who provide a comprehensive discussion of the effects of switch resistance and reactance on the performance of loaded-line phase shifters.

The switched loaded-line phase shifter shown in Figure 10.28 can also be realized in microstrip, where the required shunt reactances are provided by stubs as shown in Figure 10.29. A switch at the remote end of each stub allows it to be terminated either in an open-circuit, or a short-circuit.

The spacing between the stubs is $\lambda_s/4$ to create an electrical separation of 90° and so minimize reflections at the input. The stub lengths are made $\lambda_{ss}/8$ (λ_{ss} is the wavelength on the stubs) to satisfy Eq. (10.28). Note that the stubs may not have the same characteristic impedance of the main line and so the substrate wavelength on the stubs may not be the same as that on the main line, i.e. $\lambda_s \neq \lambda_{ss}$.

The input impedance of an open-circuited stub is given by Eq. (1.34) of Chapter 1 as

$$(Z_{in})_{O/C} = -jZ_{os}\cot\beta l,$$

where Z_{os} is the characteristic impedance of the stub. It follows that the input admittance is

$$(Y_{in})_{O/C} = jY_{os}\tan\beta l.$$

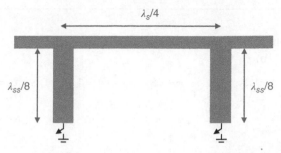

Figure 10.29 Microstrip loaded-line phase shifter.

If the stub length, l, is equal to $\lambda/8$ then

$$\beta l = \frac{\pi}{4},$$

and the input admittance becomes

$$(Y_{in})_{O/C} = -jY_{os}. \tag{10.29}$$

Similarly, using Eq. (1.33) of Chapter 1, the input admittance of a $\lambda/8$ short-circuited stub is

$$(Y_{in})_{S/C} = jY_{os}. \tag{10.30}$$

Thus, we see from Eqs. (10.29) and (10.30) that switching the stub termination between an open-circuit and a short-circuit satisfies the requirements of Eq. (10.28), and by selecting the correct characteristic impedance for the stubs we can obtain a specific phase change. The design process for a microstrip loaded-line phase shifter is illustrated in Example 10.5.

Example 10.5   Design a single bit microstrip loaded-line phase shifter to provide a phase change of 60° at 10 GHz. The phase shifter is to have a characteristic impedance of 50 Ω, and is to be fabricated on a substrate that has a relative permittivity of 9.8 and a thickness of 0.8 mm.

Solution
(i). To find the impedance of the transmission line between the two stubs.
   Using Eq. (10.25):

$$Y_T = Y_0 \sec\left(\frac{\Delta\phi}{2}\right)$$

$$= \frac{1}{50}\sec\left(\frac{60°}{2}\right)\text{S} = \frac{1}{50}\times\frac{1}{0.866}\text{S} = \frac{1}{43.30}\text{S}$$

$$Z_T = 43.30\,\Omega.$$

(ii). To find the required impedance of the stubs.
   Combining Eqs. (10.26) and (10.30):

$$jY_{os} = jB_1,$$

$$Y_{os} = Y_o \tan\left(\frac{\Delta\phi}{2}\right),$$

$$Y_{os} = \frac{1}{50}\tan\left(\frac{60°}{2}\right)\text{S},$$

$$Y_{os} = 11.55\,\text{mS},$$

$$Z_{os} = 86.58\,\Omega.$$

(Continued)

(Continued)

(iii). Using microstrip design graphs (or CAD) to find the microstrip dimensions.

Inter-stub transmission line

$$Z_T = 43.30\,\Omega \quad \Rightarrow \quad \frac{w}{h} = 1.25 \quad \Rightarrow \quad w = 1.25 \times 0.8\,\text{mm} = 1.00\,\text{mm},$$

$$\frac{w}{h} = 1.25 \quad \Rightarrow \quad \varepsilon_{r,eff} = 6.78 \quad \Rightarrow \quad \lambda_s = \frac{\lambda_o}{\sqrt{6.78}},$$

$$\lambda_s = \frac{30}{\sqrt{6.78}}\,\text{mm} = 11.52\,\text{mm} \quad l_T = \frac{\lambda_s}{4} = \frac{11.52}{4}\,\text{mm} = 2.88\,\text{mm}.$$

Stub dimensions

$$Z_{os} = 86.58\,\Omega \quad \Rightarrow \quad \frac{w_s}{h} = 0.18 \quad \Rightarrow \quad w_s = 0.18 \times 0.8\,\text{mm} = 144\,\mu\text{m},$$

$$\frac{w_s}{h} = 0.18 \quad \Rightarrow \quad \varepsilon_{r,eff} = 6.08 \quad \Rightarrow \quad \lambda_{ss} = \frac{\lambda_o}{\sqrt{6.08}},$$

$$\lambda_{ss} = \frac{30}{\sqrt{6.08}}\,\text{mm} = 12.17\,\text{mm} \quad l_{STUB} = \frac{\lambda_{ss}}{8} = \frac{12.17}{8}\,\text{mm} = 1.52\,\text{mm}.$$

50 Ω feed lines

$$Z_{os} = 50\,\Omega \quad \Rightarrow \quad \frac{w_s}{h} = 0.9 \quad \Rightarrow \quad w_s = 0.9 \times 0.8\,\text{mm} = 0.72\,\text{mm}.$$

Final design:

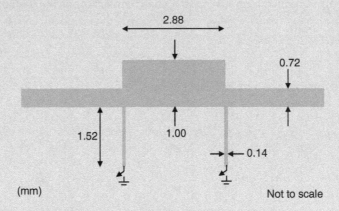

Figure 10.30 Completed design for Example 10.5.

Comment: In this design we have only considered an ideal arrangement. To obtain a precise 60° phase change the microstrip design would need to be refined to include modifications to represent the effects of discontinuities (open-ends + T-junctions + step widths), and the shunt reactances modified to include the effects of the switch impedances in the ON and OFF states. Note also that for the sake of clarity biasing circuits for the two switches have been omitted from Figure 10.30.

Whilst the microstrip switched loaded-line circuit provides a relatively simple arrangement for digital phase shifting, it is somewhat limited in the range of phase shifts that can be achieved in practice. Once the spacing of the stubs has been set to $\lambda_s/4$ and the stub lengths to $\lambda_{ss}/8$, there are only two design variables, namely the characteristic impedances of the inter-stub transmission line and those of the stubs themselves. We know from Chapter 2 that the practical range of characteristic impedance for a microstrip line is around 25–90 Ω, to avoid inconvenient line widths, and this means that the microstrip loaded-line phase shifter can normally only provide phase changes in the range 0–90°. Where larger phase changes are needed, the loaded line phase shifter is usually cascaded with an appropriate switched-line circuit.

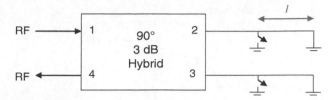

Figure 10.31   Schematic of switched reflection phase shifter.

10.3.3   Reflection-Type Phase Shifter

The most common type of switched reflection phase shifter employs a 90° 3 dB hybrid with shunt-mounted switches connected to two of the output ports, as shown schematically in Figure 10.31.

A signal applied to port 1 of the hybrid will split and emerge at ports 2 and 3 with a 90° phase difference. If the two switches are open, the signals will be reflected from the short circuits at the end of the transmission lines connected to ports 2 and 3; these reflected signals will recombine in the hybrid and emerge at port 4. In this condition the transmission phase, ϕ, will depend on the lengths of the transmission lines connected to ports 2 and 3. When the switches are closed, they will reflect the signals at earlier positions in the transmission lines and therefore decrease the transmission phase. If the transmission phase with the switches closed is ϕ_o, then with the switches open the transmission phase, ϕ, will be given by

$$\phi = \phi_o + 2\beta l, \tag{10.31}$$

where β is the phase propagation constant on the transmission lines connected to ports 2 and 3. If we wish to design a phase shifter to give a specific phase change, $\Delta\phi$, then we need to made

$$l = \frac{\Delta\phi}{2\beta}. \tag{10.32}$$

By using several switches, connected at different distances from the fixed short-circuit, a digital phase shifter can be realized. The branch line coupler (discussed in Chapter 2) is frequently used as the four-port hybrid, although it suffers from a slight disadvantage in that the bandwidth is limited to around 5–10%. A backward-wave hybrid coupler can also be used as an alternative to the branch line coupler. The backward wave coupler can be formed from two edge-coupled microstrip lines, as discussed in Chapter 2 in relation to directional couplers. Figure 10.32 shows the schematic arrangement of a reflection phase shifter using a backward wave planar structure.

As discussed in Chapter 2, the length of the coupled region is $\lambda_{av}/4$, where λ_{av} is the average of the odd and even mode wavelengths. A signal applied to port 1 will split and emerge at ports 2 and 4 with a 90° phase difference. These two signals will be reflected and recombine at port 3. When the switches are closed the transmission phase between ports 1 and 3 will be reduced by $2\beta l$. The advantage of the backward-wave coupler over other 3 dB couplers is that it provides a broad bandwidth of operation, because only the 3 dB power split is frequency dependent, and the other parameters are theoretically independent of frequency.

The reflection phase shifters described above require two switches to set each phase state. A circuit that requires only one switch to change the phase state makes use of the properties of a circulator (discussed in Chapter 6). The arrangement is shown in Figure 10.33.

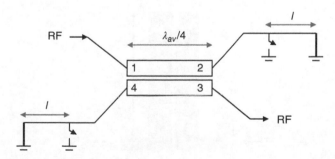

Figure 10.32   Reflection phase shifter using a 3 dB backward-wave coupler.

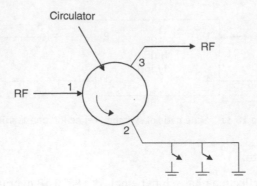

Figure 10.33 Reflective phase shifter using a circulator.

In this arrangement an RF signal applied to port 1 of the circulator will emerge at port 2, and will then be reflected either from the short-circuit terminating the line to port 2, or from any switch which is in the closed position on the line connected to port 2. The reflected signal enters the circulator at port 2, and then emerges at post 3. The transmission path through the circulator is from port 1 to port 3, with a transmission dependent on which of the switches is in the closed position. This type of digital phase shifter has the advantage of requiring only one switch for each phase change. However, this advantage is normally outweighed by the additional complexity of the circulator which requires a DC magnetic field. Also, the circuit tends to be narrowband, as the isolation between ports 1 and 3 of the circulator is frequency dependent. Consequently, the use of digital phase shifters employing circulators tends to be limited to high-power applications, with the circulator and transmission line elements formed in a metal waveguide, with high-power PIN diodes forming shunt switches.

10.3.4 Schiffman 90° Phase Shifter

The Schiffman phase shifter is named after J.B. Schiffman [16] who first observed that switching between a section of coupled transmission line with suitable terminations and a uniform through line produced a nearly constant phase change of 90° over a wide bandwidth.

The coupled line section used by Schiffman is shown in microstrip form in Figure 10.34. This behaves as an all-pass network with a transmission phase shift between ports 1 and 2 which is a function of the coupled length, θ.

When the main line is correctly terminated in a matched impedance Z_0, the input impedance at port 1 is

$$Z_0 = \sqrt{Z_{0e}Z_{0o}}, \tag{10.33}$$

with a transmission phase through the coupled section given by

$$\phi = \cos^{-1}\left(\frac{Z_{0e}/Z_{0o} - \tan^2\theta}{Z_{0e}/Z_{0o} + \tan^2\theta}\right), \tag{10.34}$$

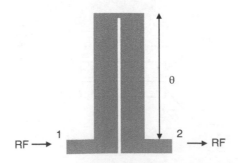

Figure 10.34 Microstrip coupled-line phase shifter.

where Z_{oe} and Z_{oo} are the even and odd mode impedances of the coupled section, as discussed in Chapter 2. The development of Eqs. (10.33) and (10.34) assumes that the velocity of propagation of the even and odd modes is the same. In Schiffman's original work, stripline was used, where these velocities are indeed the same. However, this is not the case regarding microstrip, as was pointed out in Chapter 2, and the common approach is to average the velocities of the two modes for calculation. So, in calculating the transmission phase through the coupled section the value of θ is found from

$$\theta = \beta_{av} l = \frac{2\pi}{\lambda_{av}} l, \tag{10.35}$$

where β_{av} and λ_{av} are the averaged values of the phase propagation constants and the wavelengths for the two modes, and

$$\lambda_{av} = \frac{v_{odd} + v_{even}}{2f}, \tag{10.36}$$

where v_{odd} and v_{even} are the velocities of the odd and even modes, respectively. Free and Aitchison [17] performed a more exact analysis of the coupled line phase shifter in terms of the individual velocities of the two modes and found that the transmission phase could be accurately represented by

$$\phi = \frac{\pi}{2} + \tan^{-1} \left(\frac{Z_0(Z_{oe} \cot \theta_e - Z_{oo} \tan \theta_o)}{2Z_{oe}Z_{oo} \tan \theta_o \cot \theta_e} \right), \tag{10.37}$$

where the symbols have their usual meanings.

A comparison between measured and theoretical data in [17] showed excellent agreement at *X*-band when Eq. (10.37) was used to predict the transmission phases through coupled microstrip lines of varying length.

The expression given in Eq. (10.33) for the input impedance is also an approximation if the two modes do not have the same velocity. A more exact expression for the input impedance of the coupled lines is [17]

$$Z_{in} = \frac{2Z_{oe}Z_{oo} \cot \theta_e \tan \theta_o - jZ_0(Z_{oe} \cot \theta_e - Z_{oo} \tan \theta_o)}{2Z_0 - j(Z_{oe} \cot \theta_e - Z_{oo} \tan \theta_o)}, \tag{10.38}$$

where Z_0 is the impedance terminating the coupled section.

Example 10.6   The following data apply to a 2.5 mm long microstrip coupled line having the geometry shown in Figure 10.32:

Odd mode: $Z_{oo} = 35.7\,\Omega$ $\lambda_{s,odd} = \dfrac{141}{f}$ mm, where f is in GHz,

Even move: $Z_{oe} = 77.4\,\Omega$ $\lambda_{s,even} = \dfrac{119}{f}$ mm, where f is in GHz.

Calculate the transmission phases at 10 GHz using Eqs. (10.34) and (10.37), and hence find the percentage error in transmission phase due to using the approximate expression.

Solution

$$\lambda_{s,odd} = \frac{141}{10} \text{ mm} = 14.1 \text{ mm},$$

$$\lambda_{s,even} = \frac{119}{10} \text{ mm} = 11.9 \text{ mm},$$

$$\theta_o = \beta_o l = \frac{2\pi}{\lambda_{s,odd}} l = \frac{2\pi}{14.1} \times 2.5 \text{ rad} = 1.114 \text{ rad} \equiv 63.82°,$$

$$\theta_e = \beta_e l = \frac{2\pi}{\lambda_{s,even}} l = \frac{2\pi}{11.9} \times 2.5 \text{ rad} = 1.320 \text{ rad} \equiv 75.62°,$$

$$\theta_{av} = \frac{63.82 + 75.62°}{2} = 69.72°.$$

(Continued)

(Continued)

Substituting in Eq. (10.34):

$$\phi = \cos^{-1}\left(\frac{77.4/35.7 - \tan^2 69.72°}{77.4/35.7 + \tan^2 69.72°}\right) = \cos^{-1}\left(\frac{-5.16}{9.49}\right) = 122.94°.$$

Substituting in Eq. (10.37):

$$\phi = \frac{\pi}{2} + \tan^{-1}\left(\frac{Z_o(Z_{oo}\tan\theta_o - Z_{oe}\cot\theta_e)}{2Z_{oe}Z_{oo}\tan\theta_o\cot\theta_e}\right)$$

$$= 90° + \tan^{-1}\left(\frac{50 \times (35.7 \times \tan 63.82° - 77.4 \times \cot 75.62°)}{2 \times 77.4 \times 35.7 \times \tan 63.82° \times \cot 75.62°}\right)$$

$$= 90° + \tan^{-1}\left(\frac{50 \times (72.62 - 19.84)}{2882.00}\right)$$

$$= 90° + \tan^{-1}(0.92)$$

$$= 132.61°.$$

$$\text{Error} = \frac{132.61 - 122.94°}{132.61°} \times 100\% = 7.29\%.$$

The layout of a switched microstrip Schiffman phase shifter is shown in Figure 10.35; for convenience, the biasing arrangement for the switches and the necessary DC breaks have been omitted.

The switched Schiffman phase shifter has four switches so that the RF signal can be routed through either a continuous (non-dispersive) microstrip path, or through a (dispersive) path containing a coupled line section. Schiffman observed that there was a 90° phase difference between these paths which was almost constant over a relatively wide range of frequency. Typical theoretical phase responses for the two paths are shown in Figure 10.36.

The slope and extent of the quasi-linear region in the centre of the phase response for the coupled line path can be changed by altering the value of ρ (the ratio of Z_{oe} to Z_{oo}) for the coupled lines. The slope of the phase response for the through path can be changed by altering the length (L) of this path. The useful working range shown in Figure 10.36 is effectively an indication of the bandwidth of the phase shifter, since $\theta \propto$ frequency. By increasing the value of ρ and accepting a small error in ϕ the bandwidth can be increased very significantly, although for coupled lines fabricated in a microstrip there are practical limitations imposed on the maximum value of ρ that can be used because of the difficulty in fabricating very narrow gaps between the lines. Figure 10.37 shows that the range over which the phase difference of 90° has been increased by increasing the value of ρ from 2.3 to 3.3, but the non-linearity of the coupled line response has caused small errors in the nominal 90° value.

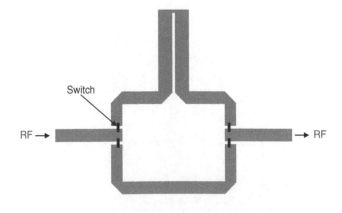

Figure 10.35 Switched microstrip Schiffman phase shifter.

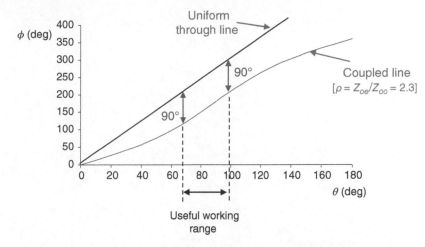

Figure 10.36 Theoretical phase responses for a Schiffman phase shifter ($\rho = 2.3$).

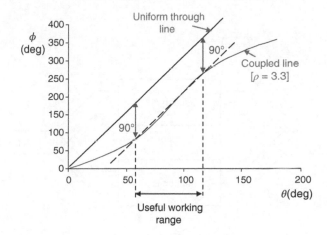

Figure 10.37 Theoretical phase responses for a Schiffman phase shifter ($\rho = 3.3$).

Although the Schiffman technique was developed some while ago, it is still widely used for obtaining broadband phase responses. A good example is the work of Wincza and Gruszczynski [18], who incorporated Schiffman couplers in a broadband microwave antenna with multi-beam capability.

10.3.5 Single-Switch Phase Shifter

All of the planar phase shifters that have been discussed so far require more than one switch to achieve each bit of phase change. Since the switches in a phase shifter are necessarily in the path of the signal, reducing the number of these switches will improve the uncertainties associated with the electrical performance of the individual switches, whilst reducing the cost of the circuit. Free and Aitchison [19] proposed a microstrip phase shifter that used only one switch to achieve each phase change, and also minimized the substrate area occupied by the phase shifter circuit. The circuit is shown in Figure 10.38.

The single-switch phase shifter consists of two coupled-line sections of microstrip, with a switch mounted at their intersection with the main 50 Ω feed line. Section A, in which the lines are connected together at the remote end to form a short circuit, contains two DC breaks. These are to permit the DC biasing of the switch. In order to satisfy the design requirements described in the following sections, the section B normally requires wide stubs. Two recesses between the wide stubs and the main line were used to minimize the discontinuities between the stubs and the feed lines. The principle of operation of the phase shifter can be explained by considering the behaviour of the circuit in the two switch states.

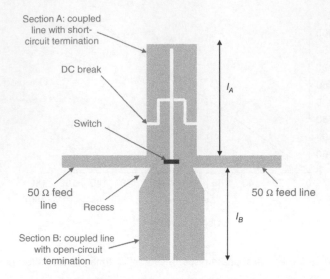

Figure 10.38 Single-bit switched phase shifter.

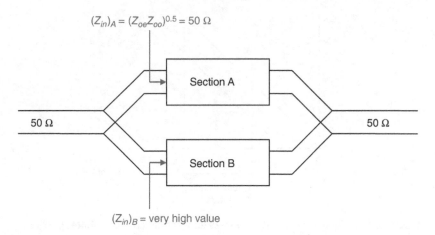

Figure 10.39 Single-bit switched phase shifter in OFF state.

Switch OFF In this state the two coupled line sections are effectively in parallel. Section B is designed to present a high impedance to the main line, so that most of the signal will be transmitted through section A, with a transmission phase determined by l_A. Note that section A is also designed to present a matched impedance to the main line, i.e. $Z_o = \sqrt{Z_{oe}Z_{oo}}$. The transmission phase through section A can be found from

$$\phi = \cos^{-1}\left(\frac{Z_{oe}/Z_{oo} - \tan^2\theta}{Z_{oe}/Z_{oo} + \tan^2\theta}\right) + \delta, \tag{10.39}$$

where δ has been added to Eq. (10.39) to represent the additional phase change introduced by the DC breaks. Thus, in the OFF state the phase shifter can be represented by the equivalent circuit shown in Figure 10.39.

With $|Z_{in}|_B \gg |Z_{in}|_B$ most of the signal will be transmitted through section A, and a good match will be maintained at the input to the phase shifter.

Switch ON In this state, sections A and B appear as stubs, connected in parallel with the 50 Ω feed line, as shown in Figure 10.40.

It should be noted that with the switch ON and so presenting a nominal short circuit across the centre of the coupler line, the formation of the odd mode is inhibited, and the stubs support only the even mode. In Figure 10.40, B_A and B_B represent the magnitudes of the susceptances, which the two coupled sections present to the main line when only the even mode is present. The design requires that in the ON state the susceptances should cancel, and the circuit behave as a simple

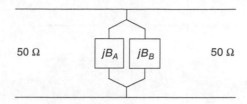

Figure 10.40 Single-bit switched phase shifter in ON state.

50 Ω through connection, i.e. we require

$$jB_A + jB_B = 0. \tag{10.40}$$

The ability to satisfy the circuit requirements in the ON and OFF states depends largely on the differences between the behaviour of the coupled line sections with both modes present, and with only the even mode present. Only small phase shifts were reported in [19] using simple microstrip circuits, but fabricating the circuit using multi-layer techniques would introduce more degrees of design freedom and extend the range of possible phase shifts.

10.4 Supplementary Problems

Q10.1 Suppose the PIN diode specified in Example 10.1 was mounted across a 300 μm gap in a 50 Ω microstrip line, which has been fabricated on a substrate that has a relative permittivity of 9.8 and a thickness of 0.75 mm. Determine the effect which the gap has on the insertion loss and the isolation at 5 GHz. Comment upon the results.

Q10.2 A packaged PIN diode has a total lead inductance of 0.24 nH. If the junction capacitance in the OFF state is 0.08 pF, determine the maximum permissible packaging capacitance if the minimum self-resonant frequency is to be 30 GHz.

Q10.3 A PIN diode has a forward resistance of 1.35 Ω and a junction capacitance in the OFF state of 0.095 pF. What is the FoM for the diode?

Q10.4 Suppose that the MEMS switch specified in Example 10.4 has a series inductance of 100 pH. Verify that this inductance has little effect on the insertion loss and isolation of the switch at 5 GHz.

Q10.5 Design a 10 GHz microstrip switched-line phase shifter that will provide 45° steps in phase over the range 0–360°. The phase shifter is to be matched to 50 Ω and is to be fabricated on a substrate that has a relative permittivity of 9.8 and a thickness of 0.5 mm. Include provision for biasing the switches, which are to be series-mounted PIN diodes. Assume that where DC breaks are required these will be provided by surface-mount lumped-element capacitors.

Q10.6 A lumped-element loaded-line phase shifter is to be designed to provide a 90° phase change at 1 GHz. Determine the required values of the shunt reactances, and give a specification for the transmission line connecting the reactances.

Q10.7 Figure 10.41 shows the layout of an ideal microstrip loaded-line phase shifter that has been fabricated on a substrate having a relative permittivity of 9.8 and a thickness of 25 mil (635 μm). Deduce:
  (i) The value of the phase change when the switches are activated together
  (ii) The frequency of operation
  (iii) The values of the dimensions a and b

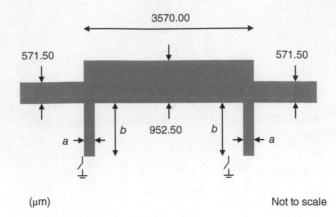

Figure 10.41 Loaded-line microstrip phase shifter for Example Q10.7.

Q10.8 (i) Show that the length of the coupled line shown in Figure 10.34 can be expressed in terms of the transmission phase, ϕ, as

$$l = \frac{\lambda_{av}}{2\pi}\tan^{-1}\left(\frac{\rho(1-\cos\phi)}{(1+\cos\phi)}\right)^{0.5},$$

where the symbols have their usual meanings.

(ii) Hence find the length of coupled line needed to give a phase change of 90° at 12 GHz given the following data for the coupled lines:

Odd mode:	$Z_{oo} = 35.7\,\Omega$ $\lambda_{s,odd} = \dfrac{141}{f}$ mm
	where f is in GHz
Even mode:	$Z_{oe} = 77.4\,\Omega$ $\lambda_{s,even} = \dfrac{119}{f}$ mm
	where f is in GHz

(iii) Find the percentage error in the 90° value due to averaging the wavelengths of the odd and even modes.

(iv) Determine the VSWR at the input to the coupled section if it is connected between 50 Ω source and load impedances.

References

1 Fukui, H. (1979). Optimal noise figure of microwave GaAs MESFETs. *IEEE Transactions on Electron Devices* ED-26 (7): 1032–1037.

2 Chang, K., Bahl, I., and Nair, V. (2002). *RF and Microwave Circuit and Component Design for Wireless Systems*. New York: Wiley.

3 Liu, D., Gaucher, B., Pfeiffer, U., and Grzyb, J. (2009). *Advanced Millimeter-Wave Technologies*. Chichester, UK: Wiley.

4 Goldsmith, C.L., Yao, Z., Eshelmen, S., and Denniston, D. (1998). Performance of low-loss RF MEMS capacitive switches. *IEEE Microwave and Guided Wave Letters* 8 (8): 269–271.

5 Rebeiz, G.M. and Muldavin, J.B. (2001). RF MEMS switches and switch circuits. *IEEE Microwave Magazine* 2: 59–71.

6 Rebeiz, G.M., Entesari, K., Reines, I.C. et al. (2009). Tuning in to RF MEMS. *IEEE Microwave Magazine* 10: 55–72.

7 Ko, C.-H., Ho, K.M.J., and Rebeiz, G. (2013). An electronically-scanned 1.8–2.1 GHz Base-station antenna using packaged high-reliability RF MEMS phase shifters. *IEEE Transactions on Microwave Theory and Techniques* 61 (2): 979–985.

8 Moran, T., Keimel, C., and Miller, T. (2016). Advances in MEMS switches for RF test applications. *Proceedings of 46th European Microwave Conference*, London, UK (October 2016), pp. 1369–1372.

9 Koul, S.K., Dey, S., Poddar A.K., and Rohde, U.L. (2016). Micromachined switches and phase shifters for transmit/receive module applications. *Proceedings of 46th European Microwave Conference*, London, UK (October 2016), pp. 971–974.

10 Borodulin, P., El-Hinnawy, N., Padilla, C.R., Ezis, A., King, M.R., Johnson, D.R., Nichols, D.T., and Young, R.M. (2017). Recent advances in fabrication and characterization of GeTe-based. Phase-change RF switches and MMICs. *Proceedings of IEEE MTT-S International Microwave Symposium Digest*, Honolulu (June 2017), pp. 285–288.

11 Wang, M., Lin, F., and Rais-Zadeh, M. (2016). A reconfigurable filter using germanium telluride phase change RF switches. *IEEE Microwave Magazine* 17 (12): 70–79.

12 Bahl, I.J. and Gupta, K.C. (1980). Design of loaded-line p-i-n diode phase shifter circuits. *IEEE Transactions on Microwave Theory and Techniques* 28 (3): 219–224.

13 Atwater, H.A. (1985). Circuit design of the loaded-line phase shifter. *IEEE Transactions on Microwave Theory and Techniques* 33 (7): 626–634.

14 Garver, R.V. (1972). Broad-band diode phase shifters. *IEEE Transactions on Microwave Theory and Techniques* 20 (5): 314–323.

15 Koul, S.K. and Bhat, B. *Microwave and Millimeter Wave Phase Shifters*, vol. 2. Norwood, MA: Artech House.

16 Schiffman, J.B. (1958). A new class of broadband microwave 90° phase shifters. *IRE Transactions on Microwave Theory and Techniques* 6: 232–237.

17 Free, C.E. and Aitchison, C.S. (1995). Improved analysis and design of coupled-line phase shifters. *IEEE Transactions on Microwave Theory and Techniques* 43 (9): 2126–2131.

18 Wincza, K. and Gruszczynski, S. (2016). Broadband integrated 8 × 8 Butler matrix utilizing quadrature couplers and Schiffman phase shifters for multibeam antennas with broadside beam. *IEEE Transactions on Microwave Theory and Techniques* 64 (8): 2596–2604.

19 Free, C.E. and Aitchison, C.S. (1985). Single PIN diode X-band phase shifter. *Electronics Letters* 21 (4): 128–129.

11

Oscillators

11.1 Introduction

Oscillators, for the generation of power, are fundamental devices in RF and microwave transmitters and receivers. The most common type of semiconductor oscillator in modern high-frequency systems is that using transistors with positive feedback, and this type of oscillator will be the main focus of this chapter. Other types of oscillators exist, such as those using microwave diodes, but they are normally restricted to relatively specialized applications. For completeness, a brief account of the two most common types of microwave diode oscillator, namely those using Gunn and Impatt diodes, has been included.

The chapter commences with a discussion of the basic conditions for oscillation in feedback circuits. Particular emphasis is given to the Colpitts oscillator, which is the most common form of oscillator for RF and low microwave frequency applications. The use of crystals and varactor diodes will be included in the discussion as they have particular relevance for oscillator stability and tuning. In addition to oscillators using discrete components, the application of feedback in distributed circuits, such as the dielectric resonator oscillator (DRO), often used in hybrid microwave circuits, will be discussed.

Many communication circuits, and much high-frequency instrumentation, derive the required oscillation frequency from very stable low-frequency (LF) sources through the use of frequency synthesizers, and a review of basic synthesizer techniques has been given.

Oscillator noise is an important issue in system design, particularly in receivers, and some general information on the characterization of noise in oscillators is presented. Also included is a brief discussion of the use of delay-line and tuned circuit techniques for measuring oscillator noise.

11.2 Criteria for Oscillation in a Feedback Circuit

An amplifier with external feedback and an open-loop gain $G(\omega)$ can be represented by the model shown in Figure 11.1, where $\beta(\omega)$ is the gain of the feedback path.

The output voltage, V_O, can be written in terms of the input voltage, V_i, as

$$V_O = G(\omega)\, V_i + \beta(\omega)\, G(\omega)\, V_O. \tag{11.1}$$

This can be rewritten as

$$\frac{V_O}{V_i} = \frac{G(\omega)}{1 + \beta(\omega)G(\omega)}. \tag{11.2}$$

It follows from Eq. (11.2) that the amplifier will exhibit oscillation at a given frequency if

$$\beta(\omega)G(\omega) = -1, \tag{11.3}$$

since this condition implies that the amplifier will produce an output with zero input.

The conditions for oscillation defined by Eq. (11.3) can be written as

$$|\beta(\omega)G(\omega)| = 1,$$

$$\angle(\beta(\omega)G(\omega)) = n\pi, \tag{11.4}$$

where n is an odd integer.

RF and Microwave Circuit Design: Theory and Applications, First Edition. Charles E. Free and Colin S. Aitchison.
© 2022 John Wiley & Sons Ltd. Published 2022 by John Wiley & Sons Ltd.
Companion website: www.wiley.com/go/free/rfandmicrowave

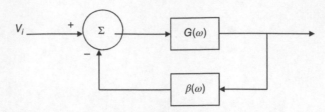

Figure 11.1 Model of an amplifier with feedback.

The conditions stated in Eq. (11.4) are referred to as the Nyquist criteria for oscillation.

11.3 RF (Transistor) Oscillators

11.3.1 Colpitts Oscillator

Figure 11.2 depicts a common-emitter transistor amplifier with the typical feedback network configuration used in RF oscillators. For clarity, the DC-biasing components have been omitted from the diagram. The feedback network consists of three impedances: Z_1, Z_2, and Z_3. It can be seen that Z_1 and Z_3 form a potential divider, allowing a portion of the output signal to be fed back to the input (base) of the transistor amplifier. The impedance Z_2, in combination with Z_1 and Z_3, determines the transfer function of the feedback network, and in particular the resonant frequency.

By replacing the transistor with its small-signal equivalent circuit it can be shown by straightforward circuit analysis [1] that the Nyquist conditions for oscillation are met when

$$Z_1 + Z_2 + Z_3 = 0 \tag{11.5}$$

and

$$\beta = \frac{Z_1}{Z_2}, \tag{11.6}$$

where β is the base–collector current gain of the transistor.

The components in the feedback network are normally pure reactances, and so Eqs. (11.5) and (11.6) can be rewritten as

$$X_1 + X_2 + X_3 = 0 \tag{11.7}$$

and

$$\beta = \frac{X_1}{X_2}. \tag{11.8}$$

Since β is always positive, we can deduce from Eq. (11.8) that X_1 and X_2 must have the same sign, i.e. the reactances must both be capacitive or both be inductive. It then follows from Eq. (11.7) that X_3 must have the opposite sign to X_1 and X_2. If X_1 and X_2 represent capacitors and X_3 an inductor, the circuit is called a Colpitts oscillator, and the basic arrangement is shown in Figure 11.3, where for clarity the DC-biasing components have been omitted.

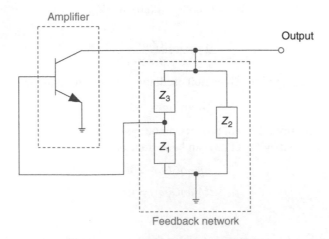

Figure 11.2 Basic transistor oscillator.

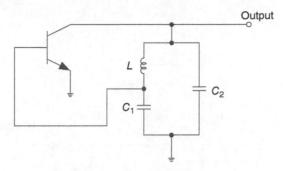

Figure 11.3 Basic Colpitts oscillator.

Oscillation will occur at the resonant frequency, f_O, of the feedback network, and this is easily found from Eq. (11.7) as

$$X_1 + X_2 + X_3 = 0,$$

$$-j\frac{1}{\omega C_1} - j\frac{1}{\omega C_2} + j\omega L = 0,$$

$$\omega L = \frac{1}{\omega C_1} + \frac{1}{\omega C_2},$$

$$\omega^2 L = \frac{1}{C_1} + \frac{1}{C_2},$$

$$\omega^2 = \frac{C_1 + C_2}{LC_1C_2},$$

i.e.

$$f_O = \frac{1}{2\pi}\sqrt{\frac{C_1 + C_2}{LC_1C_2}}. \tag{11.9}$$

The resonant frequency given by Eq. (11.9) assumes the impedance of the feedback network is unaffected by other reactive sources in a practical circuit, and should be regarded as an approximation.

Example 11.1 A 45 MHz Colpitts oscillator is to be designed using an NPN transistor with a base–collector current gain of 80, and the feedback configuration shown in Figure 11.3. If a 150 nH inductor is used, determine the values of C_1 and C_2.

Solution
Using Eq. (11.8):

$$\beta = 80 = \frac{1/\omega C_1}{1/\omega C_2} = \frac{C_2}{C_1}.$$

Using Eq. (11.9):

$$f_O = \frac{1}{2\pi}\sqrt{\frac{C_1 + C_2}{LC_1C_2}} = \frac{1}{2\pi}\sqrt{\frac{1 + C_2/C_1}{LC_2}},$$

i.e.

$$45 \times 10^6 = \frac{1}{2\pi}\sqrt{\frac{1 + 80}{150 \times 10^{-9} \times C_2}},$$

$$C_2 = 6.75 \, \text{nF},$$

$$80 = \frac{6.75 \times 10^{-9}}{C_1},$$

$$C_1 = 84.38 \, \text{pF}.$$

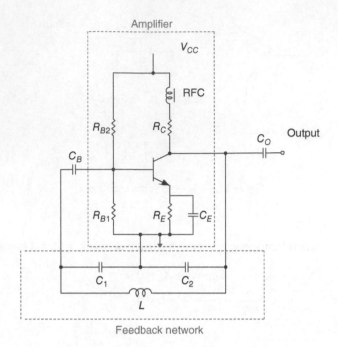

Figure 11.4 Practical Colpitts oscillator.

The arrangement of a practical Colpitts oscillator is shown in Figure 11.4, from which it is clear that there will be other reactive sources in the circuit, including those within the transistor.

Functions of the components in the circuit shown in Figure 11.4 are:

Amplifier:

R_{B1} and R_{B2} form a potential divider across the DC supply voltage, V_{CC}, and enable the DC base voltage to be set.

R_E and R_C are the emitter and collector resistors, respectively, and enable the DC operating point to be set.

RFC is an RF choke, which has a low DC resistance and a high RF resistance, and blocks the amplified RF signal from entering the supply line.

C_E is a large value by-pass capacitor; it enables the emitter to be grounded for RF signals, but allows a DC voltage to be set on the emitter.

Feedback network:

C_1, C_2, and L set the oscillating conditions as defined by Eqs. (11.7)–(11.9).

Blocking capacitors:

These are large value capacitors, usually a few μF, with the following functions:

C_B: Since the inductor will have a low DC resistance, C_B is necessary to prevent the DC voltage on the collector from being coupled to the base of the transistor through the inductor, L.

C_O: This blocks the DC collector voltage from appearing at the output of the oscillator; thus the output is just a sinusoidal signal at the required RF frequency.

11.3.2 Hartley Oscillator

The conditions for oscillation defined by Eqs. (11.7) and (11.8) can also be satisfied with X_1 and X_2 representing inductors and X_3 representing a capacitor. This gives the feedback configuration shown in Figure 11.5a, and the oscillator is then known as a Hartley oscillator. The inductors can be two discrete components as shown in Figure 11.5a, or be formed by tapping a single inductor as shown in Figure 11.5b. The advantage of using the tapping technique is that by varying the position of the tap the amount of feedback can be controlled, thus enabling the amplitude of the oscillations to be varied.

Additional points concerning the Harley oscillator:

(i) The feedback network tends to have a lower Q-value than the Colpitts oscillator due to the additional series resistance of the inductors. This reduces the stability of the oscillator and increases the noise.

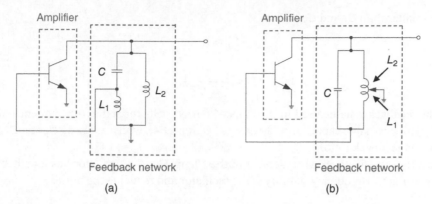

Figure 11.5 Feedback configurations for a Hartley oscillator: (a) using two discrete inductors and (b) using a single tapped inductor.

(ii) If two discrete inductors are used, some caution is needed in the practical layout of the circuit to prevent electromagnetic coupling between the two components. Any coupling will introduce mutual inductance and modify the conditions for oscillation.

(iii) The Hartley oscillator is easily tuned by replacing the single capacitor with a varactor diode (*the properties of a varactor diode are covered in a later section of this chapter*).

11.3.3 Clapp–Gouriet Oscillator

The Clapp–Gouriet oscillator is a simple modification of the Colpitts oscillator shown in Figure 11.4. The modification involves the addition of a small capacitor C_3 in series with the feedback inductor, as shown in Figure 11.6.

The frequency of oscillation of the Clapp–Gouriet oscillator is given by:

$$f_O = \frac{1}{2\pi\sqrt{LC_T}}, \tag{11.10}$$

where

$$C_T = \left(\frac{1}{C_1} + \frac{1}{C_2} + \frac{1}{C_3}\right)^{-1}. \tag{11.11}$$

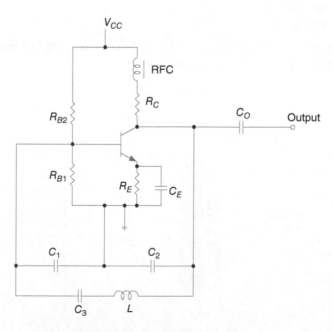

Figure 11.6 Clapp–Gouriet oscillator.

Normally C_3 is much smaller than C_1 and C_2 and so

$$C_T \approx C_3 \tag{11.12}$$

and

$$f_O \approx \frac{1}{2\pi\sqrt{LC_3}}. \tag{11.13}$$

The advantage of the feedback scheme used in the Clapp–Gouriet oscillator is that the frequency of oscillation is easily tuned by replacing C_3 with a variable capacitance diode (see Section 11.4), whilst allowing C_1 and C_2 to remain fixed, and so achieving the optimum feedback ratio.

It can also be shown that the feedback arrangement of the Clapp–Gouriet oscillator has a higher Q than the Colpitts oscillator, thereby improving the frequency stability of the oscillator and reducing the noise.

11.4 Voltage-Controlled Oscillator

The frequency of transistor oscillators with the feedback networks described in Section 11.3 can be controlled by including a voltage-controlled reactance, such as a varactor diode, in the feedback network. The varactor diode, also referred to as a varicap diode, or simply as a tuning diode, exploits the change in junction capacitance of a reverse-biased PN junction caused by a change in the bias voltage. The only difference between a varactor diode and a simple signal diode is that the microelectronic structure of the varactor diode is modified to enhance the change in junction capacitance. When a PN junction is reverse biased, electrons in the N-material diffuse across the junction to the P-material, and holes in the P-material diffuse across the junction to the N-material. Thus, recombination takes place in the vicinity of the junction, leading a region that is depleted of charge. There will be a build-up of charge of opposite sign on either side of this depletion region, which can thus be regarded as analogous to a parallel-plate capacitor, as depicted in Figure 11.7.

Thus, the capacitance of the diode can be regarded, somewhat simplistically, as equivalent to that of a parallel-plate capacitor, i.e.

$$C_j = \frac{\varepsilon A}{d}, \tag{11.14}$$

where

C_j is the diode capacitance
ε is the permittivity of the diode semiconductor material
A is the cross-sectional area of the junction
d is the width of the depletion region.

Since d varies with the magnitude of the reverse bias, the junction behaves as a voltage-controlled capacitance.

The variation of diode capacitance with bias voltage is given by [2]

$$C_j(V_R) = \frac{C_O}{\left(1 - \dfrac{V_R}{\phi}\right)^{\gamma}}, \tag{11.15}$$

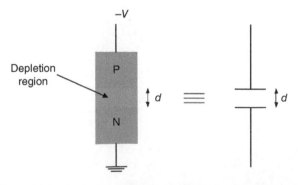

Figure 11.7 Depletion region in a reverse-biased PN junction.

where

$C_j(V_R)$ is the diode capacitance with a reverse-bias voltage, V_R

C_O is the diode capacitance with zero bias

ϕ is the diffusion potential of the junction and depends on the material from which the junction is made:

   $\phi = 1.3$ V for Gallium Arsenide (GaAs)

   $\phi = 0.7$ V for Silicon (Si)

γ is a parameter that depends on the type of PN junction:

   $\gamma = 0.5$ for an abrupt junction (*an abrupt junction is formed when there is a constant impurity density in the P and N regions*)

   $\gamma = 0.33$ for a graded junction (*a graded junction is formed when there is a linear decrease in impurity density with distance from the junction*)

   $\gamma \geq 1$ for a hyper-abrupt junction (*a hyper-abrupt junction is formed when there is a linear increase in impurity density with distance from the junction*)

Varactors with hyper-abrupt junctions are particularly useful for tuning purposes in voltage-controlled oscillators (VCOs) because they give a greater rate of change of capacitance with bias voltage when compared with other junctions, which means that less voltage change is needed to obtain a given capacitance change.

Example 11.2   Determine the value of reverse bias needed across a silicon-graded junction PN diode to produce a junction capacitance of 15 pF, given that the junction capacitance is 25 pF with zero bias voltage. Assume $\phi = 0.7$ V.

Solution

We need to use Eq. (11.15), where $\phi = 0.7$ V for silicon, and $\gamma = 0.33$ for a graded junction, i.e.

$$15 = \frac{25}{\left(1 - \dfrac{V_R}{0.7}\right)^{0.33}},$$

$$1 - \frac{V_R}{0.7} = \left(\frac{25}{15}\right)^{\frac{1}{0.33}} = 4.70.$$

$$V_R = -2.59 \text{ V}.$$

The capacitance–voltage variation of a typical varactor with a hyper-abrupt junction is shown in Figure 11.8, where C_j is the junction capacitance and V_R is the magnitude of the reverse bias voltage. Also shown is the value of junction capacitance, C_O, with zero bias.

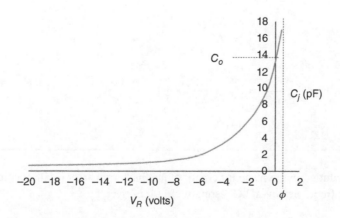

Figure 11.8   Capacitance–voltage variation of a typical hyper-abrupt varactor diode.

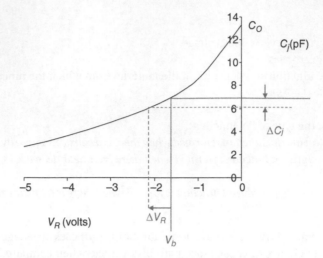

Figure 11.9 Capacitance change in a hyper-abrupt varactor diode, showing limited region of characteristic in Figure 11.8.

For the purpose of circuit tuning, the useful region of the characteristic shown in Figure 11.8 is that between $V_R = 0$ V and $V_R = -6$ V, since this region gives a significant change in junction capacitance for a small change in bias voltage. The region is shown expanded in Figure 11.9, which also shows the effect of applying a small change in bias voltage, at a fixed value, V_b, of DC bias.

The change in junction capacitance due to a change in reverse bias can be found by differentiating Eq. (11.15), i.e.

$$\frac{d}{dV_R}\{C_j(V_R)\} = C_O(-\gamma)\left\{1 - \frac{V_R}{\phi}\right\}^{-\gamma-1}\left\{-\frac{1}{\phi}\right\}$$

$$= C_O\left\{\frac{\gamma}{\phi}\right\}\left\{1 - \frac{V_R}{\phi}\right\}^{-(\gamma+1)}. \tag{11.16}$$

In order to maintain a quasi-linear relationship between the change in bias voltage and the change in capacitance, varactor diodes are normally used in situations where only a small change in capacitance is required. For small changes in capacitance at a given value of bias voltage, V_b, we may rewrite Eq. (11.16) as

$$\frac{\Delta C_j(V_R)}{\Delta V_R} = C_O\left\{\frac{\gamma}{\phi}\right\}\left\{1 - \frac{V_R}{\phi}\right\}^{-(\gamma+1)}. \tag{11.17}$$

Example 11.3   A silicon PN varactor diode with a hyper-abrupt junction ($\gamma = 1.1$) and $C_O = 22$ pF has a DC reverse-bias of -4.5 V. Find the change in reverse bias needed to produce a capacitance change of $+0.2$ pF. Assume $\phi = 0.7$ V.

Solution
Using Eq. (11.17) we have

$$\frac{0.2}{\Delta V_R} = 22 \times \left(\frac{1.1}{0.7}\right) \times \left(1 - \frac{(-4.5)}{0.7}\right)^{-(1.1+1)}.$$

$$\Delta V_R = 0.39 \text{ V}.$$

Varactor diodes are often included in the resonant circuit of feedback transistor oscillators in order to tune the frequency of the oscillator. The resonant frequency of an LC resonant circuit is given by

$$f_O = \frac{1}{2\pi\sqrt{LC}}. \tag{11.18}$$

We can determine a relationship between the change in resonant frequency and the change in capacitance by differentiating Eq. (11.18). Writing Eq. (11.18) in a more convenient form gives

$$f_O = \frac{1}{2\pi\sqrt{L}}C^{-0.5}. \tag{11.19}$$

Then,

$$\frac{df_O}{dC} = -0.5\frac{1}{2\pi\sqrt{L}}C^{-1.5} = -0.5\frac{1}{2\pi\sqrt{LC}}C^{-1}, \tag{11.20}$$

i.e.

$$\frac{df_O}{dC} = -0.5f_O C^{-1}. \tag{11.21}$$

For small changes in capacitance we may rewrite Eq. (11.21) as

$$\frac{\Delta f_O}{\Delta C} = -0.5\frac{f_O}{C}, \tag{11.22}$$

which can be rearranged to give

$$\frac{|\Delta f_O|}{f_O} = \frac{|\Delta C|}{2C}. \tag{11.23}$$

Example 11.4   A resonant circuit consists of a 50 µH inductor in parallel with a capacitor.

(i) What value of capacitor is needed for resonance at 2 MHz?
(ii) Suppose that a silicon varactor diode with a graded junction, and $C_O = 18$ pF, is connected in parallel with the capacitor. What value of DC bias is needed to reduce the resonant frequency by 3%?

Solution

(i) Using Eq. (11.18) we have

$$f_O = \frac{1}{2\pi\sqrt{LC}} \quad\Rightarrow\quad 2\times 10^6 = \frac{1}{2\pi\sqrt{50\times 10^{-6}\times C}} \quad\Rightarrow\quad C = 126.62\,\text{pF}.$$

(ii) New resonant frequency, $(f_O)_{new}$, is

$$(f_O)_{new} = 2(1 - 0.03)\,\text{MHz} = 1.94\,\text{MHz}.$$

Calculating total capacitance, C_{new}, required for resonance at 1.94 MHz:

$$1.94\times 10^6 = \frac{1}{2\pi\sqrt{50\times 10^{-6}\times C_{new}}} \quad\Rightarrow\quad C_{new} = 134.57\,\text{pF}.$$

Since the capacitor and varactor diode are in parallel we have

$$134.57 = 126.62 + C_j \quad\Rightarrow\quad C_j = 7.95\,\text{pF},$$

i.e. the varactor diode must be biased to provide a junction capacitance of 7.95 pF.
   Using Eq. (11.15) (with $\phi = 0.7$ and $\gamma = 0.33$ for the diode specified):

$$7.95 = \frac{18}{\left(1 - \dfrac{V_R}{0.7}\right)^{0.33}},$$

$$1 - \frac{V_R}{0.7} = \left(\frac{18}{7.95}\right)^{\frac{1}{0.33}} = 11.90,$$

$$V_R = -7.63\,\text{V}$$

Figure 11.10 shows practical arrangements for varactor diodes used as tuning elements in LC resonant circuits. A single reverse-biased varactor is used in Figure 11.10a, and since the varactor will only provide a small capacitance a padding

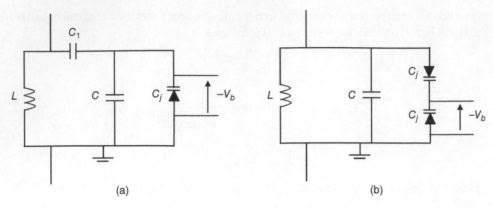

Figure 11.10 Varactor diodes used to tune resonant circuits: (a) using a single varactor and (b) using two back-to-back varactors.

capacitor, C, is included to avoid the need for large inductors. Also included in the circuit of Figure 11.10a is a blocking capacitor, C_1, which is needed to avoid the low DC resistance of the inductor shorting-out the bias voltage. Normally, $C_1 \gg C$, so that it does not affect the resonant frequency of the circuit. The need for a blocking capacitor can be avoided by using two varactor diodes connected back-to-back, as shown in Figure 11.10b. Note that in the circuit of Figure 11.10b, the low DC resistance of the inductor means that both varactor diodes are DC reverse biased equally.

The use of two back-to-back diodes, as shown in Figure 11.10, is clearly advantageous and avoids the need for a relatively high value DC blocking capacitor (C_1). For convenience, varactor diodes are often packaged in a back-to-back format. Figure 11.10 shows the normal circuit convention with the anodes of the diodes at earth potential so that positive values of V_{bias} will reverse bias the diodes.

The circuits shown in Figure 11.10 are often used to produce frequency modulated (FM) signals in which the oscillation (carrier) frequency is periodically deviated from its nominal value. This is achieved by using a bias voltage of the form

$$V_{bias} = V_b + V_m \cos \omega_m t. \tag{11.24}$$

The DC value, V_b, is used to set the carrier frequency, and V_m to set the amount of frequency deviation. The design procedure is illustrated in Example 11.5.

Example 11.5   The resonant circuit shown in Figure 11.11 is used as part of an oscillator to produce an FM signal with a carrier frequency of 5 MHz and an RMS frequency deviation of 5 kHz.

A silicon varactor diode with a graded junction and $C_O = 43$ pF is to be used. Determine the required values of V_b and V_m.

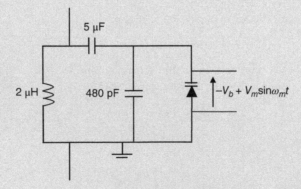

Figure 11.11   Circuit for Example 11.5.

Solution
Note that the 5 μF blocking capacitor has a large value in comparison to the padding capacitor and the varactor diode, and can therefore be neglected.

DC calculation to find V_b

Firstly we need to find the total capacitance required for resonance:

$$f_O = \frac{1}{2\pi\sqrt{LC_T}},$$

i.e.

$$5 \times 10^6 = \frac{1}{2\pi\sqrt{2 \times 10^{-6} \times C_T}} \quad \Rightarrow \quad C_T = 506.47 \text{ pF}.$$

Therefore, to achieve resonance the varactor diode must provide a capacitance of $(506.47-480)$ pF $= 26.47$ pF. Using Eq. (11.15) (with $\phi = 0.7$ and $\gamma = 0.33$ for the diode specified):

$$26.47 = \frac{43}{\left(1 - \dfrac{V_R}{0.7}\right)^{0.33}},$$

$$V_R = -2.35 \text{ V}.$$

AC calculation

The peak frequency deviation required is

$$(\Delta f_O)_{pk} = \sqrt{2} \times 5 \times 10^3 \text{ Hz} = 7.07 \times 10^3 \text{ Hz}.$$

Using Eq. (11.23) to find the peak capacitance deviation gives

$$\frac{7.07 \times 10^3}{5 \times 10^6} = \frac{(\Delta C)_{pk}}{2 \times 506.47},$$

$$(\Delta C)_{pk} = 1.43 \text{ pF}.$$

Now V_m is the peak AC voltage that will produce the peak capacitance change, and so substituting in Eq. (11.17), with $V_R = -V_b = -2.35$ V we have

$$\frac{1.43}{V_m} = 43 \times \left(\frac{0.33}{0.7}\right) \times \left(1 - \frac{(-2.35)}{0.7}\right)^{-(0.33+1)},$$

$$V_m = 499.6 \text{ mV}.$$

Example 11.6 The resonant circuit shown in Figure 11.12 is used as part of an oscillator to produce an FM signal.

If GaAs abrupt junction varactor diodes with $C_O = 27$ pF are used, determine the carrier frequency and RMS frequency deviation of the FM signal.

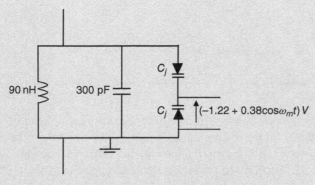

Figure 11.12 Circuit for Example 11.6.

(Continued)

(Continued)

Solution

DC calculation

Using Eq. (11.15) to find the junction capacitance of each diode with a -1.22 V DC bias (with $\phi = 1.3$ and $\gamma = 0.5$ for the diode specified):

$$C_j = \frac{27}{\left(1 - \frac{(-1.22)}{1.3}\right)^{0.5}} \text{ pF}$$

$$= 19.39 \text{ pF}.$$

The total capacitance of the parallel circuit will then be

$$C_T = \left(300 + \frac{19.39}{2}\right) \text{ pF} = 309.70 \text{ pF}.$$

And the resonant frequency (carrier frequency) will be

$$f_O = \frac{1}{2\pi\sqrt{LC_T}}$$

$$= \frac{1}{2\pi\sqrt{90 \times 10^{-9} \times 309.70 \times 10^{-12}}} \text{ Hz} = 30.14 \text{ MHz}.$$

AC calculation

Using Eq. (11.17) to find the peak capacitance change in each diode due to the 0.38 peak AC voltage:

$$\frac{(\Delta C_j)_{pk}}{0.38} = 27 \times \left(\frac{0.5}{1.3}\right) \times \left(1 - \frac{(-1.22)}{1.3}\right)^{-(0.5+1)} \text{ pF/V},$$

$$(\Delta C_j)_{pk} = 1.46 \text{ pF}.$$

Each diode will contribute a peak capacitance change of 1.46 pF, and since there are two diodes in series the total peak capacitance change will be

$$(\Delta C_T)_{pk} = \frac{1.46}{2} \text{ pF} = 0.73 \text{ pF}.$$

Using Eq. (11.23) to find the peak frequency deviation we have:

$$\frac{(\Delta f_O)_{pk}}{f_O} = \frac{(\Delta C_T)_{pk}}{2C_T},$$

$$\frac{(\Delta f_O)_{pk}}{30.14 \times 10^6} = \frac{1.46 \times 10^{-12}}{2 \times 309.70 \times 10^{-12}},$$

$$(\Delta f_O)_{pk} = 71.04 \text{ kHz}.$$

The RMS frequency deviation is then

$$(\Delta f_O)_{RMS} = \frac{71.04}{\sqrt{2}} \text{ kHz} = 50.23 \text{ kHz}.$$

We saw in Figure 11.8 that there is a non-linear relationship between the capacitance of a varactor diode and the tuning voltage. In order to obtain quasi-linear tuning of a resonant circuit varactor diodes are normally biased with small modulating voltages. This means that oscillators with varactor frequency control can only be tuned over a limited frequency range. Frequency synthesizers (described in Section 11.6) are normally used to give oscillators with wideband frequency performance.

So far we have considered varactor diodes to be ideal, and only contribute capacitance to a circuit. In reality, a varactor diode can be represented by the equivalent circuit shown in Figure 11.13, for a given value of reverse bias.

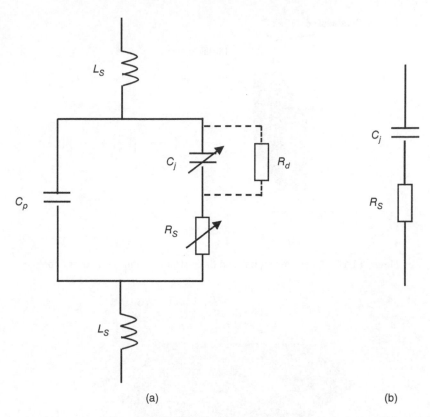

Figure 11.13 Equivalent circuits representing a reverse-biased varactor diode: (a) full equivalent circuit and (b) simplified equivalent circuit.

The variable capacitance resulting from variations in the width of the depletion layer is shown as C_j in Figure 11.13, and the resistance due to the doped P and N regions as R_S. Note that as the width of the depletion layer changes with the value of reverse bias giving a change in C_j there will also be a change in R_S since the volume of doped semiconductor will change. There will be some shunt resistance, R_d, associated with the depletion layer, but this will be very high and can normally be neglected, and so it is shown dotted in the equivalent circuit. Also shown in the full equivalent circuit are C_p to represent the packaging capacitance of the semiconductor, and L_S to represent the inductance of each lead. Neglecting the packaging parasitics, the varactor can be represented by the simplified equivalent circuit shown in Figure 11.13b, and the quality factor, Q, of the varactor at a given value of reverse bias is then defined simply as

$$Q = \frac{X_C}{R_S} = \frac{1}{R_S \omega C_j}.$$

(11.25)

11.5 Crystal-Controlled Oscillators

11.5.1 Crystals

The frequency stability of transistor oscillators can be improved significantly by including a crystal in the feedback network. The crystal acts like a resonant circuit in the feedback path, and the properties of the crystal give it a very high Q-factor and result in very high frequency stability for the oscillator with correspondingly low noise. Quartz is the most common material used for crystals, although cultured or synthetic quartz has now largely replaced natural quartz in electronic components.

The electronic component known simply as a crystal is formed from a thin slice of pure crystal with metallization on the two flat sides to form electrodes and a capacitor-like structure, as shown schematically in Figure 11.14a. Figure 11.14b shows the standard circuit symbol used to represent a crystal.

The direction of the slice with respect to the axes of the crystal is critical in determining the properties of the crystal, particularly its temperature stability. Quartz has a six-sided hexagonal structure and the most common cut for RF quartz

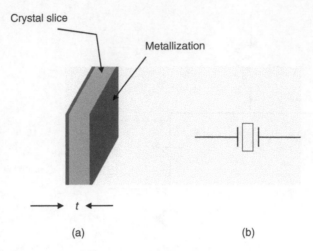

Figure 11.14 Electronic crystals: (a) basic structure and (b) circuit symbol.

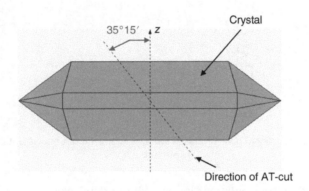

Figure 11.15 AT-cut in quartz crystal.

crystals is the AT-cut, which is made at an angle of 35°15′ to the z-axis (optical axis) as shown in Figure 11.15. Changes of only a few minutes of arc in the direction of the cut have a very significant effect on the temperature stability of a crystal oscillator.

Quartz is a material that exhibits piezoelectric properties, which means that it will vibrate mechanically when an electric signal is applied across the material. For an electronic crystal the frequency of vibration depends on the thickness of the cut, t, the stiffness of the material, and the direction of the cut. It is the need for very thin slices that determines the highest frequency for crystals, although with modern fabrication techniques crystals can be manufactured with resonant frequencies in excess of 200 MHz. Some crystals make use of overtones, or harmonics, to achieve higher resonant frequencies, but usually the quality of performance of these crystals is inferior to those using the fundamental mode of oscillation.

One of the primary reasons for using crystals as resonant circuits in electronic designs is that the resulting frequency of oscillation is very stable with changes in temperature. The temperature stability is expressed in terms of the temperature coefficient, TC, specified as

$$TC = \frac{\Delta f_O}{f_O}, \tag{11.26}$$

where f_O is the resonant frequency, and Δf_O is the change in resonant frequency for a 1 °C temperature change. TC is usually quoted in ppm (parts per million). The temperature coefficient is a cubic function of temperature, as shown in Figure 11.16.

It can be seen from Figure 11.16 that the temperature coefficient response has an inflexion point at around 25 °C, with zeros around −20 °C and 70 °C. This indicates that there is a very wide range of temperature, around 90 °C, over which the crystal has very small variations in resonant frequency.

As an example of the significance of using crystals, the TC of an LC oscillator is normally of the order of several hundred ppm, whereas a crystal-controlled oscillator usually has a TC less than 10 ppm.

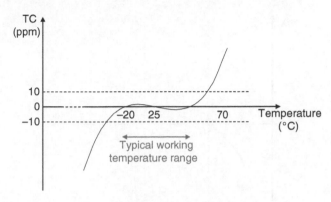

Figure 11.16 Typical variation of TC with temperature for an AT-cut quartz crystal with $\phi = 35°15'$.

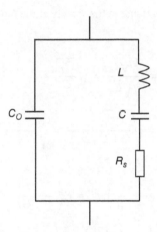

Figure 11.17 Equivalent circuit of a crystal.

The equivalent circuit of a crystal is shown in Figure 11.17.

In the equivalent circuit the values of L and C are related to the stiffness of the material, and R_s to the mechanical damping. Since the values of L, C, and R_s depend on the mechanical motion within the crystal they are also referred to as motional parameters. The static capacitance of the structure is represented by C_O. This capacitance can be viewed simply as that between the electrodes shown in Figure 11.14a, which are separated by a quartz dielectric. Any fringing capacitance across the structure is also included in C_O. Normally the value of C is very small, of the order of fF, and L is very large, typically in excess of 1H.

A crystal has two fundamental resonant frequencies, namely f_s and f_p. The series resonant frequency, f_s, where the impedance is minimum is given by

$$f_s = \frac{1}{2\pi\sqrt{LC}}. \tag{11.27}$$

The parallel resonant frequency (antiresonant frequency), f_p, at which the impedance is maximum is found by determining the impedance of the parallel circuit shown in Figure 11.17. After straightforward, but somewhat lengthy circuit analysis, it can be shown that the parallel resonant frequency is given by [1]

$$f_p = \frac{1}{2\pi\sqrt{LC_T}}, \tag{11.28}$$

where

$$C_T = \frac{CC_O}{C + C_O}. \tag{11.29}$$

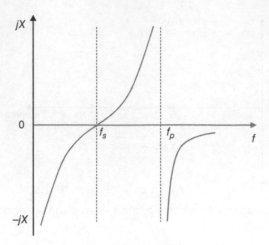

Figure 11.18 Variation of crystal reactance in the vicinity of the fundamental resonances, when $R_s = 0$.

Combining Eqs. (11.27) and (11.28) gives

$$\frac{f_s}{f_p} = \frac{1/\sqrt{C}}{1/\sqrt{C_T}} = \sqrt{\frac{C_T}{C}}$$

$$= \sqrt{\frac{CC_O}{C + C_O} \times \frac{1}{C}} = \sqrt{\frac{C_O}{C + C_O}}.$$

Therefore,

$$f_p = f_s \left(\frac{C_O}{C + C_O}\right)^{-0.5} = f_s \left(\frac{1}{1 + C/C_O}\right)^{-0.5}$$

$$= f_s \left(\left[1 + \frac{C}{C_O}\right]^{-1}\right)^{-0.5} = f_s \left(1 + \frac{C}{C_O}\right)^{0.5}.$$

Using the binomial expansion, and noting that C/C_O is small, we have

$$\left(1 + \frac{C}{C_O}\right)^{0.5} = 1 + 0.5\frac{C}{C_O},$$

giving

$$f_p = f_s \left(1 + \frac{0.5C}{C_O}\right). \tag{11.30}$$

Since C/C_O is small, f_p and f_s will be close. Figure 11.18 shows the typical variation of crystal impedance as a function of frequency, in the vicinity of the two resonant frequencies, for a crystal where we have assumed $R_s = 0$. The shape of the impedance-frequency graph follows from the reactance of a series resonant circuit being zero at the resonant frequency and very large (theoretically infinite) at the parallel resonant frequency.

It can be seen from Figure 11.18 that there is a very steep variation of inductive reactance between the series and parallel resonant frequencies, and this is the normal operating region for crystals used in crystal-controlled oscillators. Normally the frequency separation of f_s and f_p is small compared to the resonant frequencies, as shown in Example 11.7, and the Q-factor of the crystal is very large, as shown in Example 11.8.

Example 11.7   A quartz crystal with a parallel resonant frequency of 8.8 MHz has the following parameters: $C = 26$ fF, $C_O = 9.6$ pF. Calculate the difference between the series and parallel resonant frequencies.

Solution
The series and parallel resonant frequencies are related through Eq. (11.30):

$$f_p = f_s \left(1 + \frac{0.5C}{C_O}\right).$$

Substituting given data we have:

$$8.8 \times 10^6 = f_s \left(1 + \frac{0.5 \times 0.026}{9.6}\right).$$

$$f_s = 8.788 \text{ MHz}.$$

Difference between series and parallel resonant frequencies is then:

$$\Delta f = f_p - f_s = (8.8 - 8.788) \text{ MHz} = 12 \text{ kHz}.$$

Example 11.8   The following data apply to an 18.6 MHz crystal: $R = 8.2\ \Omega$, $C = 29$ fF, $C_O = 7$ pF. What is the Q of the crystal?

Solution
Using the standard definition of Q for a capacitor we have

$$Q = \frac{1}{\omega C_T R}.$$

Using Eq. (11.29)

$$C_T = \frac{C C_O}{C + C_O}.$$

Since $C \ll C_O$ we can put

$$C_T = C.$$

Then

$$Q = \frac{1}{\omega C R} = \frac{1}{2\pi \times 18.6 \times 10^6 \times 29 \times 10^{-15} \times 8.2}$$

$$= 35,978.17.$$

11.5.2   Crystal-Controlled Oscillators

Since crystals have much higher Q values than LC resonant circuits they can be incorporated in the feedback circuits of oscillators to very significantly improve the frequency stability and reduce the oscillator noise. One of the common techniques for making a crystal-controlled oscillator is to replace the inductor in the feedback circuit of a Colpitts oscillator with a suitable crystal, as shown in Figure 11.19.

In the circuit shown in Figure 11.19, the crystal operates in the parallel mode so that it presents an effective inductance. It is important to appreciate that the resonant frequency of the crystal will be affected by the load capacitance presented by the rest of the circuit. For the circuit shown in Figure 11.19 the load capacitance, C_L, is given by

$$C_L = \frac{C_1 C_2}{C_1 + C_2}. \tag{11.31}$$

When a crystal manufacturer trims a crystal to work at a particular frequency, this is done for a particular load capacitance. It is therefore essential that the circuit designer should specify the correct load capacitance when ordering a crystal for a particular application. It should further be appreciated that the values of C_1 and C_2 do not include the terminal capacitances of the transistor, and these can significantly alter the load capacitance at higher frequencies in the RF spectrum.

In some designs a variable capacitor is used in series with the crystal to tune the load capacitance to the correct value. The circuit then resembles that of the Clapp oscillator which was discussed in an earlier section.

Crystal-controlled oscillators are essentially fixed frequency devices, but a limited amount of frequency tuning is possible by including a variable capacitor, such as a varactor diode, in parallel with the crystal. This enables the antiresonant frequency to be pulled towards the series resonant value. The difference between f_s and f_p, as shown in Figure 11.18, is known

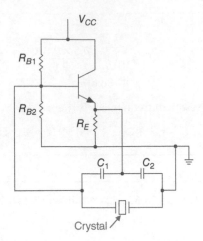

Figure 11.19 Crystal-controlled oscillator.

as the pulling range of the oscillator. From Eq. (11.30) the pulling range, $f_p - f_s$ of the oscillator can be written as

$$f_p - f_s = \frac{0.5C}{C_O} f_s. \tag{11.32}$$

Since $C \ll C_O$ the pulling range, and hence the tuning range, is normally quite small.

There are two problems associated with crystal-controlled oscillators using the parallel mode, as in Figure 11.19:

(i) It can be shown [1] that for crystal-controlled oscillators with crystals operating in the parallel mode the conditions for oscillation require

$$g_m \geq 4C_1 C_2 \pi^2 f^2 R, \tag{11.33}$$

where g_m is the transconductance of the transistor, R is the series resistance of the crystal, f is the frequency of oscillation, and C_1 and C_2 are the capacitors in the feedback network, as shown in Figure 11.19. Since the value of transconductance is fixed for a particular transistor, the condition given in Eq. (11.33) requires that C_1 and C_2 be reduced to maintain oscillation as the frequency increases. However, as C_1 and C_2 approach the values of the terminal capacitances of the transistor, the stability of the oscillator decreases because these terminal capacitances are not well defined, and so it is difficult to include consistent values in the design.

(ii) It can be seen through inspection of Figure 11.19 that the two biasing resistors, R_{B1} and R_{B2}, are effectively in parallel with the crystal for AC operation. This has the effect of placing a resistive load on the crystal, which reduces the Q of the resonant circuit, and thereby reduces the frequency stability of the oscillator.

The two problems identified above for parallel mode crystal oscillators can be overcome by connecting the crystal so that it operates in a series resonant mode. Crystal oscillators using series resonant crystals are referred to as Pierce oscillators, and most RF transistor oscillators are designed in this way. Pierce and Colpitts oscillators use similar circuits, with the main difference being in the ground location within the feedback network. To clarify this point examples of Pierce and Colpitts crystal-controlled oscillators are shown in Figure 11.20, with the feedback paths identified by the dotted lines.

It can be seen from Figure 11.20 that the crystal is directly in the feedback path of the Pierce oscillator, and so operates in the series resonant mode, whereas in the Colpitts oscillator the crystal is in parallel with the feedback path and so operates in the parallel, or antiresonant, mode. However, the current consumption tends to be higher for oscillators operated with crystals in the series resonant mode, and this must be considered when selecting a particular topology for an oscillator circuit.

11.6 Frequency Synthesizers

In Section 11.5, we saw how the frequency stability of an oscillator can be significantly improved through the use of electronic crystals. However, crystals have severe frequency limitations with an upper frequency of around 200 MHz for crystals operating in the fundamental mode. Consequently, much RF and microwave equipment and instrumentation

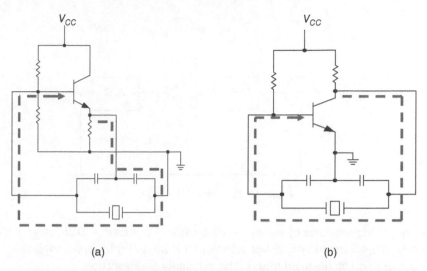

Figure 11.20 Comparison between Pierce (a) and Colpitts oscillators (b).

makes use of frequency synthesizers. A frequency synthesizer is an electronic sub-system that generates a digitally controllable frequency at its output. The output frequency is derived, or synthesized, from an extremely stable LF source in such a way that the output frequency has the same stability characteristics as the LF source. Essentially, there are three approaches to the design of modern frequency synthesizers:

Direct synthesizer (DS): The output from a reference oscillator, usually a crystal-controlled LF source, is directly processed through the use of multipliers, dividers, and mixers, to produce the required RF frequency. The multiplier and divider functions are normally digitally controlled so as to provide a programmable output frequency.

Advantages of direct frequency synthesis:

(1) *Fast frequency switching*
(2) *Very fine frequency resolution*
(3) *Low phase noise.*

Disadvantages of direct frequency synthesis:

(1) *Complex hardware is needed*
(2) *Expensive*
(3) *Spurious frequencies at the output.*

Indirect synthesizer (IDS): A phase-locked loop (PLL) is used to lock the frequency of a VCO to that of a reference oscillator, which is normally a crystal-controlled source. Indirect synthesis involves relatively little hardware, and in its simplest form just requires a single loop circuit. It overcomes many of the disadvantages associated with direct synthesis, and in particular it does not suffer from spurious frequencies in the output.

Direct digital synthesizer (DDS): In this type of synthesizer a sine wave of the required frequency is digitally constructed, using a phase accumulator and sine wave look-up table [3]. The digitized signal is then converted to an analogue format using a DA converter, and this provides the RF output.

Of the three types of synthesizers mentioned above, the digital architecture of the DDS give it a number of specific advantages:

- High-frequency agility
- Fast lock-up (low settling times between switched frequencies)
- Precise tuning
- Very fine frequency tuning, to nano-Hz values
- Low phase noise (the only significant noise being clock jitter)
- Low power consumption
- Potential for compact integrated circuit format.

These advantages contribute to the increasing popularity of DDS devices in practical systems. However, the highest output frequency available from a DDS device is limited by the speed of the digital circuitry, and in particular the highest available

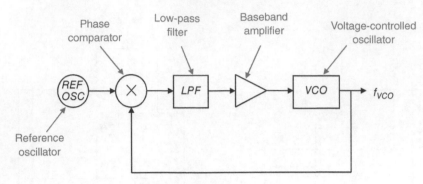

Figure 11.21 Basic PLL circuit.

clock frequency. Currently, the highest output frequency available from a commercial DDS integrated circuit is of the order of 500 MHz. Consequently, the dominant synthesizer technique for RF and microwave applications is still the indirect frequency synthesizer, and this will be the main focus of the remainder of this section.

11.6.1 The Phase-Locked Loop

The underlying principle of indirect frequency synthesizers is that of the PLL, and before considering practical IDS circuits the essential behaviour of PLL circuits will be reviewed.

11.6.1.1 Principle of a Phase-Locked Loop

The primary function of a PLL is to lock the frequency and phase of a VCO to that of a very stable reference oscillator. In its simplest form a PLL can be represented by the circuit shown in Figure 11.21, and this will be used to explain the basic operation.

In this circuit, the output of a VCO is connected to one input of a phase comparator. The output of a reference source, which is usually a very stable crystal oscillator, is connected to the other input of the phase comparator. The function of the phase comparator is to provide a voltage output proportional to the difference in phase between the two input signals. Since frequency is simply the rate of change of phase, the phase comparator will also give a voltage output proportional to the difference in frequency between the two input signals. Thus, an output voltage, V_c, will be produced which is proportional to the difference in frequency between the VCO and the reference oscillator. This voltage is sometimes referred to as the error voltage. A low-pass filter (LPF) is used to remove any unwanted frequency products from the phase comparator. The required error voltage will be present at the output of the LPF, and this is connected to the frequency control input of the VCO. An amplifier is normally included between the LPF and the VCO to increase the level of V_c, thereby increasing the sensitivity of the circuit. The polarity of V_c is chosen such that the frequency of the VCO is always driven closer to that of the reference source. Thus, any change in the frequency of the VCO will generate an error voltage which will drive the VCO frequency towards that of the reference oscillator. When the VCO frequency and phase are the same as those of the reference source, the error voltage will be zero and the circuit is said to be in lock.

11.6.1.2 Main Components of a Phase-Locked Loop

Phase Comparator   Phase detectors can be implemented using analogue or digital techniques. Whilst the majority of modern RF phase detectors are constructed using digital techniques, the essential principle can conveniently be explained by considering an ideal analogue phase detector as shown in Figure 11.22.

The phase comparator shown in Figure 11.22 consists of an ideal multiplier, with two voltage inputs v_i and v_r, followed by an LPF. The output from the multiplier will be

$$v_m = \sin(\omega_i t + \phi_i) \cos(\omega_r t + \phi_r) \text{ volts}$$
$$= 0.5[\sin(\omega_i t + \phi_i + \omega_r t + \phi_r) + \sin(\omega_i t + \phi_i - \omega_r t - \phi_r)].$$

When the loop is locked, $\omega_i = \omega_r$, then

$$v_m = 0.5\,[\sin(2\omega_i t + \phi_i + \phi_r) + \sin(\phi_i - \phi_r)]. \tag{11.34}$$

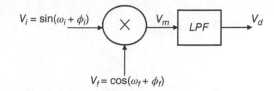

Figure 11.22 Analogue phase comparator.

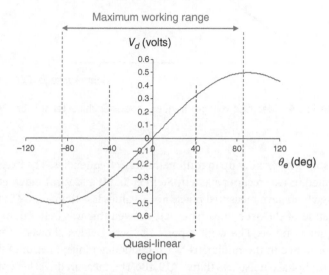

Figure 11.23 Response of analogue phase comparator.

The LPF will remove the high-frequency term, leaving

$$v_d = 0.5 \sin(\phi_i - \phi_r). \tag{11.35}$$

For small differences between ϕ_i and ϕ_r, we have

$$v_d \approx 0.5(\phi_i - \phi_r) = 0.5\,\theta_e, \tag{11.36}$$

where

$$\theta_e = \phi_l - \phi_r.$$

The response of the phase comparator is sketched in Figure 11.23, and shows the quasi-linear region in the vicinity of $v_d = 0$.

Since the PLL must be designed to provide either a positive or negative control voltage, the maximum working range will be equal to one-half of the phase detector cycle. Normally, phase detectors are designed to work over a linear, or quasi-linear, range where the response of the phase detector may be written as

$$v_d = k_{PD}\,\theta_e, \tag{11.37}$$

where k_{PD} is the transfer function or gain of the phase comparator.

Voltage-Controlled Oscillator A VCO is an oscillator whose output frequency depends on the value of an external DC bias voltage. There are various circuit techniques available for voltage-tuning an oscillator, but the two most common methods at RF and microwave frequencies are those using either varactor diodes or Yttrium Iron Garnet (YIG) resonators.

YIG is a ferrimagnetic material that can be used to make very high-Q resonators. The resonators take the form of small spheres (around 500 μm in diameter) manufactured from a single crystal of the material. The sphere is positioned in a DC magnetic field, and tuned by varying the DC current controlling the strength of the field. A YIG-tuned oscillator can be tuned over a much greater frequency range than one using a varactor, although the tuning is not as fast as with a varactor. The very high-Q of the resonator also means that YIG-tuned oscillators exhibit low phase noise, which is defined in Chapter 14. The combination of low phase noise and wideband operation make YIG-tuned oscillators very attractive for

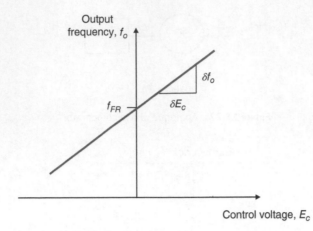

Figure 11.24 Idealized voltage–frequency transfer characteristic for a VCO.

high-quality, wideband signal sources working up to millimetre-wave frequencies. The excellent performance achievable using YIG-tuned VCOs is reflected in two recent papers by Stein et al. [4] and van Delden et al. [5]. Stein reported results for a 6–12 YIG-tuned oscillator, which gave measured phase noise values less than −130 dBc/Hz measured 100 kHz off the carrier. Stein also presented a state-of-the-art comparison that showed this was better than that of other VCOs currently working in the measured frequency range. The work reported by van Delden showed that YIG-tuned oscillators could give very good, wideband performance in the millimetre-wave frequency range; measured results showed a tuning range from 32 to 48 GHz, with measured phase noise less than −119 dBc/Hz, measured 100 kHz off the carrier, up to 42 GHz. A further interesting aspect of the work from was that the YIG had been successfully coupled between two bond pads on a SiGe MMIC.

A typical, idealized, voltage–frequency transfer characteristic for a VCO is shown in Figure 11.24. This characteristic shows a linear variation of frequency with voltage: one of the most difficult issues in VCO design is achieving the required degree of linearity.

For an oscillator with the frequency-bias characteristic shown in Figure 11.24, a positive increase in the bias voltage causes an increase in the oscillator frequency. When the bias voltage is zero, the oscillator generates its free-running frequency, f_{FR}. In this example, the transfer characteristic is linear, and the slope, k_{VCO}, of the characteristic is

$$k_{VCO} = \frac{\delta f_O}{\delta E_c}. \tag{11.38}$$

k_{VCO} is normally referred to simply as the input–output transfer function of the VCO, and has the units Hz/V, and E_c is the control voltage. Thus, in general the frequency of the oscillator may be written as

$$f_O = f_{FR} + k_{VCO}\, E_c, \tag{11.39}$$

where E_c is a particular value of bias or control voltage.

The corresponding phase, $\phi_i(t)$, is given by

$$\phi_i(t) = 2\pi \int (f_O + k_{VCO}E_c)\, dt$$
$$= 2\pi f_O t + 2\pi k_{VCO} \int E_c\, dt + \theta_o, \tag{11.40}$$

where θ_o is a constant of integration, with a value corresponding to the phase at time $t = 0$. We may assume that the phase is zero at $t = 0$, without any loss of generality. Thus, the phase deviation, $\Delta\phi_i$, due to a control voltage E_C is given by

$$\Delta\phi_i = 2\pi k_{VCO} \int E_c\, dt. \tag{11.41}$$

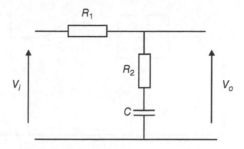

Figure 11.25 Typical low-pass loop filter.

If the control voltage is sinusoidal, then $E_c = E_{cm} \sin \omega_c t$, and

$$\Delta\phi_i = \frac{2\pi k_{VCO}}{\omega_c}(-E_{cm}\cos\omega_c t)$$

$$= \frac{2\pi k_{VCO}}{\omega_c}E_{cm}\sin(\omega_c t - 90°)$$

$$= \frac{k_{VCO}}{f_c}E_c\angle - 90°. \tag{11.42}$$

The phase sensitivity of the *VCO*, PS_{VCO}, is then defined as

$$PS_{VCO} = \frac{\Delta\phi_l}{e_c} = \frac{k_{VCO}}{f_c}\angle - 90°. \tag{11.43}$$

Low-Pass Filter The primary reason for including an LPF in the PLL is to remove unwanted frequency components at the output of the phase detector. Therefore, the bandwidth of the filter must be less than the value of the reference frequency, f_i, so as to block all the intermodulation and harmonic products involving f_i. However, the frequency response of the filter is also of crucial importance in determining the dynamic response of the loop and its stability. Making the bandwidth of the filter too small will result in poor response times, as is shown later in this chapter.

Figure 11.25 shows a typical second-order LPF used in PLL systems. This type of filter is also referred to as a lag–lead filter, from the shape of its transmission phase characteristic.

From a simple potential divider theory the voltage gain of the filter is given by

$$\frac{V_O}{V_i} = \frac{R_2 + {}^1\!/_{j\omega C}}{R_1 + R_2 + {}^1\!/_{j\omega C}}$$

$$= \frac{1 + j\omega CR_2}{1 + j\omega C(R_1 + R_2)}, \tag{11.44}$$

$$= \frac{1 + j\omega\tau_2}{1 + j\omega\tau_1}$$

where

$$\tau_1 = C(R_1 + R_2) \quad \text{and} \quad \tau_2 = CR_2.$$

A comprehensive discussion of the influence of the transmission response of the filter on the performance of a PLL is given by Smith in reference [1].

11.6.1.3 Gain of Phase-Locked Loop

A simple PLL, with the gain blocks identified, is shown in Figure 11.26, and follows from that of Figure 11.21. In Figure 11.26, k_A is the voltage gain of the broadband amplifier; the remaining gain factors have been defined in the text.

The closed-loop gain, k_L, is normally defined as the ratio of the change in VCO output frequency, Δf, to the difference in phase, $\Delta\theta$, between the phase detector inputs, i.e.

$$k_L \equiv \frac{\Delta f}{\Delta\theta}. \tag{11.45}$$

Now,

$$\Delta f = k_{VCO}v_C = k_{VCO}k_{LPF}k_A v_d \tag{11.46}$$

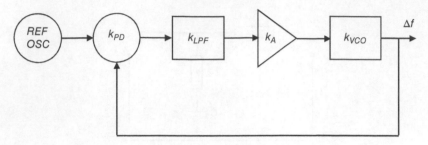

Figure 11.26 Phase-locked loop, showing gain blocks.

and

$$\Delta\theta = \frac{v_d}{k_{PD}}. \tag{11.47}$$

Thus,

$$k_L = k_{PD}k_{LPF}k_Ak_{VCO}. \tag{11.48}$$

So the loop gain is simply the product of the gain factors of the components of the loop. Some caution is needed when considering the units of k_L. Writing the equation for the loop gain with the units of each block identified gives

$$k_L = k_{PD}[\text{V/rad}] \times k_{LPF}[\text{Hz/Hz}] \times k_A[\text{V/V}] \times k_{VCO}[\text{Hz/V}], \tag{11.49}$$

from which we see that k_L has the units of Hz/rad. Consequently, the loop gain is often defined as

$$k_L = 2\pi k_{PD}k_{LPF}k_Ak_{VCO}, \tag{11.50}$$

so that k_L is dimensionless.

11.6.1.4 Transient Analysis of a Phase-Locked Loop

The interest in PLLs in this chapter is limited to their use in indirect frequency synthesizers. But since these synthesizers necessarily involve step changes in frequency and phase, transient effects within a PLL are of particular significance. A rigorous analysis of these transient effects is outside the scope of this book, and so only the main points will be summarized in this section. Transient effects within the loop can conveniently be demonstrated by considering the effect that a step change in input phase has on the phase at the output of the loop, i.e. on the phase of the signal at the output of the VCO. The analysis is most conveniently performed in the Laplacian domain, and Figure 11.27 shows a PLL with the responses of the component blocks expressed in terms of the Laplacian operator, s.

In Figure 11.27, the transfer function of the LPF is represented by $F(s)$. Also, we know from Section 11.6.1.2 that the VCO effectively integrates the phase of the signal at its input, and so we may write its transfer function as K_{VCO}/s.

The phase at the output under open-loop conditions will be

$$\theta_o(s) = \theta_i(s)K_{PC}F(s)K_A\frac{K_{VCO}}{s}, \tag{11.51}$$

and when the loop is closed this becomes

$$\theta_o(s) = \theta_i(s)K_{PC}F(s)K_A\frac{K_{VCO}}{s} - \theta_o(s)K_{PC}F(s)K_A\frac{K_{VCO}}{s}. \tag{11.52}$$

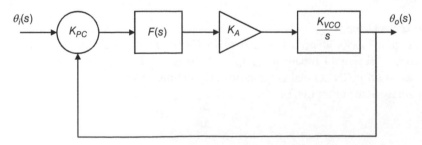

Figure 11.27 PLL in the Laplace domain.

Rearranging gives

$$\theta_o(s) + \theta_o(s)K_{PC}F(s)K_A\frac{K_{VCO}}{s} = \theta_i(s)K_{PC}F(s)K_A\frac{K_{VCO}}{s}, \tag{11.53}$$

i.e. the phase transfer function, $H_\theta(s)$, of the PLL system is

$$H_\theta(s) = \frac{\theta_o(s)}{\theta_i(s)} = \frac{K_{PC}F(s)K_A\dfrac{K_{VCO}}{s}}{1 + K_{PC}F(s)K_A\dfrac{K_{VCO}}{s}} = \frac{K_{PC}F(s)K_AK_{VCO}}{s + K_{PC}F(s)K_AK_{VCO}}. \tag{11.54}$$

If we consider the filter to have the same form as that shown in Figure 11.25, then it follows from Eq. (11.44) that the transfer function of the filter can be written as

$$F(s) = \frac{1 + \tau_2 s}{1 + \tau_1 s}. \tag{11.55}$$

Substituting for $F(s)$ in Eq. (11.54), and combining the gain factors gives

$$H_\theta(s) = \frac{K_T\dfrac{1 + \tau_2 s}{1 + \tau_1 s}}{s + K_T\dfrac{1 + \tau_2 s}{1 + \tau_1 s}}, \tag{11.56}$$

i.e.

$$H_\theta(s) = \frac{K_T(1 + \tau_2 s)}{s^2\tau_1 + s(1 + K_T\tau_2) + K_T}, \tag{11.57}$$

where

$$K_T = K_{PD}K_{LPF}K_AK_{VCO}. \tag{11.58}$$

From [6] Eq. (11.58) can be rewritten as

$$H_\theta(s) = \frac{\theta_o(s)}{\theta_i(s)} = \frac{1 + \dfrac{s}{\omega_n}\left(2\zeta - \dfrac{\omega_n}{k_T}\right)}{1 + \dfrac{s}{\omega_n}2\zeta + \left(\dfrac{s}{\omega_n}\right)^2}, \tag{11.59}$$

where

$$\omega_n = \sqrt{\frac{K_T}{\tau_1}} \tag{11.60}$$

and

$$\zeta = \frac{\omega_n}{2}\left(\frac{1}{K_T} + \tau_2\right). \tag{11.61}$$

The parameters ω_n and ζ are defined, respectively, as the loop bandwidth and damping factor. If $\frac{1}{K_T} \gg \tau_2$, then Eq. (11.61) can be written as

$$\zeta = \frac{\omega_n}{2K_T} = \frac{1}{2\tau_1\omega_n}. \tag{11.62}$$

The damping ratio and the loop bandwidth determine the transient response of the PLL system. Figure 11.28 shows normalized time domain responses of a second-order[1] PLL system, when a step change in phase is applied to the input at time $t = 0$. In this case normalization means that the values of the output phase have been adjusted so that the final, steady-state output phase has a value of one unit.

The transient responses have the form of decaying oscillatory functions, which will be familiar to readers with experience of feedback control systems. It can be seen that small values of damping factor lead to significant overshoot. If $\zeta < 1$, the system is described as being under-damped, whereas if $\zeta > 1$, the system is over-damped. An over-damped PLL system eliminates the overshoot effect, but takes a long time to settle to the steady-state output condition. It can be seen from Eq. (11.62) that a small damping factor requires a large loop bandwidth. This can be a problem in frequency synthesizers where high-frequency resolution requires a narrowband LPF in the loop, and consequently a low loop bandwidth.

---

1 A second-order system is one in which the input–output performance can be represented by a second-order linear differential equation.

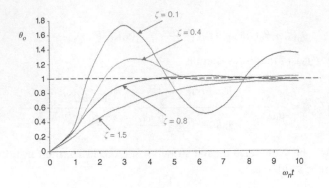

Figure 11.28 Normalized step responses of a second-order PLL system.

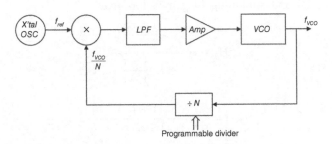

Figure 11.29 Basic indirect frequency synthesizer.

11.6.2 Indirect Frequency Synthesizer Circuits

The arrangement of a basic indirect frequency synthesizer is shown in Figure 11.29. The system is similar to the phase-locked circuit shown in Figure 11.21, but includes a frequency divider in the feedback path, so that the VCO frequency fed back to the phase detector is divided by a factor N.

In this type of synthesizer, the PLL is used to lock the output frequency of a RF or microwave VCO to that of a very stable LF reference source, which is normally a crystal-controlled oscillator (XTCO). The operation is very straightforward: the output frequency of the VCO, f_{osc}, is divided down to a lower frequency, which is then compared with the reference frequency, f_{ref}, in a phase sensitive detector (PSD). The function of the PSD is to give a DC output proportional to the difference in phase of the two input signals. The phase detector will also give a DC output if there is any difference in frequency between the two input signals. The output of the phase detector is passed through an LPF to remove any unwanted products from the PSD, to form a DC control signal for the VCO.

The synthesizer is described as being in lock when the frequency of the feedback input to the phase detector is the same as the reference frequency, i.e.

$$f_{ref} = \frac{f_{VCO}}{N}. \tag{11.63}$$

If the value of N is changed, the feedback frequency to the phase detector will change, and an error voltage will be generated at the output of the phase detector. This error voltage will drive the VCO frequency to a new value, so as to maintain the frequency fed back to the phase detector equal to f_{ref}. The synthesizer will then lock up at the new VCO frequency. The divider is normally programmable, so that desired VCO frequencies can be digitally selected.

Example 11.9   An RF VCO is to be tuned from 85 MHz to 105 MHz in 1000 steps, using the frequency synthesizer circuit shown in Figure 11.29. Determine the frequency of the crystal oscillator and the required range of N in the frequency divider.

Solution
The frequency step size will be

$$\Delta f = \frac{105 - 85}{1000} \text{ MHz} = 20 \text{ kHz}.$$

With this type of synthesizer the frequency of the crystal oscillator must equal the required step size, i.e.

$$f_{X'tal\ osc} = 20\ kHz.$$

Lower value of N:

$$N_{lower} = \frac{85\ MHz}{20\ kHz} = \frac{85 \times 10^6}{20 \times 10^3} = 4250.$$

Upper value of N:

$$N_{upper} = \frac{105\ MHz}{20\ kHz} = \frac{105 \times 10^6}{20 \times 10^3} = 5250.$$

The simple arrangement shown in Figure 11.29 requires very large values of N if an RF frequency at the output of the VCO is to be locked to a crystal reference frequency in the kHz region (as we saw in Example 11.9). Normally, for RF applications a pre-scalar is included in the feedback path as shown in Figure 11.30. In this instance the pre-scalar divides the VCO output frequency by a fixed factor of M, so that when the circuit is in lock we have

$$f_{ref} = \frac{f_{VCO}}{M \times N}. \tag{11.64}$$

The pre-scalar thus reduces the complexity and expense of the programmable divider.

Example 11.10   Repeat Example 11.9 using the circuit arrangement shown in Figure 11.30, with a pre-scalar value of 10.

Solution
The frequency of the crystal oscillator remains unchanged at 20 kHz, since this must equal the frequency step size from the VCO.

The values of N are reduced by a factor of 10, so the new range of N is 425–525.

An alternative to the pre-scalar synthesizer is to use a circuit with an offset oscillator feedback loop, as shown in Figure 11.31.

The offset oscillator mixes down the frequency of the VCO so that the programmable divider can operate at a lower frequency. Note that since the offset oscillator is included in the feedback loop, any noise from this oscillator stage is automatically suppressed by the action of the feedback loop. The circuit shown in Figure 11.31 includes an LPF following the offset oscillator to remove unwanted frequency products at the output of the mixer. Then we have

$$f_1 = f_{VCO} - f_m$$

and

$$f_{ref} = \frac{f_1}{N}. \tag{11.65}$$

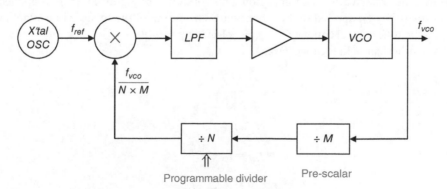

Figure 11.30   Indirect frequency synthesizer, including a pre-scalar.

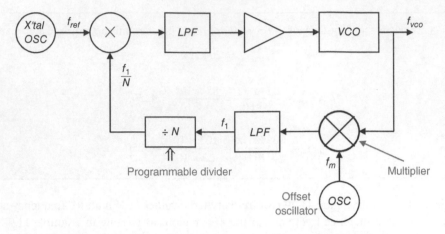

Figure 11.31 Indirect frequency synthesizer using an offset oscillator.

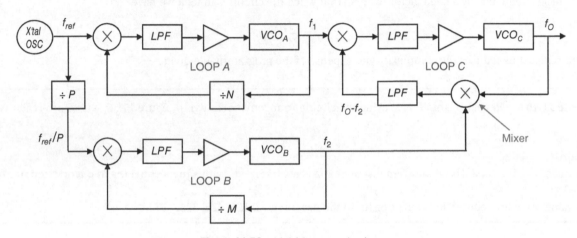

Figure 11.32 Multi-loop synthesizer.

The use of an offset oscillator is necessary when the VCO is operating at microwave frequencies, where limitations on the speed of digital circuitry preclude the use of a pre-scalar.

One drawback to the synthesizers described so far is that the minimum frequency step, i.e. the frequency resolution, is fixed by the value of the reference frequency. Improving the resolution over a given frequency range is only possible by reducing the value of the reference frequency and increasing the value of N. But small values of reference frequency lead to low loop bandwidths, and longer settling times, and also large values of N lead to large changes in dynamic range. A more versatile frequency synthesizer can be made by combining the outputs of several individual loops to form the so-called multi-loop synthesizer, an example of which is shown in Figure 11.32.

Loops A and B function as Integer-N synthesizers, and give frequency outputs f_1 and f_2. These frequencies are combined in loop C, which is sometimes referred to as a translation loop. As shown in Figure 11.32, f_2 is applied to one input of a mixer, with the output frequency f_O applied to the other input. The output of the mixer will contain all of the frequency products $mf_O \pm nf_2$, and the difference frequency $f_O - f_2$ is selected by the LPF.

When the multi-loop synthesizer is in lock we have:

Loop A:
$$f_1 = Nf_{ref},$$

Loop B:
$$f_2 = M\frac{f_{ref}}{P},$$

Loop C:
$$f_1 = f_0 - f_2,$$

giving
$$Nf_{ref} = f_0 - \frac{Mf_{ref}}{P}$$

or
$$f_O = \left(N + \frac{M}{P}\right)f_{ref}. \tag{11.66}$$

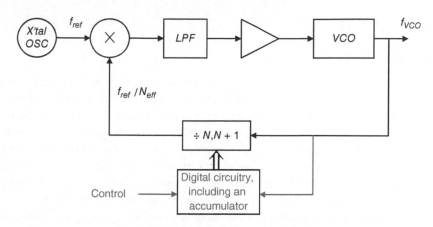

Figure 11.33   Fractional-*N* synthesizer.

Compared to single loop synthesizers, the multi-loop variant gives greater flexibility in selecting the step size, but at the cost of more complex circuitry together with a relatively slow lock-up time.

An alternative to the multi-loop synthesizer is the Fractional-*N* synthesizer, which is shown in Figure 11.33. The circuit has the same general form as a single-loop Integer-*N* synthesizer, but uses a divisor which can be digitally set to one of two values, namely *N* and *N* + 1.

The concept is that by periodically changing the divisor value from *N* to *N* + 1, and vice versa, the divider will function with an effective non-integer (fractional) value between *N* and *N* + 1. This is effectively an averaging process, and in the circuit shown in Figure 11.33 this is achieved by dividing by *N* for a certain number of VCO cycles, and then by *N* + 1 for a following number of VCO cycles. Digital circuitry is used to control the value of the divider, and includes an accumulator which is used to count the required number of VCO cycles. A control input enables the number of cycles corresponding to the two divider states to be changed. Thus, the divider functions with an effective non-integer divisor value, N_{eff}. Suppose that for *p* cycles of the VCO the divisor value is *N*, and for the following *q* cycles the value is *N* + 1; the effective divisor value will then be

$$N_{eff} = \frac{p}{p+q}N + \frac{q}{p+q}(N+1) = N + \frac{q}{p+q}. \tag{11.67}$$

The design of a Fractional-*N* synthesizer is illustrated through the following worked example.

Example 11.11   Design a Fractional-*N* synthesizer to produce an output frequency of 51.1 MHz using a reference frequency of 10 MHz.

Solution

Obviously the output frequency is not an integer multiple of the reference frequency, and so we need to create an effective divisor value equal to 5.11. Using Eq. (11.66) we have

$$5.11 = 5 + \frac{11}{100} = N + \frac{q}{p+q}.$$

It then follows that we must have

$$N = 5 \quad q = 11 \quad p + q = 100 \quad p = 89.$$

So the requirements for the divider in this Fractional-*N* synthesizer are:

$$N = 5 \text{ for } 89 \text{ VCO cycles,}$$
$$N = 6 \text{ for } 11 \text{ VCO cycles.}$$

The main advantage of the Fractional-*N* synthesizer is that it can achieve high-frequency resolution with small values of *N*, and in consequence it provides fast settling times. However, it suffers from one disadvantage, in that it tends to generate high spurious levels at the output of the VCO.

11.7 Microwave Oscillators

11.7.1 Dielectric Resonator Oscillator

A DRO is formed by including a dielectric resonator in the resonant circuit of a transistor oscillator. The DRO provides a simple, fixed-frequency, low-cost oscillator with very good frequency stability, approaching that of a phase-locked oscillator.

The dielectric resonator (sometimes referred to as a dielectric puck) is formed from low-loss, high permittivity ceramic material. A range of ceramic materials can be used to manufacture dielectric resonators, with dielectric constants ranging from around 30 to several thousand. A material often used in microwave circuits is barium titanate ($BaTiO_3$), which has a dielectric constant in the range 30–80, with the precise value depending on the grain size and processing methodology. It can provide a Q of around 9000 at 10 GHz, and has a very low temperature coefficient of expansion around ± 4 ppm/°C. The dielectric resonator is usually in the form a small solid cylinder, as shown in Figure 11.34.

Normally the resonator is used with the dominant resonant mode, denoted by $TE_{01\delta}$, where the first two suffices have the same meaning as for the air-filled cylindrical waveguides discussed in Chapter 1, and the δ indicates that there is less than one half cyclic variation in the axial direction. The resonant frequency of the structure shown in Figure 11.34 is given approximately by Larson [7] as

$$f_r(GHz) = \frac{34}{a\varepsilon_r}\left[\frac{a}{h} + 3.45\right],$$
(11.68)

where a and h are both in millimetres, and

$$0.5 \le \frac{a}{h} \le 2$$

and

$$30 \le \varepsilon_r \le 50.$$

Using Eq. (11.68) a dielectric puck of ceramic material with a relative permittivity of 40, and dimensions $a = 2.5$ mm, $h = 1$ mm, has a resonant frequency of 2.02 GHz, showing the small component size that is possible at microwave frequencies.

In practical microwave circuits, the dielectric resonator is usually placed on a substrate in close proximity to a microstrip line, so there is electromagnetic coupling between the resonator and the line. The coupling mechanism is primarily magnetic. Sometimes a spacer, usually made of quartz ($\varepsilon_r = 3.8$), is inserted between the substrate and the resonator. The spacer has the effect of reducing the induced currents in the ground plane beneath the resonator, thereby increasing its unloaded Q-factor. Figure 11.35 shows a DR coupled to a section of matched microstrip line.

The dielectric resonator is normally modelled as a parallel LCR circuit, coupled to the microstrip line through a transformer, as shown in Figure 11.36.

Komatsu and Murakami [8] showed that the dielectric resonator presents a series impedance to the microstrip line, whose impedance is given by

$$Z_s = \frac{N^2 R}{1 + j2Q_u\delta_\omega},$$
(11.69)

where N is the turns ratio of the transformer, Q_u is the unloaded Q of the resonator, and

$$\delta_\omega = \frac{\omega - \omega_0}{\omega_0},$$

where ω_0 is the resonant frequency of the DR in isolation.

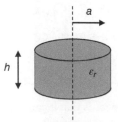

Figure 11.34 Dielectric resonator.

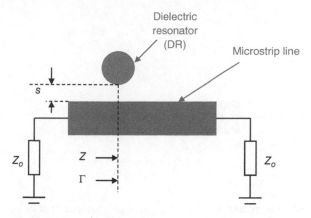

Figure 11.35 Coupling between a DR and a microstrip line.

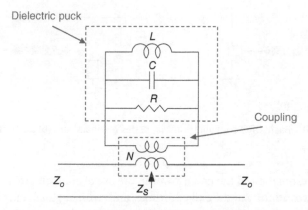

Figure 11.36 Equivalent circuit of a dielectric resonator.

Then the line impedance, Z, (see Figure 11.35) in the plane containing the resonator is

$$Z = Z_0 + Z_s = Z_0 \left(1 + \frac{N^2 R/Z_0}{1 + j2Q_u\delta_\omega} \right) = Z_0 \left(1 + \frac{\beta}{1 + j2Q_u\delta_\omega} \right), \tag{11.70}$$

where

$$\beta = \frac{N^2 R}{Z_0},$$

and β is defined as the coupling factor.

At resonance Eq. (11.70) becomes

$$Z = Z_0(1 + \beta). \tag{11.71}$$

Referring to Figure 11.35 we see that the reflection coefficient, Γ, in the same plane as Z is

$$\Gamma = \frac{Z - Z_0}{Z + Z_0}$$

$$= \frac{Z_0(1 + \beta) - Z_0}{Z_0(1 + \beta) + Z_0}$$

$$= \frac{\beta}{2 + \beta}. \tag{11.72}$$

So far we have considered the dielectric resonator to be coupled to a matched microstrip line. In some practical circuits the resonator is located $\lambda_s/4$ from an open-circuited microstrip line, as shown in Figure 11.37; the open-circuit creates a standing wave on the microstrip line, with a peak in the magnetic field in the plane of the resonator, thereby maximizing the

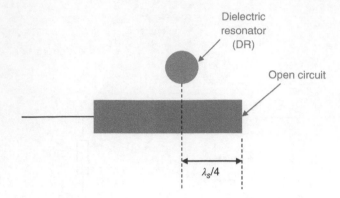

Figure 11.37 DR located $\lambda_s/4$ from the end of an open-circuited microstrip line.

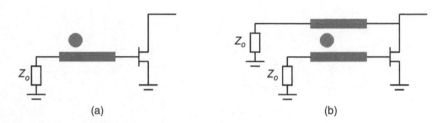

Figure 11.38 Basic DRO configurations: (a) series feedback and (b) parallel feedback.

magnetic coupling between the resonator and the line. However, the use of an open-circuited line has the minor disadvantages of creating an additional discontinuity at the end of the line, and reducing the bandwidth of the circuit. The nominal line length, $\lambda_s/4$, between the dielectric resonator and the open-circuit can shortened slightly to compensate for the stray, fringing capacitance at the open-circuit.

When used in transistor oscillators the dielectric resonator can be located in one of several possible positions. These are illustrated in Figure 11.38, where for clarity the biasing components for the FETs have been omitted.

In the case of parallel feedback the two microstrip lines are often inclined, which allows more flexibility in positioning the dielectric resonator, and hence permits some degree of circuit tuning.

A fundamental requirement for oscillation to start in a transistor network is that there should be a negative resistance created at the input and output terminals of the transistor. We will not repeat the justification for this which is presented in many textbooks, but we will show how this is achieved in the design of a DRO.

Figure 11.39 shows the overall layout of a series feedback DRO, using the conventional notation [9] for transistor oscillators.

The design of a DRO follows the well-established procedure for designing two-port transistor oscillators, and involves five steps:

(1) **Selecting a potentially unstable transistor, $K \leq 1$**
   Normally manufacturers' data sheets give the value of K along with the S-parameters to make the selection fairly straightforward.
(2) **Choosing a value of Γ_L to make $|\Gamma_{out}| \geq 1$**
   (i) Plot the input stability circle and choose Γ_L within the unstable region. *It is better not to choose a value of Γ_L with a large magnitude as this will require a large coupling coefficient between the dielectric puck and the microstrip connection to the gate of the transistor; large coupling coefficients can be difficult to achieve in practical circuits.*
   (ii) Calculate the resulting value of Γ_{out} (making use of the modified S-parameter equation, (5.12), in Chapter 5)

$$\Gamma_{out} = S_{22} + \frac{S_{12}S_{21}\Gamma_L}{1 - S_{11}\Gamma_L}. \tag{11.73}$$

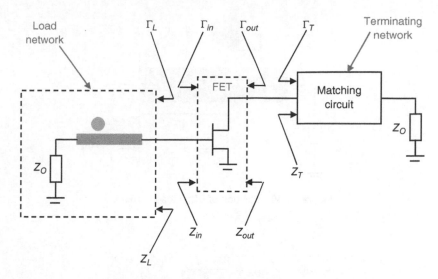

Figure 11.39 Dielectric resonator oscillator.

(3) **Calculating the value of Z_{out}**

$$Z_{out} = Z_0 \frac{1 + \Gamma_{out}}{1 - \Gamma_{out}}$$

$$= -R_{out} + jX_{out}. \tag{11.74}$$

(4) **Designing the terminating network to match Z_{out} to the output impedance, Z_o.**

For matching, the input impedance of the terminating network must be the complex conjugate of Z_{out}. However, we have so far used small-signal S-parameters in our design. But as the amplitude of the oscillations builds up in the circuit we must consider the large-signal effects within the transistor. The reactive components are essentially unaffected by the amplitude of the signals, but the magnitudes of the resistive components, R_{in} and R_{out}, will decrease. This means that R_{out} will become less negative as the oscillations increase in amplitude. It is important to ensure that the total resistance at the output of the transistor remains negative, otherwise the oscillations will cease at some point. To maintain oscillation, the normal practice is to decrease the resistive component of Z_T, i.e. make

$$R_T = n|R_{out}|, \tag{11.75}$$

where n is a fraction. Usually $n = 1/3$ is employed in practical designs, and so the required input impedance of the terminating network is

$$Z_T = \frac{R_{out}}{3} - jX_{out}. \tag{11.76}$$

The terminating network can now be designed using the matching techniques discussed in earlier chapters.

(5) **Designing the load network to give the required value of Γ_L**

In step (2) we chose a value for Γ_L, and to complete the design we need to determine the position of the dielectric resonator in the load network that will produce the required value of Γ_L. There are two design variables in the load network, namely the coupling coefficient (β) between the microstrip line and the resonator, and the distance (d_R) between the end of the microstrip line and the plane containing the resonator, as shown in Figure 11.40.

From simple transmission line theory

$$\Gamma_R = \Gamma_L e^{j2\beta_s d_R}$$

$$= |\Gamma_L|e^{j\phi_L}e^{j2\beta_s d_R} = |\Gamma_L|e^{j(\phi_L + 2\beta_L d_R)}, \tag{11.77}$$

where β_s is the phase propagation constant for the microstrip line. It follows from Eq. (11.77) that Γ_R is real (since Z is real), and therefore we must have

$$\phi_L + 2\beta_s d_R = 2\pi,$$

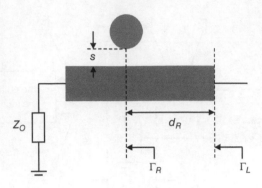

Figure 11.40 Position of DR in load network.

giving

$$d_R = \frac{2\pi - \phi_L}{2\beta_s}. \tag{11.78}$$

Using Eq. (11.72) we have

$$|\Gamma_L| = \frac{\beta}{2 + \beta},$$

i.e.

$$\beta = \frac{2|\Gamma_L|}{1 - |\Gamma_L|}. \tag{11.79}$$

Having found the value of β the spacing, s, between the edge of the microstrip puck and the edge of the microstrip line can be found using a suitable CAD package, or through published data, for example in reference [10]. *Because of the large number of variables associated with the size of the puck and its dielectric constant, there are no closed-form design equations available that relate β and s.*

Example 11.12   Design a 4 GHz DRO using a microwave high electron mobility transistor (HEMT) that has the following *S*-parameter and stability data:

$$[S] = \begin{bmatrix} 0.88\angle - 76.8° & 0.081\angle 46.7° \\ 7.94\angle 132.1° & 0.46\angle - 69.4° \end{bmatrix} \qquad K = 0.13$$

The DRO is to be fabricated on an alumina substrate that has a thickness of 0.45 mm and a relative permittivity of 9.8. The following data apply to the dielectric puck: diameter = 14.5 mm; height = 0.79 mm; $\varepsilon_r = 36.6$.

Comments: For the data given in this problem a coupling factor (β) of 6 corresponds to a spacing (s) of 1 mm between the puck and the microstrip.

Solution
We will follow the five design steps discussed in the previous section:

Step 1
A potentially unstable transistor has been specified in the question, since the stability factor given is less than 1.

Step 2
We need to plot the stability circle in the Γ_L plane. Using the equations developed in Chapter 9 we can calculate the position and radius of the required circle:

Centre of circle:

$$c = \frac{[S_{11} - DS_{22}^*]^*}{|S_{11}|^2 - |D|^2} \quad \text{where} \quad D = S_{11}S_{22} - S_{12}S_{21},$$

$$D = (0.88\angle - 76.8° \times 0.46\angle - 69.4°) - (0.081\angle 46.7° \times 7.94\angle 132.1°)$$

$$= 0.405\angle - 146.2° - 0.643\angle 178.8°$$

$$= 0.306 - j0.238$$

$$\equiv 0.388\angle - 37.875°.$$

substituting data:

$$c = \frac{[(0.88\angle - 76.8°) - (0.388\angle - 37.825° \; 0.46\angle 69.4°)]^*}{(0.88)^2 - (0.388)^2}$$

$$= 1.524\angle 87.029°.$$

Radius of circle:

$$r = \frac{|S_{12}S_{21}|}{|S_{11}|^2 - |D|^2},$$

substituting data:

$$r = \frac{0.081 \times 7.94}{0.624} = 1.031.$$

The input stability circle (i.e. stability circle in Γ_L plane) can then be plotted, as shown in Figure 11.41, and explained in Chapter 9:

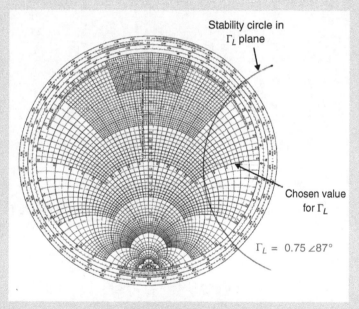

Figure 11.41 Stability circle in Γ_L plane.

Since the stability circle does not enclose the origin the unstable region will be inside the circle. We will choose a suitable point for Γ_L which is in the vicinity of the centre of the unstable region:

$$\Gamma_L = 0.75\angle 87°.$$

Using the chosen value of Γ_L we have, from Eq. (11.73):

$$\Gamma_{out} = S_{22} + \frac{S_{12}S_{21}\Gamma_L}{1 - S_{11}\Gamma_L}$$

$$= 0.46\angle - 69.4° + \frac{0.081\angle 46.7° \times 7.94\angle 132.1° \times 0.75\angle 87°}{1 - (0.88\angle - 76.8° \times 0.75\angle 87°)}$$

$$= 1.764\angle - 74.010°.$$

(Continued)

(Continued)

Step 3

We have found $\Gamma_{out} = 1.764 \angle -74.010°$ It then follows that the impedance at the output of the transistor will be

$$Z_{out} = Z_0 \frac{1 + \Gamma_{out}}{1 - \Gamma_{out}}$$

$$= 50 \times \frac{1 + (0.486 - j1.696)}{1 - (0.486 - j1.696)} \quad \Omega$$

$$= (-33.620 - j54.000) \, \Omega.$$

Step 4

Using Eq. (11.76) and $n = 1/3$ we obtain

$$Z_T = \frac{33.620}{3} + j54.000 \, \Omega$$

$$= 11.207 + j54.000 \, \Omega.$$

Normalizing with respect to 50 Ω gives

$$z_T = 0.224 + j1.08.$$

The required value of z_T can be obtained using a combination of an open-circuit shunt stub connected across the 50 Ω load, and a length (d) of microstrip line, as shown schematically in Figure 11.42.

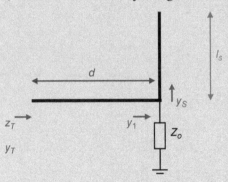

Figure 11.42 Schematic view of termination network.

The required values of l_s and d_T can be found using the Smith chart, as shown in Figure 11.43:

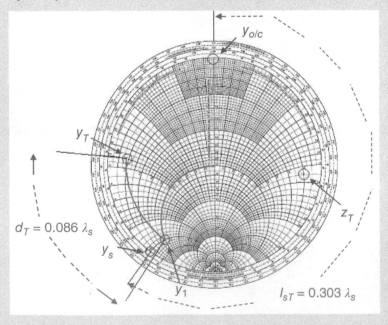

Figure 11.43 Smith chart solution for termination network.

If the microstrip lines forming the termination network each have a characteristic impedance of 50 Ω, then from the microstrip design graphs in Chapter 2 we have:

$$50\,\Omega \quad \Rightarrow \quad \frac{w}{h} = 0.9 \quad \Rightarrow \quad w = 0.9h = 0.9 \times 0.45 \text{ mm} = 0.405 \text{ mm},$$

$$\frac{w}{h} = 0.9 \quad \Rightarrow \quad \varepsilon_{r,eff} = 6.6 \quad \Rightarrow \quad \lambda_s = \frac{\lambda_o}{\sqrt{6.6}} = \frac{75}{\sqrt{6.6}} \text{ mm} = 29.19 \text{ mm}.$$

Using the electrical lengths obtained from the Smith chart in Figure 11.43 we have:

$$l_s = 0.303\lambda_s = 0.303 \times 29.19 \text{ mm} = 8.84 \text{ mm},$$

$$d_T = 0.086\lambda_s = 0.086 \times 29.19 \text{ mm} = 2.51 \text{ mm}.$$

The final circuit for the termination network is shown in Figure 11.44.

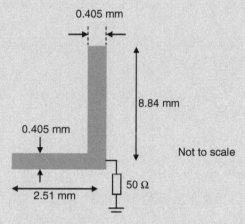

Figure 11.44 Termination network, showing designed dimensions.

Step 5

We chose $\Gamma_L = 0.75 \angle 87°$, and using Eq. (11.78) we have

$$d_R = \frac{2\pi - {}^{87\pi}\!/180}{2 \times {}^{2\pi}\!/\lambda_s},$$

i.e.

$$d_R = 0.379\lambda_s$$
$$= 0.379 \times 29.19 \text{ mm}$$
$$= 11.06 \text{ mm}.$$

Using Eq. (11.72)

$$0.75 = \frac{\beta}{2 + \beta},$$
$$\beta = 6.$$

Using data provided in the question:

$$\beta = 6 \quad \Rightarrow \quad s = 1 \text{ mm}.$$

The final design for the load network is shown in Figure 11.45.

(Continued)

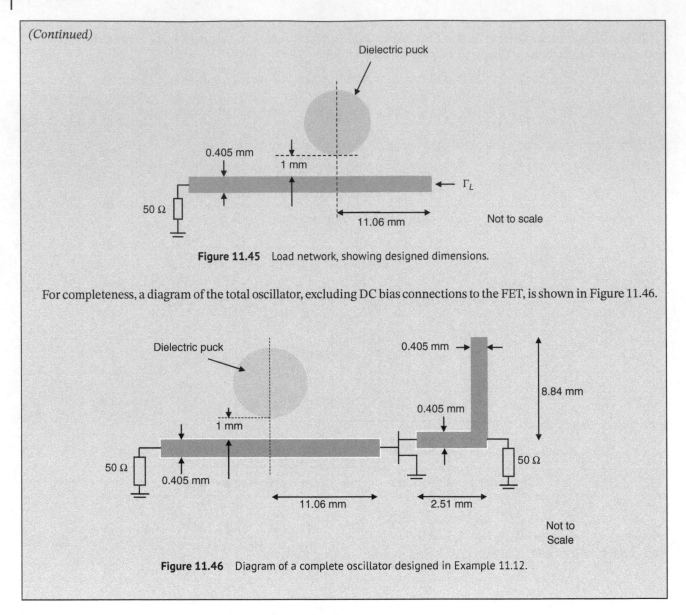

(Continued)

Figure 11.45 Load network, showing designed dimensions.

For completeness, a diagram of the total oscillator, excluding DC bias connections to the FET, is shown in Figure 11.46.

Figure 11.46 Diagram of a complete oscillator designed in Example 11.12.

11.7.2 Delay-Line Stabilized Microwave Oscillators

The transistor oscillators discussed in the earlier sections of this chapter have employed resonant circuits in the feedback path of the oscillator. An alternative method of constructing an oscillator is to use a delay-line in the feedback path, with the amount of delay such as to produce the 360° loop phase shift required by the Nyquist criterion for oscillation. The basic arrangement of a delay-line oscillator is shown in Figure 11.47, in which the output of a high-gain amplifier is fed back to its input though a delay-line. A high-gain amplifier is needed to overcome the loss in the delay line.

In most practical oscillators a surface acoustic wave (SAW) delay line is used, as this provides a compact high-performance component, with phase noise and stability comparable with crystal-controlled oscillators. Moreover, SAW oscillators can be implemented in monolithic formats as part of system-on-chip (SoC) solutions.

Surface acoustic waves are sound waves that travel on the surface of piezoelectric crystal structures that exhibit elastic properties. The amplitude of the waves decreases exponentially with depth into the material. Typical substrate materials used for RF and microwave devices are GaAs and quartz.

The typical arrangement of a SAW delay line is shown in Figure 11.48.

The SAW delay line shown in Figure 11.48 consists of two interdigitated transducers (IDTs) printed onto a piezoelectric substrate. When one of the transducers is excited by an RF or microwave signal a SAW is generated, and this travels

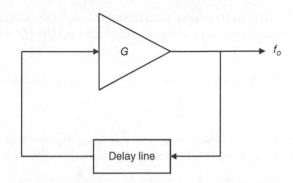

Figure 11.47 Delay-line oscillator.

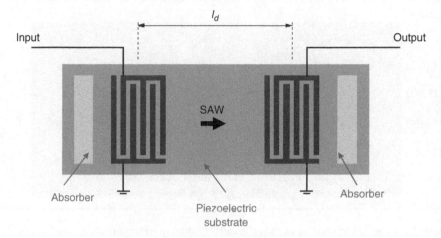

Figure 11.48 Basic SAW delay-line structure.

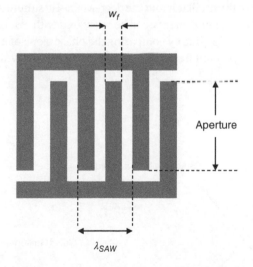

Figure 11.49 Structure of an interdigitated transducer (IDT).

along the surface of the substrate to the second IDT which converts the acoustic wave back into an electrical signal. The centre-to-centre distance between the two IDTs determines the delay time. Since the IDTs are bidirectional two absorbers are located between the IDTs and the ends of the substrate to prevent acoustic waves being reflected from the ends of the structure, thus ensuring a unique propagation path between the two IDTs. The structure of an IDT is shown in more detail in Figure 11.49.

The wavelength of the SAW generated by an IDT depends on the periodicity of the structure. Normally, the widths of the finger electrodes equal the spacing of the fingers, and so the acoustic wavelength is given by

$$\lambda_{SAW} = 4w_f, \tag{11.80}$$

where w_f is the finger width in the IDT. The operating frequency, f_O, of the SAW delay line structure is given by

$$f_O = \frac{v_{SAW}}{\lambda_{SAW}}, \tag{11.81}$$

where v_{SAW} is the velocity of propagation of the acoustic wave. Values of v_{SAW} depend on the substrate material and typically range from 2000 to 5000 m/s. For GaAs, $v_{SAW} \approx 2700$ m/s, and for quartz, $v_{SAW} \approx 3500$ m/s.

Example 11.13   A 100 MHz SAW delay line is to be fabricated using IDTs with equal finger widths and spacings. Determine the finger widths required in the IDT if the delay line is fabricated on GaAs, and if it is fabricated on quartz. Comment on the result.

Solution

GaAs: Using Eq. (11.81) $100 \times 10^6 = \dfrac{2700}{\lambda_{SAW}} \quad \Rightarrow \quad \lambda_{SAW} = 27\,\mu m.$

   Using Eq. (11.80) $27 \times 10^{-6} = 4w_f \Rightarrow w_f = 6.75\,\mu m.$

Quartz: Using Eq. (11.81) $100 \times 10^6 = \dfrac{3500}{\lambda_{SAW}} \quad \Rightarrow \quad \lambda_{SAW} = 35\,\mu m.$

   Using Eq. (11.80) $35 \times 10^{-6} = 4w_f \Rightarrow w_f = 8.75\,\mu m.$

Comment: For both materials the required finger widths are quite small. The required finger widths will decrease with increasing frequency, indicating that there is a practical upper frequency limit for SAW delay lines.

Figure 11.50 illustrates how the finger widths, w_f, within an IDT with equal finger widths and spacing vary with frequency for typical materials. Limitations in the fabrication process will set a practical upper frequency limit for the IDT.

When crystals or LC resonant circuits are used in the feedback paths of oscillators it is the Q-factor of the feedback path that determines the frequency stability of the oscillator and the degree of noise suppression. Therefore, it is useful to have an effective Q value for the delay line, so that its performance can be compared with resonant structures. Vollers and Claiborne [11] established an effective Q value for a delay line by comparing the phase slope of a delay line with the phase slope of an LC resonant circuit in the vicinity of its resonant frequency. This resulted in a delay-line Q given by

$$Q = \tau_d\, \pi f_O, \tag{11.82}$$

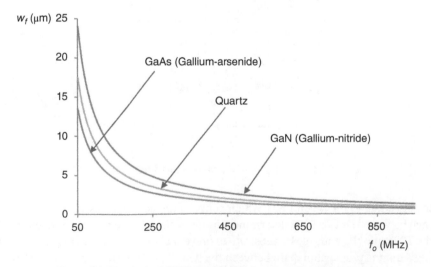

Figure 11.50 Typical IDT finger widths as a function of frequency.

where τ_d is the time delay through the line at the operating frequency f_O. In general, the Q of a typical delay line is somewhere between that of an LC resonant circuit and a crystal.

Example 11.14   The following data apply to a 300 MHz monolithic SAW delay line:

Length of delay line = 480 μm

v_{SAW} (silicon) = 4800 m/s

Determine the finger width required in the IDT, assuming equal finger-width spacing. Also, calculate the Q of the delay line and compare this with the Q of a 300 MHz LC resonant circuit, which uses a 10 nH inductor with a loss resistance of 0.8 Ω.

Solution

Using Eq. (11.81) we have:

$$\text{SAW wavelength: } \lambda_{SAW} = \frac{4800}{300 \times 10^6} \text{ m} = 16 \text{ μm}.$$

$$\text{Finger width: } w_f = \frac{16}{4} \text{ μm} = 4 \text{ μm}.$$

$$\text{Delay time: } \tau_d = \frac{480 \times 10^{-6}}{4800} \text{ s} = 0.1 \text{ μs}.$$

Using Eq. (11.82) we have:

$$Q(\text{delay line}) = \tau_d \pi f_O = 0.1 \times 10^{-6} \times 3.142 \times 300 \times 10^6 = 94.26.$$

$$Q(\text{LC resonator}) = \frac{\omega L}{R} = \frac{2\pi \times 300 \times 10^6 \times 10 \times 10^{-9}}{0.8} = 23.57.$$

Recent developments in microelectronic technology and materials have enabled the amplifying device and delay line forming a SAW oscillator to be closely integrated within a single package. For example, Lu and colleagues [12] reported results on a GaN-based 252 MHz Monolithic SAW/HEMT oscillator on silicon that exhibited very low close-carrier noise and a Q of 1000.

In general, modern SAW oscillators have excellent close-carrier noise performance and are used extensively in the space and defence industries at frequencies well into the low GHz region. In this type of oscillator the SAW delay line can also be combined with a digitally controlled phase shifter to provide a high-quality, tunable device.

11.7.3   Diode Oscillators

One of the traditional methods for constructing simple, economic microwave oscillators is to make use of special diodes, such as the Gunn diode and the IMPATT diode. Whilst many of the traditional uses for these diodes are now performed using FET-based devices, there are a number of niche applications for which microwave diode oscillators are particularly suitable, and so for completeness the basic principles of the main types of diode oscillator will be reviewed in this section.

11.7.3.1   Gunn Diode Oscillator

The Gunn diode oscillator is named after J.B. Gunn who observed oscillations from bulk semiconductors back in the 1960s [13]. Although the device is referred to as a diode, it does not contain a p–n junction but rather relies upon a transferred-electron effect, which is so-called because it involves electrons moving from a low energy band to a high energy band. The semiconductor materials normally used for Gunn diodes are GaAs and Indium Phosphide (InP). Both materials exhibit the same physical effects, although Gunn diode oscillators using GaAs require less DC bias current and voltage than those using InP. However, InP-based diode oscillators are more efficient than those using GaAs and tend to be used for higher power applications. For convenience, we will restrict the discussion to Gunn diode oscillators using GaAs, so as to illustrate the principles involved.

The energy band diagram for GaAs is shown in Figure 11.51.

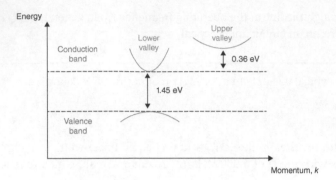

Figure 11.51 Energy band diagram for GaAs.

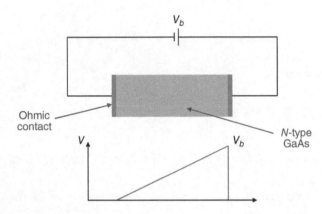

Figure 11.52 Voltage distribution across a Gunn diode with $E < E_{TH}$.

Electrons in GaAs reside in a number of energy bands, with those in the lower valley of the conduction band being responsible for the diode current at room temperature. If the energy of the electrons is increased they will move into the upper valley of the conduction band, where they have lower mobility, and higher effective mass. From Figure 11.51, we see that electrons in GaAs require 0.36 eV of energy to move from the lower valley to the upper valley of the conduction band. The energy of the electrons can be increased by increasing the electric field strength across the material, and the value of the electric field at which the electrons acquire sufficient energy to move from the lower to the upper valley of the conduction band is defined as the threshold electric field strength, E_{TH}, and the corresponding voltage across a GaAs specimen as the threshold voltage, V_{TH}.

The basic structure of a Gunn diode is shown in Figure 11.52, and consists of a slice of N-type GaAs, with two ohmic contacts that are formed by metallic coatings across the ends of the specimen; also shown in this figure is the voltage distribution across the specimen when the applied bias voltage, V_b, is less than the threshold value.

If the specimen of GaAs is homogeneous, there will be a uniform voltage distribution across the specimen, as shown in Figure 11.52, and with $V_b < V_{TH}$ the electrons will reside in the lower valley of the conduction band. If V_b is made greater than the threshold value, electrons will move into the upper valley of the conduction band, and this leads to a variation in current in the specimen as shown in Figure 11.53.

With $V_b < V_{TH}$, the slice of GaAs behaves as a normal resistive material and the current increases linearly with the value of the electric field. When the electric field exceeds the threshold value, the current begins to fall as the electrons move into the upper valley where they have less mobility. Thus, there is a region where the material exhibits negative differential conductance, i.e. the current falls with an increase in applied electric field. There will be a value of electric field, E_v, at which all of the available electrons have moved into the upper valley, and the current again begins to increase linearly. It should be noted that the slope of the J–E characteristic above E_v is less than that below E_{TH}. This is because the electrons responsible for conduction above E_v have less mobility than those below E_{TH}, which means the material exhibits lower conductivity, and consequently higher resistance.

There are two principal modes of oscillation available with Gunn diodes, namely the transit-time mode and the limited space-charge accumulation (LSA) mode. We will consider these two modes in turn.

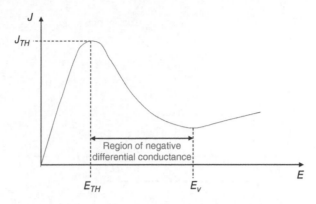

Figure 11.53 Current density (*J*) as a function of electric field strength (*E*) for GaAs.

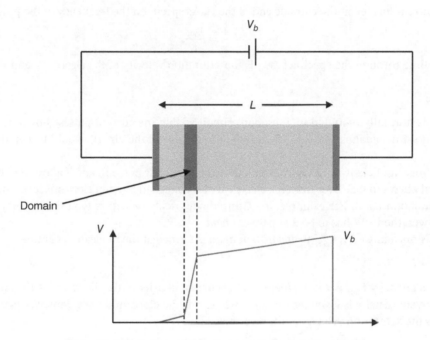

Figure 11.54 Voltage distribution across Gunn diode domain.

Transit-Time Mode of Oscillation To obtain this mode of oscillation the GaAs specimen is doped so as to create a discontinuity near the cathode end of the specimen. The presence of the discontinuity will create a non-uniform voltage distribution, with the electric field strength across the discontinuity exceeding that throughout the rest of the specimen. If the electric field strength across the discontinuity exceeds the threshold value, electrons in the vicinity of the discontinuity will move into the upper valley of the conduction band, and a domain (or layer) of low-mobility conduction electrons will be formed, as illustrated in Figure 11.54. This domain of electrons will travel through the specimen towards the anode with a reduced velocity, since the electrons in the upper valley of the conduction band have less mobility than those in the lower valley.

Since the discontinuity, which is usually referred to as a nucleation centre, is normally created near the cathode end of the specimen, the domain will travel the whole length of the specimen. Once a domain has been created, the external current will fall since the presence of the slow-moving domain will block the flow of high-mobility electrons. On reaching the anode end of the specimen, the domain will be extinguished thereby causing a pulse of current to flow in the external circuit. Once a domain has been extinguished, a new domain forms at the nucleation centre. This behaviour leads to pulses of current being observed in the external circuit, as shown in Figure 11.55. It should be noted that only one domain forms at a time, since the electric field strength outside the domain is insufficient for other electrons to move from the lower to upper valley in the conduction band.

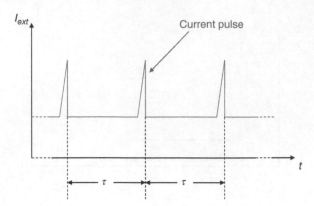

Figure 11.55 Current pulses in Gunn diode circuit.

If the nucleation centre is located at the cathode end of the GaAs specimen the frequency of the pulses will be given by

$$f_O = \frac{1}{\tau} = \frac{v_{sat}}{L}, \tag{11.83}$$

where τ is the transit time through the specimen, v_{sat} is the saturation velocity of the electrons, and L is the length of the specimen.

Other points to note:

(i) The Gunn diode is normally mounted in a resonant circuit so that the current pulses shown in Figure 11.55 stimulate the circuit into sinusoidal oscillation. The resonant frequency of the circuit should be equal to f_O, as defined in Eq. (11.83).

(ii) Gunn diodes are very inefficient, with typical efficiencies of only a few percent, and if allowed to heat up significantly will allow thermal effects to take over from the electrical bias voltage in moving electrons form the lower to the upper valley of the conduction band. For example, if a Gunn diode oscillator with a typical efficiency of 5% is producing 50 mW of CW power, then 1 W has to be dissipated as heat.

(iii) The bias voltage is normally $\sim 3 \times V_{TH}$ to ensure a domain is formed at the nucleation centre.

Comment: $V_{TH} = E_{TH} \times L$.

(iv) The bias current is given by $I_{bias} = J \times A$, where J is the current density in the diode and A the cross-sectional area of the diode. The current density is given by $J = nev_{sat}$ where n is the electron doping density, e is the unit of electronic charge, and v_{sat} is the saturation velocity of the electrons.

Example 11.15　A Gunn diode operating in the domain mode is used to make a 12 GHz oscillator with a 50 mW output. Assuming that the Gunn diode is fabricated from GaAs and has an efficiency of 5%, determine:

(i) The length of the transit region in the diode.
(ii) The required bias current.
(iii) The required cross-section of the diode.

Materials data for the GaAs material used in the Gunn diode:

$$v_{sat} = 10^7 \text{ cm/s,}$$
$$E_{TH} = 3 \text{ kV/cm,}$$
$$n = 10^{15} \text{ cm}^{-3}.$$

Make reasonable assumptions for any unspecified parameters.

Solution

(i) The transit time must be equal to one period of the 12 GHz signal.

$$\tau = \frac{1}{12 \times 10^9} \text{ s} = 83.33 \text{ ps.}$$

The length, L, of the transit region is then given by

$$L = v_{sat} \times \tau = 10^7 \times 83.33 \times 10^{-12} \text{ cm} = 8.33 \text{ μm}.$$

(ii) For a 50 mW output with 5% efficiency the DC power supplied must be

$$P_{DC} = \frac{50}{0.05} \text{ mW} = 1000 \text{ mW} = 1 \text{ W}.$$

For the specimen length calculated in part (i) the threshold voltage will be

$$V_{TH} = E \times L = 3 \times 10^5 \times 8.33 \times 10^{-6} \text{ V} = 2.5 \text{ V}.$$

Assuming the bias voltage is $3 \times V_{TH}$

$$V_{bias} = 3 \times 2.5 \text{ V} = 7.5 \text{ V}.$$

Now

$$P_{DC} = V_{bias} \times I_{bias}.$$

Therefore,

$$I_{bias} = \frac{P_{DC}}{V_{bias}} = \frac{1}{7.5} \text{ A} = 133.3 \text{ mA}.$$

(iii)

$$I_{bias} = J \times A = (nev_{sat}) \times A,$$

where J is the current density, i.e.

$$0.133 = (10^{21} \times 1.6 \times 10^{-19} \times 10^5) \times A,$$

$$A = 8.31 \times 10^{-9} \text{ m}^2 = 8.31 \times 10^{-3} \text{ mm}^2.$$

Gunn diodes are often mounted in resonant cavities formed from a rectangular waveguide, and Figure 11.56 shows a typical arrangement, with the diode mounted across the narrow dimension of the guide.

One end of the cavity is formed by a thin transverse metal plate with an iris to couple energy out of the cavity. The other end of the cavity is formed by a metal plunger which can be adjusted to change the length, L, and hence the resonant frequency, of the cavity. (*The design of waveguide resonant cavities is covered in Chapter 3.*) Figure 11.56 shows the normal practice of mounting the Gunn diode on a metal stud to ensure good heat sinking.

LSA Mode of Oscillation The LSA mode of oscillation makes use of the negative resistance properties of a Gunn diode. As with the transit-mode operation, the diode is mounted in a resonant circuit, but with the difference that the resonant frequency of the circuit is made several times the reciprocal of the transit time. The voltage bias is chosen to be between V_{TH} and V_v, where V_v is the voltage corresponding to E_v in Figure 11.53, and so the diode behaves purely as a negative resistance device, which can be modelled simply as a parallel combination of negative resistance and capacitance as shown in Figure 11.57. The load, which is the impedance of the resonant circuit, is shown as a conventional parallel L, R, C combination.

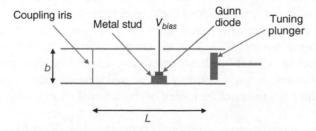

Figure 11.56 Gunn diode mounted in a waveguide cavity.

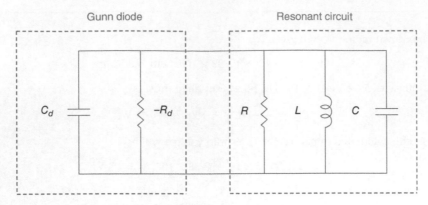

Figure 11.57 Equivalent circuit of LSA mode oscillator.

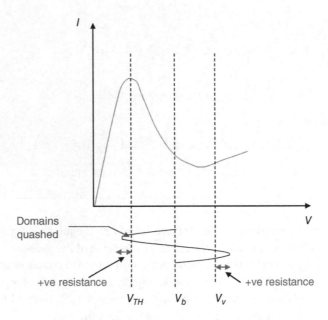

Figure 11.58 I–V characteristic showing LSA mode of oscillation.

It can be seen from Figure 11.57 that the diode negative resistance is in parallel with the resistance of the resonant circuit, so the total resistance R_T is given by

$$R_T = -\frac{R_d R}{R - R_d}. \tag{11.84}$$

The value of the load resistance, R, is normally made around 10% larger than the negative resistance of the diode, so at time $t = 0$ the total resistance will be negative, and this will ensure that oscillations start. Figure 11.58 shows the I–V characteristic of the diode and includes a typical cycle of the RF oscillation.

As the oscillations build up the maximum voltage excursions will enter regions of positive resistance, and so the total circuit resistance will become less negative. The amplitude of the RF oscillations will increase until the total resistance of the diode and resonant circuit combination is zero, and steady-state oscillation occurs. Any tendency for domains to form in the specimen is quashed partly by the voltage excursions entering the $V < V_{TH}$ region for part of a cycle, and partly by the high resonant frequency of the external circuit, which reduces the time for domain formation.

The efficiency of the LSA mode of oscillation is significantly better than that of the transit-time mode, and can typically approach 20%, and consequently higher output powers are possible using this mode.

11.7.3.2 IMPATT Diode Oscillator

IMPATT diode oscillator The acronym IMPATT stands for **Imp**act **A**valanche and **T**ransit **T**ime, and describes the principal physical mechanisms occurring within the diode. The diode is often referred to as a Read diode, after W.T. Read who first proposed a high-frequency negative-resistance diode in the 1950s.

The basic structure of an IMPATT diode is shown in Figure 11.59.

The diode consists of two heavily doped semiconductor regions, denoted by P⁺ and N⁺, separated by a thin N-doped layer and a somewhat lager intrinsic region. The purpose of the thin N-doped layer is to create a P⁺–N junction. Two end contacts are used to connect the diode to a DC bias supply. With the bias polarity shown in Figure 11.59, the P⁺–N junction will be reverse biased. The electric field distribution is also shown in Figure 11.59; there will be negligible E-field across the heavily doped regions, which are good conductors, a linear decrease across the partially conducting N-layer, and a constant E-field across the un-doped intrinsic layer.

The current–voltage characteristic for the P⁺–N junction is drawn in Figure 11.60, and shows the avalanche breakdown voltage, V_B.

When the reverse bias voltage across the P⁺–N junction exceeds the avalanche breakdown value there will be a significant reverse current through the device. It is the behaviour of the P⁺–N junction in the vicinity of breakdown that is key to the operation of the IMPATT diode oscillator. Figure 11.61 uses one cycle of the RF signal to illustrate the three basic stages in the operation.

Stage 1:   The diode is reverse biased at, or just below, the breakdown value, V_{av}. If an RF voltage is superimposed on the breakdown value, the P⁺–N junction will exhibit the avalanche effect for half the cycle. (Note that it is thermal or switching noise that causes the junction voltage to exceed the breakdown value when the device is first switched on.)

Stage 2:   When the reverse bias voltage exceeds the threshold value there is an exponential build-up of carriers at the P⁺–N junction; thus a burst of carriers is produced, with an exponential rise-time, and a sharp fall when the junction voltage falls below threshold and the avalanche process ceases.

Stage 3:   The burst of carriers (electrons) will drift across the intrinsic region towards the positive anode of the diode. The drift time, τ, will depend on the length of the i-region, and the saturation velocity of the electrons. When the burst reaches the anode it causes a pulse of current to flow in the external circuit. The precise shape of the current pulse is of no particular consequence, and will depend on the impedance of the circuit. If the diode is mounted in a resonant cavity the current pulses will stimulate the cavity into resonance, thereby giving sinusoidal oscillations.

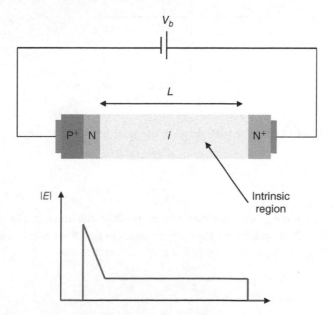

Figure 11.59   IMPATT diode structure and typical E-field distribution.

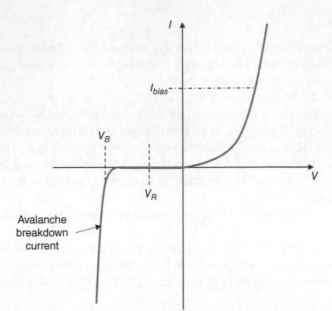

Figure 11.60 I–V characteristic for a P–N junction.

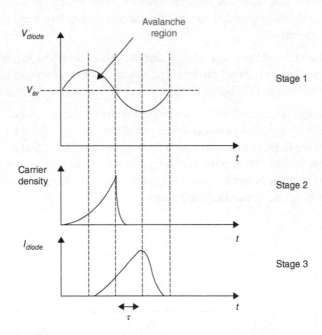

Figure 11.61 The three stages of operation of an IMPATT diode.

The drift time is given by

$$\tau = \frac{L}{v_{sat}},$$

(11.85)

where L is the length of the i-region and v_{sat} is the saturation velocity of the electrons. In Figure 11.61, the delay is shown to be one-quarter of the period of the RF signal. Under these circumstances the diode is exhibiting negative resistance, since the negative peak of the RF voltage corresponds to a positive peak of current, and

$$\tau = \frac{T}{4} = \frac{1}{4f_O} = \frac{L}{v_{sat}}$$

(11.86)

or

$$f_O = \frac{v_{sat}}{4L},$$

(11.87)

where f_O is the frequency of the current pulses, and this should equal the resonant frequency of the cavity in which the IMPATT diode is mounted.

The IMPATT diode oscillator has an efficiency similar to that of the LSA mode of the Gunn diode, but is capable of significantly more power. However, it has poor noise performance due to the noise generated by the avalanche process. Thus, IMPATT diode oscillators are used in situations where noise is less important than output power, such as pulsed radar systems, where the receiver is concerned primarily with the envelope of the received signal, rather than the quality of the CW signal.

11.8 Oscillator Noise

Oscillator noise refers to the random variations in amplitude and phase of the signal at the output of an oscillator. Variations in the amplitude of the oscillator's output are referred to as AM noise, and variations in phase as PM noise. The noise will generate unwanted sidebands around the centre frequency of the oscillator, and some of the power, $P_o(f)$, at the output of the oscillator will be distributed around the nominal carrier frequency, f_c, as shown in Figure 11.62.

The presence of noise sideband may have serious consequences in communication systems, where these sidebands mix with other frequency components in the system to produce unwanted intermodulation products within the pass-band of the system. In general, AM noise from an oscillator is significantly less than the PM noise. Random variations in the phase of the CW output from an oscillator may also be referred to as FM noise, remembering that frequency is simply the rate of change of phase. The spectra of PM noise and FM noise are very similar.

The effects of noise on the output from an oscillator may be visualized by considering a noise sinusoid modulating a CW signal, as shown by the vector diagram in Figure 11.63.

As the noise sinusoid shown in Figure 11.63 rotates about the unmodulated carrier there will be a change in the amplitude of the carrier, leading to AM noise, and similarly there will be changes in the phase of the carrier, leading to FM or PM noise.

Various models have been developed to represent the effects of phase noise in oscillators. The model that is still the most widely used is that originally developed by Leeson [14], which consisted of an amplifier with a single-resonator feedback network. This model led to the development of the following equation to represent the single-sided phase noise spectrum of an oscillator

$$L(f_n) = \frac{FkT}{2P_c}\left[1 + \left(\frac{f_c}{2Qf_n}\right)^2\right]\left(1 + \frac{f_1}{f_n}\right),\tag{11.88}$$

where $L(f_n)$ denotes a Lorentzian-shaped spectrum with the variable f_n being the frequency offset from the centre frequency of the oscillator. The other variables in the expression are:

F – the noise factor of the active device in the oscillator
k – the Boltzmann constant (1.38×10^{-23} J/K)
T – absolute temperature in Kelvin
f_c – centre frequency
P_c – output power at the centre frequency

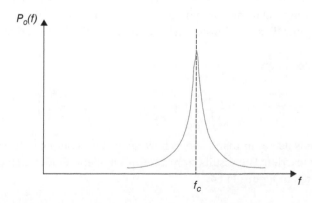

Figure 11.62   Noise power spectral density at the output of an oscillator.

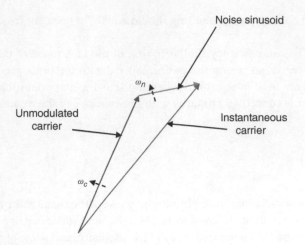

Figure 11.63 Carrier modulated by a single noise sinusoid.

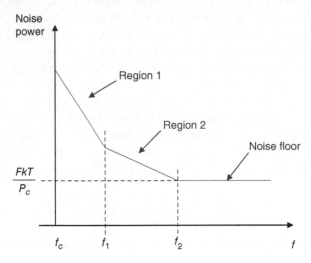

Figure 11.64 Linearized representation of oscillator output noise power.

Q – loaded Q-factor of the oscillator's resonator

f_1 – is the corner frequency for flicker noise,[2] as shown in Figure 11.64.

There are several obvious conclusions that can be drawn from Eq. (11.88), which show how the phase noise from an oscillator can be minimized: (i) a high Q resonator should be used; (ii) the active device should have a low noise factor; (iii) an active device that exhibits low flicker noise should be used. It can also be seen that the phase noise decreases with an increase in carrier power.

Equation (11.88) can be expanded to give

$$L(f_n) = \frac{FkT}{P_c} \left[1 + f_1 \left(\frac{1}{f_n} \right) + \left(\frac{f_c}{2Q} \right)^2 \left(\frac{1}{f_n^2} \right) + \left(\frac{f_c \sqrt{f_1}}{2Q} \right)^2 \left(\frac{1}{f_n^3} \right) \right]. \tag{11.89}$$

This form of Leeson's equation is useful in that it shows how the various components of oscillator noise are affected by f_n, the frequency offset from the carrier. The results from Leeson's model are often depicted in a linearized view of the oscillator noise spectrum, as shown in Figure 11.64.

---

2 Flicker noise is discussed under 'Semiconductor Noise' in Chapter 14.

There are three principal regions of the noise characteristic shown in Figure 11.64:

Region 1:   Sometimes referred to as the $1/f^3$ region, this shows the relatively sharp decrease in close carrier noise due to the term in Leeson's equation that has f^3 in the denominator.

Region 2:   As the frequency offset from the carrier increases the term in Leeson's equation with f^2 in the denominator tends to dominate the noise reduction, and so region 2 is often referred to as the $1/f^2$ region. However, this only applies to oscillators with a low-Q resonator, since the term with f^2 in the denominator also has a factor of Q^2 in the denominator, which means that with a high-Q resonator the term including $1/f$ tends to dominate, and region 2 is then referred to as the $1/f$ region.

Noise floor: With large frequency offsets from the carrier all the terms involving f in their denominators tend to zero, and the phase noise is determined simply by the components of the oscillator.

Whilst Leeson's model still remains the essential basis for the analysis of phase noise in oscillators, a number of more recent papers [15, 16] have suggested significant improvements to Leeson's model to make it more generally applicable to practical oscillators. A useful paper that discusses the effects of phase noise in relation to particular types of oscillator is that of Lee and Hajimiri [17].

11.9 Measurement of Oscillator Noise

Conventional methods of measuring oscillator noise are based on the use of a frequency discriminator. A typical measurement system using a delay-line frequency discriminator is shown in Figure 11.65.

In the system shown in Figure 11.65, the output from the oscillator is divided into two paths using a power divider (sometimes called a power splitter). One path contains a delay line which introduces a phase shift, and as the frequency of the oscillator increases the phase difference between the two inputs to the phase detector increases linearly with frequency. Thus, the voltage output from the phase detector varies linearly with changes in the oscillator frequency. The phase shifter is used to ensure the inputs to the phase detector are in phase quadrature at the carrier frequency, thus ensuring the phase detector operates in a quasi-linear region, with zero output at the carrier frequency. The output of the phase detector is passed through an LPF to remove any unwanted modulation products generated by the phase detector and applied to an LF spectrum analyzer, which will display the close-carrier phase noise as a function of the frequency offset from the carrier frequency. However, some caution is needed in interpreting the display on the spectrum analyzer. Since noise is random and has a continuous frequency spectrum the analyzer will display the value of noise integrated over the bandwidth of the analyzer, and so the values of noise displayed on the analyzer must be divided by the bandwidth of the analyzer to obtain a true reading at any particular value of frequency offset. For convenience in practical measurements, the spectrum analyzer is often replaced by a computer-based processor, which can analyze and store the noise data. One drawback to the measurement system shown in Figure 11.65 is that it is also sensitive to the amplitude noise from the oscillator, and the analyzer will display the sum of the phase and amplitude noise. The assumption is normally made that the AM noise from

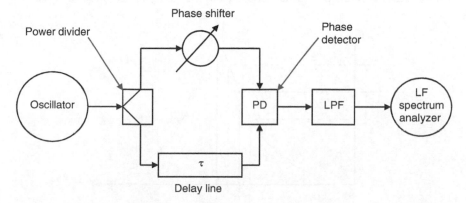

Figure 11.65   Delay-line discriminator technique for measuring oscillator phase noise.

the oscillator is significantly less than the phase noise and so can be ignored. Two significant advances have recently been made in phase-noise measurement systems using delay-line discriminators:

(i) One disadvantage of the measurement system shown in Figure 11.65 is that a phase shifter is required, which generally has to be manually tuned. Gheidi and Banai [18] developed a measurement system that still uses a delay-line discriminator, but which overcomes the need for a phase shifter. In their proposed arrangement the phase shifter is replaced by a 90° hybrid so as to create a second channel with its own phase detector. This technique effectively produces an in-phase/quadrature (I/Q) measurement system, which does not need manual tuning of a phase shifter, and is self-calibrating.

(ii) Another modification to the basic measurement system shown in Figure 11.65 has been provided by Barzegar and colleagues [19], who down-converted the output of the microwave oscillator under test to an intermediate frequency (IF) before implementing the delay-line circuit. This had the effect of reducing the delay-line loss, since the loss in delay lines tends to increase with frequency, thereby improving the sensitivity of the measurement system. Good performance has been reported from this technique, which is particularly attractive for measuring the noise from millimetre-wave oscillators, where high line losses normally prohibit the use of delay-lines at the frequency of the oscillator.

An alternative to using delay-line based circuits for measuring oscillator noise is to exploit the frequency discriminator properties of high-Q resonant circuits, where each side of the voltage–frequency Q-curve behaves as a slope detector. A typical measurement system using a tuned circuit is shown in Figure 11.66.

The voltage–frequency response (the Q-curve) of the resonant circuit is shown in Figure 11.67.

Figure 11.67 shows the Q-curve of a resonant circuit with a resonant frequency, f_O. If the value of f_O is chosen such that the frequency of the oscillator, f_c, lies around the mid-point of one of the sides of the Q-curve any deviation, Δf, in the frequency of the oscillator will be converted into a voltage deviation, Δv, at the output of the circuit. For a high-Q resonant circuit the sides of the voltage–frequency response will be very steep, and there will be a quasi-linear relationship between Δv and Δf, i.e.

$$k = \frac{\Delta v}{\Delta f},$$
(11.90)

where k is the slope of the Q-curve at frequency, f_c, in Hz/V.

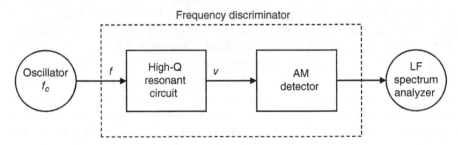

Figure 11.66 Measurement system for oscillator noise using a high-Q resonant circuit.

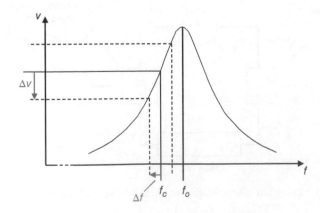

Figure 11.67 Q-curve of a resonant circuit.

Thus, the side of the Q-curve is acting as a slope detector; it can also be regarded as an FM-to-AM converter. The oscillator frequency, f_c, will still be present at the output of the resonant circuit, and a simple AM detector is required to recover values of Δv. Thus for a noisy, fixed-frequency oscillator the values of Δv will be proportional to the frequency deviations in the oscillators frequency due to FM noise. The values of Δv can conveniently be viewed on an LF spectrum analyzer, which will display a voltage proportional to the FM noise of the oscillator, as a function of the frequency offset from the carrier.

Normally, the magnitudes of the noise components at the output of an oscillator are much smaller than the carrier power, and small-signal modulation principles can be used to analyze the situation depicted in Figure 11.63. Using this technique Ondria [20] developed a useful expression for the FM noise (N_{FM}) at the output of an oscillator as

$$N_{FM} = 10\log\left[\frac{\Delta f_{RMS}}{f_n}\right]^2 \text{ dBc/Hz,} \tag{11.91}$$

where Δf_{RMS} is the RMS frequency deviation due to noise, at a frequency offset of f_n from the carrier. It is important to note the units used to specify the noise: dBc means dB relative to the carrier, and since random noise has a continuous frequency spectrum the value of N_{FM} is specified per Hertz.

Some additional points to note about the use of the tuned-circuit measurement technique:

(i) It is essential to use a high-Q resonant circuit to ensure linearity in the FM–AM conversion process. At microwave frequencies a cavity resonator is normally used, where Q-values of several thousand are easily achievable.

(ii) The spectrum analyzer shown in Figure 11.66 will display the total oscillator noise due to both frequency and amplitude fluctuations, since AM noise will pass unaffected through the resonant circuit. The AM noise can be measured independently by simply removing the resonant circuit, and then the FM noise at a particular offset frequency can be calculated using

$$\overline{\sigma_T^2} = \overline{\sigma_{AM}^2} + \overline{\sigma_{FM}^2}, \tag{11.92}$$

where $\overline{\sigma_T^2}$ is the total mean square value of noise voltage, $\overline{\sigma_{AM}^2}$ is the mean square value of AM noise voltage, and $\overline{\sigma_{FM}^2}$ is the mean square value of FM noise voltage. (Note that in Eq. (11.92) we must sum the mean square values because we are dealing with noise quantities that have a random probability distribution, and are not correlated.)

(iii) Since random noise has a continuous noise spectrum, when noise is applied to the input of a spectrum analyzer the instrument will display the noise voltage integrated over the bandwidth of the analyzer. Therefore, to obtain a correct reading of noise voltage at a particular frequency the displayed value must be divided by the bandwidth of the instrument.

Figure 11.68 shows typical responses taken from the LF analyzer shown in the measurement system of Figure 11.66; the vertical axis gives the RMS voltage at the output of the AM detector, and the horizontal axis shows the frequency off

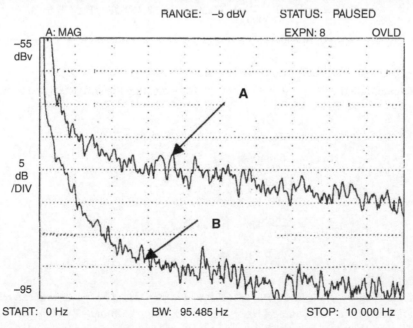

Figure 11.68 Typical LF network analyzer responses. (Trace A) With resonant circuit connected (AM+FM noise). (Trace B) Without resonant circuit connected (AM noise only).

the carrier (the oscillator's microwave frequency) from 0 to 10 kHz. In this instance the spectrum analyzer automatically selected the instrument's bandwidth, and this is shown as 95.485 Hz.

Example 11.16 Using the responses shown in Figure 11.68 determine the FM noise of the oscillator, 5 kHz off carrier, given that $k = 220$ mV/MHz for the high-Q resonator, and that the voltage efficiency of the AM detector is 69%.

Solution
As shown in Figure 11.69 we have

$$\text{AM} + \text{FM value} \approx -76 \text{ dBV} \equiv 1.585 \times 10^{-4} \text{ V},$$

$$\text{AM value} \approx -90 \text{ dBV} \equiv 3.162 \times 10^{-5} \text{ V}.$$

As previously mentioned, we must divide the voltage value displayed on the spectrum analyzer by the bandwidth of the instrument to obtain a true reading of noise voltage at a given frequency, i.e.

$$\sqrt{\sigma_T^2} = \frac{1.585 \times 10^{-4}}{95.485} \text{ V} = 1.660 \text{ μV},$$

$$\sqrt{\sigma_{AM}^2} = \frac{3.162 \times 10^{-5}}{95.485} \text{ V} = 0.331 \text{ μV}.$$

Using Eq. (11.92) we have

$$(1.660 \times 10^{-6})^2 = (0.331 \times 10^{-6})^2 + \overline{\sigma_{FM}^2},$$

$$\sqrt{\sigma_{FM}^2} = 1.627 \text{ μV}.$$

FM RMS noise voltage at the input to the AM detector will be

$$(\sigma_{FM})_{I/P} = \frac{1.627}{0.69} \text{ μV} = 2.358 \text{ μV}.$$

Using Eq. (11.90) the RMS frequency deviation at the output of the oscillator will be

$$\Delta f = \frac{2.358 \times 10^{-6}}{0.22 \times 10^{-6}} \text{ Hz} = 10.718 \text{ Hz}.$$

The FM noise 5 kHz from the oscillator's frequency is found by substituting in Eq. (11.91)

$$N_{FM} = 10 \log \left(\frac{10.718}{5000} \right)^2 \text{ dBc/Hz} = -53.377 \text{ dBc/Hz}.$$

Example 11.17 Re-calculate the FM noise in Example 11.16, neglecting the AM noise, and comment upon the result.

Solution
Neglecting the AM noise we have

$$\sigma_{FM} = 1.660 \text{ μV},$$

$$(\sigma_{FM})_{I/P} = \frac{1.660}{0.69} \text{ μV} = 2.406 \text{ μV},$$

$$\Delta f_{RMS} = \frac{2.406}{0.22} \text{ Hz} = 10.936 \text{ Hz},$$

$$N_{FM} = 10 \log \left(\frac{10.936}{5000} \right)^2 \text{ dBc/Hz} = -53.202 \text{ dBc/Hz}.$$

Comment: The AM noise makes very little difference to the calculated value of FM noise, and can reasonably be neglected. Looking at the original spectra, the AM value 5 kHz off the carrier is approximately 14 dB below the combined AM + FM value; this difference is typical of microwave oscillators and the effects of AM noise can reasonably be neglected in both tuned circuit and delay-line measurement systems.

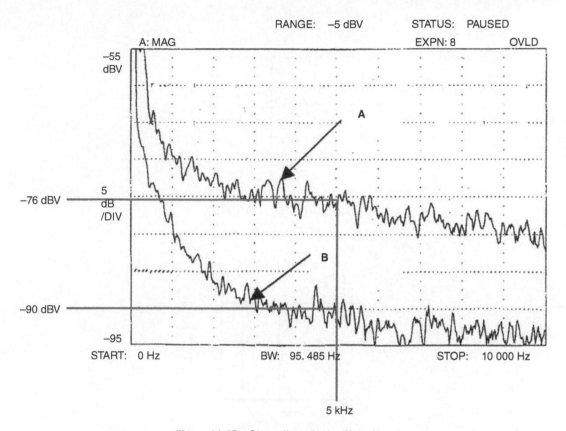

RANGE: −5 dBV STATUS: PAUSED

A: MAG EXPN: 8 OVLD

Figure 11.69 SA readings 5 kHz off carrier.

11.10 Supplementary Problems

Q11.1 The following data apply to the material used in a 5 GHz Gunn diode microwave oscillator:
Threshold electric field = 3000 V/cm
Saturated electron drift velocity = 10^7 cm/s
Doping density = 2×10^{15} cm^3
Cross-sectional area = 1.18×10^{-4} cm^2
 Estimate the maximum microwave power available from the oscillator, assuming a diode efficiency of 7%.

Q11.2 Estimate the length of the drift region of a GaAs Impatt diode used to provide a 20 GHz oscillation. Compare the thickness of the drift region in the Impatt with that of a GaAs Gunn diode used to provide a 20 GHz signal. (Saturation electron drift velocity in GaAs is 10^7 cm/s.)

Q11.3 Then following data apply to a crystal with a parallel resonant frequency of 20 MHz: $C_1 = 26$ fF; $C_O = 5.8$ pF. What is the frequency pulling range of the crystal?

Q11.4 The following data apply to a crystal: $R = 25 \, \Omega$, $C_O = 6$ pF, where the symbols have their usual meanings.
 If the anti-resonant frequency is 5 MHz, with a Q of 85 000, determine:
 (i) The series resonant frequency
 (ii) The effective inductance of the crystal.

Q11.5 The inductor in the Colpitts oscillator shown in Figure 11.4 is to be replaced by a crystal, which requires a load capacitance of 11 pF. If feedback conditions in the oscillator require C_1/C_2 to be equal to 0.02, determine the values of C_1 and C_2, given that the crystal has a C_O value of 4.3 pF.

Q11.6 The junction of a silicon ($\varepsilon_r = 11.68$) varactor diode has a cross-sectional area of 0.36 mm^2. The series resistance of the diode is 3.8 Ω.
 Estimate the Q of the diode at 98 MHz when a DC bias is applied that produces a depletion width of 55 µm.

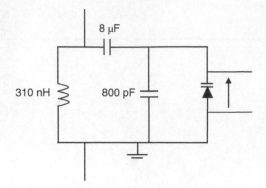

Figure 11.70 Circuit for Q11.7.

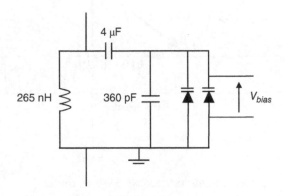

Figure 11.71 Circuit for Q11.8.

Q11.7 The resonant circuit of an oscillator is shown in Figure 11.70, and includes a silicon varactor diode having a graded junction, with $C_O = 23.5\,\text{pF}$. If the oscillator is to be used to generate an FM signal with a carrier frequency of 10 MHz, and a peak frequency deviation of 5 kHz, determine the DC bias and the peak value of the AC signal that needs to be applied across the varactor diode.

Q11.8 The resonant circuit of an oscillator is shown in Figure 11.71, and includes two identical GaAs varactor diodes with hyper-abrupt junctions ($\gamma = 1.3$), and $C_O = 27\,\text{pF}$. The oscillator is to provide an FM signal with a carrier frequency of 15.85 MHz, and a peak frequency deviation of 25 kHz. Determine the required values of V_b and V_m. What is the advantage of using two varactor diodes in parallel?

Q11.9 Figure 11.72 shows the resonant circuit of an oscillator, in which two identical GaAs diodes with a hyper-abrupt junctions (with $\gamma = 1.25$) and $C_O = 18.5\,\text{pF}$ have been used.
 (i) Determine the value of DC bias required to make the circuit resonate at 45 MHz.
 (ii) Determine the change in DC bias needed to increase the resonant frequency by 0.1%.

Q11.10 Repeat Q11.9 assuming the GaAs diodes have graded junctions with $C_O = 18.5\,\text{pF}$.

Q11.11 Figure 11.73 shows the resonant circuit of a 10 MHz oscillator used to produce frequency modulation. The varactor diodes are silicon with graded junctions, and $C_O = 28\,\text{pF}$. Determine the values of V_b and V_m required to produce an FM signal with a modulation index of 0.3.

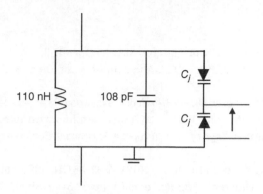

Figure 11.72 Circuit for Q11.9.

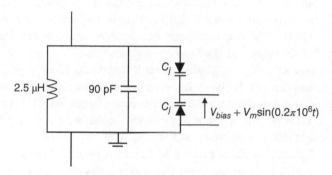

Figure 11.73 Circuit for Q11.11.

Q11.12 The circuit shown in Figure 11.66 was used to measure the noise from a microwave oscillator. The following data apply to the circuit:

Spectrum analyzer reading: −82 dBV at 10 kHz

Spectrum analyzer bandwidth: 120 Hz

Discriminator response at oscillator frequency: 450 mV/MHz

AM detector efficiency: 72%

With the discriminator removed from the system, the spectrum analyzer reading dropped to −105 dBV at 10 kHz. Determine the FM noise from the oscillator, 10 kHz off the carrier, in dBc/Hz, stating any assumptions made.

Q11.13 A PLL synthesizer using an offset oscillator (shown in Figure 11.31) is used to provide the local oscillator for a single-conversion superhet receiver (see Chapter 14) in which the IF is 10.7 MHz. The carrier signal band is 88–108 MHz, and the tuning resolution is 20 kHz. If the offset oscillator is set to 80 MHz, and the LO frequency is above the received frequency, what is the range of the divisor, N, required in the synthesizer?

Q11.14 A PLL synthesizer using an offset oscillator is to be used to generate a local oscillator, and cover the frequency range 130–140 MHz in 5 kHz steps. If the minimum divider ratio N is 500, corresponding to an output frequency of 130 MHz, determine:

  (i) The reference frequency
 (ii) The frequency of the offset oscillator
(iii) The maximum value of N.
(iv) What value of N is required to give an output frequency of 137.02 MHz?

Q11.15 Design a Fractional-N synthesizer to produce an output frequency of 114.8 MHz using a reference frequency of 8 MHz.

References

1 Smith, J. (1998). *Modern Communication Circuits*. Boston, MA: McGraw-Hill.

2 Glover, I.A., Pennock, S.R., and Shepherd, P.R. (2005). *Microwave Devices, Circuits and Subsystems*. Chichester, UK: Wiley.

3 Crawford, J.A. (1994). *Frequency Synthesizer Design Handbook*. Norwood, MA: Artech House.

4 Stein, W., Huber, F., Bildek, S., Aigle, M., and Vossiek, M. (2017). An improved ultra-low-noise tunable YIG oscillator operating in the 6–12 GHz range. *Proceedings of 47th European Microwave Conference*, Nuremberg, Germany (10–12 October 2017), pp. 767–770.

5 van Delden, M., Pohl, N., Aufinger, K., and Musch, T. (2019). A 32–48 GHz differential YIG oscillator with low phase noise based on a SiGe MMIC. *Proceedings of IEEE Radio and Wireless Symposium*, Orlando, FL (20–23 Januray 2019).

6 Buchwald, A.W., Martin, K.W., Oki, A.K., and Kobayashi, K.W. (1992). A 6 GHz integrated phase-locked loop using AlGaAs/GaAs Heterojunction bipolar transistors. *IEEE Journal of Solid-State Circuits* 27 (12): 1752–1761.

7 Larson, L.E. (ed.) (1996). *RF and Microwave Circuit Design for Wireless Applications*. Norwood, MA: Artech House.

8 Komatsu, Y. and Murakami, Y. (1983). Coupling coefficient between microstrip line and dielectric resonator. *IEEE Transactions on Microwave Theory and Techniques* 31 (1): 34–40.

9 Pozar, D.M. (2001). *Microwave and RF Design of Wireless Systems*. New York: Wiley.

10 Chaimbault, D., Verdeyme, S., and Guillon, P. (1994). Rigorous design of the coupling between a dielectric resonator and a microstrip line. *Proceedings of 24th European Microwave Conference*, Cannes, pp. 1191–1196.

11 Vollers, H.G. and Claiborne, L.T. (1974). RF oscillator control utilizing surface wave delay lines. *Proceedings of 28th IEEE Annual Symposium on Frequency Control*, Fort Monmouth, NJ, pp. 256–259.

12 Lu, X., Ma, J., Zhu, X.L., Lee, C.M., Yue, C.P., and Lau, K.M. (2012). A novel GaN-based monolithic SAW/HEMT oscillator on silicon. *Proceedings of the 2012 IEEE International Ultrasonics Symposium*, Dresden, pp. 2206–2209.

13 Gunn, J.B. (1976). The discovery of microwave oscillations in gallium arsenide. *IEEE Transactions on Electron Devices* 23 (7): 705–713.

14 Leeson, D.B. (1966). A simple model of feedback oscillator noise spectrum. *Proceedings of the IEEE* 54: 329–330.

15 Nallatamby, J.C., Prigent, M., Camiade, C., and Obregon, J. (2003). Phase noise in oscillators – Leeson formula revisited. *IEEE Transactions on Microwave Theory and Techniques* 51 (4): 1386–1394.

16 Huang, X., Tan, F., Wei, W., and Fu, W. (2007). A revisit to phase noise model of Leeson. *Proceedings of the 2007 IEEE Int Frequency Control Symposium*, pp. 238–241.

17 Lee, T.H. and Hajimiri, A. (2000). Oscillator phase noise: a tutorial. *IEEE Journal of Solid-State Circuits* 35 (3): 326–336.

18 Gheidi, H. and Banai, A. (2010). Phase-noise measurement of microwave oscillators using phase-shifterless delay-line discriminator. *IEEE Transactions on Microwave Theory and Techniques* 58 (2): 468–477.

19 Barzegar, A.S., Banai, A., and Farzaneh, F. (2016). Sensitivity improvement of phase-noise measurement of microwave oscillators using IF delay line based discriminator. *IEEE Microwave and Wireless Components Letters* 26 (7): 546–548.

20 Ondria, J. (1968). A microwave system for measurements of AM and FM noise spectra. *IEEE Transactions on Microwave Theory and Techniques* 16 (9): 767–781.

12

RF and Microwave Antennas

12.1 Introduction

This chapter will introduce the fundamental properties of RF antennas through an initial discussion of the power radiated from a simple wire conductor, leading to theoretical expressions for the electric and magnetic fields at a long distance from the conductor. The theory will be extended to consider the half-wave dipole, which is probably the most basic and most useful of all antenna types. The use of wire dipole elements in array configurations will be discussed, and we will consider how arrays can be designed and implemented in planar formats. The chapter will conclude with a discussion of traditional microwave antennas, namely those using aperture techniques and radiation from waveguide slots. In the latter case, it will be shown how modern fabrication techniques allow waveguide slot antennas to be implemented in multilayer planer substrates, using photoimageable thick-film and LTCC technologies.

12.2 Antenna Parameters

(1) An antenna is a passive loss-free structure, which converts the field pattern of the transmission line that feeds it to a field pattern in free space, and vice-versa. It can be regarded as a transformer.

(2) In principle, the available output power from an antenna equals the available input power, so the available gain is unity, as in a transformer.

(3) The *radiation pattern* of an antenna shows the power or electric field radiated from the antenna as a function of direction. Figure 12.1 illustrates a typical radiation pattern, consisting of a main lobe, side lobes, and a back lobe.
One of the important features of a radiation pattern is the beamwidth, which effectively determines spatial selectivity of the antenna. There are two common ways of specifying the beamwidth; these are the half-power beamwidth, $\theta_{3\,dB}$, and the beamwidth between the first nulls of the radiation pattern, θ_{nulls}. Both definitions of beamwidth apply to the main lobe of the radiation pattern, and are illustrated in Figure 12.2, where the radial scale of the radiation pattern shown is proportional to power.

(4) The radiation pattern is the same whether the flow of energy is from the transmission line to free space, or vice versa. So antennae can be used to receive power from free space as well as transmit power to free space.

(5) The shape of the radiation pattern depends on the geometry of the antenna.

(6) Since the performance of communication systems depends upon received power we define *power density* at a distant point from an antenna as the power passing through unit area at that point. Power density therefore has the units of W/m².

(7) The *antenna gain, G,* is the ratio of the power density, P_{den}, at a distant point due to the antenna to the power density at the same point due to a reference antenna, P_{den} (reference), fed with the same power, i.e.

$$G = \frac{P_{den}}{P_{den}(\text{reference})}. \tag{12.1}$$

(8) The reference antenna is often an isotropic antenna, i.e. an antenna that radiates uniformly in all directions, although such antennas do not exist in practice. The power density at a given distance, r, from an isotropic antenna is simply

RF and Microwave Circuit Design: Theory and Applications, First Edition. Charles E. Free and Colin S. Aitchison.
© 2022 John Wiley & Sons Ltd. Published 2022 by John Wiley & Sons Ltd.
Companion website: www.wiley.com/go/free/rfandmicrowave

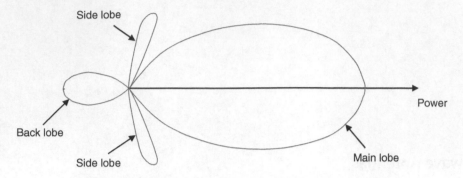

Figure 12.1 Sketch of typical antenna power radiation pattern.

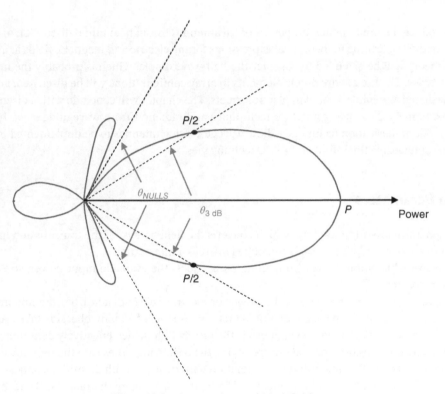

Figure 12.2 Antenna beamwidth definitions.

the radiated power, P_T, from the antenna divided by the surface area of a sphere, i.e.

$$P_{den}(\text{isotropic}) = \frac{P_T}{4\pi r^2}. \tag{12.2}$$

(9) An antenna which converts a free-space field pattern to a transmission line field pattern behaves as if it is collecting power from the free-space field pattern over an area known as the *effective aperture*, A_{eff}. We use the term effective aperture to distinguish it from the physical area of the antenna. The effective aperture, A_{eff}, is the ratio of the power received, P_R, at the transmission line output to the power density, P_{den}, at the point where the antenna is positioned, i.e.

$$A_{eff} = \frac{P_R}{P_{den}}$$

or

$$P_R = P_{den} A_{eff} \tag{12.3}$$

(10) The *radiation resistance* (R_{rad}) of an antenna is the input resistance of the transmission line input to the antenna when it is radiating into free space, assuming the antenna to be loss free. In practice, when power is supplied to an antenna some of the power will be converted into the electromagnetic field surrounding the antenna, and some will be dissipated within the antenna, mainly due to the ohmic losses of the antenna. We can define the *radiation efficiency* (η) of an antenna as the ratio of the power radiated, P_{rad}, to the power supplied, P_{in}, at the transmission line terminals of the antenna, i.e.

$$\eta = \frac{P_{rad}}{P_{in}}. \tag{12.4}$$

Since power is directly proportional to resistance we can rewrite Eq. (12.4) as

$$\eta = \frac{R_{rad}}{R_{loss} + R_{rad}}, \tag{12.5}$$

where R_{loss} is the loss resistance of the antenna. The quantity ($R_{loss} + R_{rad}$) is the total input resistance of the antenna. In practice, an antenna will exhibit some reactance (X), and we define the *antenna impedance* (Z_{ant}) at the transmission line terminals of the antenna as

$$Z_{ant} = R_{loss} + R_{rad} \pm jX. \tag{12.6}$$

(11) To calculate the performance of a one-way, free-space communication system between two antennas spaced a distance r apart it is necessary to know both the gain of the transmitting antenna (G_T) and the effective aperture ($A_{eff,R}$) of the receiving antenna. We can obtain an expression for the power received, P_R, by combining Eqs. (12.1) through (12.3). From Eq. (12.1) we have

$$G_T = \frac{P_{den}}{P_T/4\pi r^2} = \frac{4\pi r^2 P_{den}}{P_T},$$

and substituting for P_{den} from Eq. (12.3) gives

$$G_T = \frac{4\pi r^2 P_R}{P_T A_{eff,R}},$$

and rearranging gives

$$P_R = \frac{P_T G_T}{4\pi r^2} A_{eff,R}. \tag{12.7}$$

Note that Eq. (12.7) could also be written in terms of the power density at the position of the receiving antenna as

$$P_R = P_{den} A_{eff,R}, \tag{12.8}$$

where

$$P_{den} = \frac{P_T G_T}{4\pi r^2}. \tag{12.9}$$

It can be shown [1] that, at a given frequency, f, the effective aperture of any antenna is related to the gain of that antenna by the expression

$$A_{eff} = \frac{\lambda^2 G}{4\pi}, \tag{12.10}$$

where $\lambda = c/f$.

Substituting for the effective aperture in Eq. (12.7) gives

$$P_R = \frac{P_T G_T G_R \lambda^2}{16\pi^2 r^2}. \tag{12.11}$$

Equation (12.11) gives a very important relationship between the power transmitted and the power received in a free-space communication system; it is the basis of the propagation calculation in all practical wireless communication systems; this equation is also known as the Friis equation (or alternatively as the Friis transmission formula).

(12) Antennas can be subdivided into these groups: those whose relevant dimensions are small compared to the wavelength, those whose relevant dimensions are similar to the wavelength, and those whose dimensions are large compared with the wavelength. Table 12.1 shows Gain, Effective Aperture, and Radiation Resistance for an example of each type of antenna.

Table 12.1 Data on typical antennae.

Parameter	Dimensions $\ll \lambda$	Dimensions $\sim \lambda$	Dimensions $\gg \lambda$
	Hertzian dipole	*Half-wave dipole*	*Paraboloid*
Gain (G)	1.5	1.64	43 864 for 2 m diameter paraboloid at 10 GHz
Effective aperture (A_{eff})	$3\lambda^2/8\pi$ m^2	$\lambda^2/8$ m^2	$\varepsilon \times A_{phs}$, where ε is the aperture efficiency and A_{phs} is the physical area of dish
Radiation resistance (R_{rad})	$80\pi^2\left(\dfrac{\delta l}{\lambda}\right)^2$ Ω Very small if $\dfrac{\delta l}{\lambda}$ is small, e.g. if $\dfrac{\delta l}{\lambda} = 0.01$ $R_{rad} = 0.08$ Ω	73 Ω	Equal to impedance of transmission line feeding the antenna

---

Example 12.1 An antenna that has a gain of 15 dB is supplied with a power of 2 W at a frequency of 5 GHz. Determine:

(i) The power density 8 km from the antenna in the direction of maximum radiation.

(ii) The power received by an identical aerial located 20 km away from the transmitter, when both antennas are oriented for maximum reception.

Solution

$$G = 15 \text{ dB} \equiv 10^{1.5} = 31.62,$$

$$f = 5 \text{ GHz} \quad \Rightarrow \quad \lambda = \frac{3 \times 10^8}{5 \times 10^9} \text{ m} = 0.06 \text{ m}.$$

(i) Using Eq. (12.9):

$$P_{den} = \frac{2 \times 31.62}{4\pi \times (8 \times 10^3)^2} \text{ W/m}^2 = 78.62 \text{ nW/m}^2.$$

(ii) Using Eq. (12.8):

$$P_R = 78.62 \times 10^{-9} \times \frac{(0.06)^2}{4\pi} \times 31.62 \text{ W}$$

$$= 113.94 \text{ pW}.$$

12.3 Spherical Polar Coordinates

Using spherical polar coordinates the position of any point, P, in space can be described in terms of three coordinates (r, θ, ϕ) as shown in Figure 12.3.

In this system of coordinates, r is the radial distance from the origin to point P, θ is the angle measured in the vertical plane through P, and ϕ is the angle measured in the azimuthal plane. In Figure 12.3, it should be noted that point Q is the projection of point P in the horizontal plane, and that the vertical plane through P is defined by the plane $PQOR$, where O is the origin.

In antenna work, where electromagnetic waves propagate in three dimensions, it is convenient to describe the components of the propagating wave in terms of spherical polar coordinates.

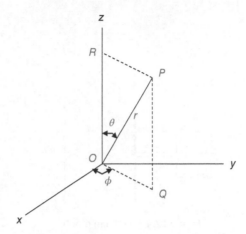

Figure 12.3 Spherical polar coordinates.

12.4 Radiation from a Hertzian Dipole

12.4.1 Basic Principles

A Hertzian dipole, shown in Figure 12.4, is an elemental length (δz) of conductor carrying a current, I, of constant magnitude, where $\delta z \ll \lambda$. It is referred to as a dipole because the ends of the conductor must be at opposite potential at any instant of time.

The Hertzian dipole does not represent a real antenna, but it is useful as a mathematical tool in analyzing real antennas whose length is an appreciable fraction of a wavelength. For example, an electrically long wire antenna may be regarded as made up of a large number of Hertzian dipoles positioned end-to-end. We will made use of this concept in analyzing the performance of a half-wave dipole in Section 12.5.

Using classical field theory it can be shown [1, 2] that the electric and magnetic fields in spherical coordinates at a distance r from a Hertzian dipole of length δz, and carrying a peak current I_O are

$$E_r = \frac{I_O \, \delta z \, 120\pi \, \cos\theta \, e^{j\omega[t-r/c]}}{\lambda} \left(\frac{c}{\omega r^2} - j\frac{c^2}{\omega^2 r^3} \right),$$

$$E_\theta = \frac{I_O \, \delta z \, 60\pi \, \sin\theta \, e^{j\omega[t-r/c]}}{\lambda} \times \left(j\frac{1}{r} + \frac{c}{\omega r^2} - j\frac{c^2}{\omega^2 r^3} \right),$$

$$E_\phi = 0,$$

$$H_r = 0,$$

$$H_\theta = 0,$$

$$H_\phi = \frac{I_O \, \delta z \, \sin\theta \, e^{j\omega[t-r/c]}}{2\lambda} \left(j\frac{1}{r} + \frac{c}{\omega r^2} \right). \tag{12.12}$$

It will also be noticed from these three equations that the time has been retarded by r/c to account for the fact that the electromagnetic field components at a distance r from the source will lag in phase with respect to the current in the dipole, due to the time taken for the wave to propagate over a distance r.

The region surrounding any radiating element can be divided into the near-field region and the far field region. The near-field components in Eq. (12.12) are those with r^2 or r^3 in the denominator, and they are known, respectively, as the

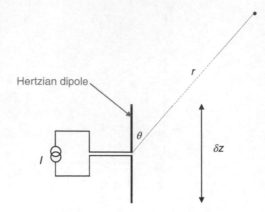

Figure 12.4 Hertzian dipole.

inductive and electrostatic components. These fields are associated with the static distribution and velocity of charges in the radiating element and decay rapidly in magnitude as r increases. The field components with r in the denominator are the far-field components, often called the radiation components, and it is these fields that are of primary interest for radio communication.

The far-field radiation components from Eq. (12.12) are

$$E_\theta = j\frac{I_O \ \delta z \ 60\pi \ \sin\theta \ e^{j\omega\left[t - r/c\right]}}{\lambda r}, \tag{12.13}$$

$$H_\phi = j\frac{I_O \ \delta z \ \sin\theta \ e^{j\omega\left[t - r/c\right]}}{2\lambda r}. \tag{12.14}$$

The peak magnitudes of the far-field components are

$$|E_\theta| = \frac{I_O \ \delta z \ 60\pi \ \sin\theta}{\lambda r}, \tag{12.15}$$

$$|H_\phi| = \frac{I_O \ \delta z \ \sin\theta}{2\lambda r}. \tag{12.16}$$

If we divide Eq. (12.15) by Eq. (12.16), we obtain the impedance of free space, Z_O, i.e.

$$Z_O = \frac{|E_\theta|}{|H_\phi|} = 120\pi. \tag{12.17}$$

The shapes of the far-field radiation patterns due to a Hertzian dipole can be deduced from Eqs. (12.15) and (12.16); the E-field polar diagram patterns are shown in Figure 12.5.

In order to determine the power radiated from a Hertzian dipole we make use of the Poynting vector. This vector, denoted by the symbol **P**, indicates the direction of power flow in a radiating electromagnetic field, and also the magnitude of the power flow. (Note that the symbol for the Poynting vector is shown in bold to denote a vector quantity.) The Poynting vector

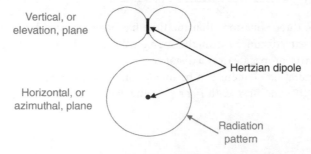

Figure 12.5 E-field polar diagrams due to a Hertzian dipole.

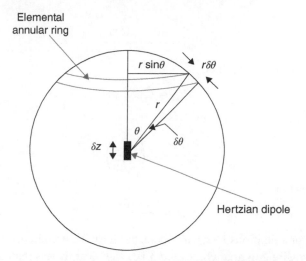

Figure 12.6 Elemental annular ring on sphere surrounding Hertzian dipole.

states that if $E_{\theta,rms}$ and $H_{\phi,rms}$ are the rms values of the electric and magnetic fields there is a power flow normal to the plane containing the electric and magnetic field with the power per unit area (P_{den}) given by

$$P_{den} = E_{\theta,rms} H_{\phi,rms} \sin \alpha, \tag{12.18}$$

where α is the angle between the electric and magnetic field components. The quantity P_{den} is also known as the power density, and has the units of W/m^2. For an electromagnetic field in free space, $\alpha = 90°$, and Eq. (12.18) reduces simply to

$$P_{den} = E_{\theta,rms} H_{\phi,rms}. \tag{12.19}$$

We can find power radiated from a Hertzian dipole by considering the dipole to be at the centre of a sphere of radius r, as shown in Figure 12.6. The total radiated power from the dipole can then be found by first determining the power flowing through an elemental annular ring, of area δS, on the surface of the sphere, and then integrating over the total surface to find the total power radiated.

Using Eq. (12.19) the power density at any point on the surface of the sphere will be

$$P_{den} = \frac{|E_\theta|}{\sqrt{2}} \times \frac{|H_\phi|}{\sqrt{2}}, \tag{12.20}$$

where E_θ and H_ϕ are given in Eq. (12.12). Hence the power flowing through the elemental annular ring will be

$$\begin{aligned}
P &= \frac{|E_\theta|}{\sqrt{2}} \times \frac{|H_\phi|}{\sqrt{2}} \times \delta S \\
&= \frac{60\pi I_O^2 (\delta z)^2 \sin^2\theta}{4\lambda^2 r^2} \times \delta S, \tag{12.21}
\end{aligned}$$

where

$$\delta S = \pi \times 2r \sin\theta \times r\delta\theta = 2\pi r^2 \sin\theta \, \delta\theta. \tag{12.22}$$

Thus, the total power radiated through the sphere will be

$$\begin{aligned}
P_{rad} &= \int_0^\pi \frac{60\pi I_O^2 (\delta z)^2 \sin^2\theta}{4\lambda^2 r^2} \times 2\pi r^2 \sin\theta \, d\theta \\
&= \frac{30\pi^2 I_O^2 (\delta z)^2}{\lambda^2} \int_0^\pi \sin^3\theta \, d\theta. \tag{12.23}
\end{aligned}$$

Substituting the standard integral

$$\int_0^\pi \sin^3\theta \, d\theta = \frac{4}{3}.$$

Equation (12.23) becomes

$$P_{rad} = \frac{40\pi^2 I_O^2 (\delta z)^2}{\lambda^2}. \qquad (12.24)$$

The radiation resistance can now be found, since

$$P_{rad} = I_{rms}^2 R_{rad}. \qquad (12.25)$$

Substituting from Eq. (12.24) into Eq. (12.25) gives

$$\frac{40\pi^2 I_O^2 (\delta z)^2}{\lambda^2} = \left(\frac{I_O}{\sqrt{2}}\right)^2 R_{rad}, \qquad (12.26)$$

and by rearranging Eq. (12.26) we obtain

$$R_{rad} = \frac{80\pi^2 (\delta z)^2}{\lambda^2}. \qquad (12.27)$$

Equation (12.27) was derived for a fictitious Hertzian dipole, but it can be used in practice to give a good approximation of the radiation resistance of any wire antenna whose length is a significantly less than a wavelength, e.g. less than $\lambda/10$. However, it should be noted that the radiation resistance of a Hertzian dipole is very small, which makes it difficult to match to a generator.

12.4.2 Gain of a Hertzian Dipole

We can find the gain of a Hertzian dipole relative to an isotropic radiator by considering each antenna to be supplied with the same power, P, and then calculating the ratio of the electric field strengths at a given distance, r, from each antenna. In the case of the Hertzian dipole, which has gain and consequently concentrates the radiated power in particular directions, we define the gain in the direction of maximum electric field strength, E_m.

The power density at distance r from the isotropic radiation is given by Eq. (12.2) as

$$P_{den} = \frac{P}{4\pi r^2}. \qquad (12.28)$$

Using Poynting vector we can write Eq. (12.28) as

$$\frac{P}{4\pi r^2} = \frac{E}{\sqrt{2}} \times \frac{H}{\sqrt{2}} = \frac{EH}{2}. \qquad (12.29)$$

Using the impedance of free space, Eq. (12.29) can be written as

$$\frac{P}{4\pi r^2} = \frac{E^2}{240\pi},$$

i.e.

$$E = \frac{\sqrt{60P}}{r}. \qquad (12.30)$$

We can find an expression for the maximum electric field due to a Hertzian dipole by first writing Eqs. (12.15) and (12.16) in the form

$$|E_\theta| = E_m \sin\theta \qquad (12.31)$$

and

$$|H_\phi| = H_m \sin\theta, \qquad (12.32)$$

where E_m and H_m are the maximum values of the electric and magnetic fields, respectively. Using this notation Eq. (12.23) can be written as

$$\begin{aligned} P_{rad} &= \int_0^\pi \frac{E_m H_m \sin^2\theta}{2} \times 2\pi r^2 \sin\theta \, d\theta \\ &= E_m H_m \pi r^2 \int_0^\pi \sin^3\theta \, d\theta = \frac{E_m^2}{120\pi} \pi r^2 \times \frac{4}{3}, \\ &= \frac{E_m^2 \, r^2}{90} \end{aligned} \qquad (12.33)$$

i.e.

$$E_m = \frac{\sqrt{90 P_{rad}}}{r}.$$ (12.34)

Now, since we have considered a loss-free antenna, the power radiated from the antenna must equal to power supplied, i.e. $P_{rad} = P$, and so Eq. (12.34) can be written as

$$E_m = \frac{\sqrt{90P}}{r}.$$ (12.35)

Thus, using Eqs. (12.35) and (12.30) we can obtain the gain, G, of the Hertzian dipole as

$$G = \left(\frac{E_m}{E}\right)^2 = \frac{90}{60} = 1.5.$$ (12.36)

We can also express the gain in dB as

$$G_{dB} = 10 \log 1.5 \text{ dB} = 1.76 \text{ dB}.$$ (12.37)

Note that since the units of electric field are volts/metre, E_m/E is equivalent to a voltage ratio, and $(E_m/E)^2$ to a power ratio.

Example 12.2   Determine the radiation resistance and radiation efficiency of a 10 cm long dipole at a frequency of 100 MHz. Assume that the loss resistance of the dipole is 1.5 Ω.

Solution

$$100 \text{ MHz} \quad \Rightarrow \quad \text{Free space wavength, } \lambda_O = \frac{3 \times 10^8}{100 \times 10^6} \text{ m} = 3 \text{ m}.$$

Since the 10 cm length of the dipole is very much less than λ we may use Eq. (12.27):

$$R_{rad} = \frac{80\pi^2 \times (0.1)^2}{(3)^2} \ \Omega = 0.88 \ \Omega.$$

Using Eq. (12.5):

$$\eta = \frac{0.88}{0.88 + 1.5} = 0.37 \equiv 37\%.$$

12.5   Radiation from a Half-Wave Dipole

12.5.1   Basic Principles

As the name implies, the half-wave dipole refers to a linear antenna whose length is $\lambda/2$, as illustrated in Figure 12.7.

The current distribution on a half-wave dipole can be approximated by a sinusoid with maximum current at the feed point and zero current at the ends of the two conductors, as shown in Figure 12.7. The magnitude of the current at a distance z from the feed point is then given by

$$I_z = I_O \cos \beta z,$$ (12.38)

where I_O is the magnitude of the current at the feed point, and β is the phase propagation constant that applies to current travelling along the antenna.

The reason for the shape of the current distribution is often explained by considering the dipole to be formed as the result of splaying-out by 90° the last $\lambda/4$ section of an open-circuited, parallel wire transmission line. The shape of the current distribution on the dipole will then correspond approximately to that of the current standing wave pattern on the end of the transmission line.

Whilst the sinusoidal current distribution is a reasonable approximation over most of the dipole's length, in practice the distribution will exhibit some deviation in the vicinity of the feed point.

For the purpose of analysis, the $\lambda/2$ dipole may be considered to be made up of a number of Hertzian dipoles, positioned end-to-end. Figure 12.8 shows the position of one such arbitrary Hertzian dipole, at a distance z from the centre.

If the current flowing through this Hertzian dipole is I_1 then using Eq. (12.12) the electric field at point P due to the Hertzian dipole will be

$$E_{\theta 1} = j \frac{I_1 \ \delta z \ 60\pi \ \sin \theta_1 \ e^{j\omega\left[t - r_1/c\right]}}{\lambda r_1},$$ (12.39)

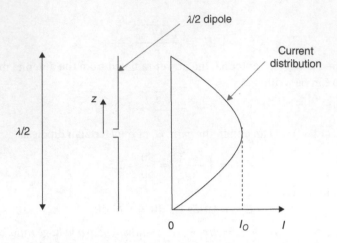

Figure 12.7 Current distribution on a half-wave dipole.

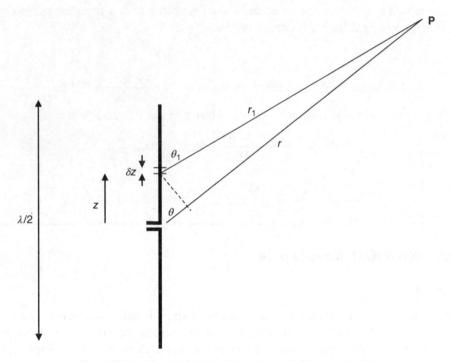

Figure 12.8 Electric field radiation from a half-wave dipole.

where

$$r_1 = r - z\cos\theta,$$

$$\theta_1 \approx \theta \quad (\text{if } r \gg \lambda)$$

and

$$I_1 = I_O \cos\beta z.$$

Thus, the electric field at point P due to the Hertzian element can be written as

$$E_{\theta 1} = j\frac{I_O \cos(\beta z)\ \delta z\ 60\pi\ \sin\theta\ e^{j\omega\left[t-(r-z\cos\theta)/c\right]}}{\lambda(r - z\cos\theta)}. \tag{12.40}$$

If $r \gg \lambda$, the difference between r and r_1 will be small and will not significantly affect the magnitude of the field, although the small difference, as a fraction of a wavelength, may be significant in terms of phase. Consequently Eq. (12.40) may be written as

$$E_{\theta 1} = j\frac{I_O \cos(\beta z)\ \delta z\ 60\pi\ \sin\theta\ e^{j\omega\left[t-(r-z\cos\theta)/c\right]}}{\lambda r}. \tag{12.41}$$

The total electric field at point P can now be found by summing the fields due to all the Hertzian elements comprising the antenna, i.e. by integrating from $z = -\lambda/4$ to $z = +\lambda/4$, giving

$$E_\theta = \int_{-\lambda/4}^{\lambda/4} j \frac{I_0 \cos(\beta z) 60\pi \ \sin\theta \ e^{j\omega\left[t - (r - z\cos\theta)/c\right]}}{\lambda r} \, dz$$

$$= j\frac{I_0 60\pi \sin\theta e^{j\omega\left[t - r/c\right]}}{\lambda r} \int_{-\lambda/4}^{\lambda/4} \cos(\beta z) e^{j\omega z \cos\theta/c} \, dz. \tag{12.42}$$

The integral in Eq. (12.42) is of a standard form, and has the solution given in Eq. (12.43)

$$\int \cos az \ e^{bz} \, dz = \frac{e^{bz}}{a^2 + b^2} (b \cos az + a \sin az). \tag{12.43}$$

Applying this standard solution to Eq. (12.42) we obtain, after lengthy but straightforward manipulation

$$E_\theta = j\frac{I_0 60 e^{j\omega\left[t - r/c\right]}}{r} \cdot \frac{\cos\left(\dfrac{\pi}{2}\cos\theta\right)}{\sin\theta} \tag{12.44}$$

or

$$|E_\theta| = \frac{60 I_0}{r} \frac{\cos\left(\dfrac{\pi}{2}\cos\theta\right)}{\sin\theta}. \tag{12.45}$$

Using Eq. (12.45) the radiation pattern of a $\lambda/2$ dipole in the elevation plane can be plotted, as shown in Figure 12.9. In the azimuthal plane the dipole appears to be a point source, and thus will radiate uniformly in all directions and have a circular radiation pattern.

The radiation resistance of the half-wave dipole may be found using the same general procedure as for the Hertzian dipole. The radiated power will be

$$P_{rad} = \int_0^\pi \frac{1}{2} \frac{|E_\theta|^2}{120\pi} 2\pi r^2 \sin\theta \, d\theta, \tag{12.46}$$

where E_θ is given by Eq. (12.45), i.e.

$$P_{rad} = \int_0^\pi \frac{\pi r^2}{120\pi} \left[\frac{60 I_0}{r} \frac{\cos\left(\dfrac{\pi}{2}\cos\theta\right)}{\sin\theta} \right]^2 \sin\theta \, d\theta, \tag{12.47}$$

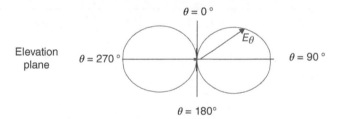

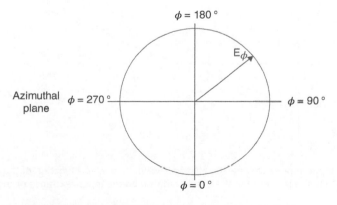

Figure 12.9 Electric field radiation patterns of a $\lambda/2$ dipole.

i.e.

$$P_{rad} = 30I_O^2 \int_0^\pi \frac{\cos^2\left(\frac{\pi}{2}\cos\theta\right)}{\sin\theta}\, d\theta. \tag{12.48}$$

The integral in Eq. (12.48) cannot be evaluated using simple functions. The results of numerical integration give

$$\int_0^\pi \frac{\cos^2\left(\frac{\pi}{2}\cos\theta\right)}{\sin\theta}\, d\theta = 1.22. \tag{12.49}$$

Using this value in Eq. (12.48) gives the power radiated from a half-wave dipole as

$$P_{rad} = 36.6I_O^2. \tag{12.50}$$

Therefore, we have (noting that I_O is a peak value)

$$36.6I_O^2 = \left(\frac{I_O}{\sqrt{2}}\right)^2 R_{rad} \quad \Rightarrow \quad R_{rad} = 73.2\ \Omega, \tag{12.51}$$

i.e.

$$R_{rad}(\text{half-wave dipole}) = 73.2\ \Omega \approx 73\ \Omega. \tag{12.52}$$

The variation of the input impedance of a typical half-wave dipole in the vicinity of resonance is shown in Figure 12.10. The nominal resonant length of the dipole at the design frequency is 0.5λ, and it can be seen from Figure 12.10 that the

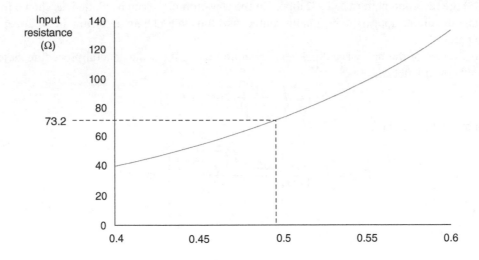

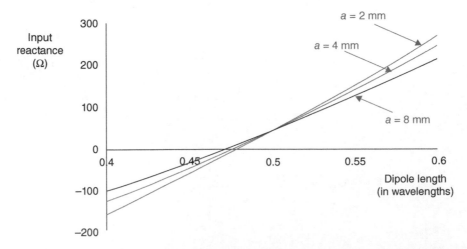

Figure 12.10 Variation of the input impedance of a 100 MHz half-wave dipole, as a function of the electrical length of the dipole, showing also the influence of the dipole radius a on the input reactance of the dipole.

input impedance is approximately $(73 + j43)\ \Omega$. The input reactance falls to zero when the dipole length is around 0.47λ, because fringing of the electromagnetic fields at the ends of the dipole make it behave electrically as slightly longer than the physical length. A length of 0.47λ is normally chosen for practical designs since the dipole can then be conveniently matched to a $75\ \Omega$ feed line, on the assumption that the loss resistance is of the order of $2\ \Omega$, giving a total input resistance of approximately $(73 + 2)\ \Omega$. The radius of the dipole element does not have a significant effect on the input resistance, but it can also be seen from Figure 12.10 that the rate of change of input reactance decreases as the radius of the dipole element increases. This indicates that the bandwidth of the dipole increases as the diameter of the element increases, and large diameter elements should always be chosen for wide bandwidth applications.

Example 12.3

(i) What is the direction of maximum radiation from a half-wave dipole?
(ii) If a half-wave dipole is supplied with an rms current of $2\ A$, find the electric and magnetic field strengths $2\ km$ from the dipole in the direction of maximum radiation.
(iii) What is the power density at the position specified in part (ii)?

Solution

(i) From inspection of Eq. (12.45) E_θ has its maximum value when $\theta = 90°$.
(ii) Using Eq. (12.45), with $\theta = 90°$:

$$|E_\theta| = \frac{60 \times 2}{2 \times 10^3}\ \text{V/m} = 60\ \text{mV/m},$$

$$\frac{E_\theta}{H_\phi} = 120\pi \quad \Rightarrow \quad H_\phi = \frac{E_\theta}{120\pi} = \frac{60 \times 10^{-3}}{120\pi}\ \text{A/m} = 159.13\ \mu\text{A/m}.$$

(iii)

$$P_{den} = \frac{|E_\theta|}{\sqrt{2}} \times \frac{|H_\phi|}{\sqrt{2}} = \frac{60 \times 10^{-3} \times 159.13 \times 10^{-6}}{2}\ \text{W/m}^2 = 4.77\ \mu\text{W/m}^2.$$

Example 12.4 Assuming the loss resistance of a half-wave dipole is $2\ \Omega$, determine the radiation efficiency.

Solution

$$\eta = \frac{R_{rad}}{R_{loss} + R_{rad}} - \frac{73}{2 + 73} = 0.97 \equiv 97\%.$$

12.5.2 Gain of a Half-Wave Dipole

The gain of the half-wave dipole is found using the same procedure as for the Hertzian dipole in Section 12.4.2. First, we write the magnitudes of the electric and magnetic fields due to the half-wave dipole as

$$|E_\theta| = E_m \frac{\cos\left(\frac{\pi}{2}\cos\theta\right)}{\sin\theta} \tag{12.53}$$

and

$$|H_\phi| = \frac{E_m}{120\pi} \frac{\cos\left(\frac{\pi}{2}\cos\theta\right)}{\sin\theta}. \tag{12.54}$$

The power radiated is then given by

$$P_{rad} = \int_0^\pi \frac{|E_\theta||H_\phi|}{2} 2\pi r^2 \sin\theta\ d\theta$$

$$= \frac{E_m^2 r^2}{120} \int_0^\pi \frac{\cos^2\left(\frac{\pi}{2}\cos\theta\right)}{\sin\theta}\ d\theta. \tag{12.55}$$

Substituting the value of the integral from numerical integration gives

$$P_{rad} = \frac{E_m^2 \, r^2}{120} \times 1.22, \tag{12.56}$$

i.e.

$$E_m = \frac{\sqrt{98.36P}}{r} \tag{12.57}$$

Using Eq. (12.36) we have the gain G of the half-wave dipole as

$$G = \frac{98.36}{60} = 1.64 \equiv 2.15 \text{ dB.} \tag{12.58}$$

12.5.3 Summary of the Properties of a Half-Wave Dipole

(i) The $\lambda/2$ dipole is extremely efficient, with a radiation efficiency approaching 100%.

(ii) The gain of a half-wave dipole is 2.15 dB.

(iii) In practice, there is some fringing of the fields at the ends of the dipole, making it behave electrically as if it is slightly longer than the physical length. To compensate for fringing the practical length of the dipole is normally made slightly less than 0.5λ, around 0.47λ.

(iv) At the design frequency, the terminal impedance of the dipole is purely resistive.

(v) If the dipole is constructed from wire elements, the bandwidth of the antenna depends on the diameter of the elements, with larger diameter elements giving a greater bandwidth.

(vi) The beamwidth of a $\lambda/2$ dipole is quite large, as seen from the radiation patterns in Figure 12.9. The 3 dB beamwidth is approximately 78°. If required, however, the beamwidth can be decreased by connecting several dipoles in an array, as demonstrated later in this chapter.

(vii) The half-wave dipole is an example of a balanced antenna, since each half of the structure has the same impedance to Earth. A common method of connecting to the dipole is to use coaxial cable, as in the case of ultra-high frequency (UHF) TV reception. But coaxial cable is an unbalanced structure, with the centre conductor having much higher impedance to Earth than to the sheath of the cable. If the properties of the half-wave dipole are to be maintained, it is important to connect it to the cable using a BALUN, which is a balanced-to-unbalanced transformer connection, and which has the effect of increasing the impedance to Earth of the sheath of the cable.

12.6 Antenna Arrays

The interconnection of a number of radiating elements to form an array is a powerful design concept at RF and microwave frequencies. Since the frequencies are high, the dimensions of the radiating elements are small, and consequently very compact antennas with high functionality can be made using the array concept. The functionality can be further increased by using active devices to control the phases of the currents in the individual elements of an array, leading to the ability to steer the radiating beam, and to control its shape.

Figure 12.11 shows a linear array of three isotropic sources, with a spacing d. The rays are shown converging on a particular point in the far field. At a distance which is much greater than d we may consider the rays to be parallel for the purpose of analysis. It can be seen from Figure 12.11 that there will be path differences equal to $d \cos \theta$ between the rays emanating from adjacent sources. These path differences will be small compared to the distance travelled to the far field, and so will have negligible effect on the magnitude of the fields, but they may have significant effect on the phases of the fields.

In the far field, we can write the resultant field, E_R, as the sum of the fields from the three individual sources

$$E_R = E_1 + E_2 + E_3. \tag{12.59}$$

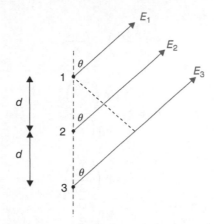

Figure 12.11 Linear array of three isotropic sources.

If each of the isotropic sources is fed with current of the same magnitude and phase, the only differences between the three field components will be the phase differences due to the path differences, and so we can write

$$E_R = E_1 + E_1 \angle - \varphi + E_1 \angle - 2\varphi, \tag{12.60}$$

where

$$\varphi = \beta d \cos \theta \tag{12.61}$$

The conventional method of summing the three vectors in Eq. (12.60) is to make use of the polygon of vectors, as shown in Figure 12.12. Each vector is drawn with the appropriate magnitude and phase, and the vector that closes the polygon represents the resultant field, E_R, both in magnitude and phase. Figure 12.12 includes a set of dotted lines at that have been added to form three isosceles triangles, each with an included angle, φ; the distance r results from the construction of the diagram and has no physical meaning.

Considering a single triangle with the included angle φ:

$$\sin\left(\frac{\varphi}{2}\right) = \frac{E_{1/2}}{r}. \tag{12.62}$$

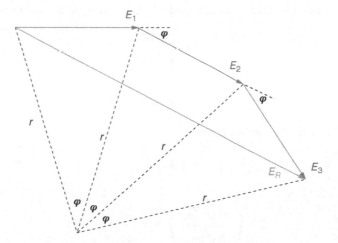

Figure 12.12 Addition of three electric field vectors from three equally separated isotropic radiators.

For the triangle with the included angle 3φ:

$$\sin\left(\frac{3\varphi}{2}\right) = \frac{E_{R/2}}{r}. \tag{12.63}$$

Combining Eqs. (12.62) and (12.63) gives

$$E_R = E_1 \frac{\sin\left(\dfrac{3\varphi}{2}\right)}{\sin\left(\dfrac{\varphi}{2}\right)}. \tag{12.64}$$

For N sources, Eq. (12.64) becomes

$$E_R = E_1 \frac{\sin\left(\dfrac{N\varphi}{2}\right)}{\sin\left(\dfrac{\varphi}{2}\right)}. \tag{12.65}$$

Now the resultant field will be a maximum when $\theta = 90°$, which means there are no path differences, and hence no phase differences between the rays from the individual sources. Thus, the maximum resultant field will be

$$E_{R,max} = NE_1, \tag{12.66}$$

and Eq. (12.65) can be rewritten as

$$E_R = \frac{E_{R,max}}{N} \frac{\sin\left(\dfrac{N\varphi}{2}\right)}{\sin\left(\dfrac{\varphi}{2}\right)}. \tag{12.67}$$

The gain, G, of an antenna array is specified in terms of the ratio of the maximum electric field from the array to the field due to a single element of the array fed with the same power, i.e.

$$G = \left(\frac{E_{R,max}}{E_1}\right)^2. \tag{12.68}$$

For a linear array of N isotropic sources, each supplied with the same current I, the power supplied to the array, P, will be

$$P = NI^2R_1, \tag{12.69}$$

where R_1 is the input resistance of an individual source, and where we have assumed there is no interaction between the sources. Then

$$I = \sqrt{\frac{P}{NR_1}}$$

and

$$E_{R,max} = Nk\sqrt{\frac{P}{NR_1}}. \tag{12.70}$$

For the same power applied to a single source we have

$$E_1 = k\sqrt{\frac{P}{R_1}}, \tag{12.71}$$

and hence the gain of the array is

$$G = \left(\frac{Nk\sqrt{P/NR_1}}{k\sqrt{P/R_1}}\right)^2 = \left(\frac{N}{\sqrt{N}}\right)^2 = N. \tag{12.72}$$

Note that since we have defined the gain as a power ratio we can express the gain in dB as

$$G_{dB} = 10\log G \text{ dB}. \tag{12.73}$$

The performance of an array is often specified in terms of the array factor, F, which is defined as

$$F = \left|\frac{E_R}{E_{R,max}}\right|, \tag{12.74}$$

and combining Eqs. (12.61), (12.67), and (12.74) gives

$$F = \left| \frac{1}{N} \frac{\sin\left(\dfrac{N\beta d \cos\theta}{2}\right)}{\sin\left(\dfrac{\beta d \cos\theta}{2}\right)} \right|.$$ (12.75)

The practical usefulness of the array factor is that we can find the radiation characteristics of a linear array of real elements, by first replacing the elements by isotropic sources to find the array factor, and then multiplying the array factor by the radiation pattern of a single real element to give the radiation pattern of the real array. This process is sometimes referred to as pattern multiplication. To find the maxima and minima of the radiation pattern we can differentiate F with respect to θ, and solve for θ using

$$\frac{\partial F}{\partial \theta} = 0.$$ (12.76)

We have so far considered the currents fed to each source to be in phase, but if there was a progressive phase change of α between the feed currents to adjacent elements, then

$$\varphi = \beta d \cos\theta + \alpha.$$ (12.77)

If the three isotropic sources shown in Figure 12.11 were fed with in-phase currents, i.e. $\alpha = 0$, the direction of the main beam would be given by $\theta = 90°$, since the three rays are in phase in this direction. Arrays with this characteristic are called broadside arrays, since the main beam is in the broadside direction.

If the three isotropic sources were fed with currents that had a progressive phase difference, βd, such that

$$I_1 = I\angle 0° \quad I_2 = I\angle \beta d \quad I_3 = I\angle 2\beta d,$$ (12.78)

then we could re-write Eq. (12.60) as

$$E_R = E_1 + E_1\angle(-\beta d \cos\theta + \beta d) + E_1\angle(-2\beta d \cos\theta + 2\beta d).$$ (12.79)

Clearly, from Eq. (12.79), all three vectors will be in phase when $\theta = 0°$ and this must therefore be the direction of the main beam. Arrays with this characteristic are called end-fire arrays, because the main beam is radiated in the end direction, relative to the line of elements.

Thus, the phases of the currents fed to the elements of an array can be used to determine the direction of the main beam. If we were to electronically vary the value of α using an appropriate phase shifter, we could electronically steer the radiated main beam. This is the basis of phased array systems, which have many RF and microwave applications, for example in radar and satellite systems.

Example 12.5 A horizontal line of four isotropic sources is shown in Figure 12.13. The sources are 0.2 m apart and fed with currents of equal magnitude.

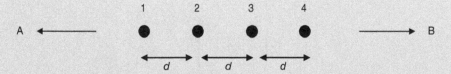

Figure 12.13 Linear array of four isotropic sources.

If the frequency is 600 MHz, determine the required phase relationships between the currents fed to the sources if the main radiated beam is to be:

(a) in direction A
(b) in direction B

(Continued)

(Continued)

Solution

$$f = 600 \text{ MHz} \quad \Rightarrow \quad \lambda_O = \frac{c}{f} = \frac{3 \times 10^8}{600 \times 10^6} \text{ m} = 0.5 \text{ m},$$

$$\phi = \beta d = \frac{2\pi}{\lambda_O} d = \frac{2\pi}{0.5} \times 0.2 \text{ rad} = 0.8\pi \text{ rad} \equiv 144°.$$

We require a progressive phase change of 144° between the currents fed to the four sources, which in the desired direction will cancel the phase differences due to the spatial separation of the sources.

(a)
$$I_1 = I\angle 0° \quad I_2 = I\angle 144° \quad I_3 = I\angle 288° \quad I_4 = I\angle 432°,$$

(b)
$$I_1 = I\angle 0° \quad I_2 = I\angle -144° \quad I_3 = I\angle -288° \quad I_4 = I\angle -432°.$$

12.7 Mutual Impedance

So far we have neglected the effects of mutual coupling between elements in an antenna array. Normally there will be an interaction between closely spaced radiating elements since the electromagnetic field due to one element will influence the magnitude and phase of the current and voltage in the other element. The coupling between elements in an array is normally specified in terms of the mutual impedance. Consider two parallel dipoles as shown in Figure 12.14, with the appropriate terminal voltages and currents.

The relationships between the complex voltages and currents can be written as

$$V_1 = I_1 Z_{11} + I_2 Z_{12}, \tag{12.80}$$

$$V_2 = I_2 Z_{22} + I_1 Z_{21}, \tag{12.81}$$

where

$Z_{11} = $ self impedance of antenna 1
$Z_{22} = $ self impedance of antenna 2
$Z_{12} = $ mutual impedance $(2 \rightarrow 1)$
$Z_{21} = $ mutual impedance $(1 \rightarrow 2)$

The mutual impedance Z_{12} is a complex quantity that represents the effect that element 2 has on element 1. Similarly, Z_{21} represents the effect that element 1 has on element 2. It follows from the principle of reciprocity that $Z_{12} = Z_{21}$ and we

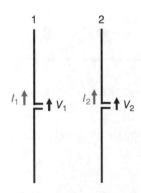

Figure 12.14 Two parallel dipoles.

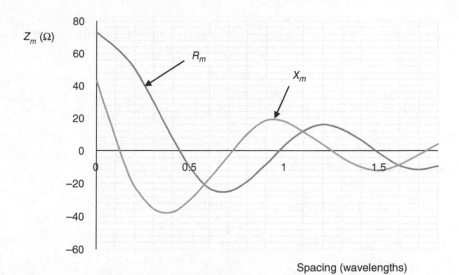

Figure 12.15 Graph showing the real (R_m) and imaginary (X_m) parts of the mutual impedance between two parallel half-wave dipoles, as a function of the spacing between the dipoles.

can write the mutual impedance simply as Z_m, and Eqs. (12.80) and (12.81) as

$$V_1 = I_1 Z_{11} + I_2 Z_m, \tag{12.82}$$

$$V_2 = I_2 Z_{22} + I_1 Z_m, \tag{12.83}$$

where

$$Z_m = \pm R_m \pm j X_m. \tag{12.84}$$

Note that the real part of Z_m can be negative, and this represents the effect of the current in one element decreasing the magnitude of the current in the other element. Figure 12.15 shows the typical variation of Z_m between two parallel $\lambda/2$ dipoles, as a function of the spacing between the dipoles.

Using Eqs. (12.82)and (12.83) we can write the impedance at the terminals of each antenna as

$$Z_1 = \frac{V_1}{I_1} = Z_{11} + \frac{I_2}{I_1} Z_m, \tag{12.85}$$

$$Z_2 = \frac{V_2}{I_2} = Z_{22} + \frac{I_1}{I_2} Z_m. \tag{12.86}$$

It should be noted that if the two dipoles are collinear, as shown in Figure 12.16, there is very little coupling between the dipoles because each dipole is in the radiation null of the other. Thus, the mutual impedance is zero for collinear dipoles.

Figure 12.16 Collinear array of two dipoles.

Example 12.6 An aerial array consists of two parallel $\lambda/2$ dipoles. The mutual impedance between the dipoles if $(50 - j22)\,\Omega$. Determine the input impedance at the terminals of each dipole, if:

  (i) They are fed with in-phase currents of the same magnitude
 (ii) They are fed with anti-phase currents of the same magnitude
(iii) They are fed with currents of the same magnitude, but in quadrature, such that $I_2 = jI_1$.

State any assumptions.

Solution Referring to Eqs. (12.85) and (12.86):

(i)
$$Z_1 = Z_{11} + \frac{I_2}{I_1}Z_m = (75 + 50 - j22)\,\Omega = (125 - j22)\,\Omega,$$

$$Z_2 = Z_{22} + \frac{I_1}{I_2}Z_m = (75 + 50 - j22)\,\Omega = (125 - j22)\,\Omega.$$

(ii)
$$Z_1 = Z_{11} - \frac{I_2}{I_1}Z_m = (75 - 50 + j22)\,\Omega = (25 + j22)\,\Omega,$$

$$Z_2 = Z_{22} - \frac{I_1}{I_2}Z_m = (75 - 50 + j22)\,\Omega = (25 + j22)\,\Omega.$$

(iii)
$$Z_1 = Z_{11} + \frac{I_2}{I_1}Z_m = (75 + j[50 - j22])\,\Omega = (97 + j50)\,\Omega,$$

$$Z_2 = Z_{22} + \frac{I_1}{I_2}Z_m = (75 - j[50 - j22])\,\Omega = (53 - j50)\,\Omega.$$

We have assumed that the loss resistance of $\lambda/2$ dipole is $2\,\Omega$, so that the total input resistance of a $\lambda/2$ dipole is $(73 + 22)\,\Omega = 75\,\Omega$.

12.8 Arrays Containing Parasitic Elements

The concept of mutual impedance can be extended to include parasitic elements; these elements are non-driven dipoles in which current is induced due to mutual coupling, and which then radiate in the same way as a driven dipole. In Section 12.5, we saw that a $\lambda/2$ dipole will produce maximum radiation in the horizontal ($\theta = \pm 90°$) directions. The purpose of the parasitic element is to enhance the radiation in one direction and produce a radiation pattern with a single main lobe, as was shown in Figure 12.1. The effectiveness of the parasitic element in achieving a radiation pattern with a single main lobe is represented by the front-to-back ratio F/B, which is defined as the ratio of the maximum magnitude radiated E-field in the main lobe (E_F) to the maximum magnitude radiated E-field in the back lobe (E_B), i.e.

$$F/B = \left|\frac{E_F}{E_B}\right| \equiv 20\log\left|\frac{E_F}{E_B}\right|\ \text{dB}. \tag{12.87}$$

A simple array consisting of a driven $\lambda/2$ dipole and a single parasitic element is shown in Figure 12.17.

Following a similar procedure to that used for an array of driven elements, we can write the mutual impedance equations for the array in Figure 12.17 as

$$V_1 = I_1 Z_{11} + I_2 Z_m, \tag{12.88}$$

$$V_2 = 0 = I_2 Z_{22} + I_1 Z_m, \tag{12.89}$$

where the symbols have their usual meanings, and where V_2 is zero for the parasitic element which has no terminals. Eq. (12.89) can be re-written as

$$I_2 = -\frac{Z_m}{Z_{22}}I_1. \tag{12.90}$$

It is clear from Eq. (12.90) that the magnitude and phase of the current I_2 induced in the parasitic element will depend on:

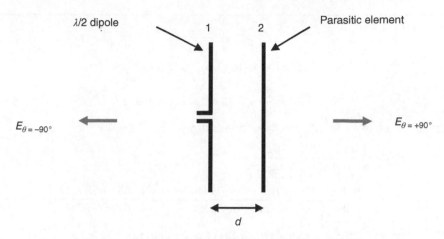

Figure 12.17 Dipole array with a single parasitic element.

(i) The magnitude and phase of the current in the driven dipole
(ii) The distance between the dipole and the parasitic, i.e. on Z_m
(iii) The self-impedance of the parasitic element.

Referring to Figure 12.17, the electric fields at a given distance from the driven element can be written as

$$E_{\theta=90°} = kI_1 + kI_2\angle + \beta d, \qquad (12.91)$$

$$E_{\theta=-90°} = kI_1 + kI_2\angle - \beta d, \qquad (12.92)$$

where k is a constant related to the distance from the array and the propagation conditions.
Substituting Eqs. (12.91) and (12.92) into Eq. (12.87) gives

$$F/B = \left| \frac{I_1 + I_2\angle + \beta d}{I_1 + I_2\angle - \beta d} \right|, \qquad (12.93)$$

and then substituting from Eq. (12.90) gives

$$F/B = \left| \frac{1 - \dfrac{Z_m}{Z_{22}}\angle + \beta d}{1 - \dfrac{Z_m}{Z_{22}}\angle - \beta d} \right| = \left| \frac{Z_{22} - Z_m\angle + \beta d}{Z_{22} - Z_m\angle - \beta d} \right|. \qquad (12.94)$$

Example 12.7 A 100 MHz antenna array is shown in Figure 12.18, and consists of a driven half-wave dipole and a parasitic element. The self-impedance of the parasitic element is $(60 - j50)\,\Omega$, and the mutual impedance between the two elements is $70\,\Omega$. Assume the loss resistance of a $\lambda/2$ dipole is $2\,\Omega$.

(i) Calculate the front-to-back ratio, in dB.
(ii) Identify the direction of maximum radiation.
(iii) Calculate the input impedance of the array.

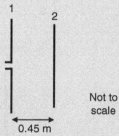

Not to
scale

0.45 m

Figure 12.18 Antenna array for Example 12.7.

(Continued)

(Continued)

Solution

(i)

$$f = 100 \text{ MHz} \quad \Rightarrow \quad \lambda = 3 \text{ m},$$

$$\beta d = \frac{2\pi}{\lambda} d = \frac{2\pi}{3} \times 0.45 \text{ rad} = 0.943 \text{ rad} \equiv 54°.$$

The front-to-back ratio can now be found directly by substituting in Eq. (12.94):

$$F/B = \left| \frac{(60 - j50) - 70\angle 54°}{(60 - j50) - 70\angle - 54°} \right|$$

$$= \left| \frac{(60 - j50) - (70\cos 54° + j70\sin 54°)}{(60 - j50) - (70\cos 54° - j70\sin 54°)} \right| = \left| \frac{18.86 - j106.63}{18.86 + j6.63} \right|,$$

$$F/B = \frac{108.29}{19.99} = 5.42 \equiv 20\log(5.42) \text{ dB} = 14.68 \text{ dB}.$$

(ii) Since we defined the front-to-back ratio as $F/B = \left| \dfrac{E_{\theta=90°}}{E_{\theta=-90°}} \right|$ our result of 5.42 means $|E_{\theta=90°}| > |E_{\theta=-90°}|$ and maximum radiation is in the $\theta = 90°$ direction.

(iii) Input impedance, $Z_{in} = \dfrac{V_1}{I_1}$

Using Eq. (12.88):

$$Z_{in} = \frac{V_1}{I_1} = Z_{11} + \frac{I_2}{I_1} Z_m = Z_{11} - \frac{Z_m^2}{Z_{22}}$$

$$= \left(75 - \frac{(70)^2}{60 - j50} \right) \Omega$$

$$= \left(75 - \frac{4900 \times (60 + j50)}{(60)^2 + (50)^2} \right) \Omega = (75 - 0.8 \times (60 + j50)) \Omega$$

$$= (27 - j40) \Omega.$$

It follows from Eq. (12.90) that if the spacing between the driven and parasitic elements is such that the mutual impedance is purely real, then the phase of I_2 relative to I_1 will be determined solely by the self-reactance of the parasitic. The spacing to give a real value of mutual impedance is around 0.15λ, as indicated by Figure 12.15. Moreover, we know that the reactance of a dipole varies with its length, as indicated by Figure 12.10, and so the phase of I_2 can be controlled by varying the length of the parasitic. If the length of the parasitic is less than $\lambda/2$, it will have a negative reactance, and it can be shown that I_2 will lag I_1. If this phase lag is made equal to βd then it can be seen from Eq. (12.91) that in the $\theta = 90°$ direction the phase lag of I_2 will cancel the phase lead of the ray from the parasitic which occurs due to the spatial separation of the elements. In these circumstances the parasitic will act as a director. Similarly, a parasitic that is longer than $\lambda/2$ will have a positive self-reactance and will act as a reflector. The behaviour of directors and reflectors is summarized in Figure 12.19.

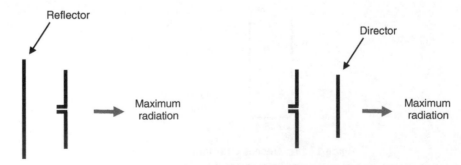

Figure 12.19 Reflectors and directors.

In addition to influencing the front-to-back ratio of an array, the use of parasitic elements will affect the gain of the array. A useful expression for the gain of an array containing parasitic elements can be found following the procedure given in Kraus and Marhefka [1]. The gain can be defined in terms of radiated electric field as

$$G = \left| \frac{E_{max}}{E_{ref}} \right|^2, \tag{12.95}$$

where E_{max} is the electric field at a given distance in the direction of maximum radiation from the antenna, and E_{ref} is the electric field at the same distance from a reference antenna. In the case of an array consisting of a driven $\lambda/2$ dipole and one or more parasitic elements, it is usual to use a $\lambda/2$ dipole as the reference antenna, since the resulting gain will be a measure of the effectiveness of adding the parasitic elements.

If we consider the simple array shown in Figure 12.17, and assume that $\theta = 90°$ is the direction of maximum radiation, then

$$\begin{aligned} E_{max} &= kI_1 + kI_2 \angle \beta d \\ &= kI_1 - kI_1 \frac{Z_m}{Z_{22}} \angle \beta d, \\ &= kI_1 \left(1 - \frac{Z_m}{Z_{22}} \angle \beta d \right) \end{aligned} \tag{12.96}$$

i.e.

$$|E_{max}| = k|I_1| \left| 1 - \frac{Z_m}{Z_{22}} \angle \beta d \right|. \tag{12.97}$$

If we let the power supplied to the array be P, then

$$P = |I_1|^2 R_{in}, \tag{12.98}$$

where R_{in} is the real part of the driving point impedance (Z_{in}) of the array.

As we found in Example 12.7, the driving point impedance is given by

$$Z_{in} = Z_{11} - \frac{Z_m^2}{Z_{22}}. \tag{12.99}$$

Combining Eqs. (12.97) and (12.98) gives

$$|E_{max}| = k \sqrt{\frac{P}{R_{in}}} \left| 1 - \frac{Z_m}{Z_{22}} \angle \beta d \right|, \tag{12.100}$$

where R_{in} is the real part of Z_{in}.

Considering the same power, P, applied to the reference antenna (half-wave dipole) we obtain

$$|E_{ref}| = \sqrt{\frac{P}{73}}, \tag{12.101}$$

since the radiation resistance of a half-wave dipole is 73 Ω.

Finally, substituting from Eqs. (12.100) and (12.101) into Eq. (12.95) gives the expression for the gain of the array as

$$G = \frac{73}{R_{in}} \left| 1 - \frac{Z_m}{Z_{22}} \angle \beta d \right|^2, \tag{12.102}$$

where

$$R_{in} = Re \left[Z_{11} - \frac{Z_m^2}{Z_{22}} \right]. \tag{12.103}$$

Example 12.8 An 800 MHz array, consisting of a driven $\lambda/2$ dipole and a parasitic element is shown in Figure 12.20, together with the relevant impedance data.
  Determine:

  (i) The driving-point impedance.
  (ii) The front-to-back ratio (in dB).
  (iii) The gain (in dB) relative to a $\lambda/2$ dipole.

(Continued)

(Continued)

(iv) The gain (in dB) relative to an isotropic radiator.

Assume the loss resistance of the $\lambda/2$ dipole is 2 Ω.

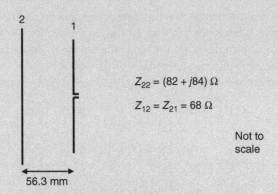

$Z_{22} = (82 + j84)\ \Omega$

$Z_{12} = Z_{21} = 68\ \Omega$

Not to scale

56.3 mm

Figure 12.20 Antenna array for Example 12.8.

Solution

(i) Using Eqs. (12.88) and (12.89)

$$Z_1 = \frac{V_1}{I_1} = Z_{11} - \frac{Z_{12}^2}{Z_{22}}$$

$$= \left(75 - \frac{(68)^2}{82 + j84}\right)\ \Omega$$

$$= \left(75 - \frac{(68)^2(82 - j84)}{(82)^2 + (84)^2}\right)\ \Omega$$

$$= \left(75 - \frac{(68)^2(82 - j84)}{13780}\right)\ \Omega = (75 - 27.52 + j28.19)\ \Omega$$

$$= (47.48 + j28.19)\ \Omega.$$

(ii)
$$800\ \text{MHz} \quad \Rightarrow \quad \lambda_O = 375\ \text{mm},$$

$$\beta d = \frac{2\pi}{375} \times 56.3\ \text{rad} \equiv 54°.$$

Using Eqs. (12.91) and (12.92) gives:

$$E_{\theta=90°} = kI_1 + kI_2\angle - 54°.$$
$$E_{\theta=-90°} = kI_1 + kI_2\angle 54°.$$

From Eq. (12.87) we have:

$$F/B = \left|\frac{E_{\theta=90°}}{E_{\theta=-90°}}\right| = \left|\frac{kI_1 + kI_2\angle - 54°}{kI_1 + kI_2\angle 54°}\right|$$

$$= \left|\frac{1 + \dfrac{I_2}{I_1}\angle - 54°}{1 + \dfrac{I_2}{I_1}\angle 54°}\right| = \left|\frac{1 - \dfrac{Z_m}{Z_{22}}\angle - 54°}{1 - \dfrac{Z_m}{Z_{22}}\angle 54°}\right|.$$

Substituting remaining data gives:

$$F/B = \left| \frac{1 - \dfrac{68}{82 + j84} \angle - 54°}{1 - \dfrac{68}{82 + j84} \angle 54°} \right| = \left| \frac{82 + j84 - (68\cos 54° - j68\sin 54°)}{82 + j84 - (68\cos 54° + j68\sin 54°)} \right|,$$

$$= \left| \frac{82 + j84 - (39.97 - j55.01)}{82 + j84 - (39.97 + j55.01)} \right| = \left| \frac{42.03 + j139.01}{42.03 + j28.99} \right| = 2.84$$

i.e.

$$F/B = 2.84 \equiv 20\log 2.84 \text{ dB} = 9.07 \text{ dB}.$$

(iii) It can be seen from the result of part (ii) that the direction of maximum radiation is $\theta = 90°$. Therefore, the gain is given by

$$G = \left| \frac{E_{\theta=90°}}{E_{ref}} \right|^2 = \left| \frac{kI_1 + kI_2 \angle - 54°}{kI_{ref}} \right|^2 = \left| \frac{I_1 \left(1 - \dfrac{Z_m}{Z_{22}} \angle - 54°\right)}{I_{ref}} \right|^2 = \left| \frac{\sqrt{\dfrac{P}{R_{in}}} \left(1 - \dfrac{Z_m}{Z_{22}} \angle - 54°\right)}{\sqrt{\dfrac{P}{75}}} \right|^2.$$

Now $R_{in} = 47.48\ \Omega$ from part (i), so substituting data we have:

$$G = \frac{75}{47.48} \times \left| 1 - \frac{68}{82 + j84} \angle - 54° \right|^2$$

$$= 1.58 \times \left| \frac{82 + j84 - 68(\cos 54° - j\sin 54°)}{82 + j84} \right|^2$$

$$= 1.58 \times \left(\frac{145.23}{117.39} \right)^2 = 2.43 \equiv 10\log 2.43 \text{ dB} = 3.86 \text{ dB}.$$

(iv) To find the gain relative to an isotropic radiator we must add the gain of the array relative to a $\lambda/2$ dipole to the gain of a $\lambda/2$ dipole relative to an isotropic radiator, i.e.
G (rel to isotropic radiator) $= 3.86 + 2.16 = 6.02$ dB.

12.9 Yagi–Uda Antenna

The Yagi–Uda antenna is the best-known antenna using parasitic elements. Named after Hidetsugu Yagi and Shintaro Uda, it was developed in the 1920s and has become one of the most common types of wire RF antenna. The basic construction is shown in Figure 12.21, and consists of a single driven half-wave dipole, and a number of parasitic elements.

Only one reflector is used, since additional reflectors would be in a low-field region and consequently have small induced currents, and so would not contribute significantly to the operation of the antenna. Additional directors, however, will always be in a high-field region, and so there will be significant interaction between these elements and the electromagnetic field. Thus, the additional directors will have significant induced currents, and these will increase the gain of the array. In the Yagi antenna shown in Figure 12.21 the lengths of the elements are tapered to ensure a progressive phase lag in the elements currents, from left to right. This will have the effect of reinforcing the radiation from the elements in the direction shown. Since the centre of each element is a voltage node, these points may be connected to a conducting boom to form a self-supporting structure.

Additional points concerning the Yagi–Uda antenna:

(i) With six elements, as shown in Figure 12.21, the gain is around 10 dB, with each additional director increasing the gain by approximately 1 dB.
(ii) Whilst only one reflector is required, the performance can be improved slightly by forming the reflector from a grid of wires, thereby decreasing back radiation, and increasing forward gain.

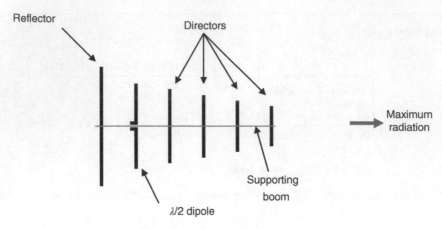

Figure 12.21 A six-element Yagi–Uda array.

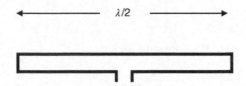

Figure 12.22 Folded dipole.

(iii) The presence of the parasitic elements adjacent to the $\lambda/2$ dipole have the effect of reducing the input impedance, as was seen in the previous worked examples. In order to have an input impedance which more closely matches the impedance of typical coaxial cable, it is usual to have a folded dipole as the driven element. A typical folded dipole is shown in Figure 12.22. The spacing between the two conductors is made very much less than a wavelength, so there is strong mutual coupling between them, and each conductor has current of the same magnitude and phase. This means that the field from the folded dipole is twice that of a single dipole, and the radiated power is greater by a factor of 4. Thus the impedance increases by 4, i.e.

$$R_{rad}(\text{folded dipole}) = 4 \times 73 \ \Omega = 292 \ \Omega. \tag{12.104}$$

Typically, the parasitic elements either side of the driven element reduce the driving point impedance by a factor of 4, and so with the folded dipole as the driven element the input impedance is reduced close to that of 75 Ω coaxial cable.

(iv) Since the Yagi–Uda array uses resonant elements it has a rather narrow bandwidth, usually only a few percent.

(v) The lengths of the parasitic elements shown in Figure 12.21 are shown progressively tapered in the intended direction of maximum radiation, in order to obtain the correct phase relationships between the element currents. But the phasing of the currents also depend on the spacing between the elements, and so Yagi–Uda antenna often have elements of a fixed length, but with the spacing between the elements chosen to give the desired phase relationships between the element currents.

(vi) Since there are a significant number of factors influencing the performance of a Yagi–Uda antenna, the design can be quite complex. A very informative discussion on the theory of Yagi–Uda arrays, and the associated design procedure is given by Balanis [2].

Example 12.9 A three-element Yagi–Uda antenna, consisting of a driven $\lambda/2$ dipole and two parasitic elements is shown in Figure 12.23.

Determine:

(i) The driving point impedance.

(ii) The front-to-back ratio, in dB.

Assume loss resistance of the $\lambda/2$ dipole is $2\,\Omega$.

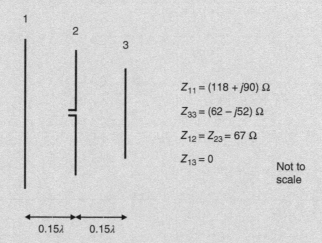

$Z_{11} = (118 + j90)\,\Omega$

$Z_{33} = (62 - j52)\,\Omega$

$Z_{12} = Z_{23} = 67\,\Omega$

$Z_{13} = 0$

Not to scale

Figure 12.23 Three-element Yagi–Uda array for Example 12.9.

Solution

(i) The mutual impedance equations are (noting that $Z_{13} = Z_{31} = 0$):

$$V_1 = 0 = I_1 Z_{11} + I_2 Z_{12},$$

$$V_2 = I_2 Z_{22} + I_1 Z_{21} + I_3 Z_{23},$$

$$V_3 = 0 = I_3 Z_{33} + I_2 Z_{32}.$$

Combining these three equations gives

$$Z_2 = \frac{V_2}{I_2} = Z_{22} - \frac{Z_{12}^2}{Z_{11}} - \frac{Z_{32}^2}{Z_{33}}.$$

Note that due to reciprocity we can write $Z_{12} = Z_{21}$ and $Z_{23} = Z_{32}$.
Substituting the impedance data gives:

$$Z_2 = \left(75 - \frac{(67)^2}{118 + j90} - \frac{(67)^2}{62 - j52}\right)\,\Omega$$

$$= (8.445 - j17.305)\,\Omega.$$

(ii) Phase shift due to spacing between elements:

$$\beta d = \frac{2\pi}{\lambda} \times 0.15\lambda = 0.3\pi \text{ rad} \equiv 54°.$$

Calculating resultant E-field in axial directions:

$$E_{R,\,\theta=90°} = kI_1 \angle -54° + kI_2 + kI_3 \angle 54°,$$

$$E_{R,\,\theta=-90°} = kI_1 \angle 54° + kI_2 + kI_3 \angle -54°.$$

Then,

$$F/B = \left| \frac{kI_1 \angle -54° + kI_2 + kI_3 \angle 54°}{kI_1 \angle 54° + kI_2 + kI_3 \angle -54°} \right| = \left| \frac{\dfrac{I_1}{I_2} \angle -54° + 1 + \dfrac{I_3}{I_2} \angle 54°}{\dfrac{I_1}{I_2} \angle 54° + 1 + \dfrac{I_3}{I_2} \angle -54°} \right|.$$

(Continued)

(Continued)

From the mutual impedance equations:

$$\frac{I_1}{I_2} = -\frac{Z_{12}}{Z_{11}} = -\frac{67}{118 + j90} = -\frac{67}{148.405\angle 37.333°} = -0.451\angle - 37.333°,$$

$$\frac{I_3}{I_2} = -\frac{Z_{32}}{Z_{33}} = -\frac{67}{62 - j52} = -\frac{67}{80.920\angle - 39.987°} = -0.828\angle 39.987°.$$

Therefore,

$$F/B = \left| \frac{-0.451\angle - 91.333° + 1 - 0.828\angle 93.987°}{-0.451\angle 16.667° + 1 - 0.828\angle - 14.013°} \right|$$

$$= \left| \frac{-0.010 + j0.451 + 1 + 0.058 - j0.826}{-0.432 - j0.129 + 1 - 0.803 + j0.200} \right| = \left| \frac{1.048 - j0.373}{-0.235 + j0.071} \right|,$$

$$F/B = \frac{1.113}{0.245} = 4.543 \equiv 13.147 \text{ dB}.$$

12.10 Log-Periodic Array

The log-periodic array is a highly directional, wideband antenna similar in appearance to the Yagi–Uda antenna described in the last section. But whereas the Yagi antenna has only one driven element, all the $\lambda/2$ dipoles forming the log-periodic are driven. A typical array is shown in Figure 12.24.

The array is fed by a twin-wire feeder at the short end. Since the elements are $\lambda/2$ dipoles there will be a number of discrete frequencies corresponding to the individual dipole resonances. If a dipole in the centre of the array is excited into resonance this will radiate strongly, as will the closely spaced elements on either side, which are also close to resonance. Thus, we can regard a region of the array as being active at a given frequency. As the frequency changes, this active region moves across the array. By scaling the lengths and spacing of the elements the properties of the active region can be made almost constant as the frequency is changed, thus giving the array its wideband characteristics. As the frequency is increased the active region moves to the left, and as the frequency decreases the active region moves to the right. The nominal bandwidth of the array is the difference between the resonant frequencies of the longest and shortest elements. If the input impedance of the array is plotted as a function of the log of the frequency, it will exhibit a periodicity, and this gives the array its name of log periodic.

The scaling factor, τ, for the array is defined as

$$\tau = \frac{l_{n+1}}{l_n} = \frac{s_{n+1}}{s_n}, \tag{12.105}$$

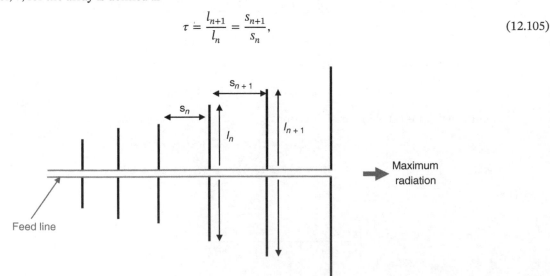

Figure 12.24 Log-periodic array.

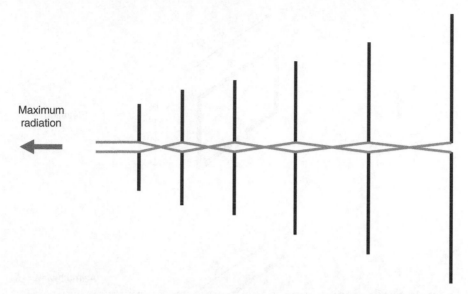

Figure 12.25 Log-periodic array with phase reversal between feeds to adjacent elements.

where l_n and s_n and the length and spacing of an element at an arbitrary position in the array, as shown in Figure 12.24.

Typically, τ has a value between 0.7 and 0.9 and the number of elements in the array is around 10.

Figure 12.24 shows a twin-wire feeder used to connect the elements. If the array is used as a transmitting antenna, there will be a progressive phase lag in the element currents, left–to–right. This will cause maximum radiation towards the right, as shown in Figure 12.24. This is not a desirable arrangement, since if the array is exited at the higher end of the frequency band the active region will be in the vicinity of the shorter elements and the radiation will be impeded by the presence of the longer elements. A better arrangement is to reverse the feed connections to adjacent elements, as shown in Figure 12.25, which will produce maximum radiation to the left. Now, even if the active region is situated on the right-hand side of the array, the presence of the shorter elements in the radiation path will only have a small adverse effect.

12.11 Loop Antenna

A simple wire loop, either of rectangular or circular cross-section, provides a useful RF antenna, primarily for receiving purposes. This type of antenna is normally constructed to be electrically small, with the dimensions of the sides less than $0.1\lambda_O$, where λ_O is the free-space wavelength at the working frequency. For an electrically small rectangular loop antenna, the radiation resistance is given by [1].

$$R_{rad} = 320\pi^4 N^2 \left(\frac{A}{\lambda_O^2}\right)^2,$$

(12.106)

where A = area of the loop, N = number of turns, λ_O = free-space wavelength. Since the dimensions of the antenna are a lot less than a wavelength, the radiation resistance is normally very small, often significantly less than 1 Ω.

Example 12.10 Find the radiation resistance at 100 MHz of a rectangular loop antenna consisting of four turns each having a width of 8 cm and a height of 5 cm.

Solution

$$f = 100 \text{ MHz} \quad \Rightarrow \quad \lambda_O = 3 \text{ m}.$$

Using Eq. (12.106);

$$R_{rad} = 320\pi^4 \times 4^2 \times \left(\frac{0.08 \times 0.05}{3^2}\right)^2$$

$$= 0.1 \ \Omega$$

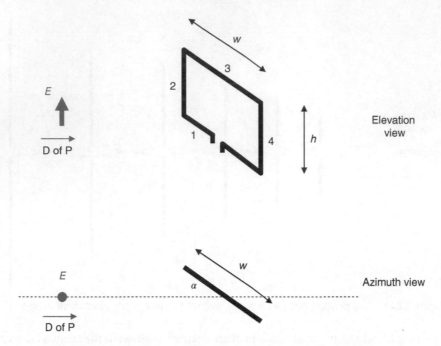

Figure 12.26 A rectangular loop antenna located in an E-field.

Figure 12.26 shows a rectangular loop located in a linearly polarized electric field, with the E-field parallel to sides 2 and 4 of the loop, and with the plane of the loop at an angle α relative to the direction of propagation (D of P) of the electric field.

The electric field will induce voltages in sides 2 and 4 of the loop, but not in sides 1 and 3 which are perpendicular to the E-field. When $\alpha = 90°$, and the plane of the loop faces the incident wave, the induced voltages will have the same magnitude and phase, and will cancel around the loop giving zero voltage across the terminals of the loop. As the loop is rotated from the 90° position the magnitudes of the induced voltages are essentially unchanged, but there will be a phase difference, ψ, given by

$$\psi = \beta \times w \cos\alpha = \frac{2\pi w \cos\alpha}{\lambda_O}. \tag{12.107}$$

Therefore, the induced voltages in the two vertical sides can be written as

$$V_2 = h \times E_m \sin\omega t, \tag{12.108}$$

$$V_4 = h \times E_m \sin(\omega t - \psi), \tag{12.109}$$

where E_m is the peak electric field strength.

The voltage across the terminals of the loop will then be

$$V_T = V_2 - V_4 = hE_m\{\sin\omega t - \sin(\omega t - \psi)\}$$
$$= 2hE_m \cos\left(\frac{2\omega t - \psi}{2}\right) \sin\left(\frac{\psi}{2}\right). \tag{12.110}$$

From Eq. (12.110) the magnitude of the terminal voltage is

$$|V_T| = 2hE_m \sin\left(\frac{\psi}{2}\right). \tag{12.111}$$

Since $w \ll \lambda_O$, ψ will be small and $\sin\left(\frac{\psi}{2}\right) \simeq \frac{\psi}{2}$. Then

$$|V_T| = 2hE\left(\frac{\psi}{2}\right) = \frac{2hE_m\pi w \cos\alpha}{\lambda_O} = \frac{2\pi AE_m}{\lambda_O}\cos\alpha, \tag{12.112}$$

where A is the area of the loop antenna.

If there are N turns on the loop, the terminal voltage becomes

$$|V_T| = \frac{2\pi NAE_m}{\lambda_O} \cos \alpha. \tag{12.113}$$

The gain, G, relative to an isotropic radiator of a small loop antenna, i.e. one whose dimensions are less than $0.1\lambda_O$, is given by [1].

$$G = \frac{1.5 R_{rad}}{R_L}, \tag{12.114}$$

where R_L is the loss resistance of the loop. For a small loop, $R_{rad} \ll R_L$ and so the gain of the loop is very small.

Example 12.11   A square loop antenna has an area of $100\,\text{cm}^2$. The antenna has four turns with a total loss resistance of $2.2\,\Omega$. Determine the gain of the antenna, relative to an isotropic radiator, at a frequency of $80\,\text{MHz}$.

Solution

$$80\,\text{MHz} \quad \Rightarrow \quad \lambda_O = \frac{3 \times 10^8}{80 \times 10^6}\,\text{m} = 3.75\,\text{m}.$$

Using Eq. (12.106) to find the radiation resistance:

$$R_{rad} = 320\pi^4 \times 4^2 \times \left(\frac{100 \times (0.01)^2}{(3.75)^2} \right)^2 \Omega = 0.252\,\Omega,$$

Using Eq. (12.114) to find the gain:

$$G = \frac{1.5 \times 0.252}{2.2} = 0.172,$$

i.e.

$$G_{dB} = 10 \log 0.172\,\text{dB} = -7.64\,\text{dB}.$$

Example 12.12   A rectangular loop antenna with 15 turns has a width of $10\,\text{cm}$ and a height of $8\,\text{cm}$ is used as a receiving antenna in a vertically polarized electric field given by

$$E = 65 \sin(6\pi \times 10^8 t)\,\mu\text{V/m}.$$

If the loop is oriented for maximum reception, determine:

(i) The value of α (as defined in Figure 12.26).
(ii) The rms value of the voltage across the terminals of the loop.
(iii) The change in voltage across the terminals of the loop if the antenna is rotated $30°$ about its vertical axis.

Solution
(i) For maximum reception the plane of the loop must be in the direction of the propagating field, i.e. $\alpha = 0°$.
(ii) From the given data we have

$$2\pi f = 6\pi \times 10^8 \quad \Rightarrow \quad f = 300\,\text{MHz}.$$

Therefore,

$$\lambda_O = \frac{c}{f} = \frac{3 \times 10^8}{300 \times 10^6} = 1\,\text{m}.$$

Substituting in Eq. (12.113) gives

$$|V_T|_{rms} = \frac{2\pi \times 15 \times (0.1 \times 0.08) \times 65/\sqrt{2}}{1}\,\mu\text{V} = 34.659\,\mu\text{V}.$$

Note that the peak value of the electric field has been divided by $\sqrt{2}$ to obtain the rms value.

(Continued)

(Continued)

(iii) $\alpha = 30° \Rightarrow \cos\alpha = 0.866$.

Therefore, $|V_T|_{rms}(\alpha = 30°) = (34.659 \times 0.866)\,\mu V = 30.015\,\mu V$.

Change in voltage across terminals

$$= (34.659 - 30.015)\,\mu V = 4.644\,\mu V.$$

The geometry of a loop antenna indicates that the reactance seen at the terminals will be primarily inductive. So we can deduce an equivalent circuit for the loop consisting of a combination of inductance and resistance, in series with a voltage source representing the induced voltage at the terminals, as shown in Figure 12.27. As was mentioned earlier, the radiation resistance of a loop antenna is very small, and therefore the resistance in the equivalent circuit will primarily be the ohmic resistance of the wires of the loop. In order to extract maximum voltage from the loop, a capacitor can be connected across the terminals of the loop to form a resonant circuit.

At resonance the voltage, V_C, across the tuning capacitor shown in Figure 12.27 will be

$$|V| = Q \times |V_T| = \frac{\omega L}{R} \times |V_T|. \tag{12.115}$$

Example 12.13 A square loop antenna whose side length is 100 mm is used to detect signals from a 100 MHz electric field that has an rms amplitude of 120 μV/m. The antenna has three turns with a total resistance of 4 Ω, and a total inductance of 250 nH. A tuning capacitor is connected across the terminals of the loop.

Determine:

(i) The required value of the tuning capacitor.
(ii) The rms voltage across the capacitor when the loop is positioned for maximum reception.

Solution

(i) At resonance

$$f_O = \frac{1}{2\pi\sqrt{LC}}$$

Therefore,

$$100 \times 10^6 = \frac{1}{2\pi\sqrt{250 \times 10^{-9} \times C}},$$

$$C = 10.13 \text{ pF}.$$

(ii) $f = 100\text{ MHz} \Rightarrow \lambda_O = 3$ m

The maximum open-circuit voltage at the antenna terminals is given by Eq. (12.113), noting that $\alpha = 0$ for maximum reception:

$$|V_T| = \frac{2\pi NAE}{\lambda_O} = \frac{2\pi \times 3 \times (0.1 \times 0.1) \times 120 \times 10^{-6}}{3}\text{ V} = 7.54\,\mu V.$$

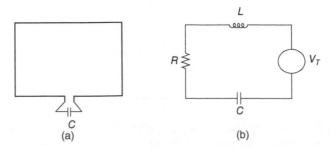

Figure 12.27 Receiving loop antenna with tuning capacitor. (a) Loop with tuning capacitor. (b) Equivalent circuit of receiving loop.

The voltage across the capacitor at resonance is given by Eq. (12.115):

$$|V_C| = Q \times |V_T| = \frac{\omega L}{R} \times |V_T|,$$

i.e.

$$|V_C| = \frac{2\pi \times 100 \times 10^6 \times 250 \times 10^{-9}}{4} \times 7.54 \ \mu V = 296.13 \ \mu V.$$

12.12 Planar Antennas

12.12.1 Linearly Polarized[1] Patch Antennas

The use of microstrip patches is one of the most common techniques for making RF and microwave planar antennas. A typical microstrip patch antenna is shown in Figure 12.28, and includes a quarter-wave impedance transformer to connect the high impedance edge of the patch to the feed line.

Patch antennas of the type shown in Figure 12.28 have a number of advantages for high-frequency applications:

(i) The radiating patch can easily be integrated into a hybrid package.
(ii) If required the patch can be formed on a conformal, curved surface.
(iii) Arrays of patches can be made in a relatively small area, leading to compact high-gain antennas.
(iv) It is easy to incorporate phase shifters into the feed lines to the patches in an array to produce a phased array.
(v) Feed lines can be made short, leading to lower line attenuation and improved noise performance.

Whilst patch antennas have the advantages mentioned above, there are some minor limitations. For good radiation, the dielectric constant of the substrate should be small, and the thickness of the substrate reasonable large. This tends to lead to relatively large size antennas at the low end of the microwave frequency band. Also, patch antennas tend to have low radiation efficiency and a low power-handling capability.

Figure 12.29 shows the planar view of a microstrip rectangular patch, together with a feed line.

When the patch is exited into axial resonance, the two end edges of the patch each form an effective slot with the ground plane, and each acts as a radiating slot of area $W \times h$, where W is the width of the patch h is the thickness of the substrate. The two slots radiate E-fields in anti-phase, and so produce an E-field radiation pattern similar to that of a half-wave dipole, with the E-field in the axial direction.

A transmission line model can be used to represent the resonant patch, and this is shown in Figure 12.30.

In the transmission line model, the radiation admittance of each slot is represented by Y_R, where

$$Y_R = G_R + jB_R, \tag{12.116}$$

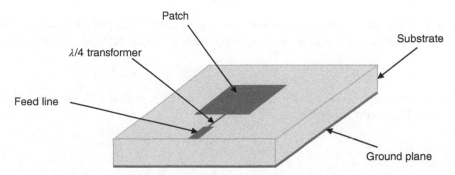

Figure 12.28 Microstrip patch antenna.

---

1 A linearly polarized antenna is one where the direction of the radiated electric field remains fixed as the wave propagates.

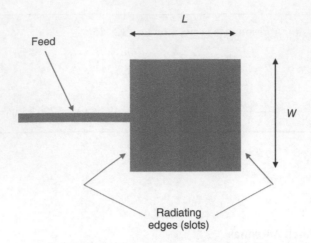

Figure 12.29 Planar view of a microstrip patch showing positions of radiating edges (slots) relative to the feed position.

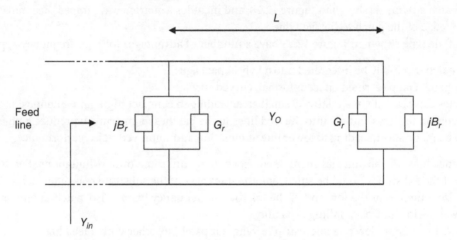

Figure 12.30 Transmission line model of a microstrip patch antenna.

and G_r represents the radiation conductance of each slot, and B_R the radiation susceptance of each slot, which is associated with the fringing field at the open end. Munson [3] gives the following expressions for G_R and B_R

$$G_R = \frac{W}{120\lambda_O} \qquad (12.117)$$

and

$$B_R = \frac{W(3.135 - 2\log\beta W)}{120\pi\lambda_O}, \qquad (12.118)$$

where $\beta = \frac{2\pi}{\lambda_s}$ and λ_s is the substrate wavelength measured along the patch. If the patch is exactly $\lambda_s/2$ long, the radiation admittance at the remote end of the patch will be reflected to the feeding end, so the total radiation admittance, Y_{in}, at the feed point will be

$$Y_{in} = 2G_R + j2B_R, \qquad (12.119)$$

and the patch will not resonate. If the length of the patch is made slightly less than $\lambda_s/2$, say around $0.48\lambda_s$, the radiation admittance reflected back, Y_{refl}, to the feed point can be made

$$Y_{refl} = G_R - jB_R, \qquad (12.120)$$

and so the patch will resonate with

$$Y_{in} = 2G_R. \qquad (12.121)$$

It should be noted that Eq. (12.119) neglects the mutual conductance between the two radiating slots, but this is normally small for practical microstrip antennas.

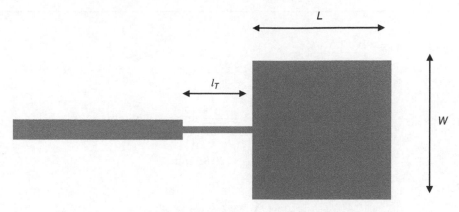

Figure 12.31 Design dimensions for a simple microstrip patch antenna.

The width of a microstrip patch is not critical, but for efficient radiation it is normally taken to be

$$W = \frac{c}{2f_O}\sqrt{\frac{2}{\varepsilon_r + 1}}, \tag{12.122}$$

where f_O is the resonant frequency.

The three design parameters, L, W, T for a front-fed microstrip patch antenna are shown in Figure 12.31.

Summary of design information:

L is the resonant length of the patch. Nominally it should be half a substrate wavelength. However, there will be fringing at each end of the patch (the open-end effect). So the actual length, A, of the patch should be

$$A = \frac{\lambda_s}{2} - 2l_{eo}, \tag{12.123}$$

where λ_s is the substrate wavelength for a microstrip line of width B, and l_{eo} is the equivalent end-effect length, l_{eo}, given by

$$l_{eo} = 0.412h\left(\frac{\varepsilon_{r,eff} + 0.3}{\varepsilon_{r,eff} - 0.258}\right)\left(\frac{W/h + 0.264}{W/h + 0.8}\right). \tag{12.124}$$

W is the width of the patch. This dimension is not critical, but is normally calculated using Eq. (12.122).

l_T is the length of the quarter-wave matching section, i.e. $\lambda_{ST}/4$, where λ_{ST} corresponds to the designed impedance of this section. The impedance, Z_{OT}, of the transformer is

$$Z_{OT} = \sqrt{R_{feed}\, R_{in}}, \tag{12.125}$$

i.e.

$$Z_{OT} = \sqrt{R_{feed} \times \frac{1}{2G_R}}, \tag{12.126}$$

where G_R is defined in Eq. (12.117).

Example 12.14 Design a microstrip single-patch antenna to work at 5 GHz. The antenna is to be fabricated on a substrate that has a relative permittivity of 2.3 and a thickness of 2.5 mm, and is to be fed from a 50 Ω microstrip line.

Solution
Using Eq. (12.122):

$$W = \frac{c}{2f_O}\sqrt{\frac{2}{\varepsilon_r + 1}} = \frac{3 \times 10^8}{2 \times 5 \times 10^9}\sqrt{\frac{2}{2.3 + 1}} \text{ m} = 23.35 \text{ mm.}$$

(Continued)

(Continued)

Using design graphs in Appendix 12.A:

$$W = 23.35 \text{ mm} \quad \Rightarrow \quad \frac{W}{h} = \frac{23.35}{2.5} = 9.34 \quad \Rightarrow \quad \varepsilon_{r.eff}^{MSTRIP} = 2.07.$$

Then,

$$\lambda_s(\text{patch}) = \frac{\lambda_O}{\sqrt{2.07}} = \frac{60}{\sqrt{2.07}} \text{ mm} = 41.70 \text{ mm},$$

$$\frac{\lambda_s(\text{patch})}{2} = 20.85 \text{ mm}.$$

Using Eq. (12.124):

$$l_{eo} = 0.412h \left(\frac{\varepsilon_{r,eff} + 0.3}{\varepsilon_{r,eff} - 0.258} \right) \left(\frac{W/h + 0.264}{W/h + 0.8} \right)$$

$$= 0.412 \times 2.5 \left(\frac{2.07 + 0.3}{2.07 - 0.258} \right) \left(\frac{9.34 + 0.264}{9.34 + 0.8} \right) \text{ mm} = 1.28 \text{ mm}.$$

Then,

$$L = (20.85 - [2 \times 1.28]) \text{ mm} = 18.29 \text{ mm}.$$

Using Eq. (12.117):

$$G_R = \frac{W}{120\lambda_O} = \frac{23.35}{120 \times 60} \text{ S} = 3.24 \text{ mS}.$$

Using Eq. (12.126):

$$Z_{OT} = \sqrt{R_{feed} \times \frac{1}{2G_r}} = \sqrt{50 \times \frac{1}{2 \times 3.24 \times 10^{-3}}} \text{ } \Omega = 87.84 \text{ } \Omega.$$

Using design graphs in Appendix 12.A:

$$87.84 \text{ } \Omega \quad \Rightarrow \quad w_T = 2.8 \text{ mm} \quad \text{and} \quad \varepsilon_{r.eff}^{MSTRIP} = 1.85.$$

Therefore,

$$l_T = \frac{\lambda_{s,T}}{4} = \frac{\lambda_O/\sqrt{1.85}}{4} = \frac{60/\sqrt{1.85}}{4} \text{ mm} = 11.03 \text{ mm}.$$

Using design graphs in Appendix 12.A:

$$50 \text{ } \Omega \quad \Rightarrow \quad \frac{w_T}{h} = 3 \quad \Rightarrow \quad w_T = 7.5 \text{ mm}.$$

The final design is shown in Figure 12.32.

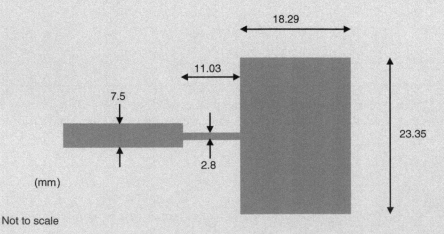

Figure 12.32 Complete design for Example 12.14.

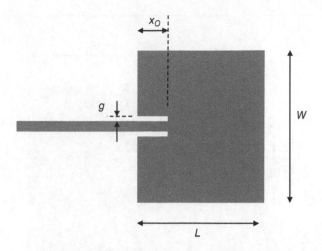

Figure 12.33 Patch antenna with inset feed point.

The previous configurations of a microstrip patch antenna have shown the feed line connected to one edge of the patch. This arrangement has the disadvantage that a quarter-wave transformer is required to match the patch to the feed line because the impedance at the edge of the patch is very high. This disadvantage can be overcome by using an inset feed point as shown in Figure 12.33. Since the patch is resonant in the axial direction, the voltage and impedance must both be zero at the centre of the patch. It can be shown [2] that the impedance along the centre line of the patch varies according to the relationship

$$R_{in}(x_O) = R_{in}(x = 0) \times \cos^2\left(\frac{\pi}{L}x\right). \tag{12.127}$$

Thus, a value of x_O can be found which matches the characteristic impedance of the feed line, and alleviates the necessity for a quarter-wave matching transformer. It can be seen from Figure 12.33 that the use of the inset feed creates two notches, of width g, either side of the feed line as it enters the patch. The value of g is not critical to the design but the two notches will introduce some coupling capacitance between the feed line and the patch. Balanis [2] suggests that this will typically change the resonant frequency of the patch by around 1%. In a detailed examination of the effect of the width of the notch on the resonant frequency, Matin and Sayeed [4] proposed the following equation to calculate the resonant frequency due to notches of width g.

$$f_r = \frac{c}{\sqrt{2\varepsilon_{r,eff}}} \frac{4.6 \times 10^{-14}}{g} + \frac{f}{1.01}. \tag{12.128}$$

Example 12.15   Repeat the design of Example 12.14 using an inset-fed patch.

Solution
The values of L and W are unchanged from the solution to Example 12.14.
  The resistance at the edge of the patch is given by

$$R_{in}(x = 0) = \frac{1}{2G_R} = \frac{1}{2 \times 3.24 \times 10^{-3}}\ \Omega = 154.32\ \Omega.$$

Using Eq. (12.127) we can find the value of x_O corresponding to a feed point impedance of 50 Ω:

$$50 = 154.32 \times \cos^2\left(\frac{\pi}{18.29}x\right).$$

$$x = 5.62\ \text{mm}.$$

We will assume that $g = 1$ mm.
The complete design is shown in Figure 12.34.

(Continued)

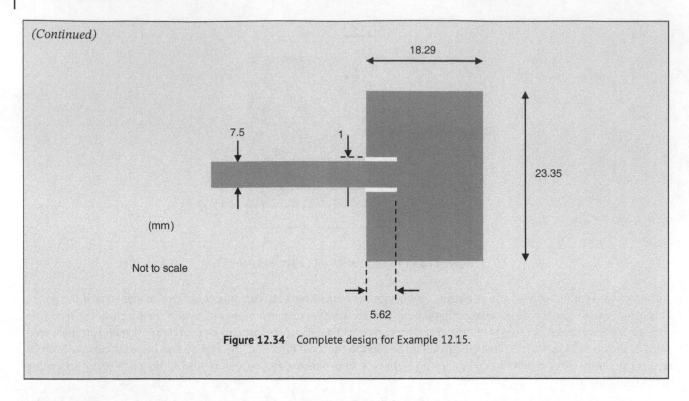

Figure 12.34 Complete design for Example 12.15.

The microstrip patch antennas that have been described so far have been front-fed, that is with the feed line on the same side of the substrate as the radiating patch. This feed method has the disadvantage that there can be significant unwanted radiation from the feed network, particularly in the case of a patch array. An alternative is to feed each radiating patch through the substrate, either by direct conduction through a probe or by electromagnetic coupling. The simplest rear-feed technique is to use a probe, formed from the end of a coaxial cable feed line, as illustrated in Figure 12.35. The probe position is designed so that it makes contact with the patch at a point where the impedance of the patch matches the impedance of the feed line.

The feed technique using aperture coupling is illustrated in Figure 12.36, which shows a multilayer structure using two different dielectrics. The radiating patch is formed on the upper surface of a low permittivity substrate that has a dielectric constant, ε_{r1}. A microstrip line on the lower substrate electromagnetically couples energy to the patch through an aperture in the buried ground plane. The position of the aperture is chosen to give the correct impedance match.

The advantage of using the multilayer structure shown in Figure 12.36 is that a low permittivity substrate can be used underneath the patch to enhance radiation, and a high permittivity substrate can be used for the lower substrate to provide a physically more compact feed network. A variant on the structure shown in Figure 12.36 is to use high permittivity for both the upper and lower dielectrics, which makes the radiating patch smaller, and to inset regions of low permittivity material under the radiating edges of the patch to enhance radiation. The modified structure is illustrated in Figure 12.37. This technique can be useful when it is desired to construct an array of patches in a small area. Thick-film printing using photo-sensitive materials, as described in Chapter 3, make the construction of this type of antenna both feasible and relatively straightforward.

One of the benefits of using patch antennas is the ease with which antenna arrays can be formed. Figure 12.38 shows a simple array of four microstrip patches, with a corporate feed network.

The array shown in Figure 12.38 incorporates phase shifters in the paths to the radiating elements. Since the array is in microstrip format, it is straightforward to include active phase shifters, as discussed in Chapter 10, thus making an array with an electronically steerable main beam.

To illustrate how the properties of new materials can be exploited in the design of high frequency planar antennas, a photograph of an actual 77 GHz fixed beam array is shown in Figure 12.39. It consists of a 16×16 array of rectangular patches on the top surface of a multilayer structure which uses polymer dielectric. Each patch is aperture-coupled to a microstrip feed line on the bottom surface using the technique illustrated in Figure 12.36. A corporate-feed arrangement was used in this structure. This array was manufactured using thin (125 μm thick) layers of flexible polymer. Thin sheets

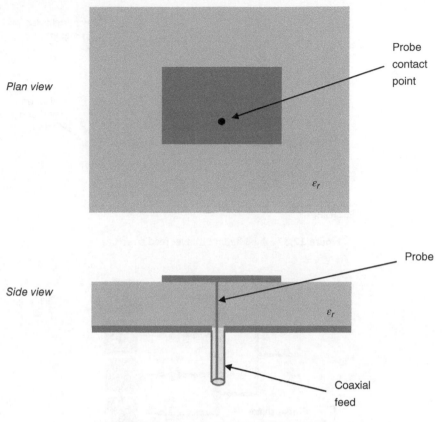

Plan view

Probe contact point

ε_r

Side view

Probe

ε_r

Coaxial feed

Figure 12.35 Probe-fed microstrip patch.

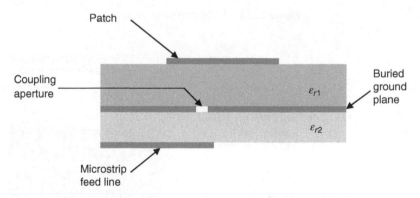

Patch

Coupling aperture

Buried ground plane

ε_{r1}

ε_{r2}

Microstrip feed line

Figure 12.36 Patch excitation through coupling aperture.

of conductor were laminated onto the polymer layers, and etched to form the required conductor patterns. The primary advantage of using thin polymer layers was that the final structure was flexible, with a total thickness of the order of 250 µm, which meant that it was compatible with mounting on curved surfaces.

12.12.2 Circularly Polarized Planar Antennas

The patch antennas described so far in this chapter have all produced linear polarization. Circular polarization can be achieved by a slight modification to the feed arrangements to the patch. Figure 12.40 shows a single square microstrip patch fed centrally on two adjacent sides, with orthogonal signals.

Referring to Figure 12.40 it can be seen that when a positive voltage peak is applied at point 1, the peak radiated electric field is in $\phi = 0°$ direction. Due to the additional $\lambda/4$ length, the positive peak of the voltage wave arrives at point 2 a quarter

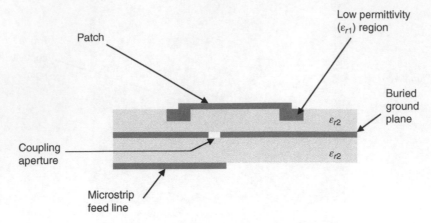

Figure 12.37 Modified multilayer feed structure.

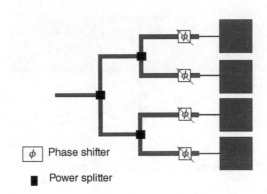

φ Phase shifter

■ Power splitter

Figure 12.38 Microstrip patch array.

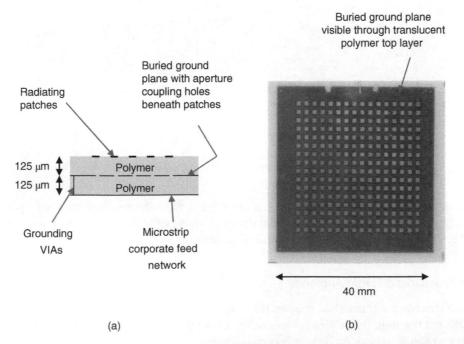

(a) (b)

Figure 12.39 A 77 GHz patch array on a multilayer polymer: (a) details of structure (not to scale) and (b) photograph of top surface showing radiating patches.

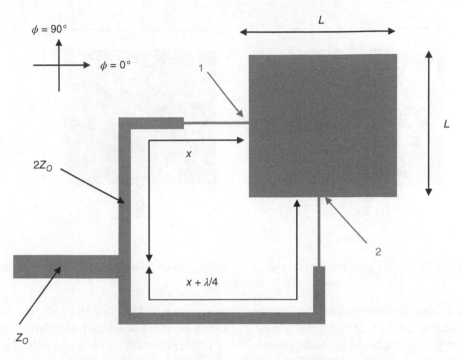

Figure 12.40 Circular polarization from a single microstrip patch.

of a cycle later producing a peak electric field in the $\phi = 90°$ direction. Thus the radiated wave will be circularly polarized, with rotation in the counter-clockwise direction.

The single patch shown in Figure 12.40 generates circular polarization because the two feeds excite identical orthogonal electromagnetic modes, which are in phase quadrature. It is also possible to achieve circular polarization with a single feed to a rectangular patch by changing some geometrical parameter of the patch to generate the two orthogonal modes with the required phase difference. A common technique is to use a single rear-fed probe, with the contact point on the diagonal of a rectangular patch. The patch cannot be square, but must be nearly so in order that the two modes have nearly identical properties. The theory underlying this technique is beyond the scope of this book, but a comprehensive discussion is presented in reference [2]. Figure 12.41 shows the probe positions necessary to produce left- and right-hand circular polarization.

In order to produce good quality circular polarization, the dimension L is normally made around 10% greater then dimension W. Another technique for producing circular polarization from a single microstrip patch is to cut slots in the patch as shown in Figure 12.42. Balanis [2] provides practical design data enabling the dimensions of the slots to be calculated, i.e.

$$\text{Slot length} = \frac{L}{2.72} \text{ and Slot width} = \frac{L}{27.2}. \tag{12.129}$$

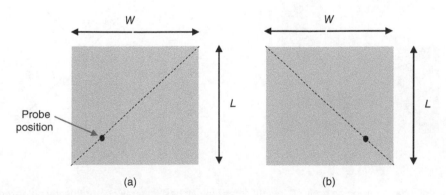

Figure 12.41 Circular polarization from probe-fed rectangular patches: (a) left-hand CP and (b) right-hand CP.

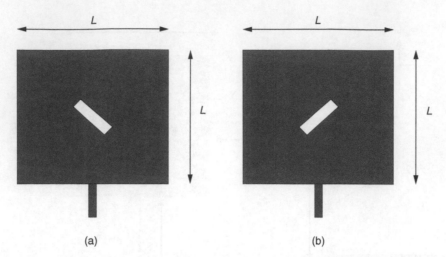

Figure 12.42 Circular polarization from front-fed microstrip patches: (a) left-hand CP and (b) right-hand CP.

Circular polarization can also be achieved using an array of linearly polarized patches. Huang [5] demonstrated excellent circular polarization using an array of four linearly polarized microstrip patches, with a particular spatial arrangement of the patches and appropriate phasing of the feed currents. The principle of Huang's array is illustrated in Figure 12.43.

The four linearly polarized patches shown in Figure 12.43a are arranged so that each patch is orthogonally oriented with respect to its neighbour. The feed points to the patches have been numbered 1, 2, 3, and 4. It has been assumed that the individual patches are fed by probes through the ground plane on the bottom side of the substrate using the technique shown in Figure 12.35. In order to generate circular polarization, the four patches should be fed with a progressive phase delay of 90°. The required phase relationships are achieved by adding extra $\lambda/4$ lengths in the feeds to each patch, as shown in Figure 12.43b, which also shows the array being fed from a single input. Figure 12.43 only gives a schematic view of the feeding arrangements, and in a practical antenna appropriate $\lambda/4$ transformers would need to be included in the feed network to match the impedances of the patches to that of the antenna input.

The principle of operation of the antenna can be explained as follows:

(i) If we consider the instant of time when a maximum positive voltage is applied at feed point 1, then due to the feed line lengths zero voltage will appear at points 2 and 4, and these two patches will not radiate. A maximum positive voltage at point 1 will result in a maximum negative voltage at point 3, due to the relative difference of $\lambda/2$ in the feed lines to these patches. But since patches 1 and 3 are oriented at 180° with respect to each other, they will radiate in phase, giving a maximum E-field in the $\phi = 90°$ direction.

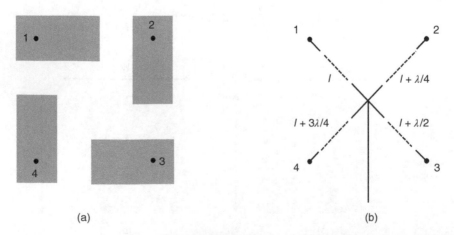

Figure 12.43 Circular polarization using an array of linearly polarized patches: (a) array of four linearly polarized patches positioned to produce circular polarization; (b) the path lengths to the feed points of the four patches.

(ii) A quarter of a cycle later there will be a maximum positive voltage at feed point 2, with a maximum negative voltage at feed point 4, and so patches 2 and 4 will radiate in phase with a maximum E-field in the $\phi = 180°$ direction. So as time progresses the maximum E-field will rotate in the counter-clockwise direction, giving left-hand circular polarization (LHCP).

Since the adjacent patches in the Huang array are orthogonally oriented there is very little mutual coupling between them, which simplifies the analysis and design. However, one limitation of this type of array is that the hand of polarization cannot be changed without changing the feed network. To achieve right-hand circular polarization (RHCP) the progressive 90° difference in the feed phases to the patched would have to be in the clockwise direction.

In an extension of the work of Huang [5], Lum and colleagues [6] used an array of linearly polarized microstrip patches with a travelling-wave fed in order to achieve polarization diversity. Using their arrangement of patches, shown in Figure 12.44, it was possible to change the hand of circular polarization by reversing the direction of travel in the slot line.

The array shown in Figure 12.44 consists of four microstrip patches, labelled 1, 2, 3, and 4, which are excited sequentially by a wave travelling in a slot in the ground plane. The separation between adjacent patches, measured around the slot, is $\lambda/4$. Thus, as the wave travels around the slot the patches are excited sequentially, and due to their orientation, the direction of the net radiated E-field rotates with time, with the time of one rotation of the field equal to the period of the wave in the slot line. In the antenna shown in Figure 12.44 the wave travels clockwise in the slot line, producing a right-hand circularly polarized (RHCP) field. The photograph in Figure 12.45 shows a practical 5 GHz antenna that employs the travelling-wave feed technique.

In this case there are eight radiating patches formed on the upper surface of a microstrip substrate. The input and output microstrip lines are on the same side as the radiating patches, and are coupled to the slot line in the ground plane through two microstrip-to-slot line transitions. Quarter-wave matching sections are included in the microstrip feed lines to match the 50 Ω impedance of the microstrip to the higher impedance of the slot line. Meander sections are included in the slot line to obtain the correct 45° phase difference between the feeds to adjacent patches. It should be noted that in the antenna shown in Figure 12.45 the wave in the slot line is travelling in the counter-clockwise direction, thus producing left-hand polarized CP. However, one disadvantage of the antenna shown in Figure 12.45 is that the microstrip feed lines are on the same side of the substrate as the radiating patches, and it is possible to have unwanted radiated from the feed lines. One solution to this problem is to use a multi-layer structure, as illustrated in Figure 12.46.

In the antenna shown in Figure 12.46 the upper surface contains only the radiating patches. The patches are fed by electromagnetic coupling from a slot line running in the ground plane, which is in the centre of the package. The microstrip feed lines are located on the lower surface and couple to the buried slot through microstrip-to-slot line transitions. The photograph of an actual antenna in Figure 12.46b includes SMA connectors to feed the signal in and out of the microstrip line. Also included in the actual antenna were a set of VIAs to connect the Earth of the connectors to the buried ground plane.

If the direction of the wave in the slot line is reversed, then the hand of polarization will be reversed. Thus, using a simple microstrip switching circuit to control the direction of the wave in the slot line, polarization diversity can be achieved.

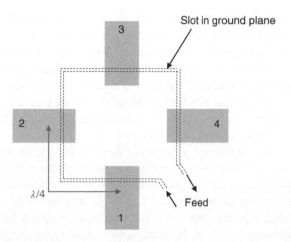

Figure 12.44 Travelling-wave-fed patch array producing circular polarization.

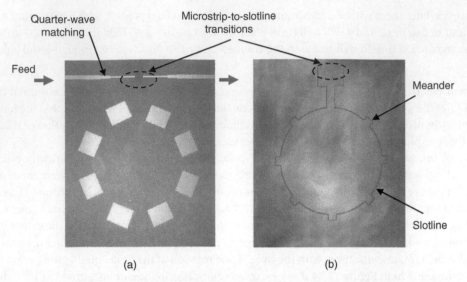

Figure 12.45 Photograph of a 5 GHz CP antenna with a travelling-wave feed: (a) top view showing eight radiating patches and microstrip input and output; (b) view of ground plane showing slotline running beneath the patches.

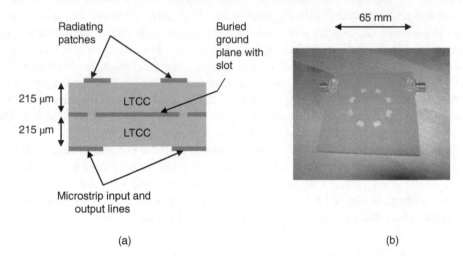

Figure 12.46 15 GHz travelling-wave-fed antenna fabricated using LTCC.

As the wave propagates around the slot line in the arrays shown in Figures 12.44 through 12.46, energy is successively coupled to each patch. Therefore, the amplitude of the signal on the slot line decreases as the wave propagates, and less energy is available for coupling to the latter patches. If the patches all have the same offset with respect to the slot line, less signal will be coupled into the latter patches, and consequently the power radiated from the four patches will not be equal and this will have an adverse effect on the quality of the circular polarization. To overcome this problem there should be a progressive increase in the degree of coupling between the slot line and the patches as the wave propagates around the slot. This is achieved by gradually increasing the offset of the patches relative to the slot line. The offset, x, is defined in Figure 12.47, which also shows the resonant length, L_p, of the patch.

At resonance there will be zero voltage at the centre of the patch, and consequently there will be little coupling between the slot line and the patch when $x = 0$. As x increases so will the degree of coupling. By choosing the correct increments for x around the array it can be arranged that all patches radiate the same power, leading to good quality circular polarization. However, once the incremental technique has been applied it is no longer possible to reverse the direction of the wave in the slot line to change the hand of polarization, since the increments are only correct for a wave travelling in one direction. In order to achieve polarization diversity Lum et al. [6] used two slot lines feeds, one for RHCP and the other for LHCP. The offsets in each slot line were optimized for the particular hand of polarization, and by switching the feed between the

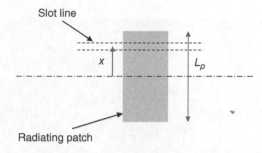

Slot line

x

L_p

Radiating patch

Figure 12.47 Slot line fed microstrip patch: (a) schematic view of construction (not to scale) and (b) photograph of an actual 15 GHz LTCC antenna.

two lines polarization diversity was achieved. Using the dual-feed technique, Lum and his co-workers [7] demonstrated excellent performance from a 5 GHz antenna, fabricated using RT/Duroid substrate. In subsequent work Min and Free [8] provided a more rigorous analysis of travelling-wave-fed patch arrays, leading to useful practical design data.

12.13 Horn Antennas

Horns have long been one of the standard ways of making high performance antennas at microwave frequencies. They are formed by flaring out the end of rectangular waveguide to form a large radiating aperture. From Huygen's principle [1] a large aperture is equivalent to an array with a large number of radiating sources, and it follows from array theory that a large aperture will provide an antenna with high directivity and high gain. The waveguide may be flared in one dimension to provide a sectorial horn as shown in Figure 12.48a and b. However, the most common type of horn is formed by flaring the waveguide in two dimensions to form a pyramidal horn as shown in Figure 12.48c.

A photograph of a typical practical pyramidal horn is shown in Figure 12.49. The horn shown has an aperture of 72×58 mm, and is fed from WR62 rectangular waveguide. It is designed to work in the 13–18 GHz microwave frequency band.

The axial length, L, of a microwave horn is defined in Figure 12.50. It is the distance measured from the throat of the flare to the face of the aperture.

The flare angle, θ, is then defined as

$$\theta = 2 \tan^{-1} \left(\frac{B/2}{L} \right) = 2 \tan^{-1} \left(\frac{B}{2L} \right).$$

(12.130)

Since the path length along the edge of the horn is greater than that along the centre there will be a phase variation across the aperture, because the centre of a propagating wave will reach the aperture before that travelling along the edge, as illustrated in Figure 12.51.

If the phase variation across the aperture is too large, it can have an adverse effect on the gain and directivity of the antenna. The normal criterion is that the phase variation should not exceed 90°, i.e. the path difference x should be less than $\lambda_O/4$, where λ_O is the free-space wavelength, and this leads to an optimum length for an air-filled horn antenna given by

$$L \simeq \frac{B^2}{2\lambda_O}.$$

(12.131)

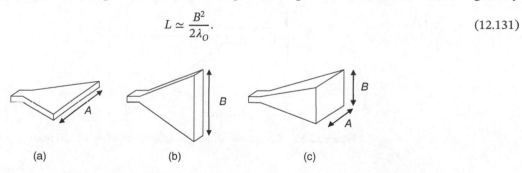

(a)   (b)   (c)

Figure 12.48 Waveguide horns: (a) *H*-plane sectorial, (b) *E*-plane sectorial, and (c) pyramidal.

Figure 12.49 Photograph of a 13–18 GHz pyramidal horn.

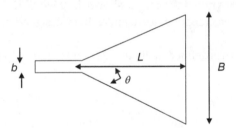

Figure 12.50 Axial length, *L*, of a pyramidal horn.

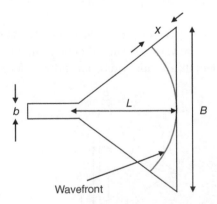

Figure 12.51 Curved wavefront across aperture of pyramidal horn.

With this optimum length the gain of a pyramidal horn antenna, relative to an isotropic radiator, is given by

$$G \approx 7.4 \times \frac{A \times B}{\lambda_O^2}.$$

(12.132)

Example 12.16   Calculate the optimum dimensions of a pyramidal horn with a square aperture that will provide 18 dB of gain at a frequency of 12 GHz.

Solution

$$f = 12 \text{ GHz} \quad \Rightarrow \quad \lambda_O = \frac{c}{f} = \frac{3 \times 10^8}{12 \times 10^9} \text{ m} = 25 \text{ mm}.$$

$$G = 18 \text{ dB} \equiv 10^{1.8} = 63.10$$

The size of the aperture of the horn is given by Eq. (12.132):

$$63.1 = 7.4 \times \frac{B^2}{(25)^2} \quad \Rightarrow \quad B = 72.95 \text{ mm}.$$

The optimum length of the horn is given by Eq. (12.131):

$$L = \frac{(72.95)^2}{2 \times 25} \text{ mm} = 106.43 \text{ mm}.$$

The restriction in the relative values of L and B indicated by Eq. (12.131) means that very long horns are needed to achieve high gains. If a short horn with a large aperture is required in order to provide high gain in a small space, a dielectric lens can be used to correct the phase variation across the aperture. The lens will convert a spherical wavefront into a plane wavefront in which all points are in phase. Normally lenses are made from low-loss plastic materials that can easily be machined or cast to provide the required profile. The material selected for a lens normally has a low dielectric constant to minimize the mismatch between air and the material at the surface of the lens. A typical convex lens is shown in Figure 12.52, and the design of the lens profile can be obtained from straightforward ray theory.

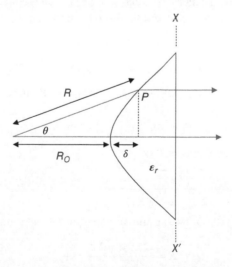

Figure 12.52   Ray paths in a convex dielectric lens.

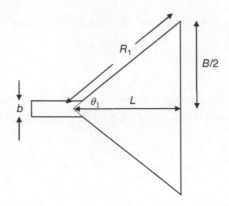

Figure 12.53 Horn dimensions for Example 12.17.

If we consider the transit time of a ray passing through an arbitrary point P on the surface of the lens shown in Figure 12.52, then it follows that all such rays will be in phase in the plane XX' if

$$\frac{R}{c} = \frac{R_O}{c} + \frac{\delta}{v_p}, \tag{12.133}$$

where $v_p = \dfrac{c}{\sqrt{\varepsilon_r}}$ and ε_r is the dielectric constant of the lens material, and δ is defined in Figure 12.52.

From Figure 12.52 it can be seen that

$$\delta = R\cos\theta - R_O.$$

Substituting into Eq. (12.133) gives

$$\frac{R}{c} = \frac{R_O}{c} + \frac{(R\cos\theta - R_O)\sqrt{\varepsilon_r}}{c}. \tag{12.134}$$

Rearranging Eq. (12.134) gives

$$R = \frac{R_O(\sqrt{\varepsilon_r} - 1)}{\sqrt{\varepsilon_r}\cos\theta - 1}. \tag{12.135}$$

Equation (12.135) enables the profile of the dielectric lens to be designed. Since $\varepsilon_r > 1$ the curved surface is hyperbolic.

Example 12.17   Determine the profile of a dielectric lens ($\varepsilon_r = 2.3$) which would need to be inserted into the pyramidal horn of Example 12.16 in order to reduce the axial length of the horn to 50 mm.

Solution
We have:
$$L = 50 \text{ mm} \qquad B/2 = 72.95/2 \text{ mm} = 36.48 \text{ mm}$$

Referring to Figure 12.53
$$R_1 = \sqrt{(50)^2 + (36.48)^2} \text{ mm} = 61.89 \text{ mm},$$

$$\theta_1 = \tan^{-1}\left(\frac{36.5}{50}\right) = 36.11°.$$

Substituting in Eq. (12.135) to obtain a value for R_O:

$$61.89 = \frac{R_O(\sqrt{2.3} - 1)}{\sqrt{2.3} \times \cos(36.11) - 1} \quad \Rightarrow \quad R_O = 26.94 \text{ mm}.$$

By substituting for R_O and ε_r in Eq. (12.135) we obtain an equation for the profile of the lens, i.e. R in terms of θ:

$$R = \frac{26.94(\sqrt{2.3} - 1)}{\sqrt{2.3} \times \cos\theta - 1},$$

i.e.

$$R = \frac{13.92}{1.52 \times \cos\theta - 1} \text{ mm}.$$

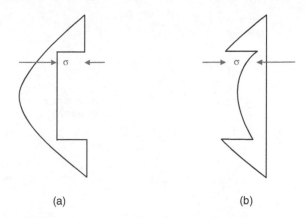

(a) (b)

Figure 12.54 Examples of stepped dielectric lenses.

One problem associated with simple lenses is that the lens can become very thick if ε_r is small, and this can lead to excess weight and uneven transmission loss through the lens. This problem can be resolved by removing sections of the lens, since the lens may be a number of wavelengths thick. Figure 12.54a shows a dielectric lens with a section of depth σ removed from the plane surface of the lens. (Strictly speaking, the depth of the step should be specified in a radial direction, but this makes very little difference in practice.)

The removal of the section will have no effect on the phase providing the difference between the transmission time in a length σ of the dielectric and a length σ of air is equal to one period of the signal, i.e.

$$\frac{\sigma}{v_p} - \frac{\sigma}{c} = \frac{1}{f} = \frac{\lambda_O}{c}. \tag{12.136}$$

Rearranging Eq. (12.136), and substituting $v_p = \dfrac{c}{\sqrt{\varepsilon_r}}$ gives

$$\sigma = \frac{\lambda_O}{\sqrt{\varepsilon_r} - 1}. \tag{12.137}$$

The step can be removed from either face of the dielectric lens; the arrangement shown in Figure 12.54b is preferable in that it is a mechanically stronger structure. Depending on the original size of the lens, a number of steps may be removed, providing the condition given by Eq. (12.137) is not violated. The use of a dielectric lens has the additional advantage of sealing the aperture of the horn against adverse weather conditions. The one minor disadvantage of using stepped dielectric lenses is that they tend to be narrowband, since the depth of the step must be a specific fraction of a wavelength.

12.14 Parabolic Reflector Antennas

Antennas which utilize the focussing properties of parabolic reflectors provide high gain and high directivity, and are popular at microwave frequencies, notably for satellite systems and point-to-point communications. The principle of operation is well known from simple optics; a radiating source at the focal point of the reflector will produce collimated (parallel) reflected rays, as illustrated in Figure 12.55.

If the parabolic reflector shown in Figure 12.55 is rotated about its central axis it generates a three-dimensional surface known as a paraboloid, and this is the form of reflector used in most practical reflector antennas. A source at the focal point of a paraboloid will theoretically produce a narrow parallel radiated beam. In practice, this is not possible since the source must have some physical size and cannot be concentrated at the focal point. But if the diameter of the paraboloid is large compared with a wavelength this type of reflector will produce a radiation pattern with a narrow, slightly divergent main beam and low sidelobes. The two most common methods of feeding paraboloid antennas are the front-feed, using a miniature dipole or horn at the focal point, or using a Cassegrain arrangement. The two feed methods are illustrated in Figure 12.56. In Figure 12.56a, a small horn is shown at the focal point; the horn aperture is usually circular so as to give a more even illumination of the reflector. The primary disadvantage of this technique is the long feed connection to the horn, which introduces attenuation and noise. A better arrangement, particularly for receiving antennas, is shown in

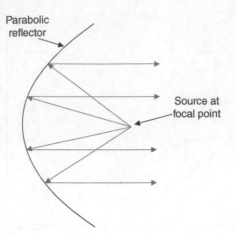

Figure 12.55 Reflection of waves at a parabolic surface.

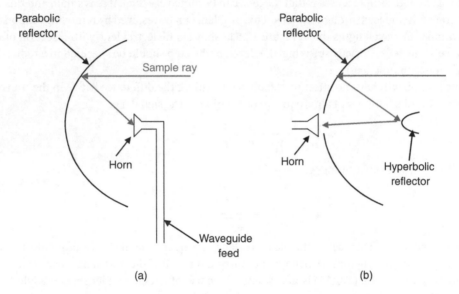

(a)   (b)

Figure 12.56 Typical feed arrangements for parabolic reflector antennas: (a) parabolic reflector with horn at focal point and (b) parabolic reflector with hyperbolic sub-reflector at focal point.

Figure 12.56b. This is known as a Cassegrain antenna and uses a hyperbolic sub-reflector to focus the rays from the main parabolic surface to a point at the centre of the main dish. The receiving horn is located at this point and enables the front-end of the receiver to be mounted behind the parabolic dish with no significant length of feeder between the antenna and the first stage of the receiver. This minimizes both the loss and noise figure of the receiving system.

The efficiency with which the feed illuminates the parabolic surface is specified as the aperture efficiency and denoted by η. Thus, we have

$$A_{eff} = \eta \times A_{phs}, \tag{12.138}$$

where A_{eff} is the effective aperture of the antenna, and A_{phs} is the physical aperture. Usually η is in the range 0.6–0.75 for a parabolic dish.

Then using Eq. (12.10) the gain of the antenna is

$$G = \frac{4\pi}{\lambda_O^2} A_{eff} = \frac{4\pi}{\lambda_O^2} \eta A_{phs} = \frac{4\pi}{\lambda_O^2} \eta \frac{\pi D^2}{4}, \tag{12.139}$$

i.e.

$$G = \eta \left(\frac{\pi D}{\lambda_O} \right)^2. \tag{12.140}$$

where D is the diameter of the parabolic reflector.

The 3 dB beamwidth of a parabolic antenna is given approximately by

$$\theta_{3\ dB} \approx 60\frac{\lambda_O}{D}, \tag{12.141}$$

where λ_O is the free-space wavelength and D is the diameter of the dish. It should be noted that the exact value of the 3 dB beamwidth for a particular antenna depends on the precise nature of the illumination of the parabolic surface by the source.

Example 12.18 A parabolic dish antenna has a diameter of 1.3 m, and an aperture efficiency of 65%. Determine the gain and 3 dB beamwidth of the antenna at 8 GHz.

Solution

$$8\ \text{GHz} \quad \Rightarrow \quad \lambda_O = 0.0375\ \text{m}.$$

Using Eq. (12.140):

$$G = 0.65 \times \left(\frac{\pi \times 1.3}{0.0375}\right)^2 = 7.71 \times 10^3 \equiv 38.87\ \text{dB}.$$

Using Eq. (12.141):

$$\theta_{3\ dB} = \left(60 \times \frac{0.0375}{1.3}\right)^\circ = 1.73^\circ.$$

Example 12.19 A communication link is established at 6 GHz between a geostationary satellite at a height of 35 800 km and a ground station. The satellite transmits 146 W using an antenna with a gain of 19 dB. Determine the required diameter a parabolic receiving dish at the ground station if the received power is to be 1.4 pW. State any assumptions.

Solution

$$6\ \text{GHz} \quad \Rightarrow \quad \lambda_O = 0.05\ \text{m},$$

$$G_{T,sat} = 19\ \text{dB} \equiv 79.43.$$

Using Eq. (12.11) to find the power received at the Earth station:

$$P_{R,Earth} = \frac{P_{T,sat} G_{T,sat} G_{R,Earth} \lambda_O^2}{16\pi^2 R^2}.$$

Substituting data:

$$1.4 \times 10^{-12} = \frac{146 \times 79.43 \times G_{R,Earth} \times (0.05)^2}{16\pi^2(35.80 \times 10^6)^2},$$

$$G_{R,Earth} = 9775.73.$$

Assuming the aperture efficiency of the antenna on the Earth is 70%, and using Eq. (12.140):

$$9775.73 = 0.7 \times \left(\frac{3.142 \times D}{0.05}\right)^2 \quad \Rightarrow \quad D = 1.88\ \text{m}.$$

The calculation assumes zero absorption losses during transmission.

Additional points relating to parabolic reflector antennas:

(i) The parabolic surface does not need to be solid, and can be made of a conducting mesh, providing the apertures in the mesh are small compared to a wavelength.

(ii) Both of the feed arrangements shown in Figure 12.56 lead to some blocking of the signal by the feed and its supporting structure. To overcome this problem the feed can be offset as shown in Figure 12.57. The principle is that a portion of the parabolic reflector is removed so that the focal point lies outside the radiated beam from the reflector. In this case

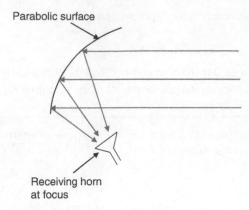

Figure 12.57 Offset feed parabolic reflector.

the feed is tilted to properly illuminate the remaining part of the reflector. Although the offset feed leads to a slightly more complicated arrangement mechanically, the absence of aperture blocking can significantly reduce the sidelobe levels.

(iii) At a given frequency the gain of a parabolic antenna increases with the square of the diameter of the dish, as shown by Eq. (12.140).

(iv) Using a feed configuration that tapers the amplitude distribution across the surface of the dish can significantly reduce the level of the side lobes. This is important when parabolic dishes are used for satellite reception, since the side lobes can pick up significant noise from the Earth. Collin [9] shows that the side lobe levels can be reduced by up to 25 dB with appropriate tapering of the amplitude distribution across the dish.

(v) By using an active feed system it is possible to achieve some degree of steerage of the main beam from a parabolic reflector.

(vi) Spillover loss can result when not all of the radiated power from the feed is intercepted by the dish. The spillover efficiency (η_S) is defined as the fraction of total power from the feed that is intercepted by the parabolic surface. Poor spillover efficiency will contribute to back lobes from the dish.

12.15 Slot Radiators

A slot antenna is formed by cutting a narrow slot in a conducting sheet whose overall dimensions are large compared with those of the slot. The feed is then connected across the centre of the slot. Babinet's principle states that the slot can be represented by a complementary dipole formed by replacing the slot with metal and feeding the resulting dipole at the centre, as shown in Figure 12.58. The radiation characteristics of the slot, such as the bandwidth and directivity, can then

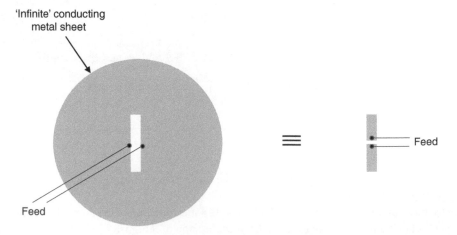

Figure 12.58 Illustration of Babinet's principle.

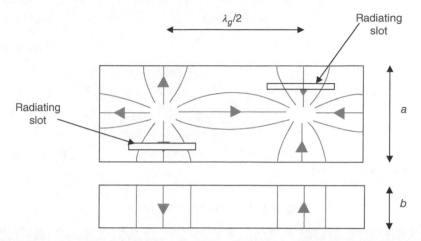

Figure 12.59 Sketch of current distribution in walls of a rectangular waveguide supporting the TE_{10} mode.

be predicted from those of the dipole, but with the E and H fields interchanged. The terminal impedances of the slot and dipole, Z_{Slot} and Z_{Dipole}, are related by

$$Z_{Dipole}Z_{Slot} = \frac{Z_O^2}{4}, \qquad (12.142)$$

where Z_O is the characteristic impedance of free space.

Slots are normally made half a wavelength long, so their properties can easily be deduced from those of the complementary half-wave dipole.

One traditional method of constructing a slot antenna is to cut an array of slots in metal rectangular waveguides. The current distribution in the walls of rectangular waveguide supporting the TE_{10} mode is sketched in Figure 12.59.

Also shown in Figure 12.59 are two radiating slots cut in the broad wall of the waveguide. The slots are cut so as to intercept the currents in the waveguide walls. It can be seen from the directions of the wall currents that the two slots will radiate in anti-phase. But since the slots are $\lambda_g/2$ apart, they will be excited in anti-phase by a wave propagating along the waveguide. The net effect is that the two slots will form an array with both slots radiating in phase. Radiating slots can also be cut in the narrow walls of the waveguide, but they must be inclined so as to intercept the current flow. The photographs in Figure 12.60 show two practical examples of slot antennas in X-band rectangular waveguide.

Although antennas formed from slots in metal waveguide have limited applications in modern microwave systems, the principles are being actively exploited in substrate integrated waveguide (SIW), whose principles were introduced in Chapter 3. The concept of the SIW slot antenna is illustrated in Figure 12.61.

Henry et al. [10] showed that a high performance 77 GHz SIW antenna using the slot array technique could be made using photoimageable thick-film technology. Figure 12.62 shows a photograph of the top surfaces of two of the 77 GHz arrays, which were fabricated using 18 layers of thick-film using the methodology described in Chapter 3.

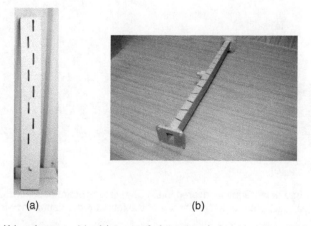

(a)   (b)

Figure 12.60 Slot antennas in X-band waveguide: (a) array of eight slots in broad wall and (b) array of 14 slots in narrow wall.

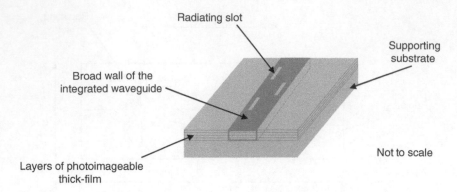

Figure 12.61 SIW antenna concept.

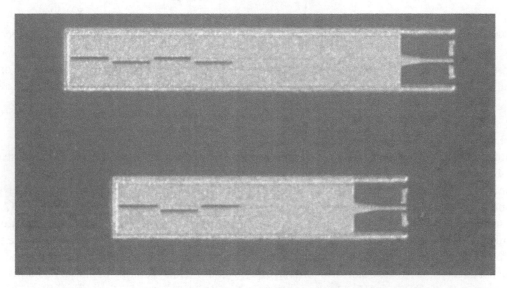

Figure 12.62 77 GHz SIW antennas.

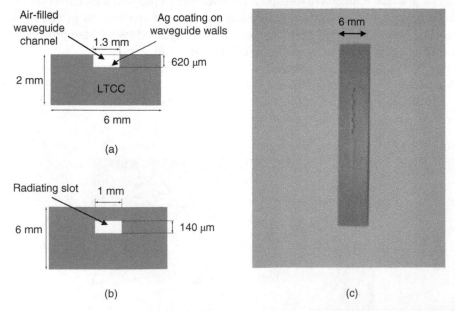

Figure 12.63 160 GHz air-filled waveguide antenna fabricated in LTCC: (a) cross-sectional dimensions (not to scale), (b) dimensions of a radiating slot (not to scale), and (c) photograph of the top of the fabricated antenna showing the eight radiating slots.

The antenna produced a well-defined radiation pattern with a return loss of around 30 dB at the design frequency.

The concept of the slot waveguide antenna was further investigated by Henry et al. [11] who showed that the technique could be applied to LTCC structures. A particular feature of this work was that the integrated waveguide was air-filled, thereby reducing feed losses. Figure 12.63 shows a photograph of the LTCC integrated air-filled waveguide together with key dimensional data.

The fabrication methodology for the antenna shown in Figure 12.63 followed that described in Chapter 3. The measured performance of the antenna at G-band was quite good, with a return loss better than 30 dB at the design frequency. The antenna had eight radiating slots which produced a narrow main beam, with a 3 dB beamwidth around 6°, which was close to the predicted value.

12.16 Supplementary Problems

Q12.1 An antenna having a gain of 8 dB radiates a power of 50 W. Determine the power density 10 km from the antenna in the direction of maximum radiation.

Q12.2 What is the effective aperture of an antenna that has a gain of 17.5 dB at a frequency of 850 MHz?

Q12.3 A line-of-sight communication link is established between two identical antennas, which are 24 km apart. Each antenna has a gain of 13 dB at a frequency of 1.5 GHz. Determine the transmission loss in dB between the two antennas at this frequency when the antennas are oriented for maximum transmission and reception.

Q12.4 Calculate the radiation resistance at 300 MHz of a 10 cm long dipole.

Q12.5 An electrically short dipole has a length of 0.08λ. What is the radiation efficiency of the dipole, assuming the loss resistance is 2 Ω?

Q12.6 Using the array factor, determine the field pattern of a broadside array of six isotropic point sources at a frequency of 500 MHz. The sources are spaced 0.3 m apart and fed with currents of equal amplitude. Assume there is no mutual coupling between the sources.

Q12.7 A linear array of eight isotropic sources have a spacing of 100 mm. The sources are fed with currents of equal amplitude and phase. There is no coupling between the sources. Plot the array factor as a function of θ, as specified in Eq. (12.75), and hence determine:
  (i) The 3 dB beam width in the main lobe
  (ii) The angle between the first nulls of the radiation pattern
  (iii) The directions of the principal side lobes.

Q12.8 Repeat Q12.7 but with 12 isotropic source with a spacing of 100 mm. Comment on the difference between the radiation patterns due to arrays formed from 8 sources and from 12 sources.

Q12.9 If a half-wave dipole is supplied with an peak current of 0.5 A, determine the maximum values of the E and H fields at a distance of 1 km from the antenna.

Q12.10 A simple communication system is established between a transmitter and receiver, using $\lambda/2$ dipoles as the antennas. In order to achieve maximum communication how should the dipoles be oriented with respect to each other?

Q12.11 Figure 12.64 shows an array of two $\lambda/2$ dipoles. The currents fed to the two dipoles are $I_1 = 2\angle0°$ A and $I_2 = 2\angle60°$ A. Determine the ratio of the magnitude of the electric field strength at a distant point in direction **B** to that in direction **A**, at a frequency of 600 MHz. Assume that there is no interaction between the two dipoles.

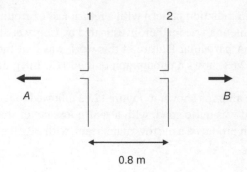

Figure 12.64 Antenna array for Q12.11.

Q12.12 An aerial array consists of two parallel $\lambda/2$ dipoles. The mutual impedance between the dipoles is $(20 - j38)\,\Omega$. Determine the input impedance at the terminals of each dipole, if:
  (i) The dipoles are fed with in-phase currents of the same magnitude
  (ii) The dipoles are fed with anti-phase currents of the same magnitude
  (iii) The dipoles are fed with currents such that $I_1 = jI_2$.
Assume the loss resistance of a $\lambda/2$ dipole is $2\,\Omega$.

Q12.13 An 800 MHz antenna consists of a driven $\lambda/2$ dipole and a parasitic dipole. The two elements of the antenna are parallel and separated by a distance of 6 cm. The mutual impedance between the elements is $(60 - j10)\,\Omega$, and the self-impedance of the parasitic element of $(70 + j40)\,\Omega$. The $\lambda/2$ dipole may be assumed to have a loss resistance of $2\,\Omega$.
Determine:
  (i) The front-to-back ratio of the antenna, in dB.
  (ii) The driving point impedance.
  (iii) The gain of the antenna relative to a $\lambda/2$ dipole.

Q12.14 Calculate the front-to-back ratio of the Yagi antenna shown in Figure 12.65, which consists of a driven half-wave dipole and two parasitic elements, with the following impedance parameters:

$$Z_{11} = (68 + j52)\,\Omega,\ Z_{33} = (47 - j41)\,\Omega,\ Z_{12} = Z_{23} = 62\,\Omega,\ Z_{13} = 0.$$

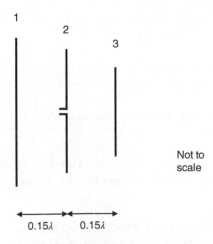

Figure 12.65 Antenna array for Q12.14.

Q12.15 A square loop antenna has 10 turns and a side length of 40 cm. The loop has a total inductance of 175 nH and a total resistance of 7 Ω. Determine the rms voltage across a tuning capacitor connected across the terminals of the antenna, when it is positioned for maximum reception in a linearly polarized E-field of frequency 50 MHz and rms field strength 6 mV/m.

Q12.16 A rectangular loop antenna is 30 cm wide and 10 cm high. There are 15 turns on the loop and the total resistance is 6 Ω and the total inductance is 280 nH. The electric field in the vicinity of the loop is

$$E = 50\sin(10^9 t) \text{ mV/m}.$$

If the loop is positioned with the narrow length parallel to the electric field, and oriented for maximum reception, determine:
  (i) The open loop rms terminal voltage
 (ii) The value of tuning capacitance needed across the terminals of the loop
(iii) The value of the rms voltage across the tuning capacitor
 (iv) The change (in dB) in the voltage across the capacitor when the loop is rotated 20° about a vertical axis.

Q12.17 A parabolic dish antenna has a diameter of 0.6 m. If the aperture efficiency of the dish is 72%, calculate the gain of the antenna at 6 GHz. Give the answer in dB.

Q12.18 A 15 km line-of-sight communication system is set up using parabolic dish antennas having the specification given in Q12.17 for transmitting and receiving. Determine the transmission loss for the link at 6 GHz, assuming there are no absorptive losses in the atmosphere.

Q12.19 How much would the diameter of the parabolic dishes used in Q12.18 need to be increased to reduce the transmission loss of the link by 10 dB? Assume that the aperture efficiency of the parabolic dishes does not change as the diameter is increased.

Q12.20 The transmitting antenna of the up-link of a geostationary satellite communications system radiates 60 W of power at 20 GHz to the satellite. The transmitting antenna is a parabolic dish, of diameter 2 m and efficiency 70%. If the receiving antenna on the satellite has a gain of 18 dB and the combined absorption losses for the antennas and the propagation path are 2.9 dB, determine the power received by the satellite. Note that the distance from Earth to a geostationary satellite is approximately 35 800 km.

Q12.21 The following data apply to the 28 GHz down-link of a geostationary satellite communication system:
Gain of the satellite transmit antenna: 22 dB
Power transmitted from satellite: 50 W
Total absorption losses: 2.3 dB
     If the required received power at the Earth station is 1.2 pW, determine the required gain (in dB) of the receiving antenna. If the receiving antenna is a parabolic dish, estimate the required diameter of the dish. Make reasonable assumptions for any unknown parameters.

Q12.22 Design a 10 GHz front edge-fed microstrip patch antenna on a substrate that has a relative permittivity of 2.3 and a thickness of 2 mm. The patch is to be fed from a 50 Ω microstrip line. (Microstrip design data are given in Figures 12.66 and 12.67.)

Q12.23 Redesign the antenna specified in Q12.22 using the inset feed technique.

Q12.24 Calculate the approximate gain at 8 GHz of a pyramidal horn, of optimized length, with an aperture whose dimensions are $A = B = 55$ mm.

Q12.25 Calculate the dimensions of an optimized 15 GHz pyramidal horn that will give a gain of 20 dB. The horn is to have a square aperture.

Q12.26 The following dimensions apply to 10 GHz E-plane sectorial horn: $L = 80$ mm and $B = 100$ mm. Design a dielectric lens ($\varepsilon_r = 2.34$) that could be inserted in the horn to correct the phase distribution across the aperture.

Q12.27 Could the lens design in Q12.26 be converted to a stepped lens?

Q12.28 A 30 GHz E-plane sectorial horn has the following dimensions: $L = 60$ mm and $B = 90$ mm. Design a stepped dielectric lens to correct the phase distribution across the aperture of the horn. The relative permittivity of the dielectric is 2.34.

Appendix 12.A Microstrip Design Graphs for Substrates with $\varepsilon_r = 2.3$

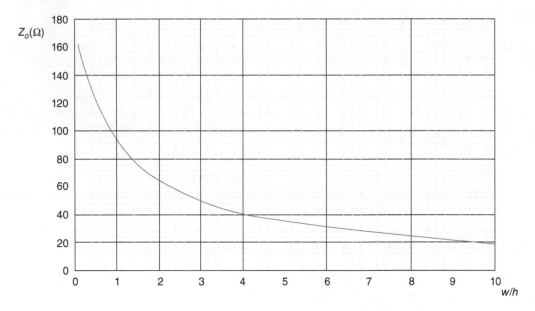

Figure 12.66 Microstrip design graph showing Z_0 as a function of w/h for $\varepsilon_r = 2.3$.

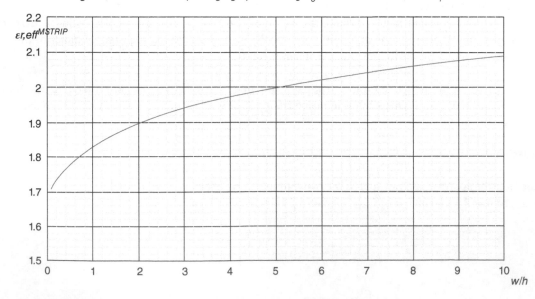

Figure 12.67 Microstrip design graph showing $\varepsilon_{r,eff}^{MSTRIP}$ as a function of w/h for $\varepsilon_r = 2.3$.

References

1 Kraus, J.D. and Marhefka, R.J. (2002). *Antennas*, 3e. New York: McGraw-Hill.

2 Balanis, C.A. (2016). *Antenna Theory: Analysis and Design*, 4e. New York: Wiley.

3 Munson, R.E. (1974). Conformal microstrip antennas and microstrip phased arrays. *Institute of Electrical and Electronics Engineers*: 74–78.

4 Matin, M.A. and Sayeed, A.I. (2010). A design rule for inset-fed rectangular microstrip patch antenna. *WSEAS Transactions* 9 (1): 63–72.

5 Huang, J. (1986). A technique for an array to generate circular polarization with linearly polarized elements. *Institute of Electrical and Electronics Engineers* AP-34 (9): 1113–1124.

6 Lum, K.M., Laohapensaeng, C., and Free, C.E. (2005). A novel travelling-wave feed technique for circularly polarized planar antennas. *IEEE Microwave and Wireless Components Letters* 15 (3): 180–182.

7 Lum, K.M., Tick, T., Free, C.E., and Jantunen, H. (2006). Design and measurement data for a microwave duel-CP antenna using a new travelling-wave feed concept. *IEEE Transactions on Microwave Theory and Techniques* 54 (6): 2880–2886.

8 Min, C. and Free, C.E. (2009). Analysis of travelling-wave-fed patch arrays. *IEEE Transactions on Antennas and Propagation* 57 (3): 664–670.

9 Collin, R.E. (1985). *Antennas and Radiowave Propagation*. New York: McGraw-Hill.

10 Henry, M., Free, C.E., Izqueirdo, B.S. et al. (2009). Millimeter wave substrate integrated waveguide antennas: design and fabrication analysis. *IEEE Transactions on Advanced Packaging* 32 (1): 93–100.

11 Henry, M., Osman, N., Tick, T., and Free, C.E. (2008). Integrated air-filled waveguide antennas in LTCC for G-band operation. *Proceedings of 2008 Asia Pacific Microwave Conference*, Hong Kong, pp. 1–4.

13

Power Amplifiers and Distributed Amplifiers

13.1 Introduction

Power amplifiers are essential devices in any RF or microwave transmission system. Their primary function is to amplify the transmitted signals to a level that will overcome losses in the transmission channel. Since a power amplifier will normally be the last active device in the transmitter chain, any distortion produced within the amplifier will be transmitted to the channel. Distortion will arise due to non-linearities within the amplifier, and consequently issues relating to non-linearity will be the main focus of this chapter, rather than the circuit design of power amplifiers which is outside the scope of the present textbook.

The issue of non-linearity is particularly important for modern, complex digital modulation schemes, which result in high peak-to-average power ratios. These schemes require a high degree of linearity within the power amplifier if error rates are to be kept at an acceptably low level.

The chapter commences with a theoretical discussion of the unwanted frequency components generated by amplifiers with a non-linear power transfer response, and in particular on the influence of third-order distortion products. This is followed by a discussion of the principal methods used to linearize the performance of a power amplifier. Since it is often not possible to achieve the desired level of output power from a single, linearized power amplifier, it may be necessary to combine the outputs of a number of such amplifiers. Consequently, a review of the techniques for combining the outputs of a number of power amplifiers has been included.

The ability of power amplifiers to deliver linear performance with high efficiency is a key design issue. The well-established Doherty amplifier, originally proposed by W.H. Doherty in 1936 [1], is one method of achieving a combination of high efficiency and linearity. In recent years, the Doherty amplifier has received a lot of attention in the technical literature for its implementation at RF and microwave frequencies, and the basic principles of the technique will be introduced in this chapter.

Also included in the chapter is a discussion of the basic principles of distributed amplifiers. This type of amplifier uses a number of FETs as the amplifying devices and employs a powerful design technique whereby the input and output capacitances of the FETs are incorporated into matched transmission lines. This has the effect of producing very wideband operation, although to achieve maximum performance the distributed amplifier requires the use of FETs with identical characteristics.

The chapter concludes with a short review of the developments in materials and packaging that are influencing the advancement of RF and microwave power amplifier devices. Two particularly significant developments are discussed, namely the use of gallium nitride (GaN) to replace gallium arsenide (GaAs), both in discrete and monolithic applications, and the use of advanced packaging techniques to improve the thermal management of high power devices.

RF and Microwave Circuit Design: Theory and Applications, First Edition. Charles E. Free and Colin S. Aitchison.
© 2022 John Wiley & Sons Ltd. Published 2022 by John Wiley & Sons Ltd.
Companion website: www.wiley.com/go/free/rfandmicrowave

13.2 Power Amplifiers

13.2.1 Overview of Power Amplifier Parameters

13.2.1.1 Power Gain

The concepts of amplifier gain were discussed in Chapter 9 in relation to small-signal amplifiers, and the same general definitions apply to power amplifiers. The transducer power gain at a given frequency is simply defined as

$$G_T = \frac{P_o}{P_{avail}}, \tag{13.1}$$

where P_o is the power delivered to the load, and P_{avail} is the power available from the source. In the case of a power amplifier, the gain is normally defined with a specific drive level at the input.

13.2.1.2 Power Added Efficiency

Since the function of a power amplifier is to convert DC power from the biasing supply into RF power at the output, the efficiency (η) of a power amplifier may be defined as:

$$\eta = \frac{P_o}{P_{DC}}, \tag{13.2}$$

where P_o is the RF power available at the output of the amplifier, and P_{DC} is the power from the DC bias supply. However, this definition takes no account of the RF drive power, which may be substantial for a power amplifier. Most power amplifiers tend to have low power gains and thus this definition tends to over-rate the performance of the amplifier. An alternative definition, usually employed by system designers who are concerned with RF levels, is the power added efficiency (PAE), where:

$$PAE = \frac{P_o - P_{avail}}{P_{DC}}, \tag{13.3}$$

i.e.

$$PAE = \frac{P_o}{P_{DC}} \left(1 - \frac{P_{avail}}{P_o} \right) = \eta \left(1 - \frac{1}{G_T} \right). \tag{13.4}$$

Since the PAE is defined in terms of the difference between the input and output RF powers, it provides a more realistic indication of the efficiency of power amplifiers with relatively high drive powers and low gains.

13.2.1.3 Input and Output Impedances

Instead of specifying power amplifier matching criteria in terms of S-parameters, which are essentially small-signal parameters, manufacturers will normally specify the large-signal input and output impedances. When designing power amplifiers, the normal practice is to design a source matching network to provide a conjugate impedance match between the source and the input of the amplifier, but in the case of the output to design a load network that will provide maximum output power (or maximum output voltage swing) at the relevant frequency, which may not correspond to an impedance match.

13.2.2 Distortion

In general, a power amplifier will have a non-linear response described by a Taylor series of the form

$$v_o(t) = a_1[f(t)] + a_2[f(t)]^2 + a_3[f(t)]^3 + \cdots, \tag{13.5}$$

where $v_o(t)$ is the signal at the output of the amplifier when the input is a function of time represented by $f(t)$, and $a_1 \ldots a_n$ are scalar coefficients. The first term on the right-hand side of Eq. (13.5) represents the linear behaviour of the amplifier, since the output is just a scalar multiple of the input. The higher-order terms represent the non-linear behaviour, and these terms will cause additional frequencies to be generated at the output. Normally, just the first three terms of the Taylor expansion shown in Eq. (13.5) are sufficient to describe the non-linear behaviour of an amplifier. The coefficients a_1 and

a_2 are usually positive, and a_3 negative. If we consider a single sinusoidal signal applied to the input of the amplifier then we can write

$$f(t) = V \cos \omega t \tag{13.6}$$

and

$$v_o(t) = a_1[V \cos \omega t] + a_2[V \cos \omega t]^2 + a_3[V \cos \omega t]^3 + \cdots . \tag{13.7}$$

Using simple trigonometric identities, Eq. (13.7) can be expanded to give

$$v_o(t) = \frac{a_2 V^2}{2} + \left(a_1 V + \frac{3a_3 V^3}{4}\right) \cos \omega t + \frac{a_2 V^2}{2} \cos 2\omega t + \frac{a_3 V^3}{4} \cos 3\omega t + \cdots . \tag{13.8}$$

It can be seen that the effect of the higher-order terms is to produce harmonics of the input signal. The resulting harmonic distortion is usually expressed as a percentage of the idealized (linear voltage gain) signal at the output of the amplifier (see Example 13.1).

Example 13.1 A sinusoidal signal with a peak amplitude of 1.5 V, and a frequency of 20 MHz, is applied to the input of an amplifier that has a response given by

$$v_o(t) = a_1[f(t)] + a_2[f(t)]^2 + a_3[f(t)]^3,$$

where $a_1 = 8.1$, $a_2 = 3.9$, $a_3 = -3.1$, and the other symbols have their usual meanings. Determine:

 (i) The frequencies at the output of the amplifier.
 (ii) The percentage second harmonic distortion at the output of the amplifier.
(iii) The percentage third harmonic distortion at the output of the amplifier.

Solution
 (i) Fundamental: 20 MHz
   Second harmonic: 40 MHz
   Third harmonic: 60 MHz
 (ii) Output amplitude with linear gain $= a_1 V = 8.1 \times 1.5 \text{ V} = 12.15 \text{ V}$,

   Output amplitude of second harmonic $= \dfrac{a_2 V^2}{2} = \dfrac{3.9 \times (1.5)^2}{2} \text{ V} = 4.39 \text{ V}$,

   Percentage second harmonic distortion $= \left| \dfrac{a_2 V^2 / 2}{a_1 V} \right| = \dfrac{4.39}{12.15} \times 100\% = 36.13\%$.

(iii) Output amplitude of third harmonic $= \dfrac{a_3 V^3}{4} = \dfrac{-3.1 \times (1.5)^3}{4} \text{ V} = -2.62 \text{ V}$,

   Percentage third harmonic distortion $= \left| \dfrac{a_3 V^3 / 4}{a_1 V} \right| = \dfrac{2.62}{12.15} \times 100\% = 21.56\%$.

Comment: Although the levels of second and third harmonic distortion are relatively high, these harmonics can normally be removed fairly easily through filtering in a practical system.

The effects of non-linearity become far more significant when more than one frequency is applied to the input of the amplifier. If we consider two frequencies, f_1 and f_2, applied to the input of an amplifier with a response represented by the first three terms of Eq. (13.5), then we have

$$f(t) = V_1 \cos \omega_1 t + V_2 \cos \omega_2 t, \tag{13.9}$$

and the signal at the output of the amplifier is given by

$$v_o(t) = a_1[V_1 \cos \omega_1 t + V_2 \cos \omega_2 t] + a_2[V_1 \cos \omega_1 t + V_2 \cos \omega_2 t]^2$$
$$+ a_3[V_1 \cos \omega_1 t + V_2 \cos \omega_2 t]^3 . \tag{13.10}$$

After lengthy, but straightforward substitution of trigonometric identities, Eq. (13.10) can be expanded in terms of the individual frequency components as

$$v_o(t) = \left[a_1 V_1 + a_3 \left(\frac{3V_1^3}{4} + \frac{3V_1 V_2^2}{2}\right)\right] \cos \omega_1 t + \left[a_1 V_2 + a_3 \left(\frac{3V_2^3}{4} + \frac{3V_1^2 V_2}{2}\right)\right] \cos \omega_2 t$$

$$+ \left[\frac{a_2 V_1^2}{2}\right] \cos 2\omega_1 t + \left[\frac{a_2 V_2^2}{2}\right] \cos 2\omega_2 t$$

$$+ \left[\frac{a_3 V_1^3}{4}\right] \cos 3\omega_1 t + \left[\frac{a_3 V_2^3}{4}\right] \cos 3\omega_2 t$$

$$+ [a_2 V_1 V_2] \cos(\omega_2 + \omega_1)t + [a_2 V_1 V_2] \cos(\omega_2 - \omega_1)t$$

$$+ \left[a_3 \frac{3V_1^2 V_2}{4}\right] \cos(2\omega_1 + \omega_2)t + \left[a_3 \frac{3V_1^2 V_2}{4}\right] \cos(2\omega_1 - \omega_2)t$$

$$+ \left[a_3 \frac{3V_1 V_2^2}{4}\right] \cos(2\omega_2 + \omega_1)t + \left[a_3 \frac{3V_1 V_2^2}{4}\right] \cos(2\omega_2 - \omega_1)t. \tag{13.11}$$

It can be seen from Eq. (13.11) that the non-linearity of the amplifier has resulted in additional frequencies at the output, which will cause distortion. These frequencies include harmonics of the input frequencies, and also intermodulation (IM) frequencies $(f_2 \pm f_1)$, $(2f_2 \pm f_1)$ and $(2f_1 \pm f_2)$. The distortion due to the IM frequencies is often referred to by the initials IMD (intermodulation distortion). For clarity, the frequencies in Eq. (13.11) and their peak amplitudes are summarized in Table 13.1.

Table 13.1 Frequency components at output of non-linear amplifier.

Frequency	Peak amplitude	Significance
f_1	$a_1 V_1 + a_3 \left(\dfrac{3V_1^3}{4} + \dfrac{3V_1 V_2^2}{2}\right)$	Input frequency
f_2	$a_1 V_2 + a_3 \left(\dfrac{3V_2^3}{4} + \dfrac{3V_1^2 V_2}{2}\right)$	Input frequency
$2f_1$	$\dfrac{a_2 V_1^2}{2}$	Second harmonic of input
$2f_2$	$\dfrac{a_2 V_2^2}{2}$	Second harmonic of input
$3f_1$	$\dfrac{a_3 V_1^3}{4}$	Third harmonic of input
$3f_2$	$\dfrac{a_3 V_2^3}{4}$	Third harmonic of input
$f_2 + f_1$	$a_2 V_1 V_2$	Second-order IM frequency
$f_2 - f_1$	$a_2 V_1 V_2$	Second-order IM frequency
$2f_1 + f_2$	$a_3 \dfrac{3V_1^2 V_2}{4}$	Third-order IM frequency
$2f_1 - f_2$	$a_3 \dfrac{3V_1^2 V_2}{4}$	Third-order IM frequency
$2f_2 + f_1$	$a_3 \dfrac{3V_1 V_2^2}{4}$	Third-order IM frequency
$2f_2 - f_1$	$a_3 \dfrac{3V_1 V_2^2}{4}$	Third-order IM frequency

Whilst some of the unwanted frequencies at the output of the amplifier can be removed through filtering, there can be a problem with the third-order IM components. For example, if the two input frequencies, f_1 and f_2, are close together the IM frequency $2f_2 - f_1$ will fall within the pass-band of the original signals and therefore cannot be removed by filtering. Consequently, one of the essential parameters used in the specifications of power amplifiers is the third-order IM performance, which is usually specified in terms of the third-order intercept point, which is defined in Section 13.2.2.2.

13.2.2.1 Gain Compression

If we consider just a single frequency signal, $f(t) = V_1 \cos \omega_1 t$, applied to the non-linear amplifier, it follows from Eq. (13.11) that the amplitude of this frequency at the output, $[v_o(t)]_{f_1}$, will be

$$[v_o(t)]_{f_1} = \left(a_1 V_1 + a_3 \frac{3V_1^3}{4} \right) \cos \omega_1 t. \tag{13.12}$$

The voltage gain, G_V, of the amplifier at a frequency f_1 is then

$$G_V = \frac{a_1 V_1 + a_3 \frac{3V_1^3}{4}}{V_1} = a_1 + a_3 \frac{3V_1^2}{4}. \tag{13.13}$$

Since a_3 is normally negative, Eq. (13.13) shows that one of the effects of the non-linearity of the amplifier is to reduce the voltage gain, with the amount of reduction increasing with an increase in the input level. This effect is known as gain compression, and is normally specified in terms of the 1 dB compression point. This is the point at which the output power, P_{out}, from the amplifier has fallen 1 dB below the extrapolated linear power output.

The typical power transfer characteristic of a non-linear amplifier is shown in Figure 13.1. For low levels of input power the amplifier has a linear response, but as the input level increases the output power begins to fall as the amplifier goes into compression. The input power level, P_{1dB}, is normally used to specify the 1 dB compression point. In addition, the effect of compression can be described by the *single-tone gain compression factor* (G_{VCF}). This is defined as the ratio of the voltage gain with distortion to that of the idealized linear voltage gain, thus we have

$$G_{VCF} = \frac{a_1 V_1 + a_3 \frac{3V_1^3}{4}}{u_1 V_1} = 1 + \frac{3a_3 V_1^2}{4a_1}. \tag{13.14}$$

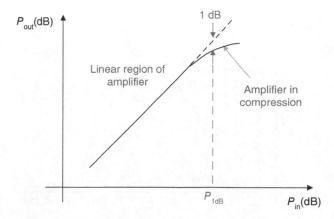

Figure 13.1 Power transfer characteristic of a non-linear amplifier showing 1 dB compression point.

Example 13.2 Determine the maximum voltage at the input to the amplifier specified in Example 13.1 if the 1 dB compression point is not to be exceeded.

Solution

1 dB ≡ 1.122 (voltage ratio)

Using Eq. (13.14): $\dfrac{1}{1.122} = 1 + \dfrac{3a_3 V_1^2}{4a_1}$.

Data from Example 13.1: $a_1 = 8.1$ $a_3 = -3.1$.

Then

$$\frac{1}{1.122} = 1 - \frac{3 \times 3.1 \times V_1^2}{4 \times 8.1},$$

$$V_1 = 0.615 \text{ V}.$$

The 1 dB power compression point is also used in specifying the *dynamic range* (DR) of a power amplifier, which is defined as

$$(DR)_{dB} = 10 \log \frac{P_{1dB}}{P_n}, \tag{13.15}$$

where P_{1dB} is the input power corresponding to the 1 dB compression point and P_n is the noise power floor of the amplifier at the input (the meaning and characteristics of noise are discussed in more detail in Chapter 14).

13.2.2.2 Third-Order Intercept Point

The third-order intercept point is a convenient figure-of-merit used to indicate the significance of third-order distortion in non-linear amplifiers, and also in other non-linear devices such as mixers. It is specified as the point at which the extrapolated third-order IMD power at the output, generated by an input of two signals of equal amplitude, is equal to the extrapolated linear output power from an amplifier, and this is illustrated in Figure 13.2. On an amplifier specification sheet, the third-order intercept point is often denoted by the IP_3 value, which is the value of input power corresponding to the third-order intercept point.

The two-tone third-order intercept specified in Figure 13.2 refers to the situation where the input signal consists only of two frequencies, f_1 and f_2. If these two frequencies have the same amplitudes at the input, i.e. $V_1 = V_2 = V_o$, then it follows from Eq. (13.11) that the power P_{IM} dissipated in a load, R, due to each of the third order intermodulation terms will be

$$P_{IM} = \left(\frac{3a_3 V_o^3/4}{\sqrt{2}} \right)^2 \times \frac{1}{R} = \frac{9a_3^2 V_o^6}{32R}. \tag{13.16}$$

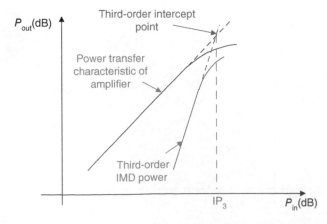

Figure 13.2 Intercept diagram showing the two-tone third-order intercept point.

Considering just the linear term of Eq. (13.11) the output power, P_o, dissipated in a load, R, due to a single tone at the input is

$$P_o = \left(\frac{a_1 V_o}{\sqrt{2}}\right)^2 \times \frac{1}{R} = \frac{a_1^2 V_o^2}{2R}. \tag{13.17}$$

Now the input power, P_i, dissipated in a load R for a single tone will be

$$P_i = \frac{V_o^2}{2R}. \tag{13.18}$$

Combining Eqs. (13.16) through (13.18) we can write P_{IM} and P_o in terms of the input power as

$$P_{IM} = \frac{9a_3^2(2RP_i)^3}{32} = \frac{9a_3^2 R^2}{4} \times P_i^3 \tag{13.19}$$

and

$$P_o = a_1^2 \times P_i. \tag{13.20}$$

From Eq. (13.19) we see that the intermodulation power is proportional to the cube of the input power, whereas from Eq. (13.20) we see that the single tone (linear) output power is directly proportional to the input power. Thus, on a dB scale the slope of the linear portion of the IM power will be three times that of the linear portion of the transfer function, as can be seen in Figure 13.2.

A further parameter often used to represent the effects of intermodulation distortion is the *intermodulation distortion ratio* (*IMR*). Considering a two-tone equal amplitude input to a power amplifier (PA) the *IMR* is defined as the ratio of the amplitude of one of the output third order intermodulation terms to one of the linear output amplitudes. Again, using the amplitudes given in Eq. (13.11) we have

$$IMR = \left|\frac{a_3 3V_1^2 V_2/4}{a_1 V_1}\right| = \left|\frac{3a_3 V_1 V_2}{4a_1}\right|, \tag{13.21}$$

and since $V_1 = V_2 = V_o$ we may write Eq. (13.21) as

$$IMR = \left|\frac{3a_3 V_o^2}{4a_1}\right|. \tag{13.22}$$

Example 13.3 The voltage output from a matched power amplifier is represented by

$$v_o(t) = 5[f(t)] + 3.8[f(t)]^2 - 0.27[f(t)]^3,$$

where $f(t)$ represents the input signal. If the input is a two-tone signal given by $f(t) = 1.3\cos(31.420 \times 10^7 t) + 1.7\cos(34.562 \times 10^7 t)$ V determine;

(i) The frequencies at the output of the amplifier.
(ii) The power in the third-order IM components at the output (assuming a 50 Ω matched load).
(iii) The *IMR*.

Solution

(i)
$$31.420 \times 10^7 = 2\pi f_1 \quad \Rightarrow \quad f_1 = 50 \text{ MHz},$$
$$34.562 \times 10^7 = 2\pi f_2 \quad \Rightarrow \quad f_2 = 55 \text{ MHz}.$$

Frequencies at output:

$$f_1: \quad 50 \text{ MHz},$$
$$f_2: \quad 55 \text{ MHz},$$
$$2f_1: \quad 100 \text{ MHz},$$
$$2f_2: \quad 110 \text{ MHz},$$
$$3f_1: \quad 150 \text{ MHz},$$
$$3f_2: \quad 165 \text{ MHz},$$

(Continued)

(Continued)

$$f_2 \pm f_1 : \quad 105 \text{ MHz} \quad 5 \text{ MHz},$$

$$2f_1 \pm f_2 : \quad 155\text{MHz} \quad 45\text{MHz},$$

$$2f_2 \pm f_1 : \quad 160 \text{ MHz} \quad 60 \text{ MHz}.$$

(ii) Power in the third order IM components. There are two sets of IM components to consider:

(a) $\qquad\qquad 2f_1 \pm f_2 \quad \Rightarrow \quad 155 \text{ MHz} \quad \text{and} \quad 45 \text{ MHz}.$

Using Eq. (13.16) we have

$$P_{IM} = \frac{a_3^2 \left(\dfrac{3V_1^2 V_2}{4} \right)^2}{2 \times R}$$

$$= \frac{(0.27)^2 \left(\dfrac{3 \times (1.3)^2 \times 1.7}{4} \right)^2}{2 \times 50} \text{ W}$$

$$= 0.0034 \text{ W}.$$

$$IMR = \left| \frac{3a_3 V_1 V_2}{4a_1} \right| = \frac{3 \times 0.27 \times 1.3 \times 1.7}{4 \times 5} = 0.09.$$

(b) $\qquad\qquad 2f_2 \pm f_1 \quad \Rightarrow \quad 160 \text{ MHz} \quad \text{and} \quad 60 \text{ MHz}.$

Again, using Eq. (13.16) we have

$$P_{IM} = \frac{a_3^2 \left(\dfrac{3V_1 V_2^2}{4} \right)^2}{2 \times R}$$

$$= \frac{(0.27)^2 \left(\dfrac{3 \times 1.3 \times (1.7)^2}{4} \right)^2}{2 \times 50} \text{ W}$$

$$= 0.0058 \text{ W}.$$

(iii) $\qquad\qquad IMR = \left| \frac{3a_3 V_1 V_2}{4a_1} \right| = \frac{3 \times 0.27 \times 1.3 \times 1.7}{4 \times 5} = 0.09.$

13.2.3 Linearization

The need for power amplifiers to operate in a linear mode is essential for many modern communication systems, particularly those carrying multi-channel digital traffic. We know from the earlier sections that power amplifiers exhibit non-linearity as the saturation level is approached, leading to increasing levels of distortion. The situation is summarized in Figure 13.3.

In practice, in order to avoid distortion the output of a power amplifier must be backed off from the non-linear region, which results in a significant loss of efficiency. Linearization is a term that refers to techniques that are used to compensate for the non-linearity in a power amplifier, and so enable it to operate in a quasi-linear mode at higher power levels approaching saturation.

The three most important techniques for achieving linearization are:

(i) *Pre-distortion*: In these systems, the behaviour of the power amplifier is already known, usually as a function of the input power, and the input signal is then pre-distorted, so as to give a linear output. The system usually employs a look-up table that has stored data on the power amplifier non-linearities. *(The technique is simple, and usually low-cost.)*

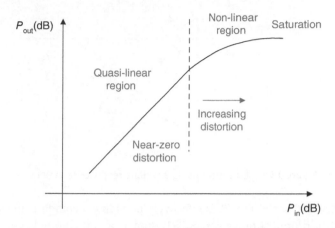

Figure 13.3 Typical non-linearity in power amplifiers.

(ii) *Negative feedback*: In these systems, the output of the power amplifier is compared to the input, and a correction applied to linearize the output. The main problem with this technique is the bandwidth limitation imposed by the feedback network. *(The technique is precise, does not require the shape of the amplifier to be known in advance, but is narrowband.)*

(iii) *Feed-forward*: In these systems, the input signal is split between the power amplifier input and a reference path. The characteristics of the signal at the output of the power amplifier are then compared with those of the signal in the reference path, and a subsequent correction applied. The key point here is that there is no feedback involved, and consequently there is no significant bandwidth limitation. *(The technique is precise and wideband.)*

The principles of the three linearization techniques are described in more detail in the following sections. For clarity, analogue circuits have been used to explain the operation of each technique, although in modern practical linearizers most of the circuit functions would be realized digitally.

13.2.3.1 Pre-Distortion

The basic principle of pre-distortion is illustrated by the circuit shown in Figure 13.4.

A directional coupler is used to take a small sample of the input signal to the PA. This sample is fed to a detector, which provides a DC output proportional to the amplitude of the RF sample. The DC signal is then fed into a PA look-up table that has stored data on the non-linear relationship between P_{in} and P_{out}. This table then provides an output, which can be used to modify the input level to the PA so as to maintain a linear response. It should be noted that a delay circuit is included before the correction unit to compensate for the delay incurred by the signal in passing through the coupler, detector, and

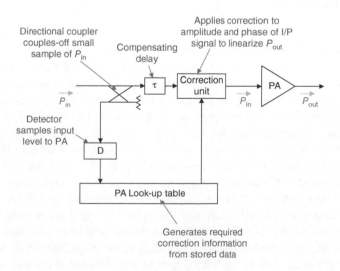

Figure 13.4 Pre-distortion circuit.

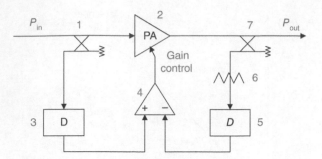

Figure 13.5 Linearizer using envelope feedback technique.

look-up unit. Thus, the circuitry in front of the PA is effectively providing a complimentary response to that of the PA, but at a lower level. The pre-distortion technique is relatively simple, and therefore low-cost to implement, but has the disadvantage that the response of the PA must be known in advance. However, simplicity and cost advantages have made digital predistortion (DPD) the dominant linearization technique.

13.2.3.2 Negative Feedback

A simple example of a feedback linearizer is shown in Figure 13.5.

The envelope feedback linearizer uses two directional couplers (1, 7) to sample the input and output powers of the power amplifier (2). The coupled ports of the directional couplers are connected to two envelope detectors (3, 5), the outputs of which are connected to a differential amplifier (4). An attenuator (6) is included in the input path to the envelope detector on the output side to compensate for the nominal gain of the PA. The output of the differential amplifier is connected to the amplitude gain control of the PA. In the linear mode of operation the two inputs to the differential amplifier are equal, and there is zero voltage applied to the gain control of the PA. But if there is any variation in the output level of the PA due to non-linearities, an error voltage will be generated at the output of the differential amplifier, and this is applied to the gain control of the PA to maintain linear operation.

The arrangement shown in Figure 13.5 will only provide amplitude correction. However, the basic method can be extended to include phase correction by using phase detectors to compare the phases at the input and output of the power amplifier, and this is known as polar loop correction. An alternative method of providing both amplitude and phase correction, without the need for phase detectors, is to use Cartesian feedback. This approach separates the complex signals at the input and output of the power amplifier into in-phase and quadrature components, often referred to as the I and Q components. The I and Q components of any signal must necessarily contain information about both the amplitude and phase of the signal. By comparing these components at the input and output of the PA, both amplitude and phase correction can be provided, without the need for phase detectors. More detailed discussions of Cartesian loop linearization systems can be found in References [2, 3].

13.2.3.3 Feed-Forward Linearization

The basic arrangement of a feed-forward linearizer is shown in Figure 13.6.

In the feed-forward linearization scheme, the input power (P_{in}) is divided by the power splitter (1) into two paths. The signal in the upper path is applied to the input of the power amplifier (2). A directional coupler (4) samples a small amount of the output from the PA and applies this through an attenuator (5) to one port of a comparator (6). The signal in the lower path from the power splitter is applied to the other port of the comparator. A delay circuit (3) is included in the lower path to compensate for the delay through the power amplifier. The signal $s_0(t)$ will then contain only the wanted input frequencies, since the signal has not passed through any non-linear device, whereas $s_1(t)$ will contain the wanted frequency plus all the unwanted frequencies resulting from the non-linear behaviour of the PA. The value of the attenuator is such that the wanted frequency terms are cancelled out in the comparator, leaving only the unwanted frequency terms in the comparator output, $s_e(t)$. This output signal is amplified (8) and reinserted into the output from the PA through a second directional coupler (9), but with a phase difference of 180° relative to the signal coming directly from the PA. Thus, the reinserted signals will cancel with those from the PA. The value of G_e is set so that the reinserted signals have the same amplitude as those coming directly from the PA, thus providing total cancellation of the unwanted frequencies generated in the PA, and effectively linearizing the performance. A delay circuit (7) is normally included to compensate for the delay through the amplifier (8).

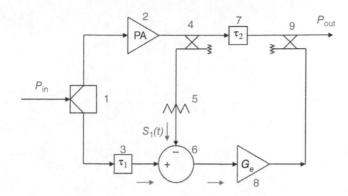

Figure 13.6 Basic feed-forward linearization.

13.2.4 Power Combining

There are many practical situations where the required RF or microwave power cannot be achieved with a single amplifier, and where it is therefore necessary to combine the outputs of a number of individual power amplifiers. Whilst combining amplifier outputs may seem a straightforward task, there are a number of requirements that must be satisfied by the combining circuit:

  (i) It should not change the loading conditions for individual amplifiers.
 (ii) It should not substantially alter the bandwidth of the individual amplifiers.
(iii) It should not worsen the match of the amplifiers when off-tune.
 (iv) It should provide effective isolation between the inputs and outputs of the amplifiers being combined.

There are three techniques normally used for PA combining at RF and microwave frequencies:

Corporate combining. This is based on the use of the Wilkinson power divider, which was discussed in Chapter 2. The arrangement for combining four power amplifiers is shown in Figure 13.7.

In the arrangement shown in Figure 13.7, three Wilkinson power dividers are used to split the input power into four paths, which provide the inputs to the four amplifiers. As described in Chapter 2, each of the Wilkinson power dividers includes two $\lambda/4$ sections. The outputs of the four amplifiers are combined using another three Wilkinson sections, connected in the opposite sense.

Quadrature combining. The performance of the basic corporate combiner can be improved by including additional $\lambda/4$ sections to form a so-called quadrature combiner, which is shown in Figure 13.8.

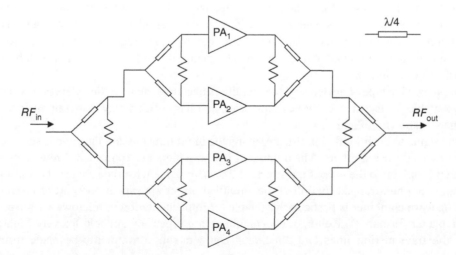

Figure 13.7 Corporate power combiner.

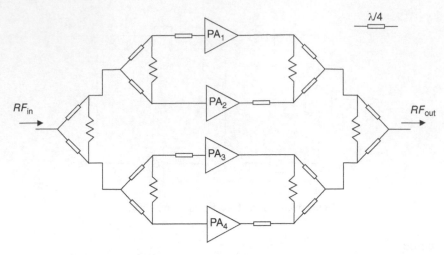

Figure 13.8 Quadrature power combiner.

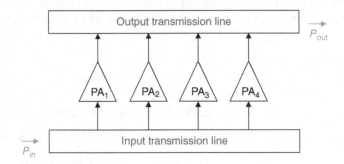

Figure 13.9 Travelling-wave power combiner.

It can be seen that the quadrature combiner has the same basic structure as the corporate combiner, but with additional quarter-wave sections at the input and output of each pair of power amplifiers. If we consider the pair of amplifiers comprising PA$_1$ and PA$_2$ we see that the inputs to the two amplifiers are now in phase quadrature. If each amplifier has the same input reflection coefficient, then the reflections from the two amplifiers will be out-of-phase across the previous Wilkinson section and will therefore cancel. Thus, the presence of the additional quarter-wave section will improve the match of the combiner. However, since the behaviour of the quarter-wave section is frequency dependent, there will only be complete cancellation at the design frequency, and the odd harmonics of the design frequency. It should also be noted that since a quarter-wave section has been added in the input to PA$_1$, a similar section must be added to the output from PA$_2$ in order to maintain the same transmission phase through each amplifier path. Whilst the quadrature combiner improves the input match, the presence of the additional, frequency dependent, $\lambda/4$ sections will have an adverse effect on the bandwidth of the combiner.

Travelling-wave combiner: This type of device uses the amplifiers to couple the travelling waves on two transmission lines. It is particularly suitable for use at microwave frequencies, and has been employed in a number of MMIC designs. A typical arrangement is shown in Figure 13.9.

In this combiner, the signal to be amplified propagates along the input transmission line, and is successively coupled into the inputs of the four power amplifiers. The outputs of the amplifiers are connected through coupling mechanisms (usually directional couplers) to the output transmission line. Through appropriate choice of coupling mechanism, and amplifier spacing, it can be arranged that all of the amplified power appears at one port of the output transmission line. The travelling-wave combiner is particularly suitable for implementation in microwave transmission lines, either in waveguide or planar formats. Travelling-wave combiners are especially suitable for very high-power microwave systems, where the transmission lines and directional couplers can conveniently be made from hollow metallic waveguide.

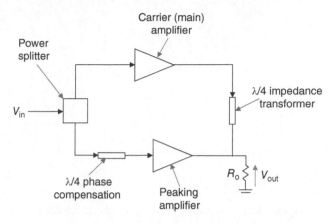

Figure 13.10 Doherty amplifier.

13.2.5 Doherty Amplifier

It was seen in earlier sections that power amplifiers become non-linear as the output power increases. In applications where linearity is of particular importance, such as mobile base stations, the power amplifiers must be backed off from their peak output power to maintain adequate linearity. This back-off results in a significant loss of efficiency. The Doherty amplifier [1], named after W.H. Doherty who first proposed the amplifier in 1936, provides high efficiency whilst maintaining good linearity. Although the Doherty amplifier has existed for some considerable time, it is only in recent years that it has achieved popularity at RF and microwave frequencies.

The basic arrangement of a Doherty amplifier is illustrated in Figure 13.10, and consists of two power amplifiers, one is known as the carrier or main amplifier, and the other as the peaking amplifier. The output of the carrier amplifier is connected to an impedance transformer, usually a $\lambda/4$ transmission line, the output of which is connected to the output of the peaking amplifier, and to the load, R_o. An additional $\lambda/4$ transmission line is included before the peaking amplifier to compensate for the phase shift introduced by the impedance transformer, thereby ensuring correct phasing of the two amplifier outputs.

At low input drive levels the peaking amplifier is cut-off and the carrier amplifier operates alone in a linear mode. As the input level increases a transition point is reached where the carrier amplifier starts to saturate and the peaking amplifier starts to switch on. The circuits are normally designed so that the transition point occurs at half of the peak output voltage of the system.

Since the outputs of the two amplifiers are connected to a common load, R_o, the loading on each amplifier will depend on the state of the other amplifier, which will change with the drive level. This is known as load modulation. For the purpose of analysis the operation of the Doherty amplifier can be represented by the arrangement shown in Figure 13.11, where each amplifier is represented by an appropriate current source [4]. The load impedances of the carrier and peaking amplifiers are Z_C and Z_p, respectively, and Z_T is the characteristic impedance of the $\lambda/4$ transformer.

Cripps [2] provides a comprehensive analysis of the Doherty amplifier, based on the circuit of Figure 13.11, leading to the following expressions for the load impedances of the carrier and peaking amplifiers.

$$Z_C = \frac{Z_T^2}{R_o \left(1 + {}^{I_p}\!/\!{}_{I_C}\right)}, \tag{13.23}$$

$$Z_P = R_o \left(1 + {}^{I_C}\!/\!{}_{I_p}\right). \tag{13.24}$$

For low input drive levels to the Doherty amplifier, where the peaking amplifier is cut off, $I_p = 0$ and the carrier load impedance is simply given by

$$Z_C = \frac{Z_T^2}{R_o}. \tag{13.25}$$

The essential behaviour of the Doherty amplifier can be understood by considering the individual responses of the carrier and peaking amplifiers, as illustrated in Figure 13.12.

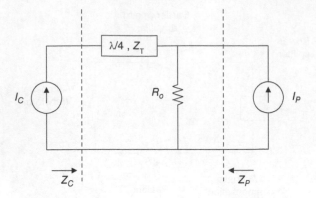

Figure 13.11 Equivalent circuit of a Doherty amplifier.

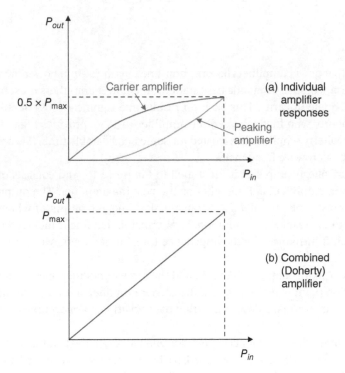

Figure 13.12 Illustration of the general principle of the Doherty amplifier.

The carrier amplifier has a linear region, and goes into saturation at high levels of P_{in}. The peaking amplifier switches on when the carrier amplifier ceases to be linear, and has an effective response which is the inverse of that the carrier amplifier in the non-linear region. Combining the two responses gives a quasi-linear response for much higher input levels than the carrier amplifier alone. The improved linearity means the Doherty amplifier can operate at much higher drive levels than a single power amplifier, with a consequent improvement in efficiency.

13.3 Load Matching of Power Amplifiers

Load matching networks for power amplifiers differ from source matching networks in that there is not a conjugate impedance match at the output. In the case of a power amplifier the criterion for the load matching network is to ensure that the output voltage and current swings are such that the amplifier can deliver maximum power. Conjugate matching of the output can, in a typical situation, result in the output power being 3 dB less than the maximum power output.

There are three techniques widely used to design an optimum load network for a power amplifier:

(i) Load-pull method

This is a well-established practical technique for determining the optimum load impedance of power amplifiers under large-signal conditions. The principal features of a load-pull measurement system are shown in Figure 13.13.

In the arrangement shown, tuning networks (6, 8) are connected to the input and output of the device-under-test (7), in this case the power amplifier. The values of the elements in the tuning networks can be varied by control signals from the controller (10). The input signal to the DUT is supplied from a high-frequency source (1), whose amplitude and frequency can be changed by data signals from the controller. An isolator (2) is connected to the output of the source to prevent changes in impedance pulling the frequency of the source off-tune, and also changing the output level from the source. A directional coupler (3) samples the input and reflected signals from the DUT. The coupled ports of the directional coupler are connected to amplitude level detectors (4, 5), the outputs of which feed level information to the controller. These levels enable the VSWR on the main path to be determined. The output power level from the DUT is monitored by a conventional power metre (9), whose DC output is connected to the controller.

The operation of the load-pull measurement system is straightforward. The controller selects the desired frequency and power level from the source. An algorithm within the controller then changes the values within the tuning circuits until the desired DUT output power, commensurate with an acceptable input VSWR, is achieved. Once the desired conditions have been achieved the tuning circuits are removed from the measurement system, and connected to a VNA. The VNA enables the impedances of the tuning circuits to be determined. Finally, these impedances can be used to realize appropriate input and output matching circuits in the required medium, either distributed or lumped. The load-pull measurement system can also be used to generate data to enable load-pull contours to be plotted. These are contours plotted on a Smith chart that show constant output power as a function of the load impedance, at a given frequency. Examples of load-pull contours for a microwave power amplifier using an FET are shown in Figure 13.14. It should be noted that if a power amplifier is operating in a non-linear region the load-pull contours will not form perfect circles, which was the case for constant noise and constant gain circles for low-power (linear) amplifiers discussed in an earlier chapter. The contours surround the optimum position, which is the load impedance resulting in maximum output power from the power amplifier.

(ii) Cripps method [1]

There are two steps to determine the required load impedance using the Cripps method. Firstly, the optimum load resistance value is found from the IV characteristics of the active device, usually an FET. The optimum resistance is the value which gives maximum voltage swing at the output of the amplifier. Secondly, the load reactance is found. The load reactance used in this method is simply the conjugate of the FETs small-signal output reactance. It should be noted that the small-signal reactance can be used since the output reactance is not a strong function of the power level. Once the required load impedance has been found, the input matching network is designed to provide a conjugate match

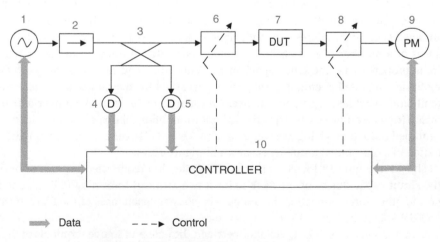

Figure 13.13 Load-pull measurement system.

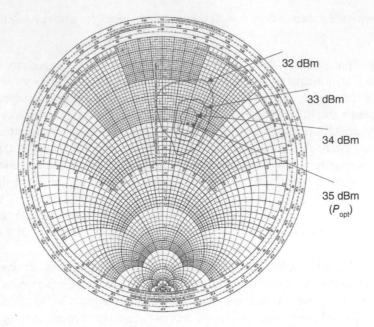

Figure 13.14 Typical load-pull contours at 12 GHz for an FET.

to the source. (Note that the input impedance of the FET will be a function of the optimum load impedance if there is any reverse transmission through the FET.)

(iii) Large-signal CAD

With this computer-based method, large-signal parameters of the FET are used in a non-linear simulation package. The technique has the advantage that the simulator can also predict harmonic and intermodulation distortion. However, some industrial users of this technique suggest that it should be used with caution, since the realization of load circuits to achieve the predicted PAE, gain, and spectral purity over a broad bandwidth is often not feasible.

13.4 Distributed Amplifiers

13.4.1 Description and Principle of Operation

The bandwidth of an amplifier that uses one or more FETs as the amplifying device is normally limited by the input and output capacitances of the FET. This limitation is overcome in the distributed amplifier by incorporating the input and output capacitances of the FETs into matched lumped, wideband artificial transmission lines. The principle is not new, and was originally proposed for implementation using valves as the active devices [5]. However, it was not until the 1980s that the technique attracted significant interest for implementation using FETs. Since the technique relies upon the input and output capacitances of a number of FETs being absorbed into matched transmission lines, it is essential that each FET has the same circuit properties, i.e. the same capacitances. Whilst this requirement is easily achievable if the FETs are formed within a monolithic integrated circuit, it is still possible to exploit the distributed amplifier principle using discrete FETs in a hybrid circuit, and to achieve high performance. An example of 1980s work on hybrid distributed amplifiers is that of Law and Aitchison [6] who reported a wideband distributed amplifier, fabricated in a hybrid form with discrete FETs interconnected with lumped inductors, giving a power gain of (4.5 ± 1.5) dB over a frequency range 2–18 GHz.

The basic configuration of a distributed amplifier is shown in Figure 13.15.

In the arrangement shown in Figure 13.15, a linear array of FETs are interconnected with inductors such that the inductors combine with the shunt capacitances of the FETs to form two artificial transmission lines, one interconnecting the gates of the FETs and the other interconnecting the drains. The shunt capacitances of the FETs are shown in the simple equivalent circuit of a FET, shown in Figure 13.16.

In the equivalent circuit shown, C_{GS} is the capacitance between the gate and source terminals of the FET, and C_{DS} is the capacitance between the drain and source terminals. This equivalent circuit neglects the series gate resistance, which is

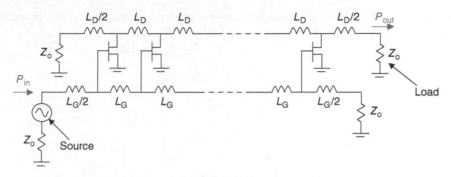

Figure 13.15 Basic configuration of a distributed amplifier.

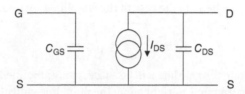

Figure 13.16 Simple equivalent circuit of an FET.

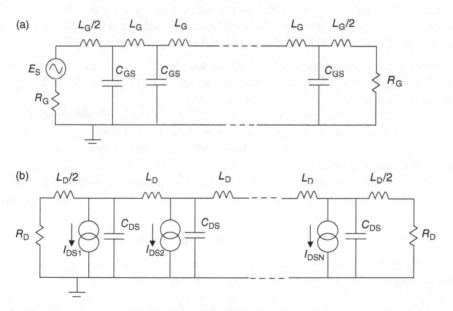

Figure 13.17 Equivalent circuits representing distributed amplifier: (a) Equivalent circuit of gate line. (b) Equivalent circuit of drain line.

normally small, and the shunt drain resistance, which is normally large. Also shown in the equivalent circuit is a current source, I_{DS}, to represent the amplification produced by the FETs, where

$$I_{DS} = v_G g_m, \tag{13.26}$$

and v_G is the gate voltage, and g_m is the transconductance of the FET.

The two artificial transmission lines representing the gate and drain lines of a distributed amplifier are shown in Figure 13.17, where L_G and L_D are the gate and drain line inductances, respectively.

Artificial transmission lines are conveniently analyzed in terms of T or π sections and have characteristic impedances Z_{oT} and $Z_{o\pi}$, respectively. The relationships between Z_o ($= \sqrt{L/C}$), and Z_{oT} and $Z_{o\pi}$ are

$$Z_{oT} = Z_o \sqrt{1 - \frac{\omega^2}{\omega_C^2}}, \tag{13.27}$$

$$Z_{o\pi} = \frac{Z_o}{\sqrt{1 - \frac{\omega^2}{\omega_C^2}}}, \tag{13.28}$$

where

$$\omega_C = \frac{2}{\sqrt{LC}}. \tag{13.29}$$

To avoid reflections the artificial transmission lines should each be terminated in the characteristic impedance of the lines, namely Z_{oT} or $Z_{o\pi}$, depending on whether the analysis is in terms of T or π sections. The values of Z_{oT} and $Z_{o\pi}$ are determined from Eqs. (13.27) and (13.28) where, in the case of the simplified FET equivalent circuit,

$$Z_o = \sqrt{\frac{L_G}{C_{GS}}} = \sqrt{\frac{L_D}{C_{DS}}}. \tag{13.30}$$

A wave travelling from left-to-right along the gate line will be successively amplified by the FETs and coupled to the drain line. This will result in a wave travelling from left-to-right along the drain line, and also a wave travelling from right-to-left along the drain line. Assuming the phase velocities on the gate and drain line have been appropriately designed, the wave travelling from left-to-right will increase in amplitude with distance, as more current is coupled from successive FETs, and the total energy will finally be dissipated in the load. Thus, in principle, the forward gain of the distributed amplifier will increase with the number FETs in the circuit. However, the reverse gain, due to the wave travelling from right-to-left along the drain line, will decrease due to the transmission phase differences caused by the path differences through the different FETs. There are some practical issues that should be noted in relation to distributed amplifiers:

(i) The simple equivalent circuits neglect resistances associated with the FETs and the inductors. In practice, there will be some resistances and hence sources of power loss, and the number of FETs that can be used will ultimately be limited by dissipative losses in the circuit.

(ii) As the wave from the source propagates along the gate line its amplitude will decrease as energy is successively absorbed by the FETs. However, since the input impedance of the FETs is normally very high, this decease will be small, and easily overcome by using additional capacitors to taper the gate line impedance so that the gate of each FET receives the signals of the same amplitude.

(iii) The amplitude of the wave propagating along the drain line will increase as it approaches the right-hand load. This means the drain voltage will be higher for those FETs near the right-hand load, and breakdown criteria will eventually limit the number of FETs being used.

Some additional points on the general layout of distributed amplifiers should also be noted:

(i) The operation of the distributed amplifier relies upon the velocities of the waves on the gate and drain lines being equal. The wave velocities on the gate and drain lines are v_G and v_D, respectively, where

$$v_G = \frac{1}{\sqrt{L_G C_{GS}}}, \tag{13.31}$$

$$v_D = \frac{1}{\sqrt{L_D C_{DS}}}. \tag{13.32}$$

Normally for an FET, $C_{GS} \gg C_{DS}$, and additional shunt capacitors are often used between the drain and source of each FET to increase the effective value of C_{DS}, thus avoiding the need for high value inductors in the drain line. High value inductors in lumped format have the disadvantage of introducing significant series resistance, and consequent series loss.

(ii) The series inductances shown in Figure 13.15 can be provided by short, narrow sections of microstrip line, rather than by lumped inductors. The use of planar inductors reduces the losses and permits monolithic implementation. Meandering of the microstrip sections can also be employed, both to reduce space and to increase the inductance for a given overall length.

(iii) The distributed amplifier is primarily used because of its very broad bandwidth. But it can also function as a medium power amplifier using power FETs. However, this results in some compromise in terms of operational bandwidth, since power FETs have greater gate widths, which lead to higher values of C_{GS}, and lower bandwidth. To some extent this problem can be overcome by capacitively coupling the gates of the FETs to the gate line (using a series capacitor connected to each gate), which has the effect of reducing the overall gate-source capacitance.

13.4.2 Analysis

In the foregoing description of the operation of the distributed amplifier we saw that travelling waves are excited on the drain line that travel in both the forward (left-to-right) and reverse (right-to-left) directions. The primary objective of the analysis is to determine expressions for the gains of the amplifier in the forward and reverse directions, and hence show that most of the significant gain occurs in the forward direction, with appropriate circuit design.

Aitchison [7] analyzed an N-section distributed amplifier in terms of π-sections to determine expressions for the forward available gain ($G_{av,F}$) and the reverse available gain ($G_{av,R}$). For convenience, the equivalent circuit of the drain line has been redrawn in Figure 13.18, and shows the forward and reverse directions.

Following the procedure in [7] superposition can be used to find the total current, I_D, due to N sections, in the forward load on the drain line as

$$I_D = \frac{1}{2}(I_{DS1}e^{-jN\phi_D} + I_{DS2}e^{-j(N-1)\phi_D} + \ldots I_{DSN}e^{-j\phi_D}), \tag{13.33}$$

where ϕ_D is the phase change through one section of the drain line. The wave propagating along the gate line will produce voltages across the gate-source capacitors given by

$$V_{GS1} = V_{in}e^{-j\phi_G}, \quad V_{GS2} = V_{in}e^{-j2\phi_G}, \cdots V_{GSN} = V_{in}e^{-jN\phi_G}, \tag{13.34}$$

where V_{in} is the voltage across the terminals of the gate line, and ϕ_G is the phase change through one section of the gate line. From the basic theory of FETs we know

$$I_{DS1} = g_m V_{GS1}, \quad I_{DS2} = g_m V_{GS2}, \cdots I_{DSN} = g_m V_{GSN}. \tag{13.35}$$

Since we have considered the gate line to be loss free, the magnitudes of the gate voltages will be equal, i.e.

$$|V_{GS1}| = |V_{GS2}| = |V_{GSN}| = V_G \tag{13.36}$$

and

$$|I_{DS1}| = |I_{DS2}| = |I_{DSN}| = I_{DS}. \tag{13.37}$$

Combining Eqs. (13.33) through (13.37) gives the current in the forward drain load as

$$I_D = \frac{V_G g_m}{2}(e^{-j(N\phi_D+\phi_G)} + e^{-j([N-1]\phi_D+2\phi_G)} + \cdots + e^{-j(\phi_D+N\phi_G)})$$
$$= \frac{V_G g_m}{2}\left(\frac{1-e^{-jN(\phi_D-\phi_G)}}{1-e^{-j(\phi_D-\phi_G)}}\right)e^{-j(N\phi_D+\phi_G)}. \tag{13.38}$$

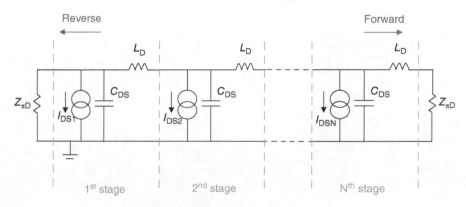

Figure 13.18 Equivalent circuit of drain line.

It can be shown [7] that Eq. (13.38) can be rewritten as

$$|I_D| = \frac{V_G g_m}{2} \left| \frac{\sin \frac{N}{2}(\phi_D - \phi_G)}{\sin \frac{1}{2}(\phi_D - \phi_G)} \right|. \tag{13.39}$$

Now, if the gate line is matched, $V_G = E_S/2$ where E_S is the open-circuit voltage from the source (see Figure 13.17a), and the power dissipated in the forward drain load, $Z_{\pi D}$, is then

$$P_{\pi D} = \frac{E_S^2 g_m^2}{16} \left(\frac{\sin \frac{N}{2}(\phi_D - \phi_G)}{\sin \frac{1}{2}(\phi_D - \phi_G)} \right)^2 Z_{\pi D}. \tag{13.40}$$

The power available from the source is

$$P_{source} = \frac{E_S^2}{4 Z_{\pi G}}, \tag{13.41}$$

and so the available forward gain is

$$G_{av,F} = \frac{P_{\pi D}}{P_{source}} = \frac{g_m^2 Z_{\pi D} Z_{\pi G}}{4} \left(\frac{\sin \frac{N}{2}(\phi_D - \phi_G)}{\sin \frac{1}{2}(\phi_D - \phi_G)} \right)^2. \tag{13.42}$$

If the wave velocities along the drain and gate lines are equal then $\phi_D = \phi_G$ and

$$G_{av,F} = \frac{g_m^2 Z_{\pi D} Z_{\pi G} N^2}{4}. \tag{13.43}$$

Equation (13.43) shows that with equal wave velocities on the gate and drain lines the forward gain becomes essentially independent of frequency. Moreover, if the gate and drain line cut-off frequencies are made sufficiently high, $Z_{\pi D}$ and $Z_{\pi G}$ can be replaced by Z_{oD} and Z_{oG}, and obtained simply from Eq. (13.30). The expression in Eq. (13.43) also shows that the available forward gain can be increased indefinitely, in principle, by increasing N, the number of sections, although this only applies for the loss-free case.

The analysis in [7] also generates an expression for the available gain in the reverse direction as

$$G_{av,R} = \frac{g_m^2 Z_{\pi D} Z_{\pi G}}{4} \left(\frac{\sin \frac{N}{2}(\phi_G + \phi_D)}{\sin \frac{1}{2}(\phi_G + \phi_D)} \right)^2. \tag{13.44}$$

Now for maximum forward gain $\phi_G = \phi_D$, and putting $\phi_G = \phi_D = \phi$ in Eq. (13.44) we have

$$G_{av,R} = \frac{g_m^2 Z_{\pi D} Z_{\pi G}}{4} \left(\frac{\sin N\phi}{\sin \phi} \right)^2. \tag{13.45}$$

The function $(\sin N\phi / \sin \phi)^2$ is plotted in Figure 13.19 for $N = 7$.

It can be seen that the values of the function $(\sin N\phi / \sin \phi)^2$ are small for a large range of ϕ centred on 90°. Thus $G_{av,R}$, the reverse gain, will be negligible over a large range of frequencies, centred on the frequency that makes ϕ equal to 90°. Within this frequency range most of the amplified power will be dissipated in the forward drain load.

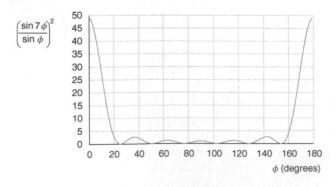

Figure 13.19 Plot of the function $(\sin N\phi/\sin \phi)^2$, for $N = 7$.

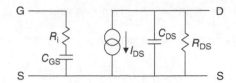

Figure 13.20 Simple equivalent circuit of FET including resistors to represent the principal sources of loss.

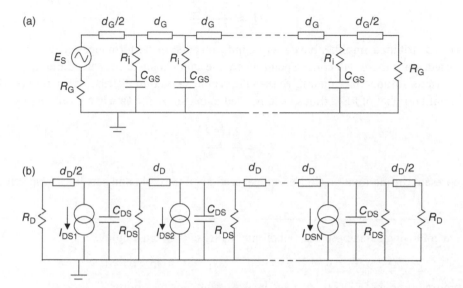

Figure 13.21 Equivalent circuits representing distributed amplifier, including principal sources of loss: (a) Equivalent circuit of gate line. (b) Equivalent circuit of drain line.

The foregoing analysis, leading to the expressions for the forward and reverse gains in Eqs. (13.43) and (13.45), is useful in explaining the basic operation of a distributed amplifier, but it is necessary to consider the sources of loss in order to obtain practical design data on the optimum number of FETs to be used, and also to determine the noise performance [7]. Figure 13.20, shows an enhanced equivalent circuit for an FET in which a resistor, R_G, represents the series loss in the gate connection, and a resistor, R_{DS}, represents the loss between drain and source. As before, the capacitance C_{GD} between gate and drain has been neglected, although this can be a source of unwanted coupling in some designs.

Using the modified equivalent circuit of an FET the equivalent circuits of the gate and drain lines shown in Figure 13.17 can be redrawn as those in Figure 13.21. In this figure the inductors in the gate and drain lines have been replaced with equivalent lengths of transmission line. Thus, d_G and d_D are the lengths of transmission line that provide the inductances L_G and L_D, respectively.

Ayasli et al [8] analyzed equivalent circuits of the form shown in Figure 13.21 and developed the following expression for the forward power gain, G_p, of a distributed amplifier in which the gate and drain lines have the same characteristic impedance, Z_o.

$$G_p = \frac{g_m^2 Z_o^2}{4} \frac{(e^{-\alpha_G d_G N} - e^{-\alpha_D d_D N})^2}{(\alpha_G d_G - \alpha_D d_D)^2}, \tag{13.46}$$

where

$$\alpha_G = \frac{R_i \omega^2 C_{GS}^2 Z_o}{2 d_G} \tag{13.47}$$

and

$$\alpha_D = \frac{Z_o}{2 R_{DS} d_D}, \tag{13.48}$$

where ω is the angular frequency of operation, N is the number of FETs, and the other symbols have been defined previously. Note that α_G and α_D represent the attenuation per unit length (in Np/m) of the gate and drain lines, respectively.

Two simplifications can be made [8] to the expression for the power gain in Eq. (13.46):

(i) When the drain line losses are small compared to the gate line losses, which often the case for practical FETs

$$G_p \approx \frac{g_m^2 N^2 Z_0^2}{4} \left(1 - \frac{(\alpha_G d_G N)}{2} + \frac{(\alpha_G d_G N)^2}{6} \right)^2.$$

(13.49)

(ii) It follows from Eq. (13.46) that if the gate line losses are also neglected

$$G_p \approx \frac{g_m^2 N^2 Z_0^2}{4}.$$

(13.50)

In practice, losses in a distributed amplifier have a very significant effect on performance and ultimately limit the number of FETs that can be used, since there will come a point where the gain obtained from an additional FET will be offset by increased FET loss. Various authors have quantified the effects of losses within distributed amplifiers and Beyer et al. [9] showed that the optimum number of FETs that should be used to maximize the gain at a given frequency is

$$N_{opt} = \frac{\ln\left(\frac{A_D}{A_G}\right)}{A_D - A_G},$$

(13.51)

where A_D and A_G represent the attenuation (in Nepers) per section on the drain and gate lines, respectively.

Example 13.4 The following data apply to a monolithic N-FET distributed amplifier

$$A_G = 0.11 \, \text{Np/section} \qquad A_D = 0.15 \, \text{Np/section}$$

What is the optimum number of FETs that should be used to provide maximum gain?

Solution

$$N_{opt} = \frac{\ln\left(\frac{A_D}{A_G}\right)}{A_D - A_G} = \frac{\ln\left(\frac{0.15}{0.11}\right)}{0.15 - 0.11} = \frac{0.31}{0.04} = 7.75.$$

Since there must be an integer number of FETs in the array, we choose the nearest whole number for N_{opt}, i.e. $N_{opt} = 8$

Recent developments in distributed amplifiers have been driven by the need for very wideband amplifiers to cope with the demands of high speed digital systems, some with pulse widths of the order of picoseconds. One of the consequences of these demands has been the development of ultra-wideband distributed amplifiers, usually in distributed monolithic formats. A particular feature of these amplifiers is that the operating frequency range is not confined to the microwave region, but encompasses frequencies from low RF values through to millimetre-wave frequencies in a single amplifier. Some representative examples reported in the literature are:

(i) Yoon and colleagues [10] in 2014 demonstrated an ultra-wideband distributed amplifier working over the frequency range 40–222 GHz. The amplifier consisted of four cascaded gain cells, giving a total measured gain of ~10 dB over the full frequency range, with a power output of ~8.5 dBm. One minor drawback to the reported performance was a relatively poor return loss of ~10 dB.

(ii) Eriksson and colleagues [11] in 2015 demonstrated the ability of a distributed amplifier to cover the frequency range from low RF through to millimetre frequencies, with the development of an amplifier working from 75 kHz through to 180 GHz. This amplifier had a reported gain greater than 10 dB, with a noise figure varying over the range 5–10 dB.

(iii) Kobayshi and colleagues [12] in 2016 reported a distributed amplifier working over the frequency range 100 MHz to 45 GHz. The gain of the amplifier was >10 dB, with a power output of 1–2 W. However, a particular feature of this work was the low noise figure of ~1.6 dB. This demonstrated the potential of distributed amplifiers to provide a combination of ultra-wideband frequency performance with low noise figure.

13.5   Developments in Materials and Packaging for Power Amplifiers

The traditional semiconductor material for high-frequency power amplifiers has been gallium arsenide (GaAs). Over the past two decades this material has gradually been replaced by gallium nitride (GaN), which offers substantial improvements in device performance.

The principal characteristics of GaN that make it attractive for RF and microwave power amplifiers are [13]:

(i) A large breakdown voltage of ~3.3 MV/cm. is ~0.4 MV/cm.) *This means GaN FET power devices can operate with higher DC voltage rails, which for a given RF output power will reduce the drain current and hence the thermal dissipation.*

(ii) A wide bandgap of ~3.39 eV. (The value for GaAs is ~1.42 eV.) *This means GaN devices can operate at higher temperatures than GaAs devices.*

(iii) A high operating power density of ~1.3 W/cm K. (The value for GaAs is ~0.43 W/cm K.) *This means devices can be manufactured with much smaller device areas.*

(iv) A high saturation velocity of ~2.5×10^7 cm/s. (The value for GaAs is ~1.0×10^7 cm/s.) *This gives GaN devices the potential to operate at very high frequencies.*

The benefits of GaN technology for power amplifiers, using both discrete and monolithic devices, have been extensively reported in the recent technical literature. Data have been presented that show GaN power devices operating with significant output power and relatively high efficiencies at frequencies extending well into the millimetre-wave region. Some representative examples of the potential of GaN technology, and indicating the current state-of-the-art are:

- The potential of GaN technology for integration of high-power microwave amplifiers was shown by the results of Tao et al. [14], who reported an MMIC device giving output powers of 56–74 W, over the frequency range 8–12 GHz. The circuit was a three-stage power amplifier using 0.25 μm GaN HEMT (high electron mobility transistor) devices. The chip area was only 13.3 mm^2, and had a PAE ~43%. The output power was achieved in pulse mode operation, with a pulse width of 100 μs, and a duty cycle of 10%.

- The ability of GaN devices to work at high millimetre frequencies has been shown by the work reported by Shaobing et al. [15]. The results of a monolithic three-stage amplifier working at W-band (75–110 GHz) were presented. The amplifier was manufactured using 0.1 μm AlGaN/GaN technology. A peak output power of 32.2 dBm (1.66 W) in CW mode was obtained at 93 GHz. A particular feature of the results presented was the high power density of 3.46 W/mm, indicating the ability of GaN-based devices to work at high temperatures.

- The use of GaN technology to provide a fully integrated high-power sub-system has been demonstrated by Heijningen et al. [16]. A high-power GaN MMIC was developed to provide a complete radar front-end, operating over the frequency range 5.2–5.6 GHz. The MMIC formed the front-end of a transceiver, and included a driver amplifier, high-power amplifier, Tx/Rx switch and low-noise amplifier. The output power the chip was reported in excess of 40 W with a PAE of 36%. As the authors point out, the use of GaN with its higher breakdown voltage means that high output power can be achieved without the need for power combiners, and without the need for additional protection circuits in the receive side of the module.

Developments in semiconductor technology, notably those using GaN materials, have enabled high-power RF and microwave devices to be manufactured in small circuit areas. One of the consequences of these developments is that thermal management, both of discrete and monolithic devices, has become a critical issue. Efficient heat dissipation must be provided to keep the temperature of the active region of the transistor low, thereby improving the stability, reliability and longevity of the device. The traditional method of heat sinking hybrid circuits has been to mount the active device on a finned metal heat sink. The metal normally used for this type of heat sink is aluminium, mainly because it is light and easy to manufacture, but it suffers from limited thermal performance [17]. In particular, because the aluminium has limited thermal conductivity, the heat distribution in the heat sink is very non-uniform with a high thermal concentration immediately beneath the active device, which can cause overheating of the component. This problem can be overcome through the use of some form of heat spreader. One material that has attracted a lot of recent attention for heat spreading is pyrolytic graphite (PG). The primary reasons why this material is attractive are that it has a thermal conductivity >400% that of copper and a density 80% that of aluminium [18]. However, PG materials have an anisotropic molecular structure and the favourable thermal conductivity is only obtained in two of the three coordinate directions. This means that the material must be carefully conditioned before being used as a heat spreader, in order to ensure that the high conductivity

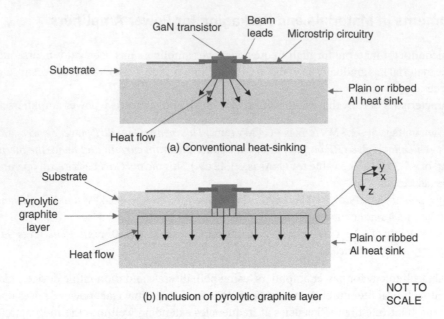

Figure 13.22 Use of pyrolytic heat spreader: (a) Conventional heat-sinking. (b) Inclusion of pyrolytic graphite layer.

is achieved in the desired direction. Figure 13.22 illustrates one method of using a sheet of PG to improve the thermal management of a GaN power amplifier.

Without the sheet of PG the heat from the transistor tends to be concentrated in the aluminium immediately below the transistor. The use of a pyrolytic sheet, configured such that the highest thermal conductivity is in the x–y plane, spreads the excess heat from the transistor over a much larger area of the aluminium, thereby improving the thermal dissipation. This technique is particularly useful for application in multilayer structures, such as LTCC.

The use of heat spreaders is just one aspect of thermal management. The methods and materials used to encapsulate the bare die within a high power transistor are also of considerable significance. New high-temperature plastic-mould compounds are now being used with GaN semiconductors, where the junction temperatures may approach 230 °C. For readers interested in more detailed information on thermal management and packaging of high power, high-frequency amplifiers, an excellent review of the techniques and materials available has been presented by Samanta [19].

References

1 Doherty, W.H. (1936). A new high efficiency power amplifier for modulated waves. *Proceedings of the IRE* 24 (9): 1163–1182.

2 Cripps, S.C. (1999). *RF Power Amplifiers for Wireless Communications*. Norwood, MA: Artech House.

3 Katz, A., Wood, J., and Chokola, D. (2016). The evolution of PA linearization. *IEEE Microwave Magazine* 17 (2): 32–40.

4 Raab, F.H. (1987). Efficiency of Doherty RF power-amplifier systems. *IEEE Transactions on Broadcasting* 33 (3): 77–83.

5 Percival, W.S. (1936). British Patent Specification No. 460562.

6 Law, C.L. and Aitchison, C.S. (1985). 2-18 GHz distributed amplifier in hybrid form. *Electronics Letters* 21 (16): 684–685.

7 Aitchison, C.S. (1985). The intrinsic noise figure of the MESFET distributed amplifier. *IEEE Transactions on Microwave Theory and Techniques* 33 (6): 460–466.

8 Ayasli, Y., Mozzi, R.L., Vorhaus, J.L. et al. (1982). A monolithic GaAs 1–13-GHz traveling-wave amplifier. *IEEE Transactions on Electron Devices* 29 (7): 1072–1077.

9 Beyer, J.B., Prasad, S.N., Becker, R.C. et al. (1984). MESFET distributed amplifier design guidelines. *IEEE Transactions on Microwave Theory and Techniques* 32 (3): 268–275.

10 Yoon, S., Lee, I., Urteaga, M. et al. (2014). A fully-integrated 40–222 GHz InP HBT distributed amplifier. *IEEE Microwave and Wireless Componcenter Letters* 24 (7): 460–462.

11 Eriksson, K., Darwazeh, I., and Zirath, H. (2015). InP DHBT distributed amplifiers with up to 235-GHz bandwidth. *IEEE Transactions on Microwave Theory and Techniques* 63 (4): 1334–1341.

12 Kobayashi, K.W., Denninghoff, D., and Miller, D. (2016). A novel 100 MHz–45 GHz input-termination-less distributed amplifier design with low-frequency low-noise and high linearity implemented with a 6 inch 0.15 μm GaN-SiC wafer process technology. *IEEE Journal of Solid-State Circuits* 51 (9): 2017–2026.

13 Mishra, U.K., Shen, L., Kazior, T.E., and Wu, Y.-F. (2008). GaN-based RF power devices and amplifiers. *Proceedings of the IEEE* 96 (2): 287–305.

14 Tao, H.-Q., Hong, W., Zhang, B., and Yu, X.-M. (2017). A compact 60W X-band GaN HEMTPower amplifier MMIC. *IEEE Microwave and Wireless Components Letters* 27 (1): 73–75.

15 Shaobing, W., Jianfeng, G., Weibo, W., and Junyun, Z. (2016). W-band MMIC PA with ultrahigh power density in 100-nm AlGaN/GaN technology. *IEEE Transactions on Electron Devices* 63 (10): 3882–3886.

16 Heijningen, M., Hek, P., Dourlens, C. et al. (2017). C-band single-Chip radar front-end in AlGaN/GaN technology. *IEEE Transactions on Microwave Theory and Techniques* 65 (11): 4428–4437.

17 Icoz, T. and Arik, M. (2010). Light weight high performance thermal management with advanced heat sinks and extended surfaces. *IEEE Transactions on Components, Packaging and Manufacturing Technology* 33 (1): 161–166.

18 Sabatino, D. and Yoder, K. (2014). Pyrolytic graphite heat sinks: a study of circuit board applications. *IEEE Transactions on Components, Packaging and Manufacturing Technology* 4 (6): 999–1009.

19 Samanta, K.K. (2016). PA thermal management and packaging. *IEEE Microwave Magazine* 17 (11): 73–81.

14

Receivers and Sub-Systems

14.1 Introduction

The superheterodyne (superhet) receiver is the most widely used receiver configuration for RF and microwave applications, and it is the main focus of this chapter. This book is concerned primarily with RF and microwave design, and consequently only the front-end aspects of receiver design will be considered. Noise is an important issue in the design of receiver circuits, and the chapter commences with a review of noise sources and their circuit specifications, leading to an analysis of the effects of noise in superhet receivers, and the use of noise budget graphs. Frequency mixers, which provide the frequency down-conversions in superhet designs, are often regarded as the critical components in the front-end of a superhet receiver, and the chapter concludes with a review of the various types of mixer used at RF and microwave frequencies.

14.2 Receiver Noise Sources

Noise refers to *random*, *unwanted* signals that occur *naturally* in an electronic system. In this section, we describe two of the most common sources of noise in electronic circuits, namely thermal noise and shot noise, and provide mathematical expressions that enable the noise to be quantified.

14.2.1 Thermal Noise

Thermal noise is due to the random motion of charge carriers in a conductor caused by thermal agitation. The random motion of the carriers produces small random currents, and consequently small random noise voltages. Thermal noise is sometimes referred to as Johnson noise, after J.B. Johnson, who observed the effect experimentally in the 1920s. Subsequently, H. Nyquist derived a theoretical expression for thermal noise based on quantum theory. The available thermal noise power from a resistive medium is obtained by considering the behaviour of a lossless transmission line, of characteristic impedance, R_O, and length, l, terminated at each end with the characteristic impedance, as shown in Figure 14.1. Also shown in this figure are two switches, that enable the terminating impedances to be shorted-out.

With the switches open P watts of noise power flow from each of the terminating resistors in a bandwidth B. At any moment the energy contained in the line is $2Pt$, where

$$2Pt = \frac{2Pl}{v_p},\tag{14.1}$$

and v_p is the velocity of propagation along the line, and t is the propagation time on the line.

With the switches closed the energy stored on the line is $2Pl/v_p$ Joules in bandwidth, B. We know from quantum mechanics that energy is stored in an infinite number of resonant modes, and that the energy (E_1) stored per resonant mode is given by Planck's law as

$$E_1 = \frac{hf}{e^{hf/kT} - 1},\tag{14.2}$$

where

f = frequency
h = Planck's constant ($= 6.6 \times 10^{-34}$ J)
k = Boltzmann's constant ($= 1.38 \times 10^{-23}$ J/K)

RF and Microwave Circuit Design: Theory and Applications, First Edition. Charles E. Free and Colin S. Aitchison.
© 2022 John Wiley & Sons Ltd. Published 2022 by John Wiley & Sons Ltd.
Companion website: www.wiley.com/go/free/rfandmicrowave

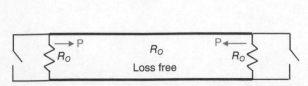

Figure 14.1 Lossless transmission line of characteristic impedance, R_O, terminated at both ends with matched impedances.

If we now consider the r_{th} resonance when the switches are closed; the wavelength associated with the r_{th} resonance is

$$\lambda_r = \frac{2l}{v_p},$$ (14.3)

and the corresponding frequency is

$$f_r = \frac{r v_p}{2l}.$$ (14.4)

The bandwidth, B_1, between the $(r+1)_{th}$ and r_{th} resonances is

$$B_1 = \frac{(r+1)v_p}{2l} - \frac{r v_p}{2l} = \frac{v_p}{2l}.$$ (14.5)

Now the number, N, of resonances within bandwidth, B, is

$$N = \frac{B}{B_1} = \frac{B}{v_p/2l} = \frac{2lB}{v_p}.$$ (14.6)

Therefore, the energy, E, associated with resonances in bandwidth, B, is

$$E = \frac{hf}{e^{hf/kT} - 1} \times \frac{2lB}{v_p}.$$ (14.7)

But from Eq. (14.1) the energy in a bandwidth, B, is also equal to $2Pl/v_p$. Thus Eq. (14.7) can be written as

$$\frac{2Pl}{v_p} = \frac{hf}{e^{hf/kT} - 1} \times \frac{2lB}{v_p},$$

i.e.

$$P = \frac{hf}{e^{hf/kT} - 1} \times B.$$ (14.8)

Now, using the standard expansion for an exponential series

$$e^{hf/kT} = 1 + \left(\frac{hf}{kT}\right) + \frac{1}{2!}\left(\frac{hf}{kT}\right)^2 + \frac{1}{3!}\left(\frac{hf}{kT}\right)^3 + \cdots.$$

Assuming $\frac{hf}{kT} \ll 1$ then $e^{hf/kT} \approx 1 + \left(\frac{hf}{kT}\right)$ and Eq. (14.8) becomes

$$P = kTB.$$ (14.9)

Equation (14.9) gives a very useful result, in that it shows that the maximum thermal noise power that can be extracted from any matched resistive source is given by kTB, and furthermore that thermal noise power can be reduced by reducing the temperature of the source and/or the operating bandwidth.

We can represent a resistor, of value R, at a temperature T, by a noise free resistor, R, in series with a noise voltage generator providing an RMS noise voltage, $\sqrt{\overline{v_n^2}}$. Figure 14.2 shows the series equivalent circuit of a resistor connected to a load resistor, R_L.

The noise power available from the resistor into a matched load, i.e. when $R_L = R$, is

$$P = \frac{\overline{v_n^2}}{4R}.$$ (14.10)

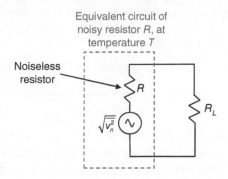

Figure 14.2 Series equivalent circuit of a noisy resistor R, at temperature T.

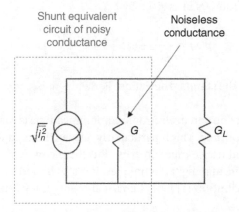

Figure 14.3 Shunt equivalent circuit of a noisy conductance at temperature T.

But we know the noise power is also given by Eq. (14.9), and so we have

$$\frac{\overline{v_n^2}}{4R} = kTB,$$

i.e.

$$\sqrt{\overline{v_n^2}} = \sqrt{4kTBR}. \tag{14.11}$$

Alternatively, we can represent the noisy resistor by a shunt equivalent circuit, consisting of an RMS current generator, $\sqrt{\overline{i_n^2}}$ in parallel with a noise free conductance, G, where $G = 1/R$. This is shown in Figure 14.3, where

$$\sqrt{\overline{i_n^2}} = \sqrt{4kTBG}. \tag{14.12}$$

Since thermal noise is the result of a large number of random events, namely the random motion of the electrons, it will have a uniform power distribution over a wide range of frequencies, and so it is often referred to as white noise (by analogy with white light). However, in communication systems the noise frequency spectrum is usually limited by filters, and more correctly the noise should be referred to as band-limited white noise.

Example 14.1   What is the open-circuit RMS thermal noise voltage from a $10\,\text{k}\Omega$ resistor measured in a bandwidth of $3\,\text{MHz}$, when the resistor is at a temperature of $20\,°\text{C}$, and what is the maximum noise power that could be extracted from the resistor?

Solution
The open circuit noise voltage is given by Eq. (14.11):

$$v_n = \sqrt{4 \times 1.38 \times 10^{-23} \times (20 + 273) \times 3 \times 10^6 \times 10^4}\,\text{V} = 22.03\,\mu\text{V}.$$

(Continued)

(Continued)

The maximum noise power is given by Eq. (14.9):

$$P_n = 1.38 \times 10^{-23} \times (20 + 273) \times 3 \times 10^6 \text{ W} = 0.012 \text{ pW.}$$

Since thermal noise is random, the probability of a particular noise level occurring needs to be specified statistically. Thermal noise has a Gaussian probability distribution (sometimes called a Normal distribution). This is based on the central limit theorem, which states that an event that depends on a large number of random individual events can be represented by a Gaussian distribution.

The Gaussian probability density function, $p(x)$, is given by

$$p(x) = \frac{1}{\sigma\sqrt{2\pi}} \exp\left(-\frac{(x-m)^2}{2\sigma^2}\right), \tag{14.13}$$

where m is the mean value of x, and σ is the standard deviation. For white noise, $m = 0$. The shape of a Gaussian probability distribution is shown in Figure 14.4.

The importance of the probability distribution in electronic communications is that it indicates the probability of a noise spike occurring that has a particular amplitude. This is particularly relevant for digital communications where a noise spike may be mistaken for a narrow pulse, and hence cause an error. But the theory is also important for systems such as pulse radar which involve a mixture of analogue and digital techniques. It is also important to note that the Gaussian distribution needs to be modified to a Rayleigh distribution [1] for systems containing filters, since the filters constrain the frequency spectrum.

Figure 14.4 shows a shaded area under the probability distribution, and this area gives the probability that noise will exceed a particular level, x_1. This can be written mathematically as

$$P(x > x_1) = \int_{x_1}^{\infty} p(x)\, \mathrm{d}x. \tag{14.14}$$

Since the total area under a probability distribution curve must be unity, we can write the probability that x will exceed x_1 as

$$P(x > x_1) = \frac{1}{2}\left(1 - \int_{-x_1}^{x} p(x)\, \mathrm{d}x\right). \tag{14.15}$$

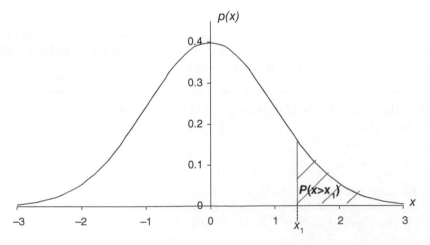

Figure 14.4 Gaussian probability distribution.

Substituting for $p(x)$ in Eq. (14.15) gives

$$P(x > x_1) = \frac{1}{2}\left(1 - \int_{-x_1}^{x_1} \frac{1}{\sigma\sqrt{2\pi}} \exp\left(-\frac{x^2}{2\sigma^2}\right) dx\right)$$

$$= \frac{1}{2}\left(1 - \frac{2}{\sigma\sqrt{2\pi}} \int_0^{x_1} \exp\left(-\frac{x^2}{2\sigma^2}\right) dx\right). \tag{14.16}$$

We can convert the integral in Eq. (14.16) into a standard form by making the following substitution

$$\frac{x^2}{2\sigma^2} = y^2. \tag{14.17}$$

Then, differentiating each side of Eq. (14.17), we have

$$\frac{2x}{2\sigma^2} dx = 2y\, dy. \tag{14.18}$$

Substituting Eq. (14.18) into Eq. (14.16), and making the appropriate change to the upper limit of the integral gives

$$P(x > x_1) = \frac{1}{2}\left(1 - \frac{2}{\sqrt{\pi}} \int_0^{x_1/\sigma\sqrt{2}} \exp(-y^2)\, dy\right). \tag{14.19}$$

The integral in Eq. (14.19) can be expressed in terms of a mathematical function known as an error function. This function is defined as follows:

$$\frac{2}{\sqrt{\pi}} \int_0^{x_1/\sigma\sqrt{2}} \exp(-y^2)\, dy = \mathrm{Erf}\left(\frac{x_1}{\sigma\sqrt{2}}\right), \tag{14.20}$$

where $\mathrm{Erf}\left(\dfrac{x_1}{\sigma\sqrt{2}}\right)$ denotes the error function whose argument is $\dfrac{x_1}{\sigma\sqrt{2}}$. The error function is a common mathematical function whose values are normally found in tabulated form. A table of error functions is given in Appendix 14.A.1.

Substituting the error function into Eq. (14.19) gives

$$P(x > x_1) = \frac{1}{2}\left[1 - \mathrm{Erf}\left(\frac{x_1}{\sigma\sqrt{2}}\right)\right]. \tag{14.21}$$

Equation (14.21) is sometimes written in terms of the complementary error function, Erfc, which is defined as

$$\mathrm{Erfc}(\alpha) = 1 - \mathrm{Erf}(\alpha). \tag{14.22}$$

So Eq. (14.21) can also be written in the form

$$P(x > x_1) = \frac{1}{2}\left[\mathrm{Erfc}\left(\frac{x_1}{\sigma\sqrt{2}}\right)\right]. \tag{14.23}$$

Example 14.2 If the RMS thermal noise in a system is 0.7 μV, what is the probability of the noise level exceeding 2 μV at any instant of time?

Solution
Using Eq. (14.21):

$$P(x > 2\,\mu V) = \frac{1}{2}\left[1 - \mathrm{Erf}\left(\frac{2}{0.7\sqrt{2}}\right)\right] = \frac{1}{2}[1 - \mathrm{Erf}(2.02)].$$

Using the error function table in Appendix 14.A.1:

$$\mathrm{Erf}[2.02] = 0.995719.$$

Therefore,

$$P(x > 2\,\mu V) = \frac{1}{2}[1 - 0.995719] = 2.14 \times 10^{-3}.$$

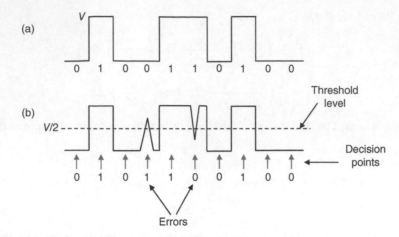

Figure 14.5 (a) Simple digital voltage waveform, and (b) Digital waveform with noise voltage spikes.

The practical significance of this noise theory can be seen by considering the detection of digital signals in the presence of thermal noise. Figure 14.5a shows a simple baseband digital signal, where the '1' state is represented by a unipolar pulse of amplitude V, and the '0' state by a space. In the receiver of a system using this type of digital format the information is recovered using a centre-sampler detector. This type of detector inspects the centre point of each pulse position to determine whether the signal level is above or below a threshold level, set at $V/2$.

Figure 14.5b shows a somewhat simplified representation of how noise spikes can cause errors in the detection process. In this simple system, errors will occur when the amplitude of the noise spike exceeds $V/2$. It can be seen that a positive noise spike at a decision point will cause a '0' to be detected as a '1', and negative spike will cause a '1' to be detected as a '0'. Since this is a baseband system without any frequency constraints, we can assume the noise has a Gaussian distribution. The probability of an error occurring in a long series of '1's and '0's is then

$$P_e = P(0) \times P\left(x > \frac{V}{2}\right) + P(1) \times P\left(x < \frac{V}{2}\right), \tag{14.24}$$

where $P(0)$ and $P(1)$ represent the probabilities of occurrence of the '0' and '1' states, respectively. For normal digital communications $P(0) = P(1) = 0.5$, and since the probability density function is symmetric about the origin

$$P\left(x > \frac{V}{2}\right) = P\left(x < \frac{V}{2}\right) = \frac{1}{2}\left[1 - \text{Erf}\left(\frac{V/2}{\sigma\sqrt{2}}\right)\right]. \tag{14.25}$$

Then

$$P_e = \frac{1}{2}\left[1 - \text{Erf}\left(\frac{V/2}{\sigma\sqrt{2}}\right)\right]. \tag{14.26}$$

The situation is slightly more complicated in RF receivers, where the digital information has been modulated onto a high-frequency carrier. The receiver will contain filters, and so we must consider the effects of band-limited Gaussian noise. Limiting the frequency range of the noise will change its characteristics, and band-limited Gaussian noise can be represented by a Rayleigh distribution, whose probability density function is given by

$$p(x) = \frac{x}{\sigma^2}\exp\left(-\frac{x^2}{2\sigma^2}\right), \tag{14.27}$$

where σ is the RMS value of the noise.

One of the simplest forms of digital modulation is the on–off keying scheme, where the '0' state is represented by a space, and the '1' state by a burst of the carrier signal; we must use different noise probability distributions to find the probability of an error occurring in each state. For the '0' state, where there is no signal, just band-limited noise we must use the Rayleigh distribution to find the probability of an error occurring. But for the '1' state, where the noise is superimposed on a carrier, we need to consider the instantaneous values of the envelope of the signal plus noise, and this is described by a Rician probability distribution [1]. The Rician probability distribution is given by

$$p(x) = \frac{x}{\sigma^2} \exp\left(-\frac{x^2 + V^2}{2\sigma^2}\right) I_O\left(\frac{xV}{\sigma^2}\right),$$

(14.28)

where V is the amplitude of the carrier, and $I_O\left(\frac{xV}{\sigma^2}\right)$ represents a modified Bessel function of zero order, whose argument is $\frac{xV}{\sigma^2}$. So it is important to use the appropriate probability distribution in error rate calculations.

A simple pulse radar system is another example of where the Rayleigh and Rician distributions can be applied in practice. This type of radar system transmits pulses of a high-frequency carrier and looks for reflections to identify targets. The noise theory principles that apply to on–off keying can be used directly to determine the probability of errors in target detection.

Example 14.3 A simple baseband digital communication system transmits data in the form of unipolar pulses of amplitude 3 mV. The information is recovered using a centre-sampler detector. If the RMS thermal noise voltage is 1.4 mV, find the probability of an error occurring in the detection process, assuming that the probabilities of occurrence of '0' and '1' are equal.

Solution
The threshold level is 1.5 mV, and an error will occur if there is a noise spike greater than this value. Hence, using Eq. (14.26):

$$P(x > 1.5\text{ mV}) = \frac{1}{2}\left[1 - \text{Erf}\left(\frac{1.5}{1.4\sqrt{2}}\right)\right] = \frac{1}{2}[1 - \text{Erf}(0.76)].$$

Using the error function table in Appendix 14.A.1:

$$\text{Erf}[0.76] = 0.717537.$$

Therefore, the probability of an error occurring is:

$$P_e = \frac{1}{2}[1 - 0.717537] = 141.23 \times 10^{-3}.$$

Example 14.4 The signal-to-noise ratio at the input to the centre-sampler detector is a simple baseband digital communication system using unipolar pulses is 6 dB. Calculate the probability of an error occurring in the detection process.

Solution
If the pulses have an amplitude of V, the average value will be $V/2$, assuming equal occurrences of the two digital states. The signal-to-noise ratio is then

$$\left(\frac{V/2}{\sigma}\right)^2 = 10^{0.6}.$$

Substituting into Eq. (14.26) gives

$$P_e = \frac{1}{2}\left[1 - \text{Erf}\left(\frac{10^{0.3}}{\sqrt{2}}\right)\right] = \frac{1}{2}[1 - \text{Erf}(1.41)].$$

Using the error function table in Appendix 14.A.1:

$$\text{Erf}[1.41] = 0.95385.$$

Therefore, the probability of an error occurring is:

$$P_e = \frac{1}{2}[1 - 0.95385] = 2.3 \times 10^{-2}.$$

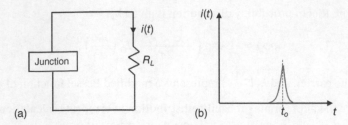

Figure 14.6 Current pulse due to shot noise: (a) PN junction connected to load resistor. (b) Current pulse.

14.2.2 Semiconductor Noise

The main source of noise in semiconductor devices is shot noise. This occurs due to the random motion of charge carriers crossing a potential barrier, such as a *p–n* junction. This gives rise to a randomly fluctuating current at the terminals of the junction. We can determine an expression for the fluctuating current by first considering the effect of a single electron flowing from cathode to anode across a semiconductor junction, assuming that here are no space charge[1] effects. Figure 14.6a shows a semiconductor junction connected to an external resistor, R.

The effect of the single electron flowing across the junction will be to produce an impulse of current, $i(t)$, in the external circuit. The impulse is sketched in Figure 14.6b.

The exact shape of the impulse is not significant, and it can be regarded as a Dirac delta function, $\delta(t-t_O)$, where t_O represents the position of the impulse in the time domain, as shown in Figure 14.6b. The impulse can then be represented by the function

$$f(t) = q\delta(t - t_O), \tag{14.29}$$

where q is the unit of electronic change $(1.6 \times 10^{-19}\ °C)$, and $\delta(t-t_O)$ is the Dirac delta function.

In general, the relationship between a function, $f(t)$, in the time domain and in the frequency domain, $g(\omega)$, is

$$g(\omega) = \int_{-\infty}^{+\infty} f(t)\, e^{-j\omega t}\, \mathrm{d}t. \tag{14.30}$$

Thus, for the current impulse we have

$$g(\omega) = \int_{-\infty}^{+\infty} q\, \delta(t - t_O)\, e^{-j\omega t}\, \mathrm{d}t$$

$$= q \int_{-\infty}^{+\infty} \delta(t - t_O)\, e^{-j\omega t}\, \mathrm{d}t$$

$$= q\, e^{-j\omega t}, \tag{14.31}$$

and so

$$|g(\omega)| = |q|. \tag{14.32}$$

We can now use Parseval's theorem to calculate the energy dissipated in resistance, R, due to current, $i(t)$, which has a transfer function, $g(\omega)$, i.e.

$$E = R \int_{-\infty}^{\infty} |g(\omega)|^2\, \mathrm{d}f$$

$$= 2R \int_{0}^{\infty} q^2\, \mathrm{d}f$$

$$= 2Rq^2 B. \tag{14.33}$$

---

1 The space-charge region at the junction of two semiconductors is an alternative name for the depletion region; the presence of the space charge will affect the flow of charge carriers across the junction.

If there are N electrons flowing from the junction in a total time, T, the power, P, dissipated in the resistor will be given by

$$P = 2Rq^2 \times \frac{N}{T}. \tag{14.34}$$

If we now consider the junction to be replaced by a generator providing an RMS current of $\sqrt{\overline{i_S^2}}$ the power dissipated in the resistor will be given by

$$P = \overline{i_S^2}R. \tag{14.35}$$

Combining Eq. (14.34) with Eq. (14.35) gives

$$\overline{i_S^2}R = 2Rq^2B \times \frac{N}{T},$$

i.e.

$$\overline{i_S^2} = 2q^2B \times \frac{N}{T}. \tag{14.36}$$

Now the DC current, I_{DC}, flowing from the junction will be

$$I_{DC} = \frac{Nq}{T}. \tag{14.37}$$

Finally, substituting the DC current into Eq. (14.36) gives the RMS shot noise current from the junction as

$$\sqrt{\overline{i_S^2}} = \sqrt{2qI_{DC}B}. \tag{14.38}$$

Two further significant sources of noise in semiconductors are partition noise and flicker noise. Partition noise occurs in semiconductors due to the random division of carriers at junctions; for example, in a bipolar junction transistor there is a random distribution of carriers between the base–emitter and base–collector junctions. Flicker noise is a low-frequency effect associated with the crystal defects in a semiconductor material. At low frequencies, the magnitude of flicker noise tends to be proportional to the inverse of the frequency, and so flicker noise is often referred to as 1/f noise.

Example 14.5 The DC current through a semiconductor junction is 10 mA. Calculate:

(i) The RMS shot noise current from the junction in an 8 MHz bandwidth.
(ii) The shot noise power dissipated in a 15 kΩ resistor shunting the junction as shown in Figure 14.6a.

Solution

(i)
$$\sqrt{\overline{i_S^2}} = \sqrt{2 \times 1.6 \times 10^{-19} \times 10 \times 10^{-3} \times 8 \times 10^6} \text{ A}$$

$$= 0.16 \text{ μA}.$$

(ii)
$$P = \overline{i_S^2}R = (0.16 \times 10^{-6})^2 \times 15 \times 10^3 \text{ W}$$

$$= 384 \text{ pW}.$$

14.3 Noise Measures

14.3.1 Noise Figure (*F*)

The noise performance of a two-port transducer is represented by the noise figure (F), sometimes referred to as the noise factor. The two terms 'noise figure' and 'noise factor' are synonymous. The fundamental definition of noise figure is that given by the Institute of Radio Engineers (IRE) Standards Committee [2]. The IRE definition can be expressed as follows:

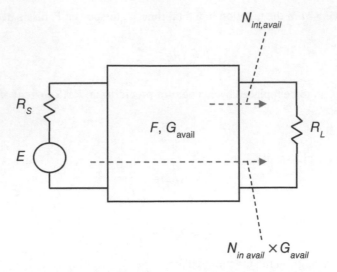

Figure 14.7 Noise components at the output of a noisy network.

At a specified input frequency the noise figure, F, is given by the following ratio:

$$F = \frac{\text{(1) The total noise power per unit bandwidth (at a corresponding output frequency) available at the output.}}{\text{(2) That portion of (1) engendered at the input frequency by the input termination at the standard noise temperature.}} \tag{14.39}$$

Four points should be noted about the definition given in Eq. (14.39):

(i) The standard noise temperature is 290 K.
(ii) For heterodyne systems there will be, in principle, more than one output frequency corresponding to a single input frequency and vice versa; for each pair of corresponding frequencies a noise figure is defined.
(iii) The phrase 'available at the output' may be replaced by 'delivered by system into an output termination'.
(iv) To characterize a system by a noise figure is only meaningful when the input termination is specified.

A two-port transducer having a noise figure, F, and an available gain, G_{avail}, is shown in Figure 14.7.

The input port is terminated by a generator, E, that has an internal resistance, R_S, at a temperature of T_O. The output of the transducer is terminated by a resistance, R_L. The noise available at the output of the transducer, $N_{out,avail}$, will be the sum of the noise generated internally in the transducer, $N_{int,avail}$, and the input noise, $N_{S,avail}$, amplified by the gain of the transducer, i.e.

$$N_{out,avail} = N_{in,avail} \times G_{avail}, \tag{14.40}$$

and $N_{in,avail}$ is the noise power available from the generator.

Applying the definition of noise figure given by Eq. (14.39) we have

$$\cdot F = \frac{N_{int,avail} + (N_{in.avail} \times G_{avail})}{(N_{in.avail} \times G_{avail})} = \frac{N_{out,avail}}{(N_{in.avail} \times G_{avail})}. \tag{14.41}$$

Now we can write G_{avail} in terms of the signal powers available at the input and output of the transducer, as,

$$G_{avail} = \frac{S_{out,avail}}{S_{in,avail}}. \tag{14.42}$$

Combining Eqs. (14.41) and (14.42) gives

$$F = \frac{(S/N)_{in,avail}}{(S/N)_{out,avail}}. \tag{14.43}$$

Where the powers are dissipated in the load, rather than available at the output, we can write Eq. (14.43) as

$$F = \frac{(S/N)_{in}}{(S/N)_{out}}. \tag{14.44}$$

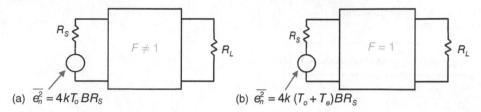

Figure 14.8 Equivalent networks using noise figure and noise temperature: (a) Network with noise. (b) Noiseless network.

Two important points to note:

(i) F is unity for a noise-free network and greater than unity for a practical network.

(ii) Equations (14.43) and (14.44) are only true if the source impedance is at a temperature of 290 K.

14.3.2 Noise Temperature (T_e)

Noise temperature is an alternative parameter to noise figure used to represent noise generated within a device or circuit; noise temperature and noise figure are directly related.

Figure 14.8a shows a two-port transducer with a resistor, R_S, at a temperature, T_O, terminating the input port. For convenience, we have represented the resistor as a series combination of a noiseless resistor, and a noise generator with a mean square output of $\overline{e_S^2}$ to represent the thermal noise produced by the resistor.

From the previous thermal noise theory

$$\overline{e_S^2} = 4kT_O BR_S. \tag{14.45}$$

We can rewrite Eq. (14.41) as

$$F = \frac{G_{avail}kT_O B + N_{int,avail}}{G_{avail}kT_O B}. \tag{14.46}$$

We can represent the noise generated within the transducer by an increase, T_e, in the source temperature, and then consider the transducer to be noiseless. This situation is depicted in Figure 14.8b, where the noiseless transducer has $F = 1$. Then Eq. (14.46) becomes

$$F = \frac{G_{avail}kT_O B + G_{avail}kT_e B}{G_{avail}kT_O B}$$

$$= 1 + \frac{T_e}{T_O} \tag{14.47}$$

or

$$T_e = (F - 1)T_O. \tag{14.48}$$

The standard source temperature is 290 K, and so Eq. (14.48) is usually written as

$$T_e = (F - 1)290 \text{ K}. \tag{14.49}$$

Example 14.6 An RF component has a noise figure of 3.42 dB. Calculate the noise temperature of the component.

Solution

$$F_{dB} = 3.42 \text{ dB} = 10 \log(F) \text{ dB} \quad \Rightarrow \quad F = 2.2.$$

Using Eq. (14.49):

$$T_e = (2.2 - 1) \times 290 \text{ K} = 348 \text{ K}.$$

Table 14.1 illustrates the relationship between F and T_e for a range of typical values.

Table 14.1 Typical values of noise figure and corresponding noise temperature.

Noise figure, F	$10\log F$ (dB)	Noise temperature, T_e (K)
1	0	0
2	3.0	290
3	4.8	580
4	6.0	870
5	7.0	1160
10	10.0	2610

Noise temperature tends to be a more numerically convenient parameter for representing noise when dealing with antennas and low-noise receiving systems.

Methods for measuring the noise figure of a two-port network are discussed in Appendix 14.A.2.

14.4 Noise Figure of Cascaded Networks

A cascade of two networks is shown in Figure 14.9. The source resistance is R_S at the standard temperature T_O, and the load resistance is R_L.

The overall noise figure, F_O, of the cascade is given by

$$F_O = \frac{N_{R_L}}{kT_O BG_{avail,1}G_{avail,2}},$$
(14.50)

where N_{R_L} is the total noise power available at the output of the second network, and where

$$N_{R_L} = N_{int,avail,2} + N_{out,avail,1}G_{avail,2}$$
$$= N_{int,avail,2} + (kT_O BG_{avail,1} + N_{int,avail,1})G_{avail,2}.$$
(14.51)

Substituting for N_{R_L} in Eq. (14.50) gives

$$F_O = \frac{N_{int,avail,2} + (kT_O BG_{avail,1} + N_{int,avail,1})G_{avail,2}}{kT_O BG_{avail,1}G_{avail,2}}.$$
(14.52)

Applying the definition of noise figure to the individual networks gives

$$F_1 = \frac{kT_O BG_{avail,1} + N_{int,avail,1}}{kT_O BG_{avail,1}} = 1 + \frac{N_{int,avail,1}}{kT_O BG_{avail,1}}$$
(14.53)

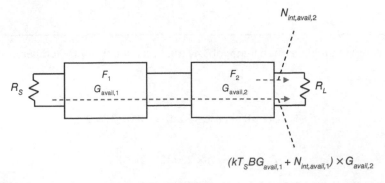

Figure 14.9 Networks in cascade.

and

$$F_2 = \frac{kT_O BG_{avail,2} + N_{int,avail,2}}{kT_O BG_{avail,2}} = 1 + \frac{N_{int,avail,2}}{kT_O BG_{avail,2}}. \tag{14.54}$$

Substituting from Eqs. (14.53) and (14.54) into Eq. (14.52) gives the overall noise figure of the two networks in cascade as

$$F_O = F_1 + \frac{F_2 - 1}{G_{avail,1}}. \tag{14.55}$$

A straightforward extension of the above analysis gives the noise figure of N networks in cascade as

$$F_{cascade} = F_1 + \frac{F_2 - 1}{G_{avail,1}} + \frac{F_3 - 1}{G_{avail,1} G_{avail,2}} + \cdots + \frac{F_N - 1}{G_{avail,1} G_{avail,2} \cdots G_{avail,(N-1)}}. \tag{14.56}$$

The expression for the overall noise figure of a number of units in cascade, given in Eq. (14.56), is very important for analyzing the noise performance of practical systems; Eq. (14.56) is also known as the Friis noise equation.

For situations in which the networks are all matched at the input and output ports, Eq. (14.56) can be written as

$$F_{cascade} = F_1 + \frac{F_2 - 1}{G_1} + \frac{F_3 - 1}{G_1 G_2} + \cdots + \frac{F_N - 1}{G_1 G_2 \cdots G_{N-1}}, \tag{14.57}$$

without specifying whether the gains are available gain, transducer gain, or power gain, since the gains are then numerically identical. (*The differences between the definitions of gain were discussed in Chapter 5.*)

Using Eq. (14.49), we can convert Eq. (14.57) into an expression for the overall noise temperature of the cascade of (matched) networks as

$$T_{e,cascade} = T_{e1} + \frac{T_{e2}}{G_1} + \frac{T_{e3}}{G_1 G_2} + \cdots + \frac{T_{eN}}{G_1 G_2 \cdots G_{N-1}}, \tag{14.58}$$

where T_{en} is the noise temperature of an individual network.

Four important points can be deduced from Eq. (14.56) through Eq. (14.58):

(i) The noise is dominated by the first units in the cascade.
(ii) High gain in the initial units will have the effect of reducing the noise contributions from subsequent units.
(iii) Ideally, the first unit should have low noise, to keep T_{e1} small, and high gain to reduce the noise of subsequent units.
(iv) The gain of the last network in the cascade does not affect the overall noise figure and temperature.

Example 14.7 The following data apply to three matched amplifiers connected in cascade:

Amplifier 1:	Gain = 8 dB	Noise figure = 1.9 dB
Amplifier 2:	Gain = 15 dB	Noise figure = 2.6 dB
Amplifier 3:	Noise figure = 3.1 dB	

Determine the overall noise figure, and the corresponding overall noise temperature.

Solution

Amplifier 1: $G_1 = 10^{0.8} = 6.31$ $\quad F_1 = 10^{0.19} = 1.55$
Amplifier 2: $G_2 = 10^{1.5} = 31.62$ $\quad F_2 = 10^{0.26} = 1.82$
Amplifier 3: $F_3 = 10^{0.31} = 2.04$

Substituting in Eq. (14.57):

$$F_{overall} = 1.55 + \frac{1.82 - 1}{6.31} + \frac{2.04 - 1}{6.31 \times 31.62} = 1.69,$$

$$F_{overall}(dB) = 10 \log(1.69) \, dB = 2.28 \, dB.$$

Using Eq. (14.49):

$$T_{e,overall} = (1.69 - 1) \times 290 \, K = 200.10 \, K.$$

14.5 Antenna Noise Temperature

Antenna noise refers to noise which is received (or picked-up) by the antenna, rather than noise which is generated within the antenna materials. The latter noise is very small since antennas are usually made from high conductivity materials. The noise which is received by an antenna is represented by an effective noise temperature, known simply as the antenna noise temperature, which is not related to the physical temperature of the antenna.

There are two main contributions to antenna noise, namely the noise from the sky and the noise from the Earth. Sky noise, represented by the sky noise temperature T_{sky}, is due to noise radiated from the galaxy and also noise originating in the atmosphere. Figure 14.10 shows the typical variation of sky noise temperature as a function of frequency. Galactic noise decreases rapidly as the frequency approaches 1 GHz, whereas atmospheric noise gradually increases above 1 GHz, with a first peak at 23 GHz, due to resonance of the water molecules in the atmosphere, and a further peak at 60 GHz, due to resonance of the oxygen molecules. The magnitudes of these peaks, particularly at 23 GHz, are very dependent on physical conditions. It can be seen from Figure 14.10 that there is a low sky noise region, approximately between 1 GHz and 10 GHz, where the sky noise temperature is of the order of 10 K. This low noise region is often referred to as the microwave window. The noise temperature within this window is very dependent on the elevation angle of the antenna's main beam, since this varies the path length in the atmosphere; at high angles of beam elevation the noise temperature within the microwave window can be as low as 5 K.

The Earth, at a typical ambient temperature of 290 K, constitutes a hot body that radiates thermal noise. The proportions of sky and Earth noise that are picked up by an antenna depend on the antenna's radiation pattern. The antenna noise temperature, T_{ae}, can be written as

$$T_{ae} = xT_{sky} + yT_{Earth}, \tag{14.59}$$

where x and y are the fractions of the antenna's radiation pattern that view the sky and Earth, respectively. If we consider the Earth to have a surface ambient temperature of 290 K, Eq. (14.59) can be rewritten as

$$T_{ae} = xT_{sky} + 290y. \tag{14.60}$$

Example 14.8   Determine the antenna noise temperature of an antenna where 95% of the polar diagram intercepts a 22 K sky, and 5% of the polar diagram intercepts the Earth.

Solution

$$T_{ae} = (0.95 \times 22 + 0.05 \times 290) \text{ K}$$
$$= (20.9 + 14.5) \text{ K}$$
$$= 35.4 \text{ K}$$

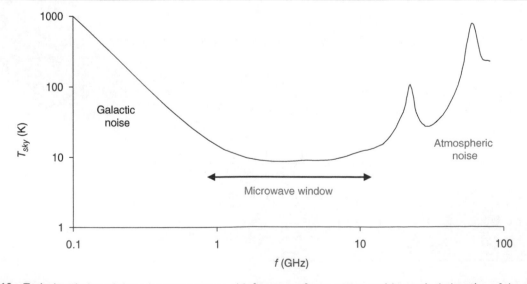

Figure 14.10   Typical variation of sky noise temperature with frequency, for an antenna with a typical elevation of the main beam.

Example 14.8 shows that although only 5% of the radiation pattern touches the Earth, this makes a major contribution to the overall noise temperature of the antenna. Clearly it is very important for low-noise wireless receivers to design and use antennas in which the side-lobes make minimum contact with the Earth.

14.6 System Noise Temperature

The system noise temperature, denoted by T_{sys}, is the sum of the noise temperature of the antenna and the noise temperature of the cascaded units that comprise the receiver, i.e.

$$T_{sys} = T_{ae} + T_{cascade}$$

$$= T_{ae} + T_{e1} + \frac{T_{e2}}{G_1} + \frac{T_{e3}}{G_1 G_2} + \cdots \frac{T_{eN}}{G_1 G_2 \cdots G_{N-1}}. \tag{14.61}$$

The noise power which is available at the output of the receiver system is then $kT_{sys}BG_{avail}$, where G_{avail} is the available gain of the network following the antenna.

14.7 Noise Figure of a Matched Attenuator at Temperature T_O

A matched attenuator is shown in Figure 14.11.

The attenuator has input and output impedances of R_O, and is connected between source and load impedances of the same value. It is assumed that the attenuator and the source and load impedances are all at the same reference temperature, T_O. The attenuator has a loss factor, α.

The noise power flowing into the matched attenuator from the source will be

$$P_{in} = kT_O B. \tag{14.62}$$

The noise power flowing out of the attenuator will be the sum of the noise due to the source, $\dfrac{kT_O B}{\alpha}$, plus the noise generated within the matched attenuator, $xkT_O B$, i.e.

$$P_{out} = \frac{kT_O B}{\alpha} + xkT_O B. \tag{14.63}$$

For thermal equilibrium the noise power flowing out must equal the noise power flowing in, i.e.

$$\frac{kT_O B}{\alpha} + xkT_O B = kT_O B \tag{14.64}$$

and so

$$x = 1 - \frac{1}{\alpha} \tag{14.65}$$

and

$$P_{out} = \frac{kT_O B}{\alpha} + \left(1 - \frac{1}{\alpha}\right) kT_O B. \tag{14.66}$$

Applying the definition of noise figure, F, given in Eq. (14.39) we have

$$F = \frac{\dfrac{kT_O B}{\alpha} + \left(1 - \dfrac{1}{\alpha}\right) kT_O B}{\dfrac{kT_O B}{\alpha}}, \tag{14.67}$$

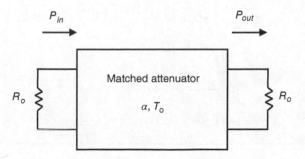

Figure 14.11 Matched attenuator.

i.e.

$$F = \alpha, \tag{14.68}$$

and so we see that the noise figure of a matched attenuator at the reference temperature, T_O, is numerically equal to the loss factor. It follows that for a matched lossy cable at the reference temperature, T_O, will have a noise figure equal to the cable loss factor.

This is an important result for receiver design, and shows that for the purpose of noise analysis the feeder cable between the antenna and the receiver must be treated as the first stage in the receiver cascade. A high feeder loss means high noise figure, or high noise temperature, for the first stage, and since this is not divided by any gain factors it will have a significant effect on the overall noise of the receiver. Moreover, having a lossy feeder as the first stage will effectively magnify the noise from subsequent stages, since the noise temperatures of these stages will be divided by the gain of the feeder, which is less than unity.

It follows from Eqs. (14.49) and (14.68) that the noise temperature, T_f, of a feeder cable whose physical temperature is 290 K will be

$$T_f = (\alpha - 1)290 \text{ K}, \tag{14.69}$$

where α is the loss factor for the cable.

Example 14.9   Figure 14.12 shows an antenna connected to a receiver through a matched feeder cable, which is at temperature T_O.

The data for the arrangement shown in Figure 14.12 are:

Antenna:	92% of radiation pattern sees a 19 K sky
	8% of radiation pattern sees a 290 K Earth
Feeder:	Loss = 3.2 dB
Receiver:	Noise figure = 6.1 dB

(i) Determine the system noise temperature.
(ii) Determine the system noise temperature if a pre-amplifier, having a gain of 12 dB and a noise figure of 2.08 dB, is inserted between the antenna and the feeder. Comment on the result.

Solution

(i) Using Eq. (14.59) to find the antenna noise temperature:

$$T_{ae} = (0.92 \times 19 + 0.08 \times 290) \text{ K} = 40.68 \text{ K}.$$

Feeder noise figure $\equiv$ feeder loss = 3.2 dB $\Rightarrow$ F = $10^{0.32}$
Using Eq. (14.69) to find the noise temperature of the feeder:

$$T_f = (10^{0.32} - 1) \times 290 \text{ K} = 315.90 \text{ K}.$$

Using Eq. (14.49) to find the receiver noise temperature:

$$T_{Rx} = (10^{0.61} - 1) \times 290 \text{ K} = 891.40 \text{ K}.$$

Using Eq. (14.61) to find the system noise temperature:

$$T_{sys} = \left(40.68 + 315.90 + \frac{891.40}{10^{-0.32}} \right) \text{ K} = 2218.98 \text{ K}.$$

Figure 14.12   Receiving system for Example 14.9.

(ii) Pre-amplifier:
$$G_{dB} = 12 \text{ dB} \quad \Rightarrow \quad G_{pa} = 10^{1.2}$$
$$F_{dB} = 2.08 \text{ dB} \quad \Rightarrow \quad T_{pa} = (10^{0.208} - 1) \times 290 \text{ K} = 178.16 \text{ K}.$$

Modifying the expression for system noise temperature to include the pre-amplifier, and noting that the pre-amplifier is inserted before the feeder:

$$T_{sys} = T_{ae} + T_{pa} + \frac{T_f}{G_{pa}} + \frac{T_{Rx}}{G_{pa}G_f}$$
$$= \left(40.68 + 178.16 + \frac{315.90}{10^{1.2}} + \frac{891.40}{10^{1.2}10^{-0.32}}\right) \text{ K}$$
$$= 356.28 \text{ K}.$$

Comment: The presence of the pre-amplifier has significantly reduced the system noise temperature. In general, an amplifier located before a lossy network will reduce the noise contribution from the network to the overall noise.

14.8 Superhet Receiver

The superheterodyne receiver (normally abbreviated to superhet receiver) is the most common form of receiver architecture used at RF and microwave frequencies. The essential principle of the superhet receiver is that the received frequency is down-converted to a lower frequency, known as the intermediate frequency (IF), where most of the amplification and detection takes place. The IF is a fixed frequency, and this enables high quality circuits to be designed, with good selectivity.

14.8.1 Single-Conversion Superhet Receiver

The essential structure of a simple, single-conversion superhet receiver is shown in Figure 14.13, with the input of the receiver connected to an antenna.

The key functions of the units shown in Figure 14.13 are:

1. A tunable, low-noise RF amplifier boosts the received signal, which has a frequency f_{RF}, and this is then applied to one input of a frequency mixer.
2. The output of a local oscillator (LO) with a frequency f_{LO} is applied to the other input of the frequency mixer.
3. The frequency mixer contains a non-linear device which generates all possible combinations of the two input frequencies, i.e.

$$f_{mixer\ o/p} = mf_{RF} \pm nf_{LO}, \tag{14.70}$$

where m and n are integers.

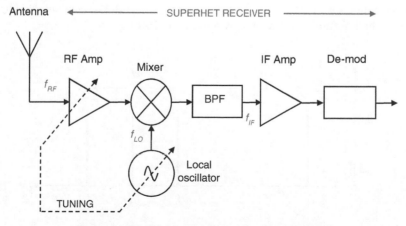

Figure 14.13 Single-conversion superhet receiver.

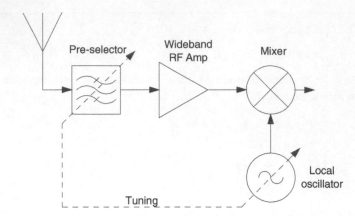

Figure 14.14 Front-end of superhet receiver modified to include pre-selector circuit.

4. The IF filter has a band-pass frequency response, and selects the difference frequency from the comb of frequencies at the mixer output, i.e.

$$f_{IF} = |f_{RF} - f_{LO}|. \tag{14.71}$$

5. The IF amplifier is a high quality, high gain, fixed-frequency amplifier that amplifies the signal prior to detection.
6. In order to receive different channels both the RF amplifier and the LO must be tuned so that the difference in their frequencies is always equal to the IF value. Often the tuning controls of these two units are linked (or ganged) so that only a single manual or electronic tuning element is needed.

The disadvantage of the simple receiver shown in Figure 14.13 is that in addition to being tunable the RF amplifier must have a relatively narrow frequency response so that the receiver can reject unwanted frequencies and channels picked up by the antenna. To overcome this disadvantage many superhet receivers include a pre-selector circuit between the antenna and the RF amplifier, as shown in Figure 14.14.

The pre-selector circuit has a band-pass filter response, whose centre frequency is tunable so as to reject unwanted frequencies from the antenna. This enables the RF amplifier to be designed at a fixed frequency, although it must have a wideband response so as to cover the desired frequency range of the receiver.

14.8.2 Image Frequency

The key aspect of a superhet receiver is that the frequency difference between f_{RF} and f_{LO} should be equal to f_{IF}. The LO can be above or below the RF frequency; normally f_{LO} is above f_{RF} because this reduces the required tuning range of the LO and RF amplifier. If we suppose that $f_{LO} > f_{RF}$ there will be another frequency on the other side of the LO frequency which will produce the correct IF when mixed with the LO signal. Since the frequency spacing between this other frequency and the LO frequency must also equal the IF value, it is known as the image frequency, and denoted by f_I. The position of the image frequency relative to the RF and LO values is shown in Figure 14.15.

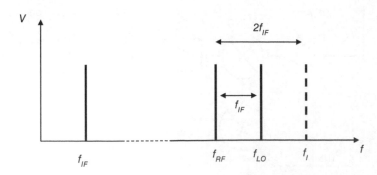

Figure 14.15 Position of the image frequency.

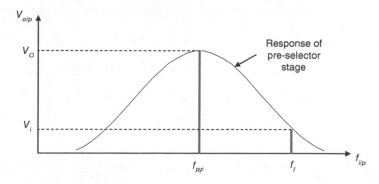

Figure 14.16 Typical response of pre-selector circuit, showing image suppression.

It is clear from Figure 14.15, that the frequency separation of the wanted RF signal and the image frequency is equal to $2f_{IF}$. One of the key functions of the pre-selector shown in Figure 14.14 is to pass the RF signal and block the image frequency. If the image frequency is allowed to reach the mixer, it will generate the correct IF frequency and cause significant interference. The pre-selector can be a narrow-band, band-pass filter, although higher suppression of the image frequency can be obtained by using a resonant circuit having a high Q-factor. Figure 14.16 shows a typical response for a pre-selector.

Figure 14.16 shows the output voltage of the pre-selector circuit plotted against the input frequency. The circuit is tuned to the f_{RF} so that the image frequency lies well down on the skirt of the response. The image suppression is then given by the expression

$$20 \log \left(\frac{V_O}{V_i} \right) \text{ dB}, \tag{14.72}$$

where V_i and V_O are the magnitudes of the input and output voltages of the pre-selector, respectively.

If a resonant circuit is used for the pre-selector stage, the image suppression can be increased by increasing the Q of the resonant circuit. But some caution is needed if a very high Q value is used, since this will also narrow the bandwidth of the circuit, and may adversely affect the sideband frequencies of the wanted signal centred on f_{RF}.

14.8.3 Key Figures of Merit for a Superhet Receiver

SENSITIVITY	The sensitivity is defined as the minimum power of the carrier at the input needed to produce a satisfactory output, which is normally taken to be a S/N ratio of 15 dB.
SELECTIVITY	This is a measure of the ability of the receiver to discriminate between adjacent channels. It can be specified in terms of the crosstalk between adjacent channels, although there is no specific value that is normally quoted for this parameter. The receiver will normally have an overall frequency response similar in shape to that of a band-pass filter, and the selectivity will depend on the quality of the roll-off at the edges of the frequency response, with a sharper roll-off leading to better selectivity.
ICRR	The Image Channel Rejection Ratio is a measure of the ability of the receiver to reject the image frequency. A value for ICRR can be calculated using Eq. (14.72).

Example 14.10 An RF superhet receiver is tuned to 145 MHz. If the IF frequency is 10.5 MHz, calculate two possible values for the LO frequency, and calculate the image frequency in each case.

Solution
LO frequency on lower side of RF frequency:

$$f_{LO} = f_{RF} - f_{IF} = 145 \text{ MHz} - 10.5 \text{ MHz} = 134.5 \text{ MHz},$$

$$f_I = f_{RF} - 2f_{IF} = 145 \text{ MHz} - 2 \times 10.5 \text{ MHz} = 124 \text{ MHz}.$$

(Continued)

(Continued)

LO frequency on upper side of RF frequency:

$$f_{LO} = f_{RF} + f_{IF} = 145 \text{ MHz} + 10.5 \text{ MHz} = 155.5 \text{ MHz},$$

$$f_I = f_{RF} + 2f_{IF} = 145 \text{ MHz} + 2 \times 10.5 \text{ MHz} = 166 \text{ MHz}.$$

Example 14.11 Figure 14.17 shows a band-pass filter used as a pre-selector circuit between the antenna and the first mixer in a superhet receiver.

The pre-selector circuit in this example has the characteristics of a resonant circuit, with the output voltage, V, at a frequency, f, being given by:

$$\frac{V}{V_O} = \left(1 + jQ \left| \frac{f}{f_O} - \frac{f_O}{f} \right| \right)^{-1},$$

where V_O is the output voltage at the resonant frequency f_O, and Q is the Q-factor of the circuit.

If the circuit is tuned to an RF frequency of 100 MHz, and the IF frequency is 10.7 MHz, determine the value of Q necessary to give an image channel rejection ratio (ICRR) of 25 dB.

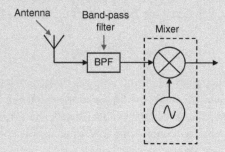

Figure 14.17 Band-pass filter used as pre-selector circuit.

Solution

Assuming $f_{LO} > f_{RF}$ then $f_i = (100 + 2 \times 10.7) \text{ MHz} = 121.4 \text{ MHz}$

$$\text{ICRR} = 25 \text{ dB} \quad \Rightarrow \quad \frac{V_O}{V} = 10^{1.25}.$$

Rearranging the given equation gives

$$\frac{V_O}{V} = 1 + jQ \left(\frac{f}{f_O} - \frac{f_O}{f} \right),$$

i.e.

$$\left| \frac{V_O}{V} \right| = \left[1 + Q^2 \left(\frac{f}{f_O} - \frac{f_O}{f} \right)^2 \right]^{0.5}.$$

Substituting data gives

$$10^{1.25} = \left[1 + Q^2 \left(\frac{121.4}{100} - \frac{100}{121.4} \right)^2 \right]^{0.5} \quad \Rightarrow \quad Q = 45.49.$$

14.8.4 Double-Conversion Superhet Receiver

The choice of f_{IF} is important in a single-conversion superhet receiver, and there are two basic requirements for its value:

(a) It should be high to allow good suppression of the image frequency.
(b) It should be low to permit easy design of high quality IF stages.

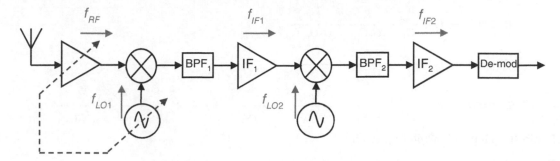

Figure 14.18 Double-conversion superhet receiver.

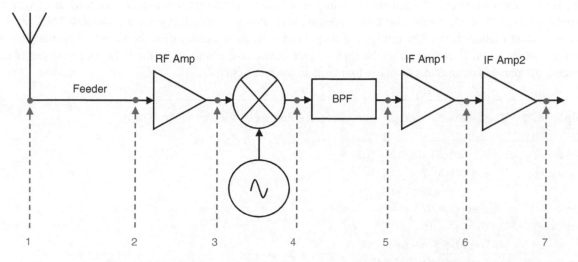

Figure 14.19 Noise temperature reference points in a superhet front-end.

Clearly these are conflicting requirements that cannot be satisfied with a single mixing stage. The double-conversion superhet receiver, which is shown in Figure 14.18, uses two mixing stages, with the majority of the IF gain in the lower-frequency IF stage.

The first IF, f_{IF1}, is chosen to be high so that the image frequency is well down on the skirt of the pre-selector's response, thereby giving good image rejection. The second IF, f_{IF2}, can then be made low to permit the design of high-quality circuits in the output stages of the receiver.

14.8.5 Noise Budget Graph for a Superhet Receiver

Noise budget graphs show how the noise and signal-to-noise ratios vary through a receiver sub-system. Figure 14.19 shows the front-end of a typical superhet wireless receiver, with seven noise reference positions identified. By considering the system noise temperatures prior to each of these positions we can determine the variation of signal-to-noise ratio through the receiver.

Using the previous theory for cascaded noise sources the system noise temperatures prior to each of the seven positions shown in Figure 14.19 are:

$$T_{sys1} = T_a,$$
$$T_{sys2} = T_a + T_f,$$
$$T_{sys3} = T_a + T_f + \frac{T_{pa}}{G_f},$$
$$T_{sys4} = T_a + T_f + \frac{T_{pa}}{G_f} + \frac{T_m}{G_f G_{pa}},$$

$$T_{sys5} = T_a + T_f + \frac{T_{pa}}{G_f} + \frac{T_m}{G_f G_{pa}} + \frac{T_{IF_{filter}}}{G_f G_{pa} G_m},$$

$$T_{sys6} = T_a + T_f + \frac{T_{pa}}{G_f} + \frac{T_m}{G_f G_{pa}} + \frac{T_{IF_{filter}}}{G_f G_{pa} G_m} + \frac{T_{IF_{amp1}}}{G_f G_{pa} G_m G_{IF_{filter}}},$$

$$T_{sys7} = T_a + T_f + \frac{T_{pa}}{G_f} + \frac{T_m}{G_f G_{pa}} + \frac{T_{IF_{filter}}}{G_f G_{pa} G_m} + \frac{T_{IF_{amp1}}}{G_f G_{pa} G_m G_{IF_{filter}}} + \frac{T_{IF_{amp2}}}{G_f G_{pa} G_m G_{IF_{filter}} G_{IF_{amp1}}}.$$

The S/N ratio at each position will be given by

$$\left(\frac{S}{N}\right)_n = \frac{P_R}{kT_{sys_n}B}, \tag{14.73}$$

where P_R is the received power at the antenna terminal, n indicates a particular position, and the remaining symbols have their usual meanings. Typical numerical calculations and the resulting noise budget graph are illustrated in Example 14.12.

In practice, what matters is the S/N ratio at the output of the receiver chain, rather than the S/N ratio at intermediate positions. However, as an introduction to the topic of receiver noise the concept of the noise budget graph is useful in emphasizing the particular significance of the first units in the receiver chain.

Example 14.12   The following data apply to the receiver front-end shown in Figure 14.19:

Received signal power at antenna terminals = 302 pW

Antenna noise:	93% of beam sees 32 K sky
	7% of beam sees 290 K Earth
Feeder:	Loss = 3.8 dB
RF pre-amplifier:	Gain = 7.1 dB
	Noise temperature = 175 K
Mixer:	Conversion loss = 1.9 dB
	Noise figure = 2.1 dB
IF filter:	Pass-band loss = 0.7 dB
First IF amplifier:	Gain = 17 dB
	Noise figure = 2.7 dB
Second IF amplifier:	Gain = 21 dB
	Noise figure = 3.6 dB

Sketch a noise budget graph showing how the S/N ratio varies through the receiver, given that the receiver bandwidth is 2 GHz.

Solution
Antenna noise (Eq. 14.59): $T_{ae} = (0.93 \times 32 + 0.07 \times 290)$ K = 50.06 K.

Feeder:	$G_f = 10^{-0.38}$	$T_f = (10^{0.38} - 1) \times 290$ K = 405.66 K
RF pre-amp:	$G_{pa} = 10^{0.71}$	$T_{pa} = 175$ K
Mixer:	$G_m = 10^{-0.19}$	$T_m = (10^{0.21} - 1) \times 290$ K = 180.32 K
IF filter:	$G_{IF_{filter}} = 10^{-0.07}$	$T_{IF_{filter}} = (10^{0.07} - 1) \times 290$ K = 50.72 K
First IF amp:	$G_{IF_{amp1}} = 10^{1.7}$	$T_{IF_{amp1}} = (10^{0.27} - 1) \times 290$ K = 250.01 K
Second IF amp:	$G_{IF_{amp2}} = 10^{2.1}$	$T_{IF_{amp2}} = (10^{0.36} - 1) \times 290$ K = 374.35 K

Calculating the intermediate system noise temperatures:

$$T_{sys1} = 50.06 \text{ K},$$

$$T_{sys2} = (50.06 + 405.66) \text{ K} = 455.72 \text{ K},$$

$$T_{sys3} = \left(455.72 + \frac{175}{10^{-0.38}}\right) \text{ K} = 875.52 \text{ K},$$

$$T_{sys4} = \left(875.52 + \frac{180.32}{10^{-0.38}10^{0.71}}\right) \text{ K} = 959.86 \text{ K},$$

$$T_{sys5} = \left(959.86 + \frac{50.72}{10^{-0.38}10^{0.71}10^{-0.19}}\right) \text{ K} = 996.60 \text{ K},$$

$$T_{sys6} = \left(996.60 + \frac{250.01}{10^{-0.38}10^{0.71}10^{-0.19}10^{-0.07}}\right) \text{ K} = 1209.39 \text{ K},$$

$$T_{sys7} = \left(1209.39 + \frac{374.35}{10^{-0.38}10^{0.71}10^{-0.19}10^{-0.07}10^{1.7}}\right) \text{ K} = 1215.75 \text{ K}.$$

Using Eq. (14.73):

$$\left(\frac{S}{N}\right)_n = \frac{P_R}{kT_{sys_n}B} = \frac{302 \times 10^{-12}}{1.38 \times 10^{-23} \times T_{sys_n} \times 2 \times 10^9} = \frac{10942.03}{T_{sys_n}}.$$

Substituting for T_{sys_n} we can obtain the S/N ratio at the seven specified positions:

Position	1	2	3	4	5	6	7
S/N	218.58	24.01	12.50	11.40	10.98	9.05	9.00
S/N (dB)	23.40	13.80	10.97	10.57	10.41	9.57	9.54

Using the data in the table we can plot the noise budget graph, showing how the S/N ratio varies with position in the receiver:

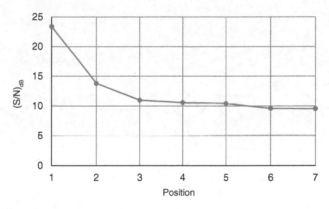

Figure 14.20   Signal-to-noise ratio budget graph for Example 14.12.

The signal-to-noise ratio budget graph plotted in Figure 14.20 emphasizes the significance of the first stages in a receiver in determining the S/N ratio at the output of the final stage. In this particular example, it can be seen that the last four stages have very little effect on the final S/N ratio.

Example 14.13   Repeat Example 14.12 with the RF pre-amplifier located between the antenna and the feeder, and compare the noise budget graphs for the two cases.

Solution
New configuration of receiver is shown in Figure 14.21.

(Continued)

(Continued)

Using the noise temperature data of individual units from Example 14.12, we can find the new intermediate system noise temperatures:

$$T_{sys1} = 50.06 \text{ K},$$

$$T_{sys2} = (50.06 + 175) \text{ K} = 225.06 \text{ K},$$

$$T_{sys3} = \left(225.06 + \frac{405.66}{10^{0.71}} \right) \text{ K} = 304.16 \text{ K},$$

$$T_{sys4} = \left(304.16 + \frac{180.32}{10^{0.71} 10^{-0.38}} \right) \text{ K} = 388.50 \text{ K},$$

$$T_{sys5} = \left(388.50 + \frac{50.72}{10^{0.71} 10^{-0.38} 10^{-0.19}} \right) \text{ K} = 425.24 \text{ K},$$

$$T_{sys6} = \left(425.24 + \frac{250.01}{10^{0.71} 10^{-0.38} 10^{-0.19} 10^{-0.07}} \right) \text{ K} = 638.03 \text{ K},$$

$$T_{sys7} = \left(638.03 + \frac{374.35}{10^{0.71} 10^{-0.38} 10^{-0.19} 10^{-0.07} 10^{1.7}} \right) \text{ K} = 644.39 \text{ K}.$$

Substituting for T_{sys_n} we can obtain the S/N ratios at the seven specified positions for the new configuration:

Position	1	2	3	4	5	6	7
S/N	218.58	48.62	35.97	28.16	25.73	17.15	16.98
S/N (dB)	23.40	16.87	15.56	14.50	14.10	12.34	12.30

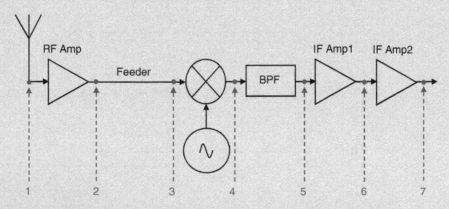

Figure 14.21 Modified receiver configuration for Example 14.13.

The noise budget graph for the modified arrangement is shown in Figure 14.22, together with that of the original arrangement.

Comment: Figure 14.22 *shows the significant improvement in S/N ratio that is obtained by positioning a pre-amplifier before a lossy feeder.*

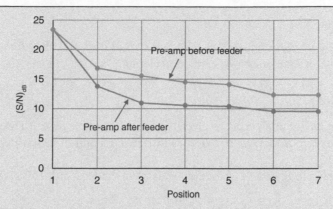

Figure 14.22 Comparison of signal-to-noise ratio budget graphs showing the influence of the position of pre-amp relative to the feeder.

Example 14.14 Suppose that the receiver specified in Example 14.12 was required to produce an output S/N ratio of at least 12 dB when the received signal power at the antenna terminals was 370 pW. What would be the maximum permissible feeder loss at room temperature to achieve this requirement?

Solution

First, we need to find the maximum system noise temperature to achieve the output requirement:

$$\left(\frac{S}{N}\right)_O = \frac{P_R}{kT_{sys}B}.$$

Therefore

$$10^{1.2} = \frac{370 \times 10^{-12}}{1.38 \times 10^{-23} \times T_{sys} \times 2 \times 10^9},$$

$$T_{sys} = 845.85 \text{ K.}$$

Using the previous expression for the system noise temperature of the complete receiver we have:

$$845.85 = T_a + T_f + \frac{T_{pa}}{G_f} + \frac{T_m}{G_f G_{pa}} + \frac{T_{IF_{filter}}}{G_f G_{pa} G_m} + \frac{T_{IF_{ampl}}}{G_f G_{pa} G_m G_{IF_{filter}}} + \frac{T_{IF_{amp2}}}{G_f G_{pa} G_m G_{IF_{filter}} G_{IF_{ampl}}}.$$

Let the loss factor of the feeder be α, then

$$\alpha = \frac{1}{G_f}.$$

Substituting data into the expression for system noise temperature, and noting that the noise temperature of the feeder is now $(\alpha - 1)290\text{K}$, gives:

$$845.85 = 50.06 + (\alpha - 1)290 + 175\alpha + \frac{180.32\alpha}{10^{0.71}} + \frac{50.72\alpha}{10^{0.71}10^{-0.19}} + \frac{250.01\alpha}{10^{0.71}10^{-0.19}10^{-0.07}}$$

$$+ \frac{374.35\alpha}{10^{0.71}10^{-0.19}10^{-0.07}10^{1.7}}.$$

Solving the equation we have:

$$\alpha = 1.79 \equiv 2.53 \text{ dB},$$

i.e. Maximum permissible feeder loss $= 2.53$ dB.

14.9 Mixers

Mixers, or rather more precisely frequency mixers, are fundamental devices providing frequency translation in superhet receivers, and also in much other RF and microwave instrumentation.

14.9.1 Basic Mixer Principles

As we have seen in the description of the superhet receiver, the fundamental purpose of a mixer is to multiply together two signals, and extract sum and difference frequencies. This is illustrated in Figure 14.23, which also shows the conventional circuit symbol for a mixer.

If the inputs to the mixer shown in Figure 14.23 are sinusoidal then the ideal output is given by

$$v_O = v_1 \times v_2 = V_1 \sin \omega_1 t \times V_2 \sin \omega_2 t$$
$$= \frac{V_1 V_2}{2} [\cos(\omega_1 - \omega_2)t - \cos(\omega_1 + \omega_2)t]. \tag{14.74}$$

Equation (14.74) shows that the ideal output consists of sum and difference frequencies; the term containing $(\omega_1 + \omega_2)$ is called the upper sideband and that containing $(\omega_1 - \omega_2)$ is called the lower sideband. Normally one of the sidebands will be removed by simple filtering. In practice, the non-linear element that performs the mixing will generate a spectrum of frequencies of the form $m\omega_1 \pm n\omega_2$ where m and n are integers, but a suitable filter can be used to isolate the required frequency at the output.

14.9.2 Mixer Parameters

Available conversion gain (G_m)

This is the ratio of the single-sideband IF output power to the RF input power.

$$(G_m)_{dB} = 10 \log \left(\frac{\text{Available IF output power}}{\text{Available RF input power}} \right) \text{ dB}. \tag{14.75}$$

In many cases, particularly for passive mixers, the IF output power will be less than the RF input power, and $(G_m)_{dB}$ will be negative, representing a conversion loss.

Noise figure (F_m)

The noise figure of a mixer is the available S/N ratio at the RF input to the mixer divided by the available S/N ratio at the IF output, i.e.

$$(F_m)_{dB} = 10 \log \left(\frac{(S/N)_{RF,avail}}{(S/N)_{IF,avail}} \right) \text{ dB}. \tag{14.76}$$

It should be noted that a mixer will down-convert noise at both the RF signal and image frequencies. If the mixer has been preceded by an RF filter to remove the image signal, then the noise figure given by Eq. (14.76) is referred to as the single-sideband noise figure.

Distortion

Mixers are necessarily non-linear devices, since their function is to provide frequency mixing. Consequently, mixers will produce relatively high levels of distortion, and they are normally regarded as the dominant distortion-producing devices in receivers. The level of distortion in a mixer can be specified in a similar manner to that used for power amplifiers (see Chapter 13), namely in terms of the IM3 intercept point, and the 1 dB compression point.

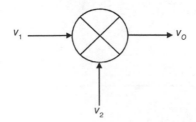

Figure 14.23 Circuit symbol for a mixer.

Bandwidth

As with all RF and microwave devices mixers will have a useable bandwidth, and this is normally determined by the couplers and matching circuits in the mixer, rather than by the active components themselves.

Isolation

This is normally specified in terms of the leakage between the RF and LO ports of a mixer. Minimizing this leakage can be very important in practical systems. For example, in a receiving system with the antenna connected to the RF port of the mixer, any leakage from the LO to RF port would result in some of the LO signal being radiated from the antenna. Typically, the leakage between the RF and LO ports should be better than 20 dB.

Dynamic range (D_m)

The dynamic range is specified in terms of the maximum and minimum powers that are supplied at the RF port.

$$D_m = 10 \log \left(\frac{P_{RF,max}}{P_{RF,min}} \right) \text{ dB,} \tag{14.77}$$

where $P_{RF,max}$ is the maximum permissible input power, and $P_{RF,min}$ is the minimum required input power, which meet the other requirements for satisfactory mixer performance. The maximum input power is normally set by the 1 dB compression point, and the minimum power is that which provides a satisfactory S/N ratio at the mixer output.

LO noise suppression

This refers to the ability of the mixer to cancel-out noise produced by the LO, and which is mixed-down to the IF frequency. This is often significant in RF and microwave receivers, where the level of noise from the LO may be comparable to the small received signal levels available from the antenna.

14.9.3 Active and Passive Mixers

Mixers fall into two broad categories, namely active mixers which provide conversion gain, and passive mixers in which there is a conversion loss. Active mixers usually employ FETs as the non-linear device; this type of mixer can provide several dB of conversion gain, but tends to have a high noise figure of the order 5–10 dB. Mixers using FETs are attractive because they are compatible with monolithic technology. Also, dual-date FETs allow input frequencies to be applied to a mixer without the need for a coupling device to maintain isolation between the input sources. In a mixer using a dual-gate FET, the RF signal is applied to one gate, and the LO signal to the other gate. Recently, high electron mobility transistor (HEMT) devices have been used for active mixers because they can provide a combination of low noise and high conversion gain at high RF frequencies. Passive mixers normally use Schottky–Barrier diodes, which have a conversion loss of a few dB, with a noise figure of a similar numerical value.

14.9.4 Single-Ended Diode Mixer

The basic arrangement of a single-ended diode mixer is shown in Figure 14.24.

In the circuit shown in Figure 14.24, the RF and LO signals are applied to the mixing diode through a forward-coupling directional coupler. The coupler provides isolation between the RF and LO sources. A DC voltage, V_{DC}, is applied to the

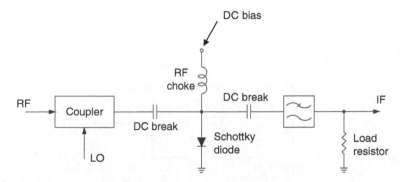

Figure 14.24   Single-ended diode mixer.

diode through an appropriate choke, and this voltage is used to select an appropriate operating point on the *I*—*V* characteristic of the diode. The total voltage, v, applied to the diode will then be given by

$$v = V_{DC} + v_{RF} + v_{LO}$$
$$= V_{DC} + V_{RF} \sin \omega_{RF} t + V_{LO} \sin \omega_{LO} t. \tag{14.78}$$

The small-signal *I*–*V* characteristic of a mixer diode is given by

$$i(v) = I_{SS} \left(\exp \left[\frac{q}{\eta k T} v \right] - 1 \right), \tag{14.79}$$

where I_{SS} is the reverse saturation current through the diode, q is the electronic charge, η is the ideality factor, k is Boltzmann's constant, and T is the absolute temperature.

The exponential term in Eq. (14.79) can be expanded as a series to give

$$i(v) = I_{SS} \left(\left[\frac{q}{\eta k T} \right] v + \frac{1}{2!} \left[\frac{q}{\eta k T} \right]^2 v^2 + \frac{1}{3!} \left[\frac{q}{\eta k T} \right]^3 v^3 + \cdots \right). \tag{14.80}$$

If we substitute v from Eq. (14.78) into Eq. (14.80) then the squared term will produce the difference frequency $\omega_{LO} - \omega_{RF}$ at the diode output. All of the remaining unwanted products at the diode output can be removed with an appropriate low-pass filter, as shown in Figure 14.24.

If the LO drive level is high, then the alternate positive and negative voltage swings from the LO source will turn the diode completely on and off during alternate cycles, and the device will perform as a switching mixer, with the LO effectively functioning as a square wave voltage source with the waveform shown in Figure 14.25, where $T = (f_{LO})^{-1}$.

The waveform shown in Figure 14.25 can be represented by a Fourier series as

$$v_{LO}(t) = \sum_{n=1}^{\infty} \frac{2V_{LO} \sin \left(\frac{n\pi}{2} \right)}{n\pi} \cos(n\omega_{LO} t). \tag{14.81}$$

The output from the mixer diode is then given by

$$v_O = v_{RF}(t) \times \sum_{n=1}^{\infty} \frac{2V_{LO} \sin \left(\frac{n\pi}{2} \right)}{n\pi} \cos(n\omega_{LO} t). \tag{14.82}$$

Expanding Eq. (14.82) and considering a sinusoidal RF input gives

$$v_O = V_{RF} \sin \omega_{RF} t \times \frac{2V_{LO}}{\pi} \left(\cos \omega_{LO} t - \frac{1}{3} \cos 3\omega_{LO} t + \cdots \right). \tag{14.83}$$

Multiplying out the trigonometric terms in Eq. (14.83) gives

$$v_O = \frac{2V_{RF} V_{LO}}{\pi} \left(\frac{\sin(\omega_{RF} \pm \omega_{LO})t}{2} - \frac{\sin 3(\omega_{RF} \pm \omega_{LO})t}{6} + \cdots \right). \tag{14.84}$$

After appropriate filtering, the term containing the difference frequency can be isolated to give

$$v_O = \frac{V_{RF} V_{LO}}{\pi} \sin(\omega_{RF} - \omega_{LO})t. \tag{14.85}$$

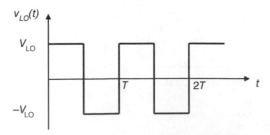

Figure 14.25 Effective LO waveform for high drive levels.

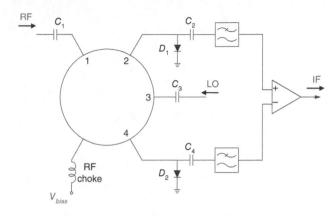

Figure 14.26 Single-balanced mixer.

14.9.5 Single Balanced Mixer

A balanced mixer refers to a mixer whose configuration is such as to remove, or balance out, one or more of the input frequencies from the output. A single-balanced mixer will remove one of the input frequencies from the output. One example of this is the use of a single balanced mixer to remove LO noise from the output of the first mixer stage in a superhet receiver. There are many methods of constructing this type of mixer, one of which is to use either a microstrip hybrid ring, or branch-line coupler, which were discussed in Chapter 2. Figure 14.26 shows the configuration of a single-balanced mixer using a hybrid ring.

In the circuit shown in Figure 14.26 ports 1 and 3 of the hybrid ring are connected to the RF input port and the LO respectively. Ports 2 and 4 are connected to mixing elements, such as semiconductor diodes. We know from the operation of the hybrid ring discussed in Chapter 2 that the RF input signal and the output of the LO will be applied together to ports 2 and 4, and these signals will be mixed in the two diodes to produce sum and difference products. The diode outputs are connected to filters which pass the required difference frequencies, and these are there applied to the inputs of a differential amplifier.

Additional points to note on the circuit of Figure 14.26:

(i) Since ports 1 and 3 of a hybrid ring are isolated from each other there is good isolation between the RF and LO sources.
(ii) The path lengths from the LO source to the two inputs of the differential amplifier are equal, and so LO noise components will be in-phase at the two inputs to the differential amplifier, and hence cancel at its output.
(iii) The RF signals leaving ports 2 and 4 of the hybrid ring will be of the same magnitude, but 180° out-of-phase. So the RF derived signal components will be out-of-phase at the two inputs to the differential amplifier, and hence in-phase at its output.

14.9.6 Double Balanced Mixer

The double-balanced mixer has several advantages over single-balanced mixers:

(i) It provides isolation between all three ports, i.e. between the RF, LO, and IF ports.
(ii) It suppresses all of the even harmonics from the RF and LO signals.
(iii) It provides a greater third-order intercept point (see Chapter 13) than a single balanced mixer.

An example of a double-balanced switching mixer is shown in Figure 14.27.

The operation of the double-balanced mixer shown in Figure 14.27 can be explained by first considering the amplitude of the LO signal to be sufficient to either turn the diodes either fully on, or fully off. Thus, when the LO signal is positive diodes D1 and D2 will be conducting, and can be replaced by a small forward resistance (r_d), and diodes D3 and D4 will be biased off and can be replaced by an open circuit. This leads to the equivalent circuit shown in Figure 14.28. Similarly, when the LO signal is negative, diodes D3 and D4 will be on, and diodes D1 and D2 will be off, leading to the equivalent circuit shown in Figure 14.29.

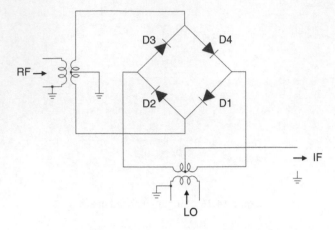

Figure 14.27 Double-balanced switching mixer.

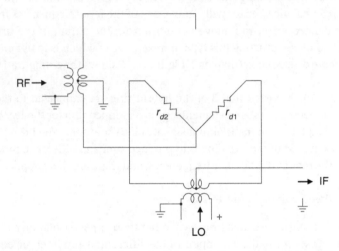

Figure 14.28 Equivalent circuit of double-balanced mixer (LO positive).

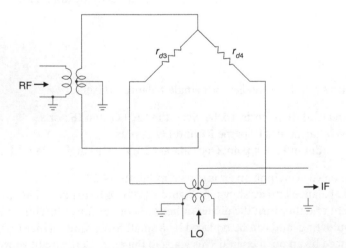

Figure 14.29 Equivalent circuit of double-balanced mixer (LO negative).

Thus, the output from the circuit of Figure 14.27 (labelled IF) will be that of an RF signal multiplied by a switching waveform, as described by Eq. (14.81), and so by appropriate filtering at the output we can obtain the difference frequency term described by Eq. (14.85).

14.9.7 Active FET Mixers

The most common type of mixer now used for RF and microwave applications is the active mixer based on the non-linear properties of an FET, usually a MESFET. The primary advantage of this type of mixer over diode mixers is that it can provide conversion gain, thus reducing the amount of additional amplification needed in a practical circuit. Also, the use of FETs rather than diodes makes the mixer easier to implement in a monolithic microwave integrated circuit (MMIC).

Various configurations are possible for FET mixers, but the most common is the *transconductance mixer*. In this configuration the LO is applied between the gate and source terminals; if a large LO voltage is used the transconductance becomes a time-varying function, $g_m(t)$, with a fundamental frequency, f_{LO}. The transconductance can be written as [3].

$$g_m(t) = \sum_{k=-\infty}^{\infty} g_k e^{jk\omega_o t},$$ (14.86)

where the Fourier coefficients of the transconductance are given by

$$g_k = \frac{1}{2\pi} \int_0^{2\pi} g_m(t) e^{-jk\omega_o t}\, d(\omega_o t)$$ (14.87)

and

$$\omega_O = 2\pi f_{LO}.$$ (14.88)

If we consider a small RF voltage, v_{RF}, superimposed on the LO voltage applied to the gate, the small-signal drain current will be

$$i_d(t) = g_m(t)\, v_{RF},$$ (14.89)

where

$$v_{RF}(t) = V_{RF} \cos \omega_{RF} t$$ (14.90)

and $|V_{RF}| \ll |V_{LO}|$.

It follows that the drain current will contain frequencies $|n\omega_O \pm \omega_{RF}|$ where n is an integer. In the case of a FET mixer used as a frequency down-converter a low-pass filter connected to the drain could be used to select the $(\omega_{RF} - \omega_O)$ frequency component.

The circuit of a simple FET mixer is shown in Figure 14.30, where for simplicity the DC bias connections have been omitted.

In the circuit shown an input RF signal is down-converted to an IF output. The hybrid circuit is used to mutually isolate the RF and LO inputs, and to combine them at the gate terminal of the FET. The hybrid will also contain a matching circuit to match the RF input to the gate of the FET. This circuit will not match the LO to the gate, since it is at a different frequency, but this is less important since the function of the LO is primarily to provide a switching function between gate and source. As previously described, the low-pass filter (LPF) is used to select the desired IF output.

One of the most important parameters used to specify the performance of an FET mixer is the available mixer conversion gain, G_c. Using the equivalent circuit of an FET, Pucel et al. [3] developed the following expression for the maximum available conversion gain

$$G_{c,max} = \frac{g_1^2 \overline{R_d}}{4\omega_{RF}^2 \overline{C}^2 R_i},$$ (14.91)

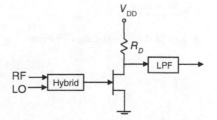

Figure 14.30 Transconductance mixer.

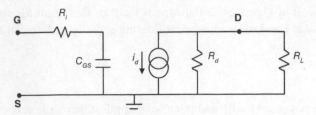

Figure 14.31 Simplified equivalent circuit of an FET.

where

g_1 = fundamental Fourier coefficient, from Eq. (14.86)
$\overline{R_d}$ = time-average value of FET drain resistance
$\omega_{RF} = 2\pi f_{RF}$ (f_{RF} = RF input frequency)
$\overline{C}$ = time-averaged value of FET gate-source capacitance
R_i = input resistance of FET.

For convenience, the parameters in Eq. (14.91) are also shown in the simplified equivalent circuit of a FET in Figure 14.31. Note that for the maximum available conversion gain specified in Eq. (14.91) $R_L = R_d$ where R_L is the load resistance.

Example 14.15   The following FET data apply to a transducer mixer used as the first stage in a 5 GHz receiver:

$$R_i = 15\,\Omega \qquad \overline{C} = 0.25\,\text{pF} \qquad \overline{R_d} = 250\,\Omega \qquad g_1 = 10\,\text{mS}.$$

Calculate the maximum available conversion gain.

Solution
Using Eq. (14.91):

$$G_{c,max} = \frac{10^{-4} \times 250}{4 \times (2\pi \times 5 \times 10^9)^2 \times (0.25 \times 10^{-12})^2 \times 15} = 6.75 \equiv 10\log(6.75)\,\text{dB} = 8.29\,\text{dB}.$$

Various configurations are possible for FET mixers. A particularly useful arrangement uses a dual-gate FET as shown in Figure 14.32.

The use of a dual-gate provides isolation between the RF and LO sources and obviates the need for a hybrid, and also enables separate matching networks to be designed for the RF and LO inputs. The theory of dual-gate FET mixers is rather different from that of the transconductance mixer and is outside the scope of the present text, but a useful in-depth analysis of the theory and performance of dual-gate mixers is given by Tsironis et al. [4]. Also, a particularly informative discussion of the various FET mixer configurations that are possible is given by S.S. Maas in Chapter 5 of reference [5].

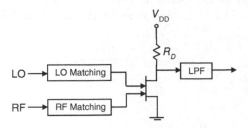

Figure 14.32   Dual-Gate FET Mixer.

14.10 Supplementary Problems

Q14.1   An RF component has a noise figure of 4.9 dB. Calculate the noise temperature of the component.

Q14.2   A length of matched transmission line at room temperature has a loss of 3.8 dB. Calculate the noise figure and the noise temperature of the line.

Q14.3   What is the open-circuit RMS thermal noise voltage across a 14 kΩ resistor measured in a bandwidth of 3 MHz, when the resistor is at a temperature of 24 °C?

Q14.4   The DC forward current through a semiconductor p–n junction is 35 mA. What is the RMS shot noise current through the diode in a bandwidth of 1 MHz.

Q14.5   What is the probability of the noise level exceeding 25 μV at any instant of time in a system in which the RMS thermal noise level is 11 μV?

Q14.6   A simple baseband digital communication system transmits data in the form of unipolar pulses in which a pulse represents a '1' and a space represents a '0'. If the pulse amplitude is 12 mV and the RMS thermal noise voltage is 7.2 mV, find the probability of an error occurring in a long message, if the probabilities of occurrence of the '1's and '0's are equal.

Q14.7   Determine the noise temperature of an antenna in which 85% of the radiation pattern sees a 45 K sky and 15% of the radiation pattern sees the Earth, which may be assumed to have a noise temperature of 290 K.

Q14.8   The available noise power at the terminals of an antenna is −110 dBm. If the antenna is connected to the input of an amplifier that has an available gain of 20 dB and a noise figure of 6 dB, what is the noise power at the output of the amplifier? Assume the antenna is impedance matched to the input of the amplifier, and that the bandwidth is 10 MHz.

Q14.9   Suppose in the arrangement specified in Q14.8 a second identical amplifier was connected in cascade with the first amplifier. What is the noise power at the output of the second amplifier? Assume that the output of the first amplifier is impedance matched to the input of the second amplifier.

Q14.10   The following data apply to two RF amplifiers;

Amplifier 1:	Available gain = 12 dB	$F_{dB} = 4.2$ dB
Amplifier 2:	Available gain = 5 dB	$F_{dB} = 2.2$ dB

In what order should the two amplifiers be connected to give minimum overall noise figure?

Q14.11   A receiving system consists of a pre-amplifier connected through a length of cable to a main receiver. The pre-amplifier has a noise figure of 6 dB, and the cable has a loss of 8 dB at room temperature. If the main receiver has a noise figure of 13 dB, find the minimum available gain required from the pre-amplifier if the overall noise figure of the receiving system is not to exceed 9 dB. Assume that all the stages of the receiving system are matched.

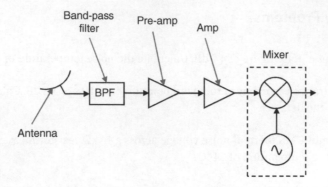

Figure 14.33 Receiver for problem Q14.12.

Q14.12 Determine the system noise temperature for the RF receiver front-end shown in Figure 14.33, where:

Antenna:	90% of beam sees 60 K sky
	10% of beam sees 290 K Earth
BPF:	Pass-band insertion loss = 0.5 dB
Pre-amp:	Gain = 8 dB
	Noise temperature = 200 K
Main amplifier:	Gain = 20 dB
	Noise figure = 3.4 dB
Mixer:	Noise figure = 4 dB

Q14.13 Determine the system noise temperature for the satellite receiver front-end shown in Figure 14.34 where:

Antenna:	88% of beam sees a 30 K sky
	12% of beam sees 290 K Earth
Feeder:	Loss = 4 dB
BPF:	Pass-band loss = 0.9 dB
Pre-amp:	Gain = 7 dB
	Noise temperature = 175 K
Main amplifier:	Gain = 18 dB
	Noise figure = 2.9 dB
Mixer:	Noise figure = 3.1 dB

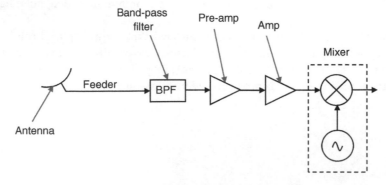

Figure 14.34 Receiver for problem Q14.13.

Q14.14 If the gain of the antenna in Q14.13 is 32 dB at a frequency of 10 GHz, and the signal power density in the vicinity of the antenna is 1 nW/m^2, determine:

   (i) The power received by the antenna.

  (ii) The S/N ratio at the output of the receiver if the bandwidth is 2 GHz.

 (iii) The change in the S/N ratio at the output of the receiver if the pre-amplifier was re-positioned between the antenna and the feeder.

 (iv) Show, on a simple noise budget graph, how the $(S/N)_O$ varies though the receiving system for parts (ii) and (iii).

Q14.15 The following data apply to the front-end of an RF receiver:

Antenna:	84% of beam sees 47 K sky
	16% of beam sees 290 K Earth
Pre-amplifier:	Noise temperature = 195 K
Feeder:	Loss = 3.2 dB
Amplifier:	Gain = 22 dB
	Noise figure = 3.8 dB
Mixer:	Noise figure = 4.5 dB

If the signal power at the antenna terminals is 3 pW, determine the gain required from the pre-amplifier to give an output S/N ratio of 9 dB, given that the bandwidth of the receiver is 80 MHz.

Q14.16 The frequency response of the pre-selector circuit in a superhet receiver is given by:

$$\frac{V}{V_O} = \frac{1}{1 + jQ \left| \dfrac{f}{f_O} - \dfrac{f_O}{f} \right|}.$$

If the circuit is tuned to an 8 MHz signal, and the IF frequency is 455 kHz, determine the value of Q required to give an ICRR (image channel rejection ratio) of 20 dB.

Q14.17 What is the 3 dB bandwidth of the pre-selector circuit specified in Q14.16?

Q14.18 Suppose that the Q value in Q14.16 was increased to give an ICRR of 25 dB. By how much would the 3 dB bandwidth change?

Appendix 14.A Appendices

14.A.1 Error Function Table

$$\text{Erf}(x) = \frac{2}{\sqrt{\pi}} \int_0^x e^{-y^2} dy.$$

x	$\text{Erf}(x)$	x	$\text{Erf}(x)$	x	$\text{Erf}(x)$	x	$\text{Erf}(x)$
0	0	0.3	0.328627	0.6	0.603856	0.9	0.796908
0.01	0.011283	0.31	0.338908	0.61	0.611681	0.91	0.801883
0.02	0.022565	0.32	0.349126	0.62	0.619411	0.92	0.806768
0.03	0.033841	0.33	0.359279	0.63	0.627046	0.93	0.811564
0.04	0.045111	0.34	0.369365	0.64	0.634586	0.94	0.816271
0.05	0.056372	0.35	0.379382	0.65	0.642029	0.95	0.820891
0.06	0.067622	0.36	0.38933	0.66	0.649377	0.96	0.825424
0.07	0.078858	0.37	0.399206	0.67	0.656628	0.97	0.82987
0.08	0.090078	0.38	0.409009	0.68	0.663782	0.98	0.834232
0.09	0.101281	0.39	0.418739	0.69	0.67084	0.99	0.838508
0.1	0.112463	0.4	0.428392	0.7	0.677801	1	0.842701
0.11	0.123623	0.41	0.437969	0.71	0.684666	1.01	0.84681
0.12	0.134758	0.42	0.447468	0.72	0.691433	1.02	0.850838
0.13	0.145867	0.43	0.456887	0.73	0.698104	1.03	0.854784
0.14	0.156947	0.44	0.466225	0.74	0.704678	1.04	0.85865
0.15	0.167996	0.45	0.475482	0.75	0.711156	1.05	0.862436
0.16	0.179012	0.46	0.484655	0.76	0.717537	1.06	0.866144
0.17	0.189992	0.47	0.493745	0.77	0.723822	1.07	0.869773
0.18	0.200936	0.48	0.50275	0.78	0.73001	1.08	0.873326
0.19	0.21184	0.49	0.511668	0.79	0.736103	1.09	0.876803
0.2	0.222703	0.5	0.5205	0.8	0.742101	1.1	0.880205
0.21	0.233522	0.51	0.529244	0.81	0.748003	1.11	0.883533
0.22	0.244296	0.52	0.537899	0.82	0.753811	1.12	0.886788
0.23	0.255023	0.53	0.546464	0.83	0.759524	1.13	0.889971
0.24	0.2657	0.54	0.554939	0.84	0.765143	1.14	0.893082
0.25	0.276326	0.55	0.563323	0.85	0.770668	1.15	0.896124
0.26	0.2869	0.56	0.571616	0.86	0.7761	1.16	0.899096
0.27	0.297418	0.57	0.579816	0.87	0.78144	1.17	0.902
0.28	0.30788	0.58	0.587923	0.88	0.786687	1.18	0.904837
0.29	0.318283	0.59	0.595936	0.89	0.791843	1.19	0.907608

x	Erf(x)	x	Erf(x)	x	Erf(x)	x	Erf(x)
1.2	0.910314	1.6	0.976348	2	0.995322	2.4	0.999311
1.21	0.912956	1.61	0.977207	2.01	0.995525	2.41	0.999346
1.22	0.915534	1.62	0.978038	2.02	0.995719	2.42	0.999379
1.23	0.91805	1.63	0.978843	2.03	0.995906	2.43	0.999411
1.24	0.920505	1.64	0.979622	2.04	0.996086	2.44	0.999441
1.25	0.9229	1.65	0.980376	2.05	0.996258	2.45	0.999469
1.26	0.925236	1.66	0.981105	2.06	0.996423	2.46	0.999497
1.27	0.927514	1.67	0.98181	2.07	0.996582	2.47	0.999523
1.28	0.929734	1.68	0.982493	2.08	0.996734	2.48	0.999547
1.29	0.931899	1.69	0.983153	2.09	0.99688	2.49	0.999571
1.3	0.934008	1.7	0.98379	2.1	0.997021	2.5	0.999593
1.31	0.936063	1.71	0.984407	2.11	0.997155	2.51	0.999614
1.32	0.938065	1.72	0.985003	2.12	0.997284	2.52	0.999635
1.33	0.940015	1.73	0.985578	2.13	0.997407	2.53	0.999654
1.34	0.941914	1.74	0.986135	2.14	0.997525	2.54	0.999672
1.35	0.943762	1.75	0.986672	2.15	0.997639	2.55	0.999689
1.36	0.945561	1.76	0.98719	2.16	0.997747	2.56	0.999706
1.37	0.947312	1.77	0.987691	2.17	0.997851	2.57	0.999722
1.38	0.949016	1.78	0.988174	2.18	0.997951	2.58	0.999736
1.39	0.950673	1.79	0.988641	2.19	0.998046	2.59	0.999751
1.4	0.952285	1.8	0.989091	2.2	0.998137	2.6	0.999764
1.41	0.953852	1.81	0.989525	2.21	0.998224	2.61	0.999777
1.42	0.955376	1.82	0.989943	2.22	0.998308	2.62	0.999789
1.43	0.956857	1.83	0.990347	2.23	0.998388	2.63	0.9998
1.44	0.958297	1.84	0.990736	2.24	0.998464	2.64	0.999811
1.45	0.959695	1.85	0.991111	2.25	0.998537	2.65	0.999822
1.46	0.961054	1.86	0.991472	2.26	0.998607	2.66	0.999831
1.47	0.962373	1.87	0.991821	2.27	0.998674	2.67	0.999841
1.48	0.963654	1.88	0.992156	2.28	0.998738	2.68	0.999849
1.49	0.964898	1.89	0.992479	2.29	0.998799	2.69	0.999858
1.5	0.966105	1.9	0.99279	2.3	0.998857	2.7	0.999866
1.51	0.967277	1.91	0.99309	2.31	0.998912	2.71	0.999873
1.52	0.968413	1.92	0.993378	2.32	0.998966	2.72	0.99988
1.53	0.969516	1.93	0.993656	2.33	0.999016	2.73	0.999887
1.54	0.970586	1.94	0.993923	2.34	0.999065	2.74	0.999893
1.55	0.971623	1.95	0.994179	2.35	0.999111	2.75	0.999899
1.56	0.972628	1.96	0.994426	2.36	0.999155	2.76	0.999905
1.57	0.973603	1.97	0.994664	2.37	0.999197	2.77	0.99991
1.58	0.974547	1.98	0.994892	2.38	0.999237	2.78	0.999916
1.59	0.975462	1.99	0.995111	2.39	0.999275	2.79	0.99992

14.A.2 Measurement of Noise Figure

Noise figure measurement is one of the key measurements employed for RF and microwave sub-systems, particularly for low-noise amplifiers used as the first stages in receivers. There are a number of methods available for measuring noise

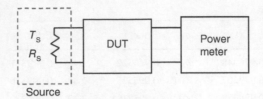

Figure 14.35 Measurement of Noise Figure: *Y*-factor method.

figure and the three most common, namely the *Y*-factor method, the noise diode method, and the discharge tube method, will be described here.

Y-factor method

The basic arrangement for a *Y*-factor measurement is shown in Figure 14.35. It consists of a source resistor, whose physical temperature can be varied, feeding the device-under-test (DUT), which is terminated with a power meter to measure the output noise.

The procedure for making a measurement is straight forward, and essentially involves measuring the noise power from the output of the DUT at two source temperatures.

Firstly, the noise power, N_1, from the output of the DUT with the source cold, i.e. at a temperature, T_{S1}, is measured. The output noise power will be the sum of the noise power from the source, amplified by the DUT, plus the noise power generated within the DUT, i.e.

$$N_1 = kT_{S1}BG_{avail} + kT_eBG_{avail}, \tag{14.92}$$

where T_e is the noise temperature of the DUT, and the other parameters have their usual meanings.

Secondly, the temperature of the source is increased to a value T_{S2}, and the new output noise power, N_2, is measured. We then have

$$N_2 = kT_{S2}BG_{avail} + kT_eBG_{avail}. \tag{14.93}$$

The *Y*-factor is defined as [6]

$$Y = \frac{N_2}{N_1}. \tag{14.94}$$

Substituting from Eqs. (14.92) and (14.93) into Eq. (14.94) gives

$$Y = \frac{T_{S2} + T_e}{T_{S1} + T_e}, \tag{14.95}$$

i.e.

$$T_e = \frac{T_{S2} - YT_{S1}}{Y - 1}. \tag{14.96}$$

Using Eq. (14.47) we have

$$F = 1 + \frac{T_e}{T_O} = 1 + \frac{T_{S2} - YT_{S1}}{(Y - 1)T_O}. \tag{14.97}$$

Rearranging Eq. (14.97) gives

$$F = \frac{(T_{S2} - T_O) + Y(T_O - T_{S1})}{(Y - 1)T_O}, \tag{14.98}$$

where T_O is the standard reference temperature, 290 K. Since the values of all the quantities on the right-hand side of Eq. (14.98) are known, or can be measured, we can determine the value of the noise figure, *F*, through measurement.

Noise diode method

This method uses an avalanche diode to provide the noise input to the DUT. The output of the DUT is connected to a precision attenuator, which is terminated by a power meter, as shown in Figure 14.36.

In Figure 14.36, the noise diode is represented by a parallel combination of a noise current generator and a conductance, G_S.

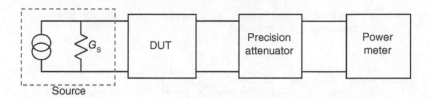

Figure 14.36   Measurement of Noise Figure: Noise diode method.

There are two stages to the measurement.

Firstly, the DC current, I_{DC}, through the diode is set to zero. The noise power, N_{S1}, available from the source can then be written as

$$N_{S1} = kT_O B + kT_e B, \tag{14.99}$$

where T_e is the noise temperature of the DUT. The noise power, N_1, available at the output of the attenuator will then be

$$N_1 = (kT_O B + kT_e B)G_{avail}L, \tag{14.100}$$

where G_{avail} is the available gain of the DUT and L represents the initial loss through the attenuator.

Secondly, the value of the attenuator is increased by 3 dB (i.e. a power ratio of 2), and the value of I_{DC} is increased to give the same noise power reading on the power meter as in stage 1. (*Note that since the same scale reading on the power meter is used for both the first and second stages of the measurement, any non-linearities associated with the power meter are eliminated.*) Increasing the value of I_{DC} causes shot noise to be generated by the diode. The shot noise power, N_S, available from the diode is given by

$$N_S = \frac{qI_{DC}BR_S}{2}, \tag{14.101}$$

where $R_S = 1/G_S$. It then follows that

$$2 \times (kT_O B + kT_e B)G_{av}L = \left(kT_O B + kT_e B + \frac{qI_{DC}BR_S}{2} \right) G_{av}L. \tag{14.102}$$

Rearranging Eq. (14.102) gives

$$kT_O B + kT_e B = \frac{qI_{DC}BR_S}{2},$$

i.e.

$$1 + \frac{T_e}{T_O} = \frac{qI_{DC}R_S}{2kT_O}. \tag{14.103}$$

Therefore, we have

$$F = \frac{qI_{DC}R_S}{2kT_O}. \tag{14.104}$$

Noting that $k = 1.38 \times 10^{-23}$ J/K, $q = 1.6 \times 10^{-19}$ C, and $T_O = 290$ K, we can write Eq. (14.104) as

$$F \approx 20I_{DC}R_S. \tag{14.105}$$

Thus, since R_S will be known in a practical arrangement, and I_{DC} can be measured, the value of F for the DUT can be determined by measurement.

Discharge tube method

At frequencies greater than ~1 GHz an avalanche diode cannot provide sufficient noise power for a measurement, and normally a gas discharge tube is used as the noise source. Gas discharge noise sources are available either with coaxial or waveguide outputs. The most common arrangement at microwave frequencies is to have an argon gas discharge tube mounted at an angle across rectangular metal waveguide that is terminated at one end with a matched load; this arrangement provides a broadband impedance match.

A typical arrangement for making a noise figure measurement using a gas discharge tube is shown in Figure 14.37, where the DUT has a noise temperature of T_e.

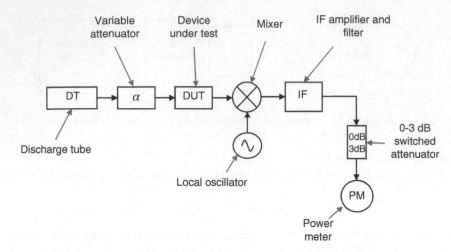

Figure 14.37 Measurement of Noise Figure: Discharge tube method.

The measurement involves two steps. The first step is to note the power meter reading with the discharge tube switched off, and the variable attenuator set to 0 dB (i.e. $\alpha = 1$), and the IF attenuator also set to 0 dB. Under these conditions the noise power available, P_n, at the output of the device under test is simply given by

$$P_n = (kT_eB + kT_OB)G_{avail}, \tag{14.106}$$

where

T_O = reference source temperature, i.e. the temperature of the discharge tube in the off state, normally 290 K
B = bandwidth of the DUT
G_{avail} = available gain of the DUT

The second step is the switch on the discharge tube so that additional noise is applied to the input of the DUT, switch the IF attenuator to 3 dB, and adjust the value of the variable attenuator, i.e. the value of α, to obtain the original reading on the power meter. Since the 3 dB attenuator is now switched in, the power available at the output of the DUT must be double the original value, i.e. we now have

$$P_n = 2 \times (kT_eB + kT_OB)G_{avail}. \tag{14.107}$$

But with the discharge tube now supplying additional noise to the input of the DUT we can write the noise available at the output of the DUT as the sum of three terms

$$P_n = \frac{kT_{tube}B}{\alpha}G_{avail} + kT_OB\left[1 - \frac{1}{\alpha}\right]G_{avail} + kT_eBG_{avail}, \tag{14.108}$$

where
$\dfrac{kT_{tube}}{\alpha}G_{avail}$ = noise from the discharge tube, which has a noise temperature of T_{tube}

$kT_OB\left[1 - \dfrac{1}{\alpha}\right]G_{avail}$ = noise from the attenuator [using Eq. (14.65)]

kT_eBG_{avail} = noise from the DUT

Equating Eqs. (14.107) and (14.108) gives

$$\frac{kT_{tube}B}{\alpha}G_{avail} + kT_OB\left[1 - \frac{1}{\alpha}\right]G_{avail} + kT_eBG_{avail} = 2 \times (kT_eB + kT_OB)G_{avail}. \tag{14.109}$$

Rearranging Eq. (14.109) yields

$$T_e = \frac{T_{tube} - T_O}{\alpha} - T_O. \tag{14.110}$$

Using Eq. (14.48) we can rewrite Eq. (14.110) in terms of the noise figure as

$$F = \frac{1}{\alpha}\left(\frac{T_{tube}}{T_O} - 1\right). \tag{14.111}$$

The quantity $\left(\dfrac{T_{tube}}{T_O} - 1\right)$ is known as the excess noise ratio (ENR) of the gas discharge tube. It is usually quoted by the manufacturers in dB, i.e.

$$\text{ENR} = 10\log\left(\frac{T_{tube}}{T_O} - 1\right)\ \text{dB}. \tag{14.112}$$

Typically, for an argon gas discharge tube ENR = 15.2 dB.

Thus we see from Eq. (14.111) that by measuring α we can determine the noise figure of the DUT.

Two particular reference sources are recommended for further information on noise measurement:

(i) Hewlett Packard[2] Application Note 57-1 [6] provides an excellent introduction to the general area of noise measurement.

(ii) J.P. Dunsmore [7] gives a very informative discussion on noise measurement using modern test equipment such as noise figure analyzers and spectrum analyzers.

References

1 Haykin, S. (1988). *Digital Communications*. Chichester, UK: Wiley.

2 IRE Standards on Electron Tubes (1957). Definition of terms. *Proceedings of the IRE* 45: 983–1010.

3 Purcel, R.A., Masse, D., and Bera, R. (1976). Performance of GaAs MESFET mixers at X band. *IEEE Transactions on Microwave Theory and Techniques* 24 (6): 351–360.

4 Tsironis, C., Meierer, R., and Stahlmann, R. (1984). Dual-gate MESFET mixers. *IEEE Transactions on Microwave Theory and Techniques* 32 (3): 248–255.

5 Larson, L.E. (1996). *RF and Microwave Circuit Design for Wireless Communications*. Norwood, MA, USA: Artech House Inc.

6 Fundamentals of RF and Microwave Noise Figure Measurements, Hewlett Packard Application Note HP-AN-57-1 1983, (Available at: http://www.hparchive.com/appnotes)

7 Dunsmore, J.P. (2012). *Handbook of Microwave Component Measurements*. Chichester, UK: Wiley.

---

2 Hewlett Packard RF and microwave test equipment is now manufactured under the name of Keysight Technologies.

Answers to Selected Supplementary Problems

Chapter 1

Q1.1 $Z = (51 + j25)\,\Omega$

Q1.2 Distance $= 0.02\lambda$

Q1.3 $\rho = 0.72 \angle -37°$ VSWR $= 6$

Q1.4 $\rho = 0.36 \angle 25°$

Q1.5 $Z_L = (232.5 - j157.5)\,\Omega$

Q1.6 $Y_L = (4 - j1.5)\,\text{mS}$ distance $= 0.191\lambda$

Q1.7 $Z_L = (7.5 - j4)\,\Omega$

Q1.8 Distance from load to stub $= 0.035$ m
Length of stub $= 0.038$ m

Q1.9 Distance from load to stub $= 0.035$ m
Length of stub $= 0.138$ m

Q1.10 Distance from load to stub $= 0.048$ m
Length of stub $= 0.034$ m

Q1.11 Length $= 16.5$ mm

Q1.12 Length $= 70.27$ mm

Q1.13 $Z_{in} = (6.5 + j18)\,\Omega$

Q1.14 $Z_{in} = (18 + j23.5)\,\Omega$

Q1.15 $Z_{in} = (24 + j2.5)\,\Omega$

Q1.16 Solution 1: $C_{series} = 64.30\,\text{pF}$ $L_{shunt} = 6.89\,\text{nH}$
Solution 2: $L_{series} = 9.98\,\text{nH}$ $C_{shunt} = 12.15\,\text{pF}$

RF and Microwave Circuit Design: Theory and Applications, First Edition. Charles E. Free and Colin S. Aitchison.
© 2022 John Wiley & Sons Ltd. Published 2022 by John Wiley & Sons Ltd.
Companion website: www.wiley.com/go/free/rfandmicrowave

Q1.17   (i) Solution 1: $C_{shunt} = 0.42\,\text{pF}$ $L_{series} = 3.18\,\text{nH}$

   Solution 2: $L_{shunt} = 4.57\,\text{nH}$ $C_{series} = 0.88\,\text{pF}$

   (ii) Solution 1: $\rho_{in} = 0.14\,\angle\,{-43°}$

   Solution 2: $\rho_{in} = 0.20\,\angle\,{-124°}$

Q1.18 L_{shunt} (nearest to load) = 2.63 nH
   L_{series} (nearest to source) = 2.39 nH

Chapter 2

Q2.3 $\lambda_s = 30.78$ mm

Q2.4

$Z_o\,(\Omega)$	$w\,(\mu\text{m})$	$\lambda_s\,(\text{mm})$	$V_p\,(\text{m/s})$
25	2159.0	10.95	1.10×10^8
50	571.5	11.68	1.17×10^8
75	190.5	12.05	1.20×10^8
100	63.5	12.27	1.23×10^8

Q2.5   17.98%

Q2.6   Minimum thickness at 1 GHz = 10.55 µm
   Minimum thickness at 10 GHz = 3.35 µm
   Minimum thickness at 100 GHz = 1.05 µm

Q2.7   Percentage error in 50 Ω line = 6.0%
   Percentage error in 70 Ω line = 11.4%

Q2.8   (a) $V_2 = 0.707\,\angle\,{-90°}$ $V_3 = 0$ $V_4 = 0.707\,\angle\,{-270°}$

   (c) Ring: $w = 175\,\mu\text{m}$ mean diameter = 5.73 mm

   Ports: $w = 450\,\mu\text{m}$

   (d) 7.99 GHz

Q2.9 $w_{transformer} = 240\,\mu\text{m}$
   $length_{transformer} = 4.96$ mm

Q2.10 $length_{transformer} = 8.24$ mm

Q2.11   First step: width = 635.0 µm   length = 2.43 mm

   Second step: width = 908.1 µm   length = 2.39 mm

   Third step: width = 1460.5 µm   length = 2.33 mm

   Fourth step: width = 1905.0 µm   length = 2.30 mm

Q2.12   First step: width = 400.0 µm   length = 1.96 mm

   Second step: width = 240.0 µm   length = 1.99 mm

   Third step: width = 150.0 µm   length = 2.01 mm

Q2.13   Bandwidth = 13.72 GHz

Q2.14   New bandwidth = 15.50 GHz
Improvement = 1.78 GHz

Q2.15 w_{ports} = 1.08 mm
$w_{transformer}$ = 420 μm
$length_{transformer}$ = 3.33 mm
Resistor = 100 Ω

Q2.16   Parallel arms: width = 360 μm   length = 1.95 mm

Series arms: width = 720 μm   length = 1.89 mm

Ports: width = 360 μm

Q2.18 Z_{oe} = 61.24 Ω Z_{oo} = 40.82 Ω

Q2.19   Frequency range = 6.78 GHz

Chapter 3

Q3.1   Tanδ = 3.03 × 10^{-3} Q = 331.13

Q3.2 ε^* = 6.38 − j0.051

Q3.3 γ = 1.34 + j10

Q3.4   Loss = 1.42 dB/m

Q3.5   Tanδ = 1.28 × 10^{-3}

Q3.6   Loss = 0.85 dB/mm

Q3.7 Q_u = 7502.95

Q3.8   (i) Length = 11.45 mm   (ii) Q_u = 17.69 × 10^3

Q3.9   (i) r = 4.27 mm   (ii) Length = 14.65 mm

Q3.10   (i) f_L = 9.312 GHz   (ii) Q_L = 3353
(iii) Tanδ = 1.74 × 10^{-4} $Q_{dielectric}$ = 5747.13

Q3.11   Length = 2.42 mm

Q3.12 Δ_{max} = 0.347 μm

Chapter 5

Q5.1 $[S] = \begin{bmatrix} 0 & 0 \\ 12.59\angle 0° & 0 \end{bmatrix}$

Q5.2 $[S] = \begin{bmatrix} 0 & 1\angle -77° \\ 1\angle -77° & 0 \end{bmatrix}$

Q5.3   (i) Γ_{in} = 0.12 ∠ 24°   (ii) Γ_{in} = 0.25 ∠ 46.8°

Q5.4 $G_{additional} = 8.12\,\text{dB}$

Q5.5 SPDT switch

Q5.6 $(S_{21})_{overall} = 24.61\,\angle -223.8°$

Q5.8 (i) $G_P = 6.58\,\text{dB}$ (ii) $G_A = 2.90\,\text{dB}$
       (iii) $G_T = 2.07\,\text{dB}$ (iv) $G_{TU} = 2.01\,\text{dB}$

Q5.9 $G_{TU,max} = 7.92\,\text{dB}$

Chapter 6

Q6.1 $f_o = 3.07\,\text{GHz}$

Q6.2 $H_o = 73.91\,\text{kA/m}$

Q6.4 (i) $\lambda_+ = 29.46\,\text{mm}$ $\lambda_- = 15.35\,\text{mm}$

       (ii) $5.61°/\text{mm}$

       (iii)   Angle $= 56.1°$ Direction: negative

Q6.5 (i) VSWR $= 3.76$

       (ii) VSWR $= 1.16$

Q6.6 VSWR $= 1.11$

Chapter 8

Q8.1 Nineth order

Q8.2 π-configuration:
     $C_1 = 3.28\,\text{pF}$ $L_2 = 21.46\,\text{nH}$
     $C_3 = 10.61\,\text{pF}$ $L_4 = L_2 \; C_5 = C_1$

Q8.3 (i) Fifth order   (ii) Eighth order

Q8.4 π-configuration:
     $L_1 = 32.19\,\text{nH}$ $C_2 = 2.19\,\text{pF}$
     $L_3 = 9.95\,\text{nH}$ $C_4 = C_2$ $L_5 = L_1$

Q8.5 Seventh order

Q8.6 π-network:

     First parallel resonant circuit: $C_1 = 333.76\,\text{pF}$ $L_1 = 14.66\,\text{nH}$

     Series resonant circuit: $C_2 = 3.79\,\text{pF}$ $L_2 = 1.29\,\mu\text{H}$

     Second parallel resonant circuit: $C_3 = 333.76\,\text{pF}$ $L_3 = 14.66\,\text{nH}$

Q8.7 π-configuration:

First parallel resonant circuit:	$C_1 = 26.33\,\text{pF}$	$L_1 = 1.98\,\text{nH}$
Series resonant circuit:	$C_2 = 0.29\,\text{pF}$	$L_2 = 180.40\,\text{nH}$
Second parallel resonant circuit:	$C_3 = 104.27\,\text{pF}$	$L_3 = 0.50\,\text{nH}$
Series resonant circuit:	$C_2 = 0.18\,\text{pF}$	$L_2 = 289.34\,\text{nH}$
Third parallel resonant circuit:	$C_3 = 104.27\,\text{pF}$	$L_3 = 0.50\,\text{nH}$
Series resonant circuit:	$C_2 = 0.29\,\text{pF}$	$L_2 = 180.40\,\text{nH}$
Fourth parallel resonant circuit:	$C_1 = 26.33\,\text{pF}$	$L_1 = 1.98\,\text{nH}$

Q8.8 10.16 dB

Q8.9 (Using microstrip data from Figures 2.24 and 2.25)

Wide section:	width = 2.8 mm	length = 2.29 mm
Narrow section:	width = 0.08 mm	length = 6.16 mm
Wide section:	width = 2.8 mm	length = 9.55 mm
Narrow section:	width = 0.08 mm	length = 6.16 mm
Wide section:	width = 2.8 mm	length = 2.29 mm

Q8.10 π-network:
$C_1 = 0.50\,\text{pF}$ $L_2 = 1.93\,\text{nH}$ $C_3 = 0.50\,\text{pF}$

Q8.11 (Using microstrip data from Figures 2.24 and 2.25)

Wide section:	width = 8.00 mm	length = 0.87 mm
Narrow section:	width = 0.24 mm	length = 2.22 mm
Wide section:	width = 8.00 mm	length = 0.87 mm

Q8.12 (Using microstrip data from Figures 2.24 and 2.25)
Refer to Figure 8.24:

First stub:	width = 0.22 mm	length = 2.53 mm
Second stub:	width = 0.50 mm	length = 2.48 mm
Third stub:	width = 0.22 mm	length = 2.53 mm
Stub separation:	width = 0.22 mm	length = 2.52 mm

Port widths = 1.47 mm

Chapter 9

Q9.1 (i) $K = 4.26$ Transistor is stable

(ii) $G_{TU,max} = 14.4\,\text{dB}$

Q9.2 Choosing $w_{wide} = 2\,\text{mm}$ $l_{wide} = 1.94\,\text{mm}$

Choosing $w_{narrow} = 50\,\mu\text{m}$ $l_{narrow} = 2.19\,\text{mm}$

Q9.3   VSWR = 4.1

Q9.4   Transformer: $w = 90\,\mu\text{m}$ $l = 2.0\,\text{mm}$

   Stub: $w = 225\,\mu\text{m}$ $l = 1.42\,\text{mm}$

Q9.5 $\Gamma = 0.87 \angle -152°$

Q9.6 $\Gamma_S = 0.39 \angle -110.3°$
   $\Gamma_L = 0.66 \angle -144.2°$

Q9.8   (i) SOURCE   Transformer: $w = 410\,\mu\text{m}$ $l = 7.47\,\text{mm}$

           Stub: $w = 900\,\mu\text{m}$ $l = 11.91\,\text{mm}$

      LOAD   Transformer: $w = 170\,\mu\text{m}$ $l = 7.62\,\text{mm}$

           Stub: $w = 900\,\mu\text{m}$ $l = 12.41\,\text{mm}$

   (ii) $F = 3.47\,\text{dB}$

   (iii) Configuration:

        Source.. $L_{SERIES,1}$ .. $L_{SHUNT,1}$ ..FET.. $L_{SHUNT,2}$ .. $L_{SERIES,2}$ ..Load

        $L_{SERIES,1} = 1.79\,\text{nH}$ $L_{SHUNT,1} = 12.43\,\text{nH}$

        $L_{SERIES,2} = 1.53\,\text{nH}$ $L_{SHUNT,2} = 3.26\,\text{nH}$

Q9.9 $G_{TU} = 11.42\,\text{dB}$

Chapter 10

Q10.1   Insertion loss = 0.166 dB
   Isolation = 11.43 dB

Q10.2 $C_{pk} = 0.037\,\text{pF}$

Q10.3 $FoM = 1.24\,\text{THz}$

Q10.6   Shunt reactance = 50 Ω
   Impedance of interconnecting line, $Z_T = 35.36\,\Omega$
   Length of interconnecting line = 0.25 λ_T

Q10.7   (i) $\phi = 75.63°$

   (ii) $f_o = 8\,\text{GHz}$

   (iii) $a = 304.8\,\mu\text{m}$ $b = 1.86\,\text{mm}$

Q10.8   (ii) $l = 1.68\,\text{mm}$

   (iii) 1.6%

   (iv) VSWR = 1.05

Chapter 11

Q11.1 $P_{RF} = 478.8$ mW

Q11.2   Length (Impatt) = 1.25 μm
Length (Gunn) = 5 μm

Q11.3   Pulling range = 43.9 kHz

Q11.4   (i) $f_S = 4.994$ MHz
(ii) $L = 67.7$ mH

Q11.5 $C_1 = 341.7$ pF $C_2 = 6.83$ pF

Q11.6 $Q = 628.41$

Q11.7 $V_b = -1.20$ V $V_m = 278.91$ mV

Q11.8 $V_b = -1.45$ V $V_m = 124.48$ mV

Q11.9   (i) $V_b = -0.62$ V
(ii) $\Delta V_b = 62.16$ mV

Q11.10   (i) $V_b = -4.37$ V
(ii) $\Delta V_b = 696.97$ mV

Q11.11 $V_b = -0.64$ V $V_m = 219.03$ mV

Q11.12   FM noise = -73.81 dBc/Hz

Q11.13 N: 935 – 1935

Q11.14   (i) $f_{ref} = 5$ kHz
(ii) $f_{offset\ osc} = 127.5$ MHz
(iii) $N_{max} = 2500$
(iv) $N = 1904$

Q11.15 $N = 14$ for 65 VCO cycles
$N = 15$ for 35 VCO cycles

Chapter 12

Q12.1 $P_{den} = 251.03$ mW/m^2

Q12.2 $A_{eff} = 0.56$ m^2

Q12.3   Tx loss = 97.57 dB

Q12.4 $R_{rad} = 7.90\,\Omega$

Q12.5 $\eta = 72\%$

Q12.7 (i) $\theta_{3dB} \approx 12.5°$

(ii) $\theta_{1st\,NULLS\,dB} \approx 27.5°$

(iii) $90° \pm 20.0°$, $90° \pm 35.9°$, $90° \pm 55.9°$

Q12.8 (i) $\theta_{3dB} \approx 8.3°$

(ii) $\theta_{1st\,NULLS\,dB} \approx 17.9°$

(iii) $90° \pm 13.3°$, $90° \pm 22.9°$, $90° \pm 33.6°$, $90° \pm 42.7°$, $90° \pm 60.3°$

Q12.9 $E_{\theta,max} = 30\,\text{mV/m}$ $H_{\phi,max} = 79.6\,\mu\text{V/m}$

Q12.11 $|E_B/E_A| = 3.59$

Q12.12 (i) $Z_1 = (95 - j38)\,\Omega$ $Z_2 = (95 - j38)\,\Omega$

(ii) $Z_1 = (55 + j38)\,\Omega$ $Z_2 = (55 + j38)\,\Omega$

(iii) $Z_1 = (37 - j20)\,\Omega$ $Z_2 = (113 + j20)\,\Omega$

Q12.13 (i) $F/B = 11.10\,\text{dB}$

(ii) $Z_{in} = (44.69 + j34.46)\,\Omega$

(iii) $G\,(\text{rel. }\lambda/2) = 4.56\,\text{dB}$

Q12.14 $F/B = 4.56\,\text{dB}$

Q12.15 $V_c = 79\,\text{mV}$

Q12.16 (i) $V_T = 52.90\,\text{mV}$

(ii) $C = 3.57\,\text{pF}$

(iii) $V_c = 2.47\,\text{V}$

(iv) Change $= -0.54\,\text{dB}$

Q12.17 $G = 30.10\,\text{dB}$

Q12.18 Tx loss $= 71.33\,\text{dB}$

Q12.19 Increase in diameter $= 78.3\%$

Q12.20 $P_R = 0.27\,\text{pW}$

Q12.21 Diameter $= 2.74\,\text{m}$

Q12.22 Patch: width $= 11.68\,\text{mm}$ length $= 8.56\,\text{mm}$

Matching transformer: width $= 2.36\,\text{mm}$ length $= 5.51\,\text{mm}$

Q12.23 Patch dimensions as Q12.22; length of inset $= 2.63\,\text{mm}$

Q12.24 $G = 12.02$ dB

Q12.25 Aperture: $73.52\,\text{mm} \times 73.52\,\text{mm}$ Length: $135.13\,\text{mm}$

Q12.27 No

Q12.28 Stepped design possible, with $\sigma = 18.88\,\text{mm}$

Chapter 14

Q14.1 $T_e = 606.2$ K

Q14.2 $F = 3.8\,\text{dB}$ $T_e = 405.7$ K

Q14.3 $v_n(\text{rms}) = 26.24\,\mu\text{V}$

Q14.4 $i_n(\text{rms}) = 0.11\,\mu\text{A}$

Q14.5 Prob $= 11.4 \times 10^{-3}$

Q14.6 $P(e) = 2.02 \times 10^{-1}$

Q14.7 $T_{antenna} = 81.75$ K

Q14.8 $P_n = 12.93$ pW

Q14.9 $P_n = 1.3$ nW

Q14.10 $2 - 1$

Q14.11 $G = 15$ dB

Q14.12 $T_{sys} = 404.82$ K

Q14.13 $T_{sys} = 1380.98$ K

Q14.14 (i) $P_R = 113.5$ pW

        (ii) $S/N = 4.74\,\text{dB}$

        (iii) Increase of $4.16\,\text{dB}$

Q14.15 $G_{PreAmp} = 12.81$ dB

Q14.16 $Q = 46.06$

Q14.17 $B_{3\text{dB}} = 176$ MHz

Q14.18 Decrease by 45.5%

Index

RF and Microwave Circuit Design: Theory and Applications, First Edition. Charles E. Free and Colin S. Aitchison.
© 2022 John Wiley & Sons Ltd. Published 2022 by John Wiley & Sons Ltd.
Companion website: www.wiley.com/go/free/rfandmicrowave